A TREATISE
ON THE
ANALYTICAL DYNAMICS
OF PARTICLES AND RIGID BODIES

A TREATISE
ON THE
ANALYTICAL DYNAMICS
OF PARTICLES AND RIGID BODIES

WITH AN INTRODUCTION TO THE
PROBLEM OF THREE BODIES

BY

E. T. WHITTAKER
Professor of Mathematics in the University of Edinburgh

FOURTH EDITION

CAMBRIDGE
AT THE UNIVERSITY PRESS
1952

First Edition 1904
Second Edition 1917
Third Edition 1927
Fourth Edition 1937

First American Printing 1944

By special arrangement with the
Cambridge University Press and The Macmillan Co.

PREFACE TO THE FOURTH EDITION

References to work published since 1927 have been inserted, and some errors corrected. For the detection of these I must thank many correspondents.

E. T. WHITTAKER

EDINBURGH
August, 1936

CONTENTS

CHAPTER I

KINEMATICAL PRELIMINARIES

CHAPTER II

THE EQUATIONS OF MOTION

CHAPTER III

PRINCIPLES AVAILABLE FOR THE INTEGRATION

CHAPTER IV

THE SOLUBLE PROBLEMS OF PARTICLE DYNAMICS

CHAPTER V

THE DYNAMICAL SPECIFICATION OF BODIES

CHAPTER VI

THE SOLUBLE PROBLEMS OF RIGID DYNAMICS

CHAPTER VII

THEORY OF VIBRATIONS

CHAPTER VIII

NON-HOLONOMIC SYSTEMS. DISSIPATIVE SYSTEMS

CHAPTER IX

THE PRINCIPLES OF LEAST ACTION AND LEAST CURVATURE

CHAPTER X

HAMILTONIAN SYSTEMS AND THEIR INTEGRAL-INVARIANTS

CHAPTER XI

THE TRANSFORMATION-THEORY OF DYNAMICS

CHAPTER XII

PROPERTIES OF THE INTEGRALS OF DYNAMICAL SYSTEMS

CHAPTER XIII

THE REDUCTION OF THE PROBLEM OF THREE BODIES

CHAPTER XIV

THE THEOREMS OF BRUNS AND POINCARÉ

CHAPTER XV

THE GENERAL THEORY OF ORBITS

CHAPTER XVI

INTEGRATION BY SERIES

CHAPTER I

KINEMATICAL PRELIMINARIES

1. *The displacements of rigid bodies.*

The name *Analytical Dynamics* is given to that branch of knowledge in which the motions of material bodies, considered as due to the mutual interactions of the bodies, are discussed by the aid of mathematical analysis.

It is natural to begin this discussion by considering the various possible types of motion in themselves, leaving out of account for a time the causes to which the initiation of motion may be ascribed; this preliminary enquiry constitutes the science of *Kinematics.* The object of the present chapter is to establish a number of kinematical theorems which will be required in the rest of the work.

Kinematics is in itself an extensive subject, for a complete account of which the student is referred to treatises dealing exclusively with it, e.g. that of Koenigs (Paris, 1897). In what follows we shall confine our attention to theorems which are of utility in the applications of Kinematics to Dynamics.

We shall say that a material body is *rigid* when the mutual distance of every pair of specified points in it is invariable, so that the body does not expand or contract or change its shape in any way, although it may change its position with reference to surrounding objects.

If a rigid body is moved from one position to another, the change of position is called a *displacement* of the body. Certain special kinds of displacement have received specific names; thus if the position in space of every point of the body which lies on some straight line L is unchanged, the displacement is called a *rotation about the line* L; if the position in space of some point P of the body is unchanged, the displacement is called a *rotation about the point* P; and if the lines joining the initial and final positions of each of the points of the body are a set of parallel straight lines of length l, so that the orientation of the body in space is unaltered, the displacement is called a *translation parallel to the direction of the lines, through a distance* l.

2. *Euler's theorem on rotations about a point**.

Consider a rigid body, one of whose points is made immoveable by some attachment; suppose that the body is free to turn about this point in any manner, and let any two possible configurations of the body be taken: for convenience we shall call these the configuration P and the configuration Q. We shall now shew that it is possible to bring the body from the configuration P to the configuration Q by simply rotating it about some definite line through the fixed point, i.e. that *a rotation about a point is always equivalent to a rotation about a line through the point.*

To establish this result (which was first given by Euler), denote the fixed point by O; let OA, OB be the positions, in the configuration P, of two lines through the fixed point which are fixed in the body and move with it; let OA', OB' be the positions of the same lines in the configuration Q. Draw the plane which is perpendicular to the plane AOA' and bisects the angle AOA'; and draw also the plane which is perpendicular to the plane BOB' and bisects the angle BOB'. Let OC be the line of intersection of these twc planes, supposing them to be not coincident; if they are coincident, we denote by OC the line of intersection of the planes OAB and $OA'B'$.

Then clearly in either case the line OC is related to the lines OA', OB' in exactly the same way as it is related to the lines OA and OB; that is to say, the angles AOC and BOC are respectively equal to the angles $A'OC$ and $B'OC$. It follows that if the system $OABC$ is rotated about O in such a way that the lines OA and OB come into the positions OA' and OB' respectively, then OC will retain its position unchanged. The line OC is therefore unaffected by the displacement in question, and so the displacement can be represented by a rotation through some angle round OC; which proves the theorem.

When a body is continuously moving round one of its points, which is fixed in space, the displacement from its position at time t to its position at time $t + \Delta t$, can, by Euler's theorem, be obtained by rotating the body about some definite line through the fixed point. The limiting position of this line, when the interval Δt is indefinitely diminished, is called the *instantaneous axis of rotation* of the body at the time t.

When a body is continuously moving round one of its points, which is fixed, the locus of the instantaneous axis in the body is a cone, whose vertex is at the fixed point: the locus of the instantaneous axis in space is also a cone whose vertex is at the fixed point. Shew that the actual motion of the body can be obtained by making the former of these cones (supposed to be rigidly connected with the body) roll on the latter cone (supposed to be fixed in space). (Poinsot.)

A similar proof shews that *if any two positions of a plane figure in the same plane are given, the displacement from one position to the other can be*

* *Novi Comment. Petrop.* xx. (1776), p. 189, § 25.

regarded as a rotation about some point in the plane. This point is called the *centre of rotation.*

When the body is regarded as continuously moving, the small displacement from one position to the position which succeeds it after an infinitesimal interval of time can therefore be accomplished by a rotation round a point; this point is called the *instantaneous centre of rotation.*

Example 1. A lamina moves in any manner in its plane. Prove that the locus at any instant of points which are at inflexions of their paths is a circle, which touches the loci in the lamina and in space of the centre of instantaneous rotation. (Coll. Exam.)

Example 2. A rigid body in two dimensions is subjected successively to two finite displacements in its plane. If D_2 be the line joining the centres of displacement, and if D_1 be the line which is brought into the position D_2 by half the first displacement (i.e. by rotation through half the angle), and if D_3 be the position to which D_2 is brought by half the second displacement, shew that the centre of the total displacement of the rigid body is the intersection of D_1 and D_3. (Coll. Exam.)

3. *The theorem of Rodrigues and Hamilton*.*

Any two successive rotations about a fixed point can be compounded into a single rotation by means of a theorem, which may be stated as follows:

Successive rotations about three concurrent lines fixed in space, through twice the angles of the planes formed by them, restore a body to its original position.

For let the lines be denoted by OP, OQ, OR. Draw Op, Oq, Or perpendicular to the planes QOR, ROP, POQ respectively. Then if a body is rotated through two right angles about Oq, and afterwards through two right angles about Or, the position of OP is on the whole unaffected, while Oq is moved to the position occupied by its image in the line Or; the effect is therefore the same as that of a rotation round OP through twice the angle between the planes PR and PQ, which we may call the angle RPQ. It follows that successive rotations round OP, OQ, OR through twice the angles RPQ, PQR, QRP, respectively, are equivalent to successive rotations through two right angles about the lines Oq, Or, Or, Op, Op, Oq; but the latter rotations will clearly on the whole produce no displacement; which establishes the theorem.

4. *The composition of equal and opposite rotations about parallel axes.*

A case of special interest is that in which a body is subjected in turn to two rotations of equal amount in opposite senses about two parallel axes.

In neither displacement is any point of the body displaced in a direction parallel to the axes, and this is therefore true of the total displacement. Moreover, if any line be taken in the body in a plane perpendicular to the

* O. Rodrigues, *Journ. de Math.* v. (1840), p. 380; Hamilton, *Lectures on Quaternions*, § 344; the proof here given is due to Burnside, *Acta Math.* xxv. (1902).

axes, this line in the first displacement will be turned through an angle equal to the angle of rotation, and in the second displacement will be turned back through the same angle; so its final position will be parallel to its original position; which evidently can be the case for every line without exception, only when the total displacement is equivalent to a simple translation. It follows that two successive equal and opposite rotations about parallel axes are equivalent to a translation in a direction perpendicular to the axes; or, in other words, *a rotation about any axis is equivalent to a rotation through the same angle about any axis parallel to it, together with a simple translation in a direction perpendicular to the axis.*

The converse of this, namely the theorem that *a rotation of a rigid body about any axis, preceded or followed by a translation in a direction perpendicular to the axis, are together equivalent to a rotation of the body about a parallel axis,* is also true, being essentially the same as the result stated in § 2, that any displacement in a plane can be regarded as a rotation round some point in the plane. By considering the angle between the initial and final positions of any line which is perpendicular to the axis and moves with the body, we see that the angles of rotation round the two axes are equal.

5. *Chasles' theorem on the most general displacement of a rigid body*.*

We shall now consider displacements of a more general character. It is evident that a free rigid body can be moved from any one selected configuration P in space to any other Q by first moving some selected point of the body from its position in the configuration P to its position in the configuration Q, each of the other points of the body being moved by a simple translation parallel to this (so that the body is oriented in the same way after the operation as before), and secondly rotating the body about this point into the configuration Q. By Euler's theorem, this latter operation can be performed by simply rotating the body about a line through the point; so we see that *the most general displacement of a rigid body can be obtained by first translating the body, and then rotating it about a line.*

We shall now shew that *the line about which the rotation takes place can be so chosen, that the motion of translation is parallel to this line.* For let A be the initial position of any point of the body, and B the position to which this point is brought by the motion of translation. Let AK be the line through A parallel to the line round which the rotation takes place, and let K be the foot of the perpendicular from B on AK. Then the motion of translation can evidently be accomplished in two stages, the first of which is a translation parallel to the line about which the rotation takes place,

* Mozzi, *Discorso matematico sopra il rotamento momentaneo dei corpi*, Naples, 1763; Cauchy, *Exercices de Math.* II. (Paris, 1827), p. 87; *Oeuvres*, (2) VII. p. 94; Chasles, *Bulletin Univ. des Sciences* (Férussac), XIV. (1830), p. 321; *Comptes Rendus de l'Acad.* XVI. (1843), p. 1420.

bringing the point A to the position K, and the second of which is a translation perpendicular to the line about which the rotation takes place, bringing the point K to the position B. But by § 4, the second translation, together with the rotation which follows it, are together equivalent to a simple rotation about a new axis parallel to the first one. If therefore any point on this axis be taken as base-point, the whole displacement can be accomplished by a translation of the body parallel to a certain line through this point, together with a rotation about this line; this establishes the theorem.

This combination of a translation and a rotation round a line parallel to the direction of translation is called a *screw*; the ratio of the distance of translation to the angle of rotation is called the *pitch* of the screw. It is clear that in a screw displacement, the order in which the translation and rotation take place is indifferent.

6. *Halphen's theorem on the composition of two general displacements.*

Halphen has shewn* how to determine geometrically the resultant of any two screw-displacements as a screw-displacement.

Let A_1 and A_2 denote the axes of the two screws, and A_{12} their common perpendicular. Let B_1 be the line which is brought to the position A_{12} by half the first displacement (i.e. half the translation, and rotation through half the angle), and let B_2 be the line to whose position A_{12} is brought by half the second displacement; let C denote the common perpendicular to the lines B_1 and B_2. Halphen's result is that *the axis of the resultant screw-displacement is C, and the displacement is twice that which brings the line B_1 to the position B_2.*

For let D_1 and D_2 be lines such that half the given displacements will bring A_{12} to the position D_1 and D_2 to the position A_{12} respectively, and let C' be the common perpendicular to D_1 and D_2.

The figure thus obtained, and that which is obtained from it by rotating it through two right angles about A_{12}, evidently coincide; whence we have the relations:

Intercept made on B_1 by A_1 and C = Intercept made on D_1 by A_1 and C',
Intercept made on B_2 by A_2 and C = Intercept made on D_2 by A_2 and C',
Intercept made on C by B_1 and B_2 = Intercept made on C' by D_1 and D_2,
Angle between the planes A_1B_1, B_1C = Angle between the planes A_1D_1, D_1C',
Angle between the planes A_2B_2, B_2C = Angle between the planes A_2D_2, D_2C',
Angle between B_1 and B_2 = Angle between D_1 and D_2.

It follows that the screw about A_1 brings C to the position of C' produced, the intersection of B_1 and C being brought to the position of the intersection of D_1 and C'; and then the screw about A_2 brings C' to the position of C produced, the intersection of D_2 and C' being brought to the intersection of B_2 and C; so C is the axis of the resultant screw, and the amount of the translation is twice the intercept made on C by B_1 and B_2. Also the line B_1, which by the first screw is brought to the position D_1, is by the second brought to a position making the same angle with B_2 that B_2 makes with B_1; and therefore

* *Nouvelles Annales de Math.* (3) I. p. 298 (1882). The proof given here is due to Burnside, *Mess. of Math.* XIX. p. 104 (1889).

the rotation of the resultant screw is twice the angle between B_2 and B_1. This establishes Halphen's theorem.

Example. Shew that any infinitesimal displacement of a rigid body can be obtained by the composition of two infinitesimal rotations round lines, and that one of these lines can be arbitrarily chosen.

7. *Analytic representation of a displacement.*

We shall now see how any displacement of a rigid body can be represented analytically.

Let rectangular axes $Oxyz$ be taken, fixed in space: these will be supposed to form a right-handed system, i.e. if the axes are so placed that Oz is directed vertically upwards and Oy is directed to the northern horizon, then Ox will be directed to the east. Let the displacement considered be equivalent to a rotation through an angle ω about a line whose direction-angles are (α, β, γ), and which passes through a point A whose coordinates are (a, b, c), together with a translation through a distance d parallel to this line. The angle ω must be taken with its appropriate sign, the sign being positive when the line (α, β, γ) being directed vertically upwards, the rotation from the southern horizon to the northern is round by the east. Let the point P whose coordinates are (x, y, z) be brought by the displacement to the position of the point $Q\,(X, Y, Z)$; and let the point P be brought by the translation alone to the position of the point $R\,(\xi, \eta, \zeta)$; then we have evidently

$$\xi = x + d\cos\alpha, \qquad \eta = y + d\cos\beta, \qquad \zeta = z + d\cos\gamma.$$

Let K be the foot of the perpendicular from R (or Q) on the axis of rotation, and let L be the foot of the perpendicular from Q on KR. Then we have

$$X - \xi = \text{projection of the broken line } RLQ \text{ on the axis } Ox,$$

it being understood that projections have their appropriate signs, so that the projection of a line AB on the axis of x is $(x_B - x_A)$, not $(x_A - x_B)$.

Now the projection of KR on the axis Ox is

$$\xi - a - (\text{projection of } AK \text{ on the axis } Ox)$$

or

$$\xi - a - \cos\alpha\,\{(\xi - a)\cos\alpha + (\eta - b)\cos\beta + (\zeta - c)\cos\gamma\},$$

and as $RL = -(1 - \cos\omega)\,KR$, it follows that the projection of RL on the axis Ox is

$$-(1 - \cos\omega)\,[\xi - a - \cos\alpha\,\{(\xi - a)\cos\alpha + (\eta - b)\cos\beta + (\zeta - c)\cos\gamma\}].$$

Moreover, the line LQ is normal to the plane RKA, and its direction-cosines are therefore proportional to the quantities

$$(\zeta - c)\cos\beta - (\eta - b)\cos\gamma, \qquad (\xi - a)\cos\gamma - (\zeta - c)\cos\alpha,$$
$$(\eta - b)\cos\alpha - (\xi - a)\cos\beta,$$

and since the sum of the squares of these three quantities, divided by the expression $\{(\xi-a)^2+(\eta-b)^2+(\zeta-c)^2\}$, represents the quantity $\sin^2 R\hat{A}K$, it follows that the sum of the three squares is equal to KR^2, and the three quantities themselves are the projections on the axes of a length $\pm KR$ measured along the line LQ. Since $LQ = \pm KR \sin\omega$, the projection of LQ on the axis Ox is therefore

$$\pm \sin\omega\,\{(\zeta-c)\cos\beta-(\eta-b)\cos\gamma\}.$$

On considering a special case, e.g. supposing that the axis of rotation is the axis Oz, we see that the upper sign is correct; and thus we have

$$\begin{aligned} X-\xi = -(1-\cos\omega)\,\{&(\xi-a)-\cos^2\alpha\,(\xi-a)\\ &-\cos\alpha\cos\beta\,(\eta-b)-\cos\alpha\cos\gamma\,(\zeta-c)\}\\ &+\sin\omega\,\{\cos\beta\,(\zeta-c)-\cos\gamma\,(\eta-b)\}. \end{aligned}$$

Substituting for ξ, η, ζ their values in terms of x, y, z, we have

$$\begin{aligned} X = x + d\cos\alpha-(1-\cos\omega)\,\{&(x-a)\sin^2\alpha\\ &-\cos\alpha\cos\beta\,(y-b)-\cos\alpha\cos\gamma\,(z-c)\}\\ &+\sin\omega\,\{\cos\beta\,(z-c)-\cos\gamma\,(y-b)\}. \end{aligned}$$

Similarly we have

$$\begin{aligned} Y = y + d\cos\beta-(1-\cos\omega)\,\{&(y-b)\sin^2\beta\\ &-\cos\beta\cos\gamma\,(z-c)-\cos\beta\cos\alpha\,(x-a)\}\\ &+\sin\omega\,\{\cos\gamma\,(x-a)-\cos\alpha\,(z-c)\} \end{aligned}$$

and

$$\begin{aligned} Z = z + d\cos\gamma-(1-\cos\omega)\,\{&(z-c)\sin^2\gamma\\ &-\cos\gamma\cos\alpha\,(x-a)-\cos\gamma\cos\beta\,(y-b)\}\\ &+\sin\omega\,\{\cos\alpha\,(y-b)-\cos\beta\,(x-a)\}. \end{aligned}$$

These equations give the new coordinates X, Y, Z in terms of the coordinates x, y, z of the original position of the point and the quantities which define the displacement.

8. *The composition of small rotations.*

We shall now apply the last result to the case in which the rotation is infinitesimal, the axis of rotation passing through the origin and there being no motion of translation. We shall write $\delta\psi$ for ω, where $\delta\psi$ is a small quantity whose square can be neglected. The equations of the last article now become

$$\begin{cases} X = x + (z\cos\beta - y\cos\gamma)\,\delta\psi,\\ Y = y + (x\cos\gamma - z\cos\alpha)\,\delta\psi,\\ Z = z + (y\cos\alpha - x\cos\beta)\,\delta\psi. \end{cases}$$

But these are the equations which we should obtain if we successively (in any order) subjected the body to infinitesimal rotations $\cos\alpha\,.\,\delta\psi$ about Ox, $\cos\beta\,.\,\delta\psi$ about Oy, and $\cos\gamma\,.\,\delta\psi$ about Oz. It follows that *any small rotation $\delta\psi$ about a line OK is equivalent to successive small rotations $\delta\psi\,.\cos K\hat{O}x$*

about Ox, $\delta\psi \,.\cos K\hat{O}y$ about Oy, and $\delta\psi \,.\cos K\hat{O}z$ about Oz, where Ox, Oy, Oz are any three mutually perpendicular lines which intersect OK in one of its points, O.

9. *Euler's parametric specification of rotations round a point**.

The analytic expressions for the translational part of a displacement are, as we have seen, extremely simple; but the expressions for the rotational part are not so simple, and these will now be further considered. Suppose then that a rigid body is rotated through an angle ω about a line through the origin, whose direction-angles are α, β, γ. By § 7, the coordinates (X, Y, Z) of the new position of a point whose original coordinates were (x, y, z) are given by the equations

$$\begin{cases} X = x - 2\sin^2 \tfrac{1}{2}\omega\,(x\sin^2\alpha - y\cos\alpha\cos\beta - z\cos\alpha\cos\gamma) \\ \qquad\qquad + 2\sin\tfrac{1}{2}\omega\cos\tfrac{1}{2}\omega\,(z\cos\beta - y\cos\gamma), \\ Y = y - 2\sin^2 \tfrac{1}{2}\omega\,(y\sin^2\beta - z\cos\beta\cos\gamma - x\cos\beta\cos\alpha) \\ \qquad\qquad + 2\sin\tfrac{1}{2}\omega\cos\tfrac{1}{2}\omega\,(x\cos\gamma - z\cos\alpha), \\ Z = z - 2\sin^2 \tfrac{1}{2}\omega\,(z\sin^2\gamma - x\cos\gamma\cos\alpha - y\cos\gamma\cos\beta) \\ \qquad\qquad + 2\sin\tfrac{1}{2}\omega\cos\tfrac{1}{2}\omega\,(y\cos\alpha - x\cos\beta). \end{cases}$$

Now introduce parameters ξ, η, ζ, χ, defined by the equations

$$\xi = \cos\alpha\sin\tfrac{1}{2}\omega, \quad \eta = \cos\beta\sin\tfrac{1}{2}\omega, \quad \zeta = \cos\gamma\sin\tfrac{1}{2}\omega, \quad \chi = \cos\tfrac{1}{2}\omega;$$

these parameters evidently satisfy the identical relation

$$\xi^2 + \eta^2 + \zeta^2 + \chi^2 = 1,$$

and the above equations can be written in the form

$$\begin{cases} X = (\xi^2 - \eta^2 - \zeta^2 + \chi^2)\,x + 2(\xi\eta - \zeta\chi)\,y + 2(\xi\zeta + \eta\chi)\,z, \\ Y = 2(\xi\eta + \zeta\chi)\,x + (-\xi^2 + \eta^2 - \zeta^2 + \chi^2)\,y + 2(\eta\zeta - \xi\chi)\,z, \\ Z = 2(\xi\zeta - \eta\chi)\,x + 2(\eta\zeta + \xi\chi)\,y + (-\xi^2 - \eta^2 + \zeta^2 + \chi^2)\,z. \end{cases}$$

If therefore the coordinate axes are denoted by $OXYZ$, and if moveable axes which originally coincide with these are brought into the position $Oxyz$ by the given rotation, the direction-cosines of the two sets of axes with reference to each other are given by the following scheme:

	X	Y	Z
x	$\xi^2-\eta^2-\zeta^2+\chi^2$	$2(\xi\eta+\zeta\chi)$	$2(\xi\zeta-\eta\chi)$
y	$2(\xi\eta-\zeta\chi)$	$-\xi^2+\eta^2-\zeta^2+\chi^2$	$2(\eta\zeta+\xi\chi)$
z	$2(\xi\zeta+\eta\chi)$	$2(\eta\zeta-\xi\chi)$	$-\xi^2-\eta^2+\zeta^2+\chi^2$

* *Novi Comment. Petrop.* xx. (1776), p. 208, § 6 sqq.

It is readily seen that the parameters $(\xi'', \eta'', \zeta'', \chi'')$, corresponding to the resultant of two successive displacements $(\xi', \eta', \zeta', \chi')$ and (ξ, η, ζ, χ), are given by the equations

$$\begin{cases} \xi'' = \xi\chi' + \eta\zeta' - \zeta\eta' + \chi\xi', \\ \eta'' = -\xi\zeta' + \eta\chi' + \zeta\xi' + \chi\eta', \\ \zeta'' = \xi\eta' - \eta\xi' + \zeta\chi' + \chi\zeta', \\ \chi'' = \chi\chi' - \xi\xi' - \eta\eta' - \zeta\zeta'. \end{cases}$$

These formulae (which were discovered independently at different times by Gauss, Rodrigues, Hamilton, and Cayley) really constitute the theorem for the *multiplication of quaternions*. For χ, ξ, η, ζ may be regarded as the components of a quaternion* $\chi + \xi i + \eta j + \zeta k$, where i, j, k satisfy the equations

$$i^2 = j^2 = k^2 = -1, \quad ij = -ji = k, \quad jk = -kj = i, \quad ki = -ik = j;$$

and the above formulae are then all comprehended in the single equation

$$\chi'' + \xi'' i + \eta'' j + \zeta'' k = (\chi + \xi i + \eta j + \zeta k)(\chi' + \xi' i + \eta' j + \zeta' k).$$

The reader who is acquainted with quaternions will observe that the effect of the rotation on any vector ρ is to convert it into the vector $q\rho q^{-1}$, where q denotes the quaternion $\chi + \xi i + \eta j + \zeta k$; the quaternion itself is *not* the rotational operator.

10. *The Eulerian angles.*

The most practically useful of the various methods of representing parametrically the displacement of a rigid body due to a rotation round a fixed point is likewise due to Euler†: it has the disadvantage of being unsymmetrical, but is otherwise very simple and convenient.

Let O be the fixed point round which the rotation takes place, and let $OXYZ$ be a right-handed system of rectangular axes fixed in space. Let $Oxyz$ be rectangular axes fixed relatively to the body and moving with it, and such that before the displacement the two sets of axes $OXYZ$ and $Oxyz$ are coincident in position. Let OK be perpendicular to the plane zOZ, drawn so that if OZ is directed to the vertical and the projection of Oz perpendicular to OZ is directed to the south, then OK is directed to the east. Denote the angles $z\hat{O}Z$, $Y\hat{O}K$, $y\hat{O}K$ by θ, ϕ, ψ, respectively: these are known as the three *Eulerian angles* defining the position of the axes $Oxyz$ with reference to the axes $OXYZ$.

In order to find the direction-cosines of Ox, Oy, Oz, with respect to OX, we observe that these are equal to the projections on Ox, Oy, Oz, respectively, of a unit length measured along OX. Now this unit length has projections $\cos\phi$ along OL and $-\sin\phi$ along OK, where OL is the intersection of the planes XOY and ZOz; but a length $\cos\phi$ along OL has projections $\cos\phi\sin\theta$ along Oz and $\cos\phi\cos\theta$ along OM, where OM is the intersection of the planes xOy and ZOz; and a length $\cos\phi\cos\theta$ along OM has projections $\cos\phi\cos\theta\cos\psi$ along Ox and $-\cos\phi\cos\theta\sin\psi$ along Oy; also, a length $-\sin\phi$ along OK has projections $-\sin\phi\sin\psi$ along Ox and $-\sin\phi\cos\psi$

* This quaternion will have its tensor equal to unity.

† *Novi Comment. Petrop.* xx. (1776), p. 189.

along Oy. Hence finally the projections on Ox, Oy, Oz respectively of the unit length measured on OX are

$$\begin{cases} \cos\phi\cos\theta\cos\psi - \sin\phi\sin\psi & \text{along } Ox \\ -\cos\phi\cos\theta\sin\psi - \sin\phi\cos\psi & \text{along } Oy, \\ \cos\phi\sin\theta & \text{along } Oz. \end{cases}$$

Proceeding in this way, we obtain for the direction-cosines of the two sets of axes $OXYZ$ and $Oxyz$ with respect to each other the following scheme:

	X	Y	Z
x	$\cos\phi\cos\theta\cos\psi - \sin\phi\sin\psi$	$\sin\phi\cos\theta\cos\psi + \cos\phi\sin\psi$	$-\sin\theta\cos\psi$
y	$-\cos\phi\cos\theta\sin\psi - \sin\phi\cos\psi$	$-\sin\phi\cos\theta\sin\psi + \cos\phi\cos\psi$	$\sin\theta\sin\psi$
z	$\cos\phi\sin\theta$	$\sin\phi\sin\theta$	$\cos\theta$

11. *Connexion of the Eulerian angles with the parameters* ξ, η, ζ, χ.

The relations between the Eulerian angles θ, ϕ, ψ and the parameters ξ, η, ζ, χ of § 9 may be obtained by comparing the schemes of direction-cosines which have been given in §§ 9 and 10; they may however be obtained directly as follows:

Let $OXYZ$ and $Oxyz$ be the fixed axes and the axes derived from these by the rotation ω round a line OR, whose direction-angles are (α, β, γ). Draw a sphere of unit radius with the point O as centre, so that planes passing through O intersect the sphere in great circles, and lines intersect the sphere in points. Then in the spherical triangle RZz, the sides are γ, γ, θ, and the angle at R is ω; whence we have the relation

$$\sin\tfrac{1}{2}\theta = \sin\gamma\sin\tfrac{1}{2}\omega.$$

Moreover, let ν denote the angle $R\hat{Z}Y$, so $R\hat{Z}z = \tfrac{1}{2}\pi - \phi - \nu$. Then the arc RZ is brought to the position Rz by successive rotations ϕ about Z, θ about the pole of Zz, and ψ about z; but the first of these transforms RZ into an arc making an angle $\tfrac{1}{2}\pi - \phi - \nu + \phi$ or $\tfrac{1}{2}\pi - \nu$ with Zz, at Z; the second rotation transforms this into an arc making the same angle $\tfrac{1}{2}\pi - \nu$ with Zz, but passing through z; and the third rotation transforms it into an arc making an angle $\tfrac{1}{2}\pi - \nu + \psi$ with Zz, at z. But this angle must be equal to $\pi - R\hat{z}Z$, or $\pi - R\hat{Z}z$, or $\pi - (\tfrac{1}{2}\pi - \phi - \nu)$, or $\tfrac{1}{2}\pi + \phi + \nu$; so we have

$$\tfrac{1}{2}\pi + \phi + \nu = \tfrac{1}{2}\pi - \nu + \psi,$$

or

$$\nu = \tfrac{1}{2}(\psi - \phi).$$

Hence, since in the spherical triangle RZX the sides are α, γ, $\frac{1}{2}\pi$, and the angle at Z is $\frac{1}{2}\pi - \nu$ or $\frac{1}{2}(\pi - \psi + \phi)$, we have

$$\cos\alpha = \sin\gamma \sin\tfrac{1}{2}(\psi - \phi)$$

Substituting for $\sin\gamma$ from the equation already found, this gives

$$\cos\alpha \sin\tfrac{1}{2}\omega = \sin\tfrac{1}{2}\theta \sin\tfrac{1}{2}(\psi - \phi),$$

or

$$\xi = \sin\tfrac{1}{2}\theta \sin\tfrac{1}{2}(\psi - \phi).$$

Similarly from the spherical triangle RZY we have

$$\cos\beta = \sin\gamma \cos\tfrac{1}{2}(\psi - \phi),$$

and again eliminating $\sin\gamma$, we have

$$\cos\beta \sin\tfrac{1}{2}\omega = \sin\tfrac{1}{2}\theta \cos\tfrac{1}{2}(\psi - \phi),$$

or

$$\eta = \sin\tfrac{1}{2}\theta \cos\tfrac{1}{2}(\psi - \phi).$$

Moreover, since we have shewn that in the spherical triangle RZz the sides are γ, γ, θ, and the angles are $\frac{1}{2}(\pi - \psi - \phi)$, $\frac{1}{2}(\pi - \psi - \phi)$, ω, we have the relations

$$\cos\tfrac{1}{2}\omega = \cos\tfrac{1}{2}\theta \cos\tfrac{1}{2}(\psi + \phi),$$

and

$$\sin\tfrac{1}{2}\omega \cos\gamma = \cos\tfrac{1}{2}\theta \sin\tfrac{1}{2}(\psi + \phi),$$

or

$$\chi = \cos\tfrac{1}{2}\theta \cos\tfrac{1}{2}(\psi + \phi),$$

$$\zeta = \cos\tfrac{1}{2}\theta \sin\tfrac{1}{2}(\psi + \phi).$$

The four parameters ξ, η, ζ, χ are thus expressed in terms of the Eulerian angles θ, ϕ, ψ by the relations

$$\begin{cases} \xi = \sin\tfrac{1}{2}\theta \sin\tfrac{1}{2}(\psi - \phi), \\ \eta = \sin\tfrac{1}{2}\theta \cos\tfrac{1}{2}(\psi - \phi), \\ \zeta = \cos\tfrac{1}{2}\theta \sin\tfrac{1}{2}(\psi + \phi), \\ \chi = \cos\tfrac{1}{2}\theta \cos\tfrac{1}{2}(\psi + \phi). \end{cases}$$

12. *The connexion of rotations with homographies; the Cayley-Klein parameters.*

Consider now a sphere, on the surface of which any figures (which we shall call S) are drawn. Let these figures be stereographically projected on a plane (e.g. by taking the highest point of the sphere as vertex of projection and the tangent-plane at the lowest point of the sphere as the plane): we shall call the projected figures P. Now let the sphere be rotated through a definite angle about some axis through its centre, so that the figures on its surface are shifted to new positions: let the figures in their new positions be called S'; and let the stereographic projections of the figures S' (with the same vertex and plane of projection as before) be called P'. Then corresponding to the rotation of the sphere, which changes S to S', we have a *transformation* in the plane, which changes the figures P into the figures P'. We shall now examine this transformation more closely.

If one of the figures P is a circle in the plane, we know that the corresponding figure S must be a circle traced on the sphere, since by stereographic projection a circle is changed into a circle: therefore S' must also be a circle; and hence P' must also be a circle. Thus we see that *the transformations of the plane, which correspond to rotations of the sphere, must be such as to change any circle in the plane into another circle in the plane.*

It may be shewn* that any transformation of this kind may be represented analytically in the following way:

Let $z = x + y\sqrt{-1}$, where x and y are the rectangular coordinates of any point in the plane; so that to this point there corresponds a definite value of the complex variable z. Similarly let $z' = x' + y'\sqrt{-1}$, where x' and y' refer to the point into which the point (x, y) is changed by the transformation. Then *any one-to-one transformation of the plane, which changes all circles into circles*†, *may be defined by an equation of the type*

$$z' = \frac{az+b}{cz+d},$$

where a, b, c, d are (real or complex) constants; or else by a transformation of this latter kind combined with a reflexion in one of the axes of coordinates.

A transformation represented by an equation of the type

$$z' = \frac{az+b}{cz+d}$$

is called a *homographic* transformation, or *homography*. It appears therefore that homographies in a plane correspond to rotations of a solid body about a fixed point, in such a way that if two homographies correspond respectively to two rotations, the homography compounded of these corresponds to the rotation compounded of the two rotations‡.

We shall now see how the connexion between rotations and homographies may be represented analytically.

Let us replace the parameters ξ, η, ζ, χ by new parameters $\alpha, \beta, \gamma, \delta$, defined by the equations

$$\xi = \frac{\beta-\gamma}{2}, \quad \eta = \frac{\beta+\gamma}{2i}, \quad \zeta = \frac{\alpha-\delta}{2i}, \quad \chi = \frac{\alpha+\delta}{2},$$

so that they are connected with the Eulerian angles θ, ϕ, ψ by the equations

$$\alpha = \cos\frac{\theta}{2} \cdot e^{\frac{i}{2}(\phi+\psi)}, \qquad \gamma = i\sin\frac{\theta}{2} \cdot e^{\frac{i}{2}(\psi-\phi)},$$

$$\beta = i\sin\frac{\theta}{2} \cdot e^{\frac{i}{2}(\phi-\psi)}, \qquad \delta = \cos\frac{\theta}{2} \cdot e^{\frac{i}{2}(-\phi-\psi)}.$$

These "Cayley-Klein" parameters clearly satisfy the relation

$$\alpha\delta - \beta\gamma = 1;$$

and replacing the quantities ξ, η, ζ, χ in the scheme of direction-cosines given in § 9 by their values in terms of $\alpha, \beta, \gamma, \delta$, we have for the values of the direction-cosines in terms of $\alpha, \beta, \gamma, \delta$ the following scheme:

	X	Y	Z
x	$\frac{1}{2}(\alpha^2+\beta^2+\gamma^2+\delta^2)$	$\frac{i}{2}(-\alpha^2-\beta^2+\gamma^2+\delta^2)$	$i(\alpha\gamma+\beta\delta)$
y	$\frac{i}{2}(\alpha^2-\beta^2+\gamma^2-\delta^2)$	$\frac{1}{2}(\alpha^2-\beta^2-\gamma^2+\delta^2)$	$-\alpha\gamma+\beta\delta$
z	$-i(\alpha\beta+\gamma\delta)$	$-\alpha\beta+\gamma\delta$	$\alpha\delta+\beta\gamma$

* Cf. L. R. Ford, *An introduction to the theory of automorphic functions* (London, 1915).

† A straight line is to be regarded as a particular kind of circle.

‡ Klein, *Math. Ann.* IX. (1875), p. 183; Cayley, *Math. Ann.* XV. (1879), p. 238.

It may readily be shewn that the parameters $(\alpha'', \beta'', \gamma'', \delta'')$ corresponding to the resultant of two successive displacements $(\alpha', \beta', \gamma', \delta')$ and $(\alpha, \beta, \gamma, \delta)$ are given by the equations

$$\alpha''=\alpha'\alpha+\gamma'\beta, \qquad \beta''=\alpha\beta'+\beta\delta',$$
$$\gamma''=\gamma\alpha'+\delta\gamma', \qquad \delta''=\gamma\beta'+\delta\delta'.$$

These equations shew that the transformation

$$z'=\frac{\alpha''z+\beta''}{\gamma''z+\delta''}$$

is the result of performing in succession the two substitutions

$$z'=\frac{\alpha'z+\beta'}{\gamma'z+\delta'} \quad \text{and} \quad z'=\frac{\alpha z+\beta}{\gamma z+\delta},$$

and the connexion between rotations and homographic transformations is thus evident analytically.

One advantage of the Cayley-Klein parameters, as compared with the parameters (ξ, η, ζ, χ), is that they retain some of the simplicity of the quaternion calculus, while using the $\sqrt{-1}$ of ordinary algebra instead of the i, j, k of Hamilton's quaternions.

Example 1. Let (θ, ϕ, ψ) denote the Eulerian angles. Suppose that a point in space which is carried about with the axes $Oxyz$ has the vectorial angles (θ_1, ϕ_1) (referred to the fixed axes $OXYZ$) before the motion, and (θ_1', ϕ_1') after the motion. Denoting $e^{i\phi_1}\tan\frac{1}{2}\theta_1$ by ζ_1, and $e^{i\phi_1'}\tan\frac{1}{2}\theta_1'$ by ζ_1', shew that

$$\zeta_1 e^{i\psi}=\frac{\zeta_1' e^{-i\phi}\cos\frac{1}{2}\theta-\sin\frac{1}{2}\theta}{\zeta_1' e^{-i\phi}\sin\frac{1}{2}\theta+\cos\frac{1}{2}\theta}.$$

Example 2. If from the equations

$$X_1=\alpha x_1+\beta x_2,$$
$$X_2=\gamma x_1+\delta x_2,$$

the quantities X_1^2, X_2^2, X_1X_2 are formed, and if these quantities are regarded as umbral symbols and the quantities X_1^2, X_2^2, X_1X_2, x_1^2, x_2^2, x_1x_2 are replaced by $-Y+iX$, $Y+iX$, Z, $-y+ix$, $y+ix$, z, respectively, shew that the equations obtained are

$$\begin{cases} -Y+iX=\alpha^2(-y+ix)+2\alpha\beta z+\beta^2(y+ix), \\ Y+iX=\gamma^2(-y+ix)+2\gamma\delta z+\delta^2(y+ix), \\ Z=\alpha\gamma(-y+ix)+(\alpha\delta+\beta\gamma)z+\beta\delta(y+ix), \end{cases}$$

and that these are the three equations connecting the coordinates (X, Y, Z) of a point referred to the axes $OXYZ$ with its coordinates (x, y, z) referred to the axes $Oxyz$.

Example 3. If

$$-y+ix : y+ix : z=\lambda\lambda' : 1 : \tfrac{1}{2}(\lambda+\lambda'),$$

and

$$-Y+iX : Y+iX : Z=\lambda_1\lambda_1' : 1 : \tfrac{1}{2}(\lambda_1+\lambda_1'),$$

shew that

$$\lambda_1=\frac{\alpha\lambda+\beta}{\gamma\lambda+\delta} \quad \text{and} \quad \lambda_1'=\frac{\alpha\lambda'+\beta}{\gamma\lambda'+\delta}.$$

13. *Vectors.*

We now proceed to consider the essential features involved in the displacement by simple translation of a rigid body.

The operation of translation in itself, considered apart from the body translated, evidently possesses the following properties:

1°. It can be specified completely by any one of the equal and parallel lines of space which have a given length (viz. the distance of the translation) and given direction (viz. the direction of the translation); since such a line furnishes all the data which describe the operation.

2°. If AB be one of these lines, and $ACDE...KB$ be a broken line joining its extremities, then the operation represented by AB is equivalent to the sum of the operations represented by AC, CD, DE, ... KB.

These properties 1° and 2° are possessed by a large number of operations and quantities other than the operation of translation; an operation or quantity which possesses them is called a *vector quantity*.

By 2°, a vector AB is equivalent to the sum of three vectors AK, KL, LB, respectively parallel to three given rectangular axes, and forming a broken line joining the points A and B. These three vectors are called the *components* of the vector AB along the given axes. If l be the length and (α, β, γ) the direction-angles of AB, the lengths of the component vectors are clearly $(l\cos\alpha, l\cos\beta, l\cos\gamma)$, being in fact the projections of AB on the axes.

A single vector which is equivalent to any number of given vectors is called their *resultant*.

If a vector is conceived as varying in dependence on a parameter (e.g. the time), the difference between the vectors corresponding to any two values of the parameter is also a vector, and hence the rate of change of the vector with respect to the parameter is also a vector, whose components are the rates of change of the corresponding components. This is called the *flux* of the vector with respect to the parameter.

14. *Velocity and acceleration; their vectorial character.*

Consider now a body which is being continuously translated (though not necessarily always in the same direction) without any change of orientation. Its total translation to any time t is a vector quantity and hence the rate at which this changes with the time, i.e. its time-flux, is also a vector quantity, which is called the *velocity* of the body; if x, y, z are the coordinates referred to fixed axes of any point fixed in the body and moving with it, then the components of the velocity referred to these axes are the rates of change of x, y, z, i.e. are $\dot{x}$, $\dot{y}$, $\dot{z}$ (where dots denote differentiations with respect to the time t).

Similarly the rate of change of the velocity is again a vector, whose components are $\ddot{x}$, $\ddot{y}$, $\ddot{z}$ (two dots indicating second derivatives with respect to the time); this vector is called the *acceleration* of the body.

It is clear that if P and Q are two moving points, the vector which represents the translation (or velocity, or acceleration) of Q is the sum of the vector which represents the translation (or velocity, or acceleration, as the case may be) of P and the vector which represents the translation (or velocity, or acceleration) of Q *relative to* P, i.e. of Q referred to axes whose origin moves with P, and whose directions are invariable.

15. *Angular velocity: its vectorial character.*

Consider next a body which is rotating continuously about a line. Let θ denote the angle turned through at any time t: then $\dot{\theta}$ represents the speed of turning at the time t. If from any point on the line round which the rotation takes place a segment whose length represents $\dot{\theta}$ is measured along the line, this segment will evidently furnish a complete specification of the nature of the rotation at the instant t, or (as it is generally expressed) of the *angular velocity* of the body. The direction in which the segment is measured from the base-point is to be connected with the sense of rotation by the usual convention, namely that when the segment is directed vertically upwards the rotation from the southern horizon to the northern is round by the east.

An angular velocity is therefore represented by a line of definite length and direction. Now by § 8, if a body one of whose points O is fixed experiences a small rotation $\delta\psi$ round any line OK, this displacement is equivalent to successive small rotations $\delta\psi\cos\alpha$ round Ox, $\delta\psi\cos\beta$ round Oy, and $\delta\psi\cos\gamma$ round Oz, where Ox, Oy, Oz are any three mutually perpendicular lines passing through O and (α, β, γ) are the direction-angles of OK with reference to $Oxyz$. From this it is clear that we can regard an angular velocity represented by a length $\dot{\psi}$ measured on OK as equivalent to angular velocities represented by lengths $\dot{\psi}\cos\alpha$, $\dot{\psi}\cos\beta$, $\dot{\psi}\cos\gamma$, measured along Ox, Oy, Oz, respectively.

But this is essentially the fundamental property of vectors, and can be expressed by the statement that *angular velocities can be resolved and compounded according to the vectorial law.*

It must be observed however that an angular velocity does not fulfil all the conditions which enter into the definition of a vector, for an angular velocity about one line is not equivalent to an angular velocity of the same magnitude about a parallel line. Angular velocity must therefore be regarded as a vector which is *localised* along a definite line.

Example. *A right circular cone of semi-vertical angle β rolls without sliding on a plane. To find its instantaneous axis of rotation, and to determine its angular velocity about this axis in terms of the angular velocity of the line of contact in the plane.*

Since all points of the generator which is in contact with the plane are instantaneously at rest (for there is no sliding), this generator is the instantaneous axis of rotation of the cone. Let ω denote the angular velocity of the cone about this generator, and let $\dot{\theta}$ denote the angular velocity of the line of contact in the plane. Then the motion of the axis of the cone can be represented by an angular velocity $\dot{\theta}$ round the normal to the plane, and the whole motion of the cone is compounded of this together with a rotation round the axis of the cone. It follows that the component of angular velocity of the cone about a line through the vertex of the cone perpendicular to the axis is $\dot{\theta}\cos\beta$; but this must equal the resolved part of ω in this direction, which is $\omega\sin\beta$. We have therefore

$$\omega=\theta\cot\beta,$$

which is the required relation between ω and $\dot{\theta}$.

16. *Determination of the components of angular velocity of a system in terms of the Eulerian angles, and of the symmetrical parameters.*

The position at any time of a rigid body which is continuously moving about a fixed point O is most conveniently described by taking two sets of rectangular axes, of which one set $OXYZ$ are fixed in space, while the other set $Oxyz$ are fixed relatively to the body, and move with it; the position of the body being then specified by the three Eulerian angles θ, ϕ, ψ, which define the position of the axes $Oxyz$ relatively to the axes $OXYZ$. We shall now determine the components, along the moving axes, of the angular velocity of the body at any instant.

Let OK denote the line of intersection of the planes XOY and xOy; the angular velocity of the system is evidently compounded of angular velocities $\dot{\theta}$ about OK, $\dot{\phi}$ about OZ, and $\dot{\psi}$ about Oz. Of these, the first can be replaced according to the vectorial law by angular velocities $\dot{\theta}\sin\psi$ about Ox and $\dot{\theta}\cos\psi$ about Oy; and the second can be resolved into $-\dot{\phi}\sin\theta\cos\psi$ about Ox, $\dot{\phi}\sin\theta\sin\psi$ about Oy, and $\dot{\phi}\cos\theta$ about Oz. So finally if ω_1, ω_2, ω_3 denote the components of angular velocity of the body about the axes Ox, Oy, Oz, respectively, we have

$$\begin{cases} \omega_1 = \dot{\theta}\sin\psi - \dot{\phi}\sin\theta\cos\psi, \\ \omega_2 = \dot{\theta}\cos\psi - \dot{\phi}\sin\theta\sin\psi, \\ \omega_3 = \dot{\psi} + \dot{\phi}\cos\theta. \end{cases}$$

From these expressions we can at once deduce the values of ω_1, ω_2, ω_3 in terms of the symmetrical parameters ξ, η, ζ, χ, of § 9; for we have

$$\begin{aligned} \dot{\phi} &= \frac{d}{dt}\left(\frac{\psi+\phi}{2}\right) - \frac{d}{dt}\left(\frac{\psi-\phi}{2}\right) \\ &= \frac{d}{dt}\left(\tan^{-1}\frac{\zeta}{\chi}\right) - \frac{d}{dt}\left(\tan^{-1}\frac{\xi}{\eta}\right) \\ &= \frac{\xi\dot{\eta}-\eta\dot{\xi}}{\xi^2+\eta^2} + \frac{\chi\dot{\zeta}-\zeta\dot{\chi}}{\zeta^2+\chi^2}. \end{aligned}$$

Similarly we have

$$\dot{\psi} = \frac{-\xi\dot{\eta}+\eta\dot{\xi}}{\xi^2+\eta^2} + \frac{\chi\dot{\zeta}-\zeta\dot{\chi}}{\zeta^2+\chi^2},$$

and we have $\cos\theta = -\xi^2 - \eta^2 + \zeta^2 + \chi^2$.

Substituting these values in the equation $\omega_3 = \dot{\psi} + \dot{\phi}\cos\theta$, we have

$$\omega_3 = 2(\eta\dot{\xi} - \xi\dot{\eta} + \chi\dot{\zeta} - \zeta\dot{\chi}).$$

The values of ω_1 and ω_2 can be at once obtained from this by the principle of symmetry; and thus we have the components of angular velocity given by the equations

$$\begin{cases} \omega_1 = 2(\chi\dot{\xi} + \zeta\dot{\eta} - \eta\dot{\zeta} - \xi\dot{\chi}), \\ \omega_2 = 2(-\zeta\dot{\xi} + \chi\dot{\eta} + \xi\dot{\zeta} - \eta\dot{\chi}), \\ \omega_3 = 2(\eta\dot{\xi} - \xi\dot{\eta} + \chi\dot{\zeta} - \zeta\dot{\chi}). \end{cases}$$

17. *Time-flux of a vector whose components relative to moving axes are given.*

Suppose now that a vector quantity is specified by its components ξ, η, ζ at any instant t with reference to the instantaneous position of a right-handed system of axes $Oxyz$ which are themselves in motion: and let it be required to find the vector which represents the rate of change of the given vector.

Let $\omega_1, \omega_2, \omega_3$ denote the components of the angular velocity of the system $Oxyz$, resolved along the instantaneous position of the axes Ox, Oy, Oz themselves.

The time-flux of the given vector is the (vector) sum of the time-fluxes of the components ξ, η, ζ, taken separately. But if we consider the vector ξ it is increased in length to $\xi + \dot{\xi}dt$ in the infinitesimal interval of time dt, and at the same time is turned by the motion of the axes, so that (owing to the angular velocity round Oy) it is displaced through an angle $\omega_2 dt$ from its position in the original plane zOx, in the direction away from Oz, and also (owing to the angular velocity round Oz) it is displaced through an angle $\omega_3 dt$ from its position in the original plane xOy, towards Oy. The coordinates of its extremity at the end of the interval of time dt, referred to the positions of the axes at the commencement of the interval dt, are therefore (neglecting infinitesimals of order higher than the first)

$$\xi + \dot{\xi}dt, \qquad \omega_3 \xi dt, \qquad -\omega_2 \xi dt,$$

and so the components of the vector which represents the time-flux of ξ are

$$\dot{\xi}, \qquad \omega_3 \xi, \qquad -\omega_2 \xi.$$

Similarly the components of the vectors which represent the time-fluxes of the vectors η and ζ are respectively

$$-\omega_3 \eta, \qquad \dot{\eta}, \qquad \omega_1 \eta,$$

and

$$\omega_2 \zeta, \qquad -\omega_1 \zeta, \qquad \dot{\zeta}.$$

Adding these, *we have finally the components of the time-flux of the given vector in the form*

$$\begin{cases} \dot{\xi} - \eta\omega_3 + \zeta\omega_2, \\ \dot{\eta} - \zeta\omega_1 + \xi\omega_3, \\ \dot{\zeta} - \xi\omega_2 + \eta\omega_1. \end{cases}$$

This result can be immediately applied to find the velocity and acceleration of a point whose coordinates (x, y, z) at time t are given with reference to axes moving with an angular velocity whose components along the axes themselves at time t are $(\omega_1, \omega_2, \omega_3)$.

For substituting in the above formulae, we see that the components of the velocity are

$$\dot{x} - y\omega_3 + z\omega_2, \qquad \dot{y} - z\omega_1 + x\omega_3, \qquad \dot{z} - x\omega_2 + y\omega_1.$$

Now applying the same formulae to the case in which the vector whose time-flux is sought is the velocity, we have the components of the acceleration of the point in the form

$$\begin{cases} \dfrac{d}{dt}(\dot{x} - y\omega_3 + z\omega_2) - \omega_3(\dot{y} - z\omega_1 + x\omega_3) + \omega_2(\dot{z} - x\omega_2 + y\omega_1), \\ \dfrac{d}{dt}(\dot{y} - z\omega_1 + x\omega_3) - \omega_1(\dot{z} - x\omega_2 + y\omega_1) + \omega_3(\dot{x} - y\omega_3 + z\omega_2), \\ \dfrac{d}{dt}(\dot{z} - x\omega_2 + y\omega_1) - \omega_2(\dot{x} - y\omega_3 + z\omega_2) + \omega_1(\dot{y} - z\omega_1 + x\omega_3). \end{cases}$$

In the case in which the motion takes place in a plane, which we may take as the plane Oxy, there will be only two coordinates (x, y), and only one component of angular velocity, namely $\dot{\theta}$, where θ is the angle made by the moving axes with their positions at some fixed epoch; the components of velocity are therefore (putting z, ω_1, ω_2 each equal to zero in the above expressions)

$$\dot{x} - y\dot{\theta} \text{ and } \dot{y} + x\dot{\theta},$$

and the components of acceleration are

$$\ddot{x} - 2\dot{y}\dot{\theta} - y\ddot{\theta} - x\dot{\theta}^2 \text{ and } \ddot{y} + 2\dot{x}\dot{\theta} + x\ddot{\theta} - y\dot{\theta}^2.$$

Example. Prove that in the general case of motion of a rigid body there is at each instant one *definite* point at a *finite* distance which regarded as invariably connected with the body has no acceleration at the instant, provided the axis of the body's screwing motion be not instantaneously stationary in direction. (Coll. Exam.)

18. *Special resolutions of the velocity and acceleration.*

The results obtained in the last article enable us to obtain formulae, which are frequently of use, relating to the components of the velocity and acceleration of a moving point in various special directions.

(i) *Velocity and acceleration in polar coordinates.*

Let the position of a point be defined by its polar coordinates r, θ, ϕ, connected with the coordinates (X, Y, Z) of the point referred to fixed rectangular axes $OXYZ$ by the equations

$$\begin{cases} X = r \sin\theta \cos\phi, \\ Y = r \sin\theta \sin\phi, \\ Z = r\cos\theta; \end{cases}$$

and let it be required to determine the components of velocity and acceleration of the point in the direction of the radius vector r, in the direction which is perpendicular to r and lies in the plane containing r and OZ (this plane is generally called the *meridian plane*), and in the direction perpendicular to the meridian plane; these three directions are frequently described as the directions of r *increasing*, θ *increasing*, and ϕ *increasing*, respectively. Take a line through the origin O, parallel to the direction of θ increasing, as a moving axis Ox; and take a line through O, parallel to the direction of ϕ increasing, as axis Oy, and a line along the direction of r increasing as axis Oz. The three Eulerian angles which determine the position of the moving axes $Oxyz$ with reference to the fixed axes $OXYZ$ are $(\theta, \phi, 0)$; so (§ 16) the components of angular velocity of the system $Oxyz$, resolved along the axes Ox, Oy, Oz themselves, are

$$\omega_1 = -\dot{\phi}\sin\theta, \qquad \omega_2 = \dot{\theta}, \qquad \omega_3 = \dot{\phi}\cos\theta.$$

The coordinates of the moving point, referred to the moving axes, are $(0, 0, r)$; and so by § 17 the components of velocity of the point resolved parallel to the moving axes are

$$r\dot{\theta}, \qquad r\dot{\phi}\sin\theta, \qquad \dot{r},$$

and the components of acceleration in the directions of θ increasing, ϕ increasing, and r increasing (again using the formulae of § 17) are

$$\begin{cases} \dfrac{d}{dt}(r\dot{\theta}) - r\dot{\phi}^2\sin\theta\cos\theta + \dot{r}\dot{\theta}, \text{ or } r\ddot{\theta} + 2\dot{r}\dot{\theta} - r\dot{\phi}^2\sin\theta\cos\theta, \\ \dfrac{d}{dt}(r\dot{\phi}\sin\theta) + \dot{r}\dot{\phi}\sin\theta + r\dot{\theta}\dot{\phi}\cos\theta, \text{ or } \dfrac{1}{r\sin\theta}\dfrac{d}{dt}(r^2\sin^2\theta\dot{\phi}), \\ \text{and} \qquad \ddot{r} - r\dot{\theta}^2 - r\dot{\phi}^2\sin^2\theta. \end{cases}$$

If the motion of the point is in a plane, we can take the initial line in this plane as axis Oz, and the quantities denoted by r and θ in these formulae become ordinary polar coordinates in the plane; since $\dot{\phi}$ is now zero, the components of velocity and acceleration in the directions of r increasing and θ increasing are

$$(\dot{r}, \quad r\dot{\theta}),$$

and

$$(\ddot{r} - r\dot{\theta}^2, \quad r\ddot{\theta} + 2\dot{r}\dot{\theta}).$$

(ii) *Velocity and acceleration in cylindrical coordinates.*

Consider now a point whose position is defined by its *cylindrical coordinates* z, ρ, ϕ, connected with the coordinates (X, Y, Z) of the point referred to fixed rectangular axes $OXYZ$ by the equations

$$X = \rho\cos\phi, \qquad Y = \rho\sin\phi, \qquad Z = z;$$

and let it be required to find the components of the velocity and acceleration of the point in the direction parallel to the axis of z, in the direction of the line drawn from the axis of z to the point, perpendicular to the axis of z, and in the direction perpendicular to these two lines. These three directions are generally called the *direction of z increasing*, the *direction of ρ increasing*, and the *direction of ϕ increasing*; and the coordinate ϕ is called the *azimuth* of the point.

In this case we take moving axes Ox, Oy, Oz passing through the origin and parallel respectively to the directions of ρ increasing, ϕ increasing, and z

increasing. The components of angular velocity of the system $Oxyz$, resolved along the axes $Oxyz$ themselves, are clearly

$$\omega_1 = 0, \qquad \omega_2 = 0, \qquad \omega_3 = \dot{\phi},$$

and the coordinates of the moving point, referred to the moving axes, are $(\rho, 0, z)$. It follows by § 17 that the components of velocity of the point in these directions are

$$(\dot{\rho}, \qquad \rho\dot{\phi}, \qquad \dot{z}),$$

and the components of acceleration are

$$(\ddot{\rho} - \rho\dot{\phi}^2, \qquad \rho\ddot{\phi} + 2\dot{\rho}\dot{\phi}, \qquad \ddot{z}).$$

(iii) *Velocity and acceleration in arc-coordinates.*

Another application of the formulae of § 17 is to the determination of the components of velocity and acceleration of a point which is moving in any way in space, resolved along the tangent, principal normal, and binormal to its path.

Consider first the case of a particle moving in a plane: and take lines through a fixed point O, parallel respectively to the tangent and inward normal to the path, as moving axes Ox and Oy. These axes are rotating round O with angular velocity $\dot{\phi}$, where ϕ is the angle made by the tangent to the path with some fixed line in the plane. If v denotes the velocity of the point, s the arc of the path described at time t, and ρ the radius of curvature of the path at the point, we have

$$v = \frac{ds}{dt}, \qquad \rho = \frac{ds}{d\phi},$$

and the angular velocity of the axes can therefore be written in the form v/ρ.

Since the components of the velocity parallel to the moving axes are $(v, 0)$, it follows from § 17 that the components of the acceleration parallel to the same axes are $\left(\dot{v}, v \cdot \dfrac{v}{\rho}\right)$. Since

$$\dot{v} = \frac{dv}{dt} = \frac{ds}{dt}\frac{dv}{ds} = v\frac{dv}{ds},$$

it follows that the acceleration of the moving point in the direction of the tangent to its path is $v\dfrac{dv}{ds}$, and the acceleration in the direction of the inward normal is $\dfrac{v^2}{\rho}$.

Now the velocity of a moving point is determined by the knowledge of two consecutive positions of the moving point, and the acceleration is therefore determined by the knowledge of three consecutive positions; so even if the path of the point is not plane, it can for the purpose of determining its acceleration at any instant be regarded as moving in the osculating plane of its path, since this plane contains three consecutive positions of the point. Hence *the components of acceleration of the point, in the directions of the tangent, principal normal, and binormal to its path, are*

$$\left(v\frac{dv}{ds}, \quad \frac{v^2}{\rho}, \quad 0\right).$$

(iv) *Acceleration along the radius and tangent.*

The acceleration of a point which is in motion in a plane may be expressed in the following form*; let r be the radius vector to the point from a fixed origin in the plane, p the perpendicular from the origin on the tangent to the path, s the arc of the path described at time t, ρ the radius of curvature of the path at the point, and v or $\dot{s}$ the velocity of the point at time t; and let h denote the product pv. Then *the acceleration of the point can be resolved into components* $\frac{h^2 r}{p^3 \rho}$ *along the radius vector to the origin and* $\frac{h}{p^2}\frac{dh}{ds}$ *along the tangent to the path.*

For the acceleration can be resolved into components $v dv/ds$ along the tangent and v^2/ρ along the normal; now a vector F directed outwards along the radius vector can be resolved into vectors $-Fp/r$ along the inward normal and $F dr/ds$ along the tangent, so a vector v^2/ρ along the inward normal can be resolved into $\frac{rv^2}{\rho p}$ inwards along the radius vector and $\frac{rv^2}{\rho p}\frac{dr}{ds}$ along the tangent. The acceleration is therefore equivalent to components

$$v\frac{dv}{ds}+\frac{rv^2}{\rho p}\frac{dr}{ds} \text{ along the tangent,}$$

and $\frac{rv^2}{\rho p}$ inwards along the radius vector.

The latter component is $\frac{h^2 r}{p^3 \rho}$, and the former can be written

$$\frac{1}{2}\frac{dv^2}{ds}+\frac{v^2}{p}\frac{dp}{ds}, \quad \text{or} \quad \frac{1}{2p^2}\frac{d\,(v^2p^2)}{ds}, \quad \text{or} \quad \frac{h}{p^2}\frac{dh}{ds};$$

which establishes Siacci's result.

Example 1. *Determine the meridian, normal, and transverse components of the acceleration of a point moving on the surface of the anchor-ring*

$$x=(c+a\sin\theta)\cos\phi, \quad y=(c+a\sin\theta)\sin\phi, \quad z=a\cos\theta.$$

Let P be the point (θ, ϕ), and let O be the centre of the anchor-ring and C the centre of the meridian cross-section on which P lies. The polar coordinates of C relative to O are (c, ϕ), and the polar coordinates of P relative to C are (a, θ, ϕ); so the components of acceleration of C relative to O are

$c\ddot{\phi}$ transverse

and $-c\dot{\phi}^2$ outwards from the axis, i.e. $-c\dot{\phi}^2\sin\theta$ along the normal,

and $-c\dot{\phi}^2\cos\theta$ along the meridian.

* Due to Siacci, *Atti della R. Acc. di Torino*, XIV. p. 750.

The components of acceleration of P relative to C are

$$\begin{cases} a\ddot{\theta} - a\dot{\phi}^2 \sin\theta\cos\theta & \text{along the meridian,} \\ \dfrac{a}{\sin\theta}\dfrac{d}{dt}(\sin^2\theta\,.\,\dot{\phi}) & \text{transverse,} \\ -a\dot{\theta}^2 - a\dot{\phi}^2\sin^2\theta & \text{normal.} \end{cases}$$

Thus finally the components of acceleration of P in space are

$$a\ddot{\theta} - (c + a\sin\theta)\,\dot{\phi}^2\cos\theta \quad \text{along the meridian,}$$

$$-a\dot{\theta}^2 - a\dot{\phi}^2\sin^2\theta - c\dot{\phi}^2\sin\theta \quad \text{normal,}$$

and $$c\ddot{\phi} + \frac{a}{\sin\theta}\frac{d}{dt}(\sin^2\theta\,.\,\dot{\phi}) \quad \text{transverse.}$$

Example 2. *If the tangential and normal components of the acceleration of a point moving in a plane are constant, shew that the point describes a logarithmic spiral.*

In this case

$$v\frac{dv}{ds} = a, \text{ where } a \text{ is a constant,}$$

so $$v^2 = as.$$

Also $$\frac{v^2}{\rho} = c, \text{ where } c \text{ is a constant,}$$

so $$s = C\rho, \text{ where } C \text{ is a constant,}$$

or $$s = C\frac{ds}{d\phi}, \text{ where } \phi \text{ is the angle made by the tangent with a fixed line.}$$

Integrating this equation, we have

$$s = Ae^{B\phi},$$

where A and B are constants: and this is the intrinsic equation of the logarithmic spiral.

Example 3. *To find the acceleration of a point which describes a logarithmic spiral with constant angular velocity about the pole.*

By Siacci's theorem, the components of acceleration are $\dfrac{h^2 r}{p^3\rho}$ along the radius vector and $\dfrac{h}{p^2}\dfrac{dh}{ds}$ along the tangent; but if ω is the constant angular velocity, we have $h = \omega r^2$: so the components of acceleration are

$$\frac{\omega^2 r^5}{p^3\rho} \text{ and } \frac{2\omega^2 r^3}{p^2}\frac{dr}{ds}.$$

Since $\dfrac{r}{p}$, $\dfrac{r}{\rho}$, and $\dfrac{dr}{ds}$ are constant in the spiral, we see that each of these components of acceleration varies directly as the radius vector.

MISCELLANEOUS EXAMPLES.

1. If the instantaneous axis of rotation of a body moveable about a fixed point is fixed in the body, shew that it is also fixed in space, i.e. the motion is a rotation round a fixed axis.

2. A point is referred to rectangular axes Ox, Oy revolving about the origin with angular velocity ω; if there be an acceleration to $x = a$, $y = 0$, of amount $n^2\omega^2 \times$ (distance), shew that the path relative to the axes can be constructed by taking (i) a point

$x=n^2a/(n^2-1)$, (ii) a uniform circular motion with angular velocity $(n-1)\omega$ about this, and (iii) a uniform circular motion with angular velocity $(n+1)\omega$, but in the opposite sense, about this last. (Coll. Exam.)

3. The velocity of a point moving in a plane is the resultant of a velocity v along the radius vector to a fixed point and a velocity v' parallel to a fixed line. Prove that the corresponding accelerations are

$$\frac{dv}{dt}+\frac{vv'}{r}\cos\theta, \text{ and } \frac{dv'}{dt}+\frac{vv'}{r},$$

θ being the angle that the radius vector makes with the fixed direction. (Coll. Exam.)

4. A point moves in a plane, and is referred to Cartesian axes making angles α, β with a fixed line in the plane, where α, β are given functions of the time. Shew that the component velocities of the point are

$$\dot{x}-x\dot{\alpha}\cot(\beta-\alpha)-y\dot{\beta}\operatorname{cosec}(\beta-\alpha), \qquad \dot{y}+y\dot{\beta}\cot(\beta-\alpha)+x\dot{\alpha}\operatorname{cosec}(\beta-\alpha),$$

and obtain expressions for the component accelerations. (Coll. Exam.)

5. A point is moving in a plane: θ is the logarithm of the ratio of its distances from two fixed points in the plane, and ϕ is the angle between them: also $2k$ is the distance between the fixed points. Shew that the velocity of the point is

$$\frac{k\sqrt{\dot{\theta}^2+\dot{\phi}^2}}{\cosh\theta-\cos\phi}.$$

(Coll. Exam.)

6. If in two different descriptions of a curve by a moving point, the product of the velocities at corresponding places in the two descriptions is constant, shew that the accelerations at corresponding places in the two descriptions are as the squares of the velocities, and that their directions make equal angles with the normal to the curve, in opposite senses. (J. von Vieth.)

7. A point is moving in a parabola of latus rectum $4a$, and when its distance from the focus is r, the velocity is v; shew that its acceleration is compounded of accelerations R and N, along the radius vector and normal respectively, where

$$R=v\frac{dv}{dr}, \qquad N=\frac{a^{\frac{1}{2}}}{2r^{\frac{3}{2}}}\frac{d}{dr}(v^2r).$$

(Coll. Exam.)

8. Shew that if the axes of x and y rotate with angular velocities ω_1, ω_2 respectively, and ψ is the angle between them, the component accelerations of the point (x, y) parallel to the axes are

$$\ddot{x}-x\omega_1{}^2-(x\dot{\omega}_1+2\dot{x}\omega_1)\cot\psi-(y\dot{\omega}_2+2\dot{y}\omega_2)\operatorname{cosec}\psi,$$

and

$$\ddot{y}-y\omega_2{}^2+(x\dot{\omega}_1+2\dot{x}\omega_1)\operatorname{cosec}\psi+(y\dot{\omega}_2+2\dot{y}\omega_2)\cot\psi.$$

(Coll. Exam.)

9. The velocity of a point is made up of components u, v in directions making angles θ, ϕ with a fixed line. Prove that the components f, f' in these directions of the acceleration of the point will be given by

$$f=\dot{u}-u\dot{\theta}\cot\chi-v\dot{\phi}\operatorname{cosec}\chi,$$
$$f'=\dot{v}+u\dot{\theta}\operatorname{cosec}\chi+v\dot{\phi}\cot\chi,$$

χ being the inclination of the two directions.

Being given that the lines joining a moving point to two fixed points are r, s in length and θ, ϕ in inclination to the line joining the two fixed points, determine the acceleration of the point in terms of ω, ω', the rates of increase of θ, ϕ. (Coll. Exam.)

10. If A, B, C be three fixed points, and the component velocities of a moving point P along the directions PA, PB, PC be u, v, and w; shew that the accelerations in the same directions are

$$\dot{u}+uv\left(\frac{1}{PB}-\frac{\cos APB}{PA}\right)+uw\left(\frac{1}{PC}-\frac{\cos APC}{PA}\right),$$

and two similar expressions. (Coll. Exam.)

11. The movement of a plane lamina is given by the angular velocity ω and the component velocities u, v of the origin O resolved along axes Ox, Oy traced on the lamina. Find the component velocities of any point (x, y) of the lamina. Shew that the equations

$$\frac{d}{dt}\tan^{-1}\left(\frac{u-y\omega}{v+x\omega}\right)=\pm\omega$$

represent circular loci on the lamina; one being the locus of those points which are passing cusps on their curve loci in space and the other being the locus of the centres of curvature of the envelopes in space of all straight lines of the lamina. (Coll. Exam.)

12. Shew that when a point describes a space-curve, its acceleration can be resolved into two components, of which one acts along the radius vector from the projection of a fixed point on the osculating plane, and the other along the tangent; and that these are respectively

$$\frac{r}{p^3}\frac{T^2}{\rho},$$

and
$$\frac{T}{p^2}\frac{dT}{ds}+\frac{T^2}{p^4}\frac{q\,dq}{ds},$$

where ρ is the radius of curvature, q the distance of the fixed point from its projection on the osculating plane, r and p are the distances of this projection from the moving point and the tangent, T is an arbitrary function (equal to the product of p and the velocity) and s is the arc. (Siacci.)

13. A circle, a straight line, and a point lie in one plane, and the position of the point is determined by the lengths t of its tangent to the circle and p of its perpendicular to the line. Prove that, if the velocity of the point is made up of components u, v in the directions of these lengths and if their mutual inclination be θ, the component accelerations will be

$$\dot{u}-uv\cos\theta/t, \qquad \dot{v}+uv/t. \qquad \text{(Coll. Exam.)}$$

14. A particle moves in a circular arc. If r, r' are the distances of the particle at P from the extremities A, B of a fixed chord, shew that the accelerations along AP, BP are respectively

$$\frac{dv}{dt}+\frac{vv'}{rr'}(r-r\cos a), \text{ and } \frac{dv'}{dt}+\frac{vv'}{rr'}(r'-r\cos a),$$

where v, v' are the velocities in the directions of r, r', and a is the angle APB.

A point describes a semicircle under accelerations directed to the extremities of a diameter, which are at any point inversely as the radii vectores r, r' to the extremities of the diameter. Shew that the accelerations are

$$\frac{4a^4V^2}{r^3r'^2}, \qquad \frac{4a^4V^2}{r^2r'^3},$$

where a is the radius of the circle and V the velocity of the point parallel to the diameter. (Coll. Exam.)

15. The motion of a rigid body in two dimensions is defined by the velocity (u, v) of one of its points C and its angular velocity ω. Determine the coordinates relative to C of the point I of zero velocity, and shew that the direction of motion of any other point P is perpendicular to PI

Find the coordinates of the point J of zero acceleration, and express the acceleration of P in terms of its coordinates relative to J. (Coll. Exam.)

16. A point on a plane is moving with constant velocity V relative to it, the plane at the same time turning round a fixed axis perpendicular to it with angular velocity ω. Shew that the path of the point is given by the equation

$$\frac{V\theta}{\omega} = \sqrt{r^2 - a^2} + \frac{V}{\omega} \cos^{-1} \frac{a}{r};$$

r and θ being referred to fixed axes, and a being the shortest distance of the point from the axis of rotation. (Coll. Exam.)

17. The acceleration of a moving point Q is represented at any instant by ωa, where ω is a fixed point and a describes uniformly a circle whose centre is ω. Prove that the velocity of Q at any instant is represented by Op, where O is a fixed point and p describes a circle uniformly; and determine the path described by Q.

(Camb. Math. Tripos, Part I, 1902.)

18. A point moves along the curve of intersection of the ellipsoid $\frac{x^2}{a^2} + \frac{y^2}{b^2} + \frac{z^2}{c^2} = 1$ and the hyperboloid of one sheet $\frac{x^2}{a^2 - \lambda} + \frac{y^2}{b^2 - \lambda} + \frac{z^2}{c^2 - \lambda} = 1$, and its velocity at the point where the curve meets the hyperboloid of two sheets $\frac{x^2}{a^2 - \mu} + \frac{y^2}{b^2 - \mu} + \frac{z^2}{c^2 - \mu} = 1$ is

$$h \left\{ \frac{\mu(\mu - \lambda)}{(a^2 - \mu)(b^2 - \mu)(c^2 - \mu)} \right\}^{\frac{1}{2}},$$

where h is constant. Prove that the resolved part of the acceleration of the point along the normal to the ellipsoid is

$$\frac{h^2 abc\,(\mu - \lambda)}{(a^2 - \mu)(b^2 - \mu)(c^2 - \mu)\sqrt{\lambda\mu}}.$$ (Coll. Exam.)

19. A rigid body is rolling without sliding on a plane, and at any instant its angular velocity has components ω_1, ω_2 along the tangent to the lines of curvature at the point of contact, and ω_3 along the normal: shew that the point of the body which is at the point of contact has component accelerations

$$-R_2\omega_1\omega_3, \qquad -R_1\omega_2\omega_3, \qquad R_1\omega_2^2 + R_2\omega_1^2,$$

where R_1, R_2 are the principal radii of curvature of the surface of the body at the point of contact. (Coll. Exam.)

CHAPTER II

THE EQUATIONS OF MOTION

19. *The ideas of rest and motion.*

In the previous chapter we have frequently used the terms "fixed" and "moving" as applied to systems. So long as we are occupied with purely kinematical considerations, it is unnecessary to enter into the ultimate significance of these words; all that is meant is, that we consider the displacement of the "moving" systems, so far as it affects their configuration with respect to the systems which are called "fixed," leaving on one side the question of what is meant by absolute "fixity."

When however we come to consider the motion of bodies as due to specific causes, this question can no longer be disregarded.

In popular language the word "fixed" is generally used of terrestrial objects to denote invariable position relative to the surface of the earth at the place considered. But the earth is rotating on its axis, and at the same time revolving round the Sun, while the Sun in turn, accompanied by all the planets, is moving with a large velocity along some not very accurately known direction in space. It seems hopeless therefore to attempt to find anything which can be really considered to be "at rest."

In the nineteenth century it was supposed that the aether of space (the vehicle of light and of electric and magnetic actions) was (apart from small vibratory motions) stagnant, and so was capable of providing a basis for absolute fixity. But this doctrine has been subverted by the modern *Principle of Relativity**, which asserts that even in the domain of electromagnetic phenomena it is impossible to distinguish absolute rest from a state of uniform translatory motion common to all the members of a system.

Accordingly in dynamics, although when we speak of the motion of bodies we always imply that there is some set of axes, or *frame of reference* as it may be called, with reference to which the motion is regarded as taking place, and to which we apply the conventional word "fixed," yet it must not be supposed that absolute fixity has thereby been discovered. When we are considering

* Cf. Whittaker's *History of the Theories of Aether and Electricity*, ch. XII. (London, 1910); or Conway's *Relativity* (London, 1915).

the motion of terrestrial bodies at some place on the earth's surface, we shall take the frame of reference to be fixed with reference to the earth, and it is then found that the laws which will presently be given are sufficient to explain the phenomena with a sufficient degree of accuracy; in other words, the earth's motion does not exercise a sufficient disturbing influence to make it necessary to allow for its effects in the majority of cases of the motion of terrestrial bodies.

It is also necessary to consider the meaning to be attached to the word "time," which in the previous chapter stood merely for any parameter varying continuously with the configuration of the systems considered. The Principle of Relativity reveals the great difficulties that attend any attempt to elucidate the idea of time: in particular, it is by no means easy to define *simultaneity*, i.e. to explain what is meant by saying that two events at different points of space happen "at the same time." However, a system of time-measurement which is intelligible from the point of view of ordinary instrumental work, and which is sufficient for our present purpose, is the following: we suppose that the angle through which the earth has rotated on its axis (measured with reference to the fixed stars, whose small motions we can for this purpose neglect), in the interval between two events, measures the *time* elapsed between the events in question. This angular measure can be converted into the ordinary measure in terms of mean solar hours, minutes, and seconds at the rate of 360 degrees to $24 \times 365\frac{1}{4}/366\frac{1}{4}$ hours.

20. *The laws which determine motion*.*

Considering now the motion of terrestrial objects, and taking the earth as the frame of reference, it is natural to begin by investigating the motion of a very small material body, or *particle* as we shall call it, when moving in vacuo and entirely unconnected with surrounding objects. The paths described by such a particle under various circumstances of projection may be observed, and the methods of the preceding chapter enable us, from the knowledge thus acquired, to calculate the acceleration of the particle at any point of any particular observed path. It is found that for all the paths the acceleration is of constant amount, and is always directed vertically downwards. This acceleration is known as *gravity*, and is generally denoted by the letter g; its amount is, in Great Britain, about 981 centimetres per second per second.

The knowledge of this experimental fact is theoretically sufficient to enable us to calculate the path of any free terrestrial particle in vacuo, when the circumstances of its projection are known: the actual calculation will not be given here, as it belongs more properly to a later chapter.

The case of motion which is next in simplicity is that of two particles which are connected together by an extremely light inextensible thread, and

* The laws of motion are due to Newton: *Principia*, p. 12 (ed. 1687).

are free to move in vacuo at the earth's surface. So long as the thread is slack, each particle moves with the acceleration gravity, just as if the other were not present. But when the thread is taut, the two particles influence each other's motion. We can now as before observe the path of one of the particles, and hence calculate the acceleration by which at any instant its motion is being modified. We thereby arrive at the experimental fact, that *this acceleration can be represented at any instant by the resultant of two vectors, of which one represents the acceleration g and the other is directed along the instantaneous position of the thread.*

The influence of one particle on the motion of the other consists therefore in superposing on the acceleration due to gravity another acceleration, which acts along the line joining the particles and which is compounded with gravity according to the vectorial law of composition of accelerations. Denoting the particles by A and B, we can at any instant calculate, from the observed paths, the magnitudes of the accelerations f_1 and f_2 thus exerted by B on A and by A on B respectively; and this calculation immediately yields the result that *the ratio of f_1 to f_2 does not vary throughout the motion.* On investigating the motions which result from various modes of projection, at various temperatures etc., we are led to the conclusion that *this ratio is an invariable physical constant of the pair of bodies A and B**.

On consideration of the motion of more complex systems it is found that the experimental laws just stated can be generalised so as to form a complete basis for all dynamics, whether terrestrial or cosmic. This generalised statement is as follows: *If any set of mutually connected particles are in motion, the acceleration with which any one particle moves is the resultant of the acceleration with which it would move if perfectly free, and accelerations directed along the lines joining it to the other particles which constrain its motion. Moreover, to the several particles A, B, C, ..., numbers m_A, m_B, m_C, ... can be assigned, such that the acceleration along AB due to the influence of B on A is to the acceleration along BA due to the influence of A on B in the ratio $m_B : m_A$. The ratios of these numbers m_A, m_B, ... are invariable physical constants of the particles.*

The evidence for the truth of this statement is to be found in the universal agreement of the calculations based on it, such as those given later in this book, with the results of observation.

It will be noticed that only the *ratios* of the numbers m_A, m_B, m_C, ... are determined by the law; it is convenient to take some definite particle A as a standard, calling it the *unit of mass*, and then to call the numbers m_B/m_A, m_C/m_A, ... the *masses* of the other particles m_B, m_C, ...

* The ratio is in fact equal to the ratio of the weight of B to the weight of A; the ratio of the weights of two terrestrial bodies, as observed at the same place on the earth's surface, is a perfectly definite quantity, and does not vary with the place of observation.

The mass of the compound particle formed by uniting two or more particles is found to be equal to the sum of the masses of the separate particles. Owing to this additive property of mass, we can speak of the mass of a finite body of any size or shape; and it will be convenient to take as our *unit of mass* the mass of the $\frac{1}{1000}$th part of a certain piece of platinum known as the *standard kilogramme*; this unit will be called a *gramme*, and the number representing the ratio of the mass of any other body to this unit mass is called the *mass of the body in grammes.*

21. *Force.*

We have seen that in every case of the interaction of two particles A and B, the mutual influence consists of an acceleration f_A on A and an acceleration f_B on B, these accelerations being vectors directed along AB and BA respectively, and being inversely proportional to the masses m_A and m_B. It follows that the vector quantity $m_A f_A$ is equal to the vector quantity $m_B f_B$, but has the reverse direction. The vector $m_A f_A$ is called the *force* exerted by the particle B on the particle A; and similarly the vector $m_B f_B$ is called the force exerted by the particle A on the particle B.

With this terminology, the law of the mutual action of a connected system of particles can be stated in the form: *the forces exerted on each other by every pair of connected particles are equal and opposite.* This is often called the *Law of Action and Reaction.*

If the various forces which act on a particle A as a result of its connexion with other particles are compounded according to the vectorial law, the resultant force gives the total influence exerted by them on the particle A; this force divided by m_A is the acceleration induced in A by the other particles; and the resultant of this acceleration and the acceleration which the particle A would have if entirely free (due to such causes as gravitation) is the actual acceleration with which the particle A moves.

In general, if an acceleration represented by a vector f is induced in a particle of mass m by any agency, the vector mf is called the *force** due to this cause acting on the particle; and the resultant of all the forces due to various agencies is called the *total force acting on the particle.* It follows that if (X, Y, Z) are the components parallel to fixed rectangular axes of the total force acting on the particle at any instant, and $(\ddot{x}, \ddot{y}, \ddot{z})$ are the components of the acceleration with which its path is being described at that instant, then we have the equations

$$m\ddot{x} = X, \quad m\ddot{y} = Y, \quad m\ddot{z} = Z.$$

Two other terms which are frequently used may conveniently be defined at this point.

* Force is the *vis motrix* of Newton's *Principia*, I. def. 8.

The product of the number which represents the magnitude of the component of a given force perpendicular to a given line L and the number which represents the perpendicular distance of the line of action of the force from the line L is called the *moment* of the force about the line L.

If the three components (X, Y, Z) of the force acting on a single free particle are given functions of the coordinates (x, y, z) of the particle, they are said to define a *field of force.*

22. *Work.*

Consider now any system of particles, whose motion is either quite free or restricted by given connexions between the particles, or constraints due to other particles which are not regarded as forming part of the system. Let m be the mass of any one of the particles, whose coordinates referred to fixed rectangular axes in any selected configuration of the system are (x, y, z); and let (X, Y, Z) be the components, parallel to the axes, of the total force acting on the particle in this configuration.

Let $(x + \delta x, y + \delta y, z + \delta z)$ be the coordinates of any point very near to the point (x, y, z), such that the displacement of the particle m from one point to the other does not violate any of the constraints (for instance, if m is constrained to move on a given surface, the two points must both be situated on the surface). Then the quantity

$$X\delta x + Y\delta y + Z\delta z$$

is called the *work** done on the particle m by the forces acting on it in the infinitesimal displacement from the position (x, y, z) to the position

$$(x + \delta x, y + \delta y, z + \delta z).$$

This expression can evidently be interpreted physically as being the product of the distance through which the particle is displaced and the component of the force (X, Y, Z) along the direction of this displacement.

Since forces obey the vectorial law of composition, the sum of the components in a given direction of any number of forces acting together on a particle is equal to the component in this direction of their resultant: and hence *the work done by a force which acts on a particle in a given displacement is equal to the sum of the quantities of work done in the same displacement by any set of forces into which this force can be resolved.*

Suppose now that in the course of a motion of the system, the particle m is gradually displaced from any position (which we can call its *initial* position) to some other position at a finite distance from the first (which we can call its *final* position). The *work done on the particle by the forces which act on*

* Newton defined the *Actio Agentis* as the product of the velocity into the component of force along the direction of motion; it is evidently the time-flux of the work done. Cf. *Principia*, I. p. 25 (ed. 1687).

it during this finite displacement is defined to be the sum of the quantities of work done in the successive infinitesimal displacements by which we can regard the finite displacement as achieved. The work done in a finite displacement is therefore represented by the integral

$$\int\left(X\frac{dx}{ds}+Y\frac{dy}{ds}+Z\frac{dz}{ds}\right)ds,$$

where the integration is taken between the initial and final positions along the arc s described in space by the particle during displacement.

These definitions can now be extended to the whole set of particles which form the system considered; the system being initially in any given configuration, we consider any mode of displacing the various particles of the system which is not inconsistent with the connexions and constraints; the sum of the quantities of work performed on all the particles of the system in the displacement is called the *total work done on the system* in the displacement by the forces which act on it.

23. *Forces which do no work.*

There are certain classes of forces which frequently occur in dynamical systems, and which are characterised by the feature that during the motion they do no work on the system.

Among these may be mentioned

1°. The reactions of fixed smooth surfaces: the term *smooth* implies that the reaction is normal to the surface, and therefore in each infinitesimal displacement the point of application of the reaction is displaced in a direction perpendicular to the reaction, so that no work is done.

2°. The reactions of fixed perfectly rough surfaces: the term *perfectly rough* implies that the motion of any body in contact with the surface is one of pure rolling without sliding, and therefore the point of application of the reaction is (to the first order of small quantities) not displaced in each infinitesimal displacement, so that no work is done.

3°. The mutual reaction of two particles which are rigidly connected together: for if (x_1, y_1, z_1) and (x_2, y_2, z_2) are the coordinates of the particles, and (X, Y, Z) are the components of the force exerted by the first particle on the second, so that $(-X, -Y, -Z)$ are the components of the force exerted by the second particle on the first, the total work done by these forces in an arbitrary displacement is

$$X(\delta x_2-\delta x_1)+Y(\delta y_2-\delta y_1)+Z(\delta z_2-\delta z_1).$$

But since the distance between the particles is invariable, we have

$$\delta\{(x_2-x_1)^2+(y_2-y_1)^2+(z_2-z_1)^2\}=0,$$

or $$(x_2-x_1)(\delta x_2-\delta x_1)+(y_2-y_1)(\delta y_2-\delta y_1)+(z_2-z_1)(\delta z_2-\delta z_1)=0,$$

and since the force acts in the direction of the line joining the particles, we have

$$X : Y : Z = (x_2 - x_1) : (y_2 - y_1) : (z_2 - z_1).$$

Combining the last two equations, we have

$$X(\delta x_2 - \delta x_1) + Y(\delta y_2 - \delta y_1) + Z(\delta z_2 - \delta z_1) = 0,$$

and therefore no work is done in the aggregate by the mutual forces between the particles.

4°. A *rigid body* is regarded from the dynamical point of view as an aggregate of particles, so connected together that their mutual distances are invariable. It follows from 3° that the reactions between the particles which are called into play in order that this condition may be satisfied (or *molecular* forces as they are called, to distinguish them from *external* forces such as gravity) do, in the aggregate, no work in any displacement of the body.

5°. The reactions at a fixed pivot about which a body of the system can turn, or at a fixed hinge, or at a joint between two bodies of the system, are similarly seen to belong to the category of forces which do no work.

In estimating the total work done by the forces acting on a dynamical system in any displacement of the system, we can therefore neglect all forces of the above-mentioned types.

24. *The coordinates of a dynamical system.*

Any material system is regarded from the dynamical point of view as constituted of a number of particles, subject to interconnexions and constraints of various kinds; a rigid body being regarded as a collection of particles, which are kept at invariable distances from each other by means of suitable internal reactions.

When the constitution of such a system (i.e. the shape, size, and mass of the various parts of which it is composed, and the constraints which act on them) is given, its configuration at any time can be specified in terms of a certain number of quantities which vary when the configuration is altered, and which will be called the *coordinates* of the system; thus, the position of a single free particle in space is completely defined by its three rectangular coordinates (x, y, z) with reference to some fixed set of axes; the position of a single particle which is constrained to move in a fixed narrow tube, which has the form of a twisted curve in space, is completely specified by one coordinate, namely the distance s measured along the arc of the tube to the particle from some fixed point in the tube which is taken as origin; the position of a rigid body, one of whose points is fixed, is completely determined by three coordinates, namely the three Eulerian angles θ, ϕ, ψ of § 10; the position of two particles which are connected by a taut inextensible string can be defined by five coordinates, namely the three rectangular coordinates of one of the particles and two of the direction-cosines of the string (since when these five quantities are known, the position of the second particle is uniquely determined); and so on.

Example. State the number of independent coordinates required to specify the configuration at any instant of a rigid bodv which is constrained to move in contact with a given fixed smooth surface.

We shall generally denote by n the number of coordinates required to specify the configuration of a system, and shall suppose the systems considered to be such that n is finite. The coordinates will generally be denoted by $q_1, q_2, \ldots q_n$. If the system contains moving constraints (e.g. if it consists of a particle which is constrained to be in contact with a surface which in turn is made to rotate with constant angular velocity round a fixed axis), it may be necessary to specify the time t in addition to the coordinates $q_1, q_2, \ldots q_n$, in order to define completely a configuration of the system.

The quantities $\dot{q}_1, \dot{q}_2, \ldots \dot{q}_n$ are frequently called the *velocities* corresponding to the coordinates $q_1, q_2, \ldots q_n$.

A heavy flexible string, free to move in space, is an example of a dynamical system which is excluded by the limitation that n is to be finite; for the configuration of the string cannot be expressed in terms of a finite number of parameters.

25. *Holonomic and non-holonomic systems.*

It is now necessary to call attention to a distinction between two kinds of dynamical systems, which is of great importance in the analytical discussion of their motion: this distinction may be illustrated by a simple example.

If we consider the motion of a sphere of given radius, which is constrained to move in contact with a given fixed plane, which we can take as the plane of xy, the configuration of the sphere at any instant is completely specified by five coordinates, namely the two rectangular coordinates (x, y) of the centre of the sphere and the three Eulerian angles θ, ϕ, ψ of § 10, which specify the orientation of the sphere about its centre. The sphere can take up any position whatever, so long as it is in contact with the plane; the five coordinates $(x, y, \theta, \phi, \psi)$ can therefore have any arbitrary values.

If now the plane is smooth, the displacement from any position, defined by the coordinates $(x, y, \theta, \phi, \psi)$, to any adjacent position, defined by the coordinates $(x + \delta x, y + \delta y, \theta + \delta\theta, \phi + \delta\phi, \psi + \delta\psi)$, where $\delta x, \delta y, \delta\theta, \delta\phi, \delta\psi$ are arbitrary independent infinitesimal quantities, is a *possible* displacement, i.e. the sphere can perform it without violating the constraints of the system. But if the plane is perfectly rough, this is no longer the case when $\delta x, \delta y, \delta\theta, \delta\phi, \delta\psi$ are arbitrary; for now the condition that the displacement of the point of contact is zero (to the first order of small quantities) must be satisfied, and this implies that the quantities $\delta x, \delta y, \delta\theta, \delta\phi, \delta\psi$ are no longer independent, but are mutually connected (in fact, they must be such as to satisfy two non-integrable linear equations); so that *in the case of the sphere on the perfectly rough plane, a displacement represented by arbitrary infinitesimal changes in the coordinates is not necessarily a possible displacement.*

A dynamical system for which a displacement represented by arbitrary infinitesimal changes in the coordinates is in general a possible displacement (as in the case of the sphere on the smooth plane) is said to be *holonomic*; when this condition is not satisfied (as in the case of the sphere on the rough plane) the system is said to be *non-holonomic.*

If $(\delta q_1, \delta q_2, \ldots \delta q_n)$ are arbitrary infinitesimal increments of the coordinates in a dynamical system, these will define a possible displacement if the system is holonomic, while for non-holonomic systems a certain number, say m, of equations must be satisfied between them in order that they may correspond to a possible displacement. The number $(n - m)$ is called the *number of degrees of freedom* of the system. Holonomic systems are therefore characterised by the fact that the number of degrees of freedom is equal to the number of independent coordinates required to specify the configuration of the system.

26. *Lagrange's form of the equations of motion of a holonomic system*.*

We shall now consider the motion of a holonomic system with n degrees of freedom. Let $(q_1, q_2, \ldots q_n)$ be the coordinates which specify the configuration of the system at the time t.

Let m_i typify the mass of one of the particles of the system, and let (x_i, y_i, z_i) be its coordinates, referred to some fixed set of rectangular axes. These coordinates of individual particles are (from our knowledge of the constitution of the system) known functions of the coordinates $q_1, q_2, \ldots q_n$ of the system, and possibly of t also; let this dependence be expressed by the equations

$$\begin{cases} x_i = f_i(q_1, q_2, \ldots, q_n, t), \\ y_i = \phi_i(q_1, q_2, \ldots, q_n, t), \\ z_i = \psi_i(q_1, q_2, \ldots, q_n, t). \end{cases}$$

Let (X_i, Y_i, Z_i) be the components of the total force (external and molecular) acting on the particle m_i; then the equations of motion of this particle are

$$m_i\ddot{x}_i = X_i, \qquad m_i\ddot{y}_i = Y_i, \qquad m_i\ddot{z}_i = Z_i.$$

Multiply these equations by

$$\frac{\partial f_i}{\partial q_r}, \quad \frac{\partial \phi_i}{\partial q_r}, \quad \frac{\partial \psi_i}{\partial q_r},$$

* Lagrange, *Mécanique Analytique* (1788), Seconde Partie, Section IV. The equations were first suggested in one of his earlier papers, *Miscell. Taurin.* II. (1760).

respectively, add them, and sum for all the particles of the system. We thus have

$$\sum_i m_i \left(\ddot{x}_i \frac{\partial f_i}{\partial q_r} + \ddot{y}_i \frac{\partial \phi_i}{\partial q_r} + \ddot{z}_i \frac{\partial \psi_i}{\partial q_r} \right) = \sum_i \left(X_i \frac{\partial f_i}{\partial q_r} + Y_i \frac{\partial \phi_i}{\partial q_r} + Z_i \frac{\partial \psi_i}{\partial q_r} \right),$$

where the symbol Σ denotes summation over all the particles of the system; this can be either an integration (if the particles are united into rigid bodies) or a summation over a discrete aggregate of particles.

But we have

$$\frac{\partial \dot{x}_i}{\partial \dot{q}_r} = \frac{\partial}{\partial \dot{q}_r} \left(\frac{\partial f_i}{\partial q_1} \dot{q}_1 + \frac{\partial f_i}{\partial q_2} \dot{q}_2 + \ldots + \frac{\partial f_i}{\partial q_n} \dot{q}_n + \frac{\partial f_i}{\partial t} \right) = \frac{\partial f_i}{\partial q_r},$$

so

$$\begin{aligned} \ddot{x}_i \frac{\partial f_i}{\partial q_r} &= \ddot{x}_i \frac{\partial \dot{x}_i}{\partial \dot{q}_r} \\ &= \frac{d}{dt} \left(\dot{x}_i \frac{\partial \dot{x}_i}{\partial \dot{q}_r} \right) - \dot{x}_i \frac{d}{dt} \left(\frac{\partial x_i}{\partial q_r} \right) \\ &= \frac{d}{dt} \left(\dot{x}_i \frac{\partial \dot{x}_i}{\partial \dot{q}_r} \right) - \dot{x}_i \left(\frac{\partial^2 f_i}{\partial q_1 \partial q_r} \dot{q}_1 + \frac{\partial^2 f_i}{\partial q_2 \partial q_r} \dot{q}_2 + \ldots + \frac{\partial^2 f_i}{\partial q_n \partial q_r} \dot{q}_n + \frac{\partial^2 f_i}{\partial t \partial q_r} \right) \\ &= \frac{d}{dt} \left(\dot{x}_i \frac{\partial \dot{x}_i}{\partial \dot{q}_r} \right) - \dot{x}_i \frac{\partial \dot{x}_i}{\partial q_r} \\ &= \frac{d}{dt} \left\{ \frac{\partial}{\partial \dot{q}_r} \left(\tfrac{1}{2} \dot{x}_i^2 \right) \right\} - \frac{\partial}{\partial q_r} \left(\tfrac{1}{2} \dot{x}_i^2 \right), \end{aligned}$$

and therefore we have

$$\sum_i m_i \left(\ddot{x}_i \frac{\partial f_i}{\partial q_r} + \ddot{y}_i \frac{\partial \phi_i}{\partial q_r} + \ddot{z}_i \frac{\partial \psi_i}{\partial q_r} \right)$$
$$= \tfrac{1}{2} \sum_i m_i \frac{d}{dt} \left\{ \frac{\partial}{\partial \dot{q}_r} (\dot{x}_i^2 + \dot{y}_i^2 + \dot{z}_i^2) \right\} - \tfrac{1}{2} \sum_i m_i \frac{\partial}{\partial q_r} (\dot{x}_i^2 + \dot{y}_i^2 + \dot{z}_i^2).$$

Now the quantity

$$\tfrac{1}{2} \Sigma m_i (\dot{x}_i^2 + \dot{y}_i^2 + \dot{z}_i^2)$$

represents the sum of the masses of the particles of the system, each multiplied by half the square of its velocity; this is called the *Kinetic Energy* of the system* From our knowledge of the constitution of the system, the kinetic energy can be calculated† as a function of

$$q_1, q_2, \ldots q_n, \dot{q}_1, \dot{q}_2, \ldots \dot{q}_n, t;$$

we shall denote it by

$$T(\dot{q}_1, \dot{q}_2, \ldots \dot{q}_n, q_1, q_2, \ldots q_n, t),$$

and shall suppose that T is a known function of its arguments. Since

$$\dot{x}_i = \frac{\partial f_i}{\partial q_1} \dot{q}_1 + \frac{\partial f_i}{\partial q_2} \dot{q}_2 + \cdots + \frac{\partial f_i}{\partial q_n} \dot{q}_n + \frac{\partial f_i}{\partial t},$$

* The mass of a particle multiplied by the square of its velocity was called the *vis viva* by Leibnitz (*Acta erud.*, 1695).

† The methods of performing this calculation for rigid bodies are given in Chapter V.

and $\dot{y}_i$ and $\dot{z}_i$ are likewise linear functions of $\dot{q}_1, \dot{q}_2, \ldots \dot{q}_n$, we see that T is a quadratic function of $\dot{q}_1, \dot{q}_2, \ldots \dot{q}_n$; if the functions f, ϕ, ψ do not involve the time explicitly (as is generally the case if there are no moving constraints in the system), the quantities $\dot{x}, \dot{y}, \dot{z}$ are *homogeneous* linear functions of $\dot{q}_1, \dot{q}_2, \ldots \dot{q}_n$, and then T is a homogeneous quadratic function of $\dot{q}_1, \dot{q}_2, \ldots \dot{q}_n$.

From the definition it follows that the kinetic energy of a system is essentially positive; T is therefore a *positive definite* quadratic form in $\dot{q}_1, \dot{q}_2, \ldots \dot{q}_n$, and so satisfies the conditions that its discriminant and the principal minors of every order of its discriminant are positive.

We have thus derived from the equations of motion the equation

$$\frac{d}{dt}\left(\frac{\partial T}{\partial \dot{q}_r}\right) - \frac{\partial T}{\partial q_r} = \sum_i \left(X_i \frac{\partial f_i}{\partial q_r} + Y_i \frac{\partial \phi_i}{\partial q_r} + Z_i \frac{\partial \psi_i}{\partial q_r}\right),$$

and the expression on the left-hand side of this equation does not involve the individual particles of the system, except in so far as they contribute to the kinetic energy T. We have now to see if the right-hand side of the equation can also be brought to a form in which the individuality of the separate particles is lost.

For this purpose, consider that displacement of the system in which the coordinate q_r is changed to $q_r + \delta q_r$, while the coordinates

$$q_1, q_2, \ldots q_{r-1}, q_{r+1}, \ldots q_n$$

and the time (so far as this is required for the specification of the system) are unaltered. Since the system is holonomic, this can be effected without violating the constraints. In this displacement, the coordinates of the particle m_i are changed to

$$x_i + \frac{\partial f_i}{\partial q_r}\delta q_r, \qquad y_i + \frac{\partial \phi_i}{\partial q_r}\delta q_r, \qquad z_i + \frac{\partial \psi_i}{\partial q_r}\delta q_r;$$

and therefore the total work done in the displacement by all the forces which act on the particles of the system is

$$\sum_i \left(X_i \frac{\partial f_i}{\partial q_r} + Y_i \frac{\partial \phi_i}{\partial q_r} + Z_i \frac{\partial \psi_i}{\partial q_r}\right)\delta q_r.$$

Now of the forces which act on the system, there are several kinds which do no work. Among these are, as was seen in § 23,

1°. The molecular forces which act between the particles of the rigid bodies contained in the system:

2°. The pressures of connecting-rods of invariable length, the reactions at fixed pivots, and the tensions of taut inextensible strings:

3°. The reactions of any fixed smooth surfaces or curves with which bodies of the system are constrained to remain in contact; or of perfectly rough surfaces, so far as these can enter into holonomic systems:

4°. The reactions of any smooth surfaces or curves with which bodies of the system are constrained to remain in contact, when these surfaces or curves are forced to move in some prescribed manner; for the displacement considered above is made on the supposition that t, so far as it is required for the specification of the system, is not varied, i.e. that such surfaces or curves are not moved during the displacement; so that this case reduces to the preceding.

The forces acting on the system, other than these which do no work, are called the *external* forces. It follows that the quantity

$$\sum_i \left(X_i \frac{\partial f_i}{\partial q_r} + Y_i \frac{\partial \phi_i}{\partial q_r} + Z_i \frac{\partial \psi_i}{\partial q_r} \right) \delta q_r$$

is the work done by the external forces in the displacement which corresponds to a change of q_r to $q_r + \delta q_r$, the other coordinates being unaltered. This is a quantity which (from our knowledge of the constitution of the system, and of the forces at work) is a known function of $q_1, q_2, \ldots q_n, t$; we shall denote it by

$$Q_r(q_1, q_2, \ldots q_n, t)\, \delta q_r.$$

We have therefore

$$\frac{d}{dt}\left(\frac{\partial T}{\partial \dot{q}_r}\right) - \frac{\partial T}{\partial q_r} = Q_r.$$

This equation is true for all values of r from 1 to n inclusive; we thus have n ordinary differential equations of the second order, in which $q_1, q_2, \ldots q_n$ are the dependent variables and t is the independent variable; as the number of differential equations is equal to the number of dependent variables, the equations are theoretically sufficient to determine the motion when the initial circumstances are given. We have now arrived at a result which may be thus stated:

Let T denote the kinetic energy of a dynamical system, and let

$$Q_1 \delta q_1 + Q_2 \delta q_2 + \ldots + Q_n \delta q_n$$

denote the work done by the external forces in an arbitrary displacement $(\delta q_1, \delta q_2, \ldots \delta q_n)$, *so that* $T, Q_1, Q_2, \ldots Q_n$ *are, from our knowledge of the constitution of the system, known functions of* $\dot{q}_1, \dot{q}_2, \ldots \dot{q}_n, q_1, q_2, \ldots q_n, t$; *then the equations which determine the motion of the system may be written*

$$\frac{d}{dt}\left(\frac{\partial T}{\partial \dot{q}_r}\right) - \frac{\partial T}{\partial q_r} = Q_r, \qquad (r = 1, 2, \ldots n).$$

These are known as *Lagrange's equations of motion.* It will be observed that the unknown reactions (e.g. of the constraints) do not enter into these equations. The determination of these reactions forms a separate branch of mechanics, which is known as *Kineto-statics**: so we can say that *in Lagrange's equations the kineto-statical relations of the problem are altogether eliminated.*

* Cf. Heun, *Deutsche Math. Ver.* IX. (Heft 2) (1900), p. 1.

27. *Conservative forces: the Kinetic Potential.*

Certain fields of force have the property that the work done by the forces of the field in a displacement of a dynamical system from one configuration to another depends only on the initial and final configurations of the system, being the same whatever be the sequence of infinitesimal displacements by which the finite displacement is effected.

Gravity is a conspicuous example of a field of force of this character; the work done by gravity in the motion of a particle of mass m from one position at a height h to another position at a height k above the earth's surface is $mg(h-k)$, and this does not depend in any way on the path by which the particle is moved from one position to the other.

Fields of force of this type are said to be *conservative.*

Let the configuration of any dynamical system be specified by n coordinates $q_1, q_2, \ldots q_n$. Choose some configuration of the system, say that for which

$$q_r = \alpha_r, \qquad (r = 1, 2, \ldots n),$$

as a standard configuration; then if the external forces acting on the system are conservative, the work done by these forces in a displacement of the system from the configuration $(q_1, q_2, \ldots q_n)$ to the standard configuration is a definite function of $q_1, q_2, \ldots q_n$, not depending on the mode of displacement. Let this function be denoted by $V(q_1, q_2, \ldots q_n)$; it is called the *Potential Energy** of the system in the configuration $(q_1, q_2, \ldots q_n)$. In this case the work done by the external forces in an arbitrary displacement

$$(\delta q_1, \delta q_2, \ldots \delta q_n)$$

is evidently equal to the infinitesimal decrease in the function V, corresponding to the displacement, i.e. is equal to the quantity

$$-\frac{\partial V}{\partial q_1}\delta q_1 - \frac{\partial V}{\partial q_2}\delta q_2 - \ldots - \frac{\partial V}{\partial q_n}\delta q_n;$$

Lagrange's equations of motion therefore take the form

$$\frac{d}{dt}\left(\frac{\partial T}{\partial \dot{q}_r}\right) - \frac{\partial T}{\partial q_r} = -\frac{\partial V}{\partial q_r}, \qquad (r = 1, 2, \ldots n).$$

If we introduce a new function L of the variables $q_1, q_2, \ldots q_n, \dot{q}_1, \ldots \dot{q}_n, t$, defined by the equation

$$L = T - V,$$

then Lagrange's equations can be written

$$\frac{d}{dt}\left(\frac{\partial L}{\partial \dot{q}_r}\right) - \frac{\partial L}{\partial q_r} = 0, \qquad (r = 1, 2, \ldots n).$$

* The Potential-function was introduced by Lagrange in 1773 (*Oeuvres*, VI. p. 335). The name *Potential* is due to Green (1828).

The function L is called the *Kinetic Potential*, or *Lagrangian function*; this single function completely specifies, so far as dynamical investigations are concerned, a holonomic system for which the forces are conservative.

28. *The explicit form of Lagrange's equations.*

We shall now shew how the second derivatives of the coordinates with respect to the time can be found explicitly from Lagrange's equations.

Let the configuration of the dynamical system considered be specified by coordinates $q_1, q_2, \ldots q_n$; we shall suppose that the configuration can be completely specified in terms of these coordinates alone, without t, so that the kinetic energy of the system is a homogeneous quadratic function of $\dot{q}_1, \dot{q}_2, \ldots \dot{q}_n$. As was seen in § 26, this is always the case when the constraints are independent of the time, but not in general when the constraints have forced motions (as for instance in the case of a particle constrained to move on a wire which is made to rotate in a given way).

Suppose then that the kinetic energy is

$$T = \tfrac{1}{2} \sum_{k=1}^{n} \sum_{l=1}^{n} a_{kl} \dot{q}_k \dot{q}_l,$$

where $a_{kl} = a_{lk}$, and where the coefficients a_{kl} are known functions of

$$q_1, q_2, \ldots q_n.$$

The Lagrangian equations of motion for the system are

$$\frac{d}{dt}\left(\frac{\partial T}{\partial \dot{q}_r}\right) - \frac{\partial T}{\partial q_r} = Q_r, \qquad (r = 1, 2, \ldots n),$$

or

$$\frac{d}{dt}\left(\sum_{s=1}^{n} a_{sr}\dot{q}_s\right) - \tfrac{1}{2} \sum_{k=1}^{n} \sum_{l=1}^{n} \frac{\partial a_{kl}}{\partial q_r} \dot{q}_k \dot{q}_l = Q_r, \qquad (r = 1, 2, \ldots n),$$

or

$$\sum_{s=1}^{n} a_{rs}\ddot{q}_s + \sum_{l=1}^{n} \sum_{m=1}^{n} \begin{bmatrix} l\,m \\ r \end{bmatrix} \dot{q}_l \dot{q}_m = Q_r, \qquad (r = 1, 2, \ldots n),$$

where the symbol $\begin{bmatrix} l\,m \\ r \end{bmatrix}$, which is called a *Christoffel's symbol**, denotes the expression

$$\frac{1}{2}\left(\frac{\partial a_{lr}}{\partial q_m} + \frac{\partial a_{mr}}{\partial q_l} - \frac{\partial a_{lm}}{\partial q_r}\right).$$

These equations, being linear in the accelerations, can be solved for the quantities $\ddot{q}_s$. In fact, let D denote the determinant

$$\begin{vmatrix} a_{11} & a_{12} & a_{13} \ldots a_{1n} \\ a_{21} & a_{22} & a_{23} \;\ldots\ldots \\ a_{31} & a_{32} & \ldots\ldots\ldots\ldots \\ \vdots & & \\ a_{n1} & \ldots\ldots\ldots\ldots & a_{nn} \end{vmatrix},$$

* It was introduced by Christoffel, *Journal für Math.* LXX. (1869), and is of importance in the theory of quadratic differential forms.

and let A_{rs} be the minor of a_{rs} in this determinant. Multiply the n equations of the above system by $A_{1\nu}, A_{2\nu}, \ldots A_{n\nu}$, respectively, and add them: remembering that the quantity $\sum_{r=1}^{n} A_{r\nu} a_{rs}$ is zero when s is different from ν, and has the value D when s is equal to ν, we have

$$D\ddot{q}_\nu + \sum_{l=1}^{n} \sum_{m=1}^{n} \sum_{r=1}^{n} A_{r\nu} \begin{bmatrix} l\ m \\ r \end{bmatrix} \dot{q}_l \dot{q}_m = \sum_{r=1}^{n} A_{r\nu} Q_r,$$

or

$$\ddot{q}_\nu = -\frac{1}{D} \sum_{l=1}^{n} \sum_{m=1}^{n} \sum_{r=1}^{n} A_{r\nu} \begin{bmatrix} l\ m \\ r \end{bmatrix} \dot{q}_l \dot{q}_m + \frac{1}{D} \sum_{r=1}^{n} A_{r\nu} Q_r.$$

This equation is true for all values of ν from 1 to n inclusive; and these n equations, in which $\ddot{q}_1, \ddot{q}_2, \ldots \ddot{q}_n$ are given explicitly as functions of $\dot{q}_1, \dot{q}_2, \ldots \dot{q}_n, q_1, q_2, \ldots q_n$, can be regarded as replacing Lagrange's equations of motion.

29. *Motion of a system which is constrained to rotate uniformly round an axis.*

In many dynamical systems, some part of the system is compelled by an external agency to revolve with constant angular velocity ω round a given fixed axis; the motion of a bead on a wire which is made to rotate in this way is a simple example. There is, as we have seen, no objection to the direct application of Lagrange's equations to such cases, provided the system is holonomic; but it is often more convenient to use a theorem which we shall now obtain, and which reduces the consideration of systems of this kind to that of systems in which no forced rotation about the given axis takes place.

Suppose that, independently of the prescribed motion round the axis, the system has n degrees of freedom, so that if the given axis is taken as axis of z, and any plane through this axis and turning with the prescribed angular velocity is taken as the plane from which the azimuth ϕ is measured, the cylindrical coordinates of any particle m of the system can be expressed in terms of n coordinates $q_1, q_2, \ldots, q_n$, these expressions not involving the time t. Then if the kinetic energy of the system in the actual motion be T, and if the work done by the external forces in an arbitrary infinitesimal displacement be $Q_1\delta q_1 + Q_2\delta q_2 + \ldots + Q_n\delta q_n$, where $Q_1, Q_2, \ldots, Q_n$ will be supposed to depend only on the coordinates $q_1, q_2, \ldots, q_n$, and if the kinetic energy of the system when the forced angular velocity is replaced by zero be denoted by T_1, we have

$$T = \tfrac{1}{2}\Sigma m \{\dot{z}^2 + \dot{r}^2 + r^2(\dot{\phi} + \omega)^2\},$$
$$T_1 = \tfrac{1}{2}\Sigma m \{\dot{z}^2 + \dot{r}^2 + r^2\dot{\phi}^2\}.$$

Now the quantity $\frac{1}{2}\Sigma mr^2$ will be a function of $q_1, q_2, \ldots, q_n$, which is determined by our knowledge of the constitution of the system: denote it by W. The quantity $\Sigma mr^2\dot{\phi}$ will also be a known function of $q_1, q_2, \ldots, q_n$,

$\dot{q}_1, \ldots, \dot{q}_n$, being linear in $\dot{q}_1, \dot{q}_2, \ldots, \dot{q}_n$; it will be zero if, when ω is zero, the motion of every particle has no component in the direction of ϕ increasing; while if n is equal to unity, so that there is only one coordinate q, it will be the perfect differential with respect to t of a function of q: these are the two cases of most frequent occurrence, and we shall include them both by assuming that $\Sigma mr^2\dot{\phi}$ is of the form $\frac{dY}{dt}$, where Y is a given function of the coordinates $q_1, q_2, \ldots, q_n$.

We have therefore

$$T = T_1 + \omega\frac{dY}{dt} + \omega^2 W,$$

and the Lagrangian equations

$$\frac{d}{dt}\left(\frac{\partial T}{\partial \dot{q}_r}\right) - \frac{\partial T}{\partial q_r} = Q_r, \qquad (r = 1, 2, \ldots, n)$$

can be written in the form

$$\frac{d}{dt}\left(\frac{\partial T_1}{\partial \dot{q}_r}\right) + \frac{d}{dt}\left(\omega\frac{\partial Y}{\partial q_r}\right) - \frac{\partial T_1}{\partial q_r} - \omega\frac{d}{dt}\left(\frac{\partial Y}{\partial q_r}\right) - \omega^2\frac{\partial W}{\partial q_r} = Q_r, \quad (r = 1, 2, \ldots, n)$$

or

$$\frac{d}{dt}\left(\frac{\partial T_1}{\partial \dot{q}_r}\right) - \frac{\partial T_1}{\partial q_r} = -\frac{\partial}{\partial q_r}(-\omega^2 W) + Q_r, \quad (r = 1, 2, \ldots, n).$$

These equations shew that, subject to the assumption already mentioned, *the motion is the same as if the prescribed angular velocity were zero, and the potential energy were to contain an additional term* $-\frac{1}{2}\Sigma mr^2\omega^2$. In this way, by modifying the potential energy, we are enabled to pass from a system which is constrained to rotate about the given axis to a system for which this rotation does not take place. The term *centrifugal forces* is sometimes used of the imaginary forces introduced in this way to represent the effect of the enforced rotation.

30. *The Lagrangian equations for quasi-coordinates.*

In the form of Lagrange's equations given in § 26, the variables are n coordinates $q_1, q_2, \ldots, q_n$, and the time t; the knowledge of these quantities, together with a knowledge of the constitution of the system, is sufficient to determine the position of any particle in any configuration of the system, which may be expressed by saying that $q_1, q_2, \ldots, q_n$ are *true coordinates* of the system. We shall now find the form which is taken by the equations when the variables used are no longer restricted to be true coordinates of the system*.

Consider a system defined by n true coordinates $q_1, q_2, \ldots, q_n$, the kinetic energy being T and the work done by the external forces in a

* Particular cases of the theorem of this article were known to Lagrange and Euler: the general form of the equations is due to Boltzmann (*Wien. Sitzungsberichte*, 1902) and Hamel (*Zeitschrift für Math. u. Phys.* 1904).

displacement $(\delta q_1, \delta q_2, \ldots, \delta q_n)$ being $Q_1\delta q_1 + Q_2\delta q_2 + \ldots + Q_n\delta q_n$, so that the Lagrangian equations of motion of the system are

$$\frac{d}{dt}\left(\frac{\partial T}{\partial \dot{q}_\kappa}\right) - \frac{\partial T}{\partial q_\kappa} = Q_\kappa \qquad (\kappa = 1, 2, \ldots, n)\ldots(1).$$

Let $\omega_1, \omega_2, \ldots, \omega_n$ be n independent linear combinations of the velocities $\dot{q}_1, \dot{q}_2, \ldots, \dot{q}_n$, defined by the relations

$$\omega_r = \alpha_{1r}\dot{q}_1 + \alpha_{2r}\dot{q}_2 + \ldots + \alpha_{nr}\dot{q}_n \qquad (r = 1, 2, \ldots, n)\ldots(2),$$

where $\alpha_{11}, \alpha_{21}, \ldots, \alpha_{nn}$ are given functions of $q_1, q_2, \ldots, q_n$; and let $d\pi_1, d\pi_2, \ldots, d\pi_n$ be n linear combinations of the differentials $dq_1, dq_2, \ldots, dq_n$, defined by the relations

$$d\pi_r = \alpha_{1r}dq_1 + \alpha_{2r}dq_2 + \ldots + \alpha_{nr}dq_n \qquad (r = 1, 2, \ldots, n),$$

where the coefficients α are the same as in the previous set of equations.

These last equations would be immediately integrable if the relations $\dfrac{\partial \alpha_{\kappa r}}{\partial q_m} = \dfrac{\partial \alpha_{mr}}{\partial q_\kappa}$ were satisfied for all values of κ, r, and m, and in that case variables π_r would exist which would be true coordinates; we shall not however suppose the equations to be necessarily integrable, so that $d\pi_1, d\pi_2, \ldots, d\pi_n$ will not necessarily be the differentials of coordinates $\pi_1, \pi_2, \ldots, \pi_n$; we shall call the quantities $d\pi_1, d\pi_2, \ldots, d\pi_n$ *differentials of quasi-coordinates*.

Suppose that the relations (2), when solved for $\dot{q}_1, \dot{q}_2, \ldots, \dot{q}_n$, give the equations

$$\dot{q}_\kappa = \beta_{\kappa 1}\omega_1 + \beta_{\kappa 2}\omega_2 + \ldots + \beta_{\kappa n}\omega_n \qquad (\kappa = 1, 2, \ldots, n)\ldots(3).$$

Multiplying the Lagrangian equations (1) by $\beta_{1r}, \beta_{2r}, \ldots, \beta_{nr}$, respectively, and adding, we obtain the equation

$$\sum_\kappa \beta_{\kappa r}\left\{\frac{d}{dt}\left(\frac{\partial T}{\partial \dot{q}_\kappa}\right) - \frac{\partial T}{\partial q_\kappa}\right\} = \sum_\kappa \beta_{\kappa r} Q_\kappa.$$

Now $\sum\limits_\kappa Q_\kappa \delta q_\kappa$ is the work done by the external forces on the system in an arbitrary displacement, so $\sum\limits_\kappa \beta_{\kappa r} Q_\kappa \delta\pi_r$ is the work done in a displacement in which all the quantities $\delta\pi$ are zero except $\delta\pi_r$. If therefore the work done by the external forces on the system in an arbitrary infinitesimal displacement $(\delta\pi_1, \delta\pi_2, \ldots, \delta\pi_n)$ is $\Pi_1\delta\pi_1 + \Pi_2\delta\pi_2 + \ldots + \Pi_n\delta\pi_n$, we have

$$\sum_\kappa \beta_{\kappa r}\left\{\frac{d}{dt}\left(\frac{\partial T}{\partial \dot{q}_\kappa}\right) - \frac{\partial T}{\partial q_\kappa}\right\} = \Pi_r.$$

By means of equations (3) we can eliminate $\dot{q}_1, \dot{q}_2, \ldots, \dot{q}_n$ from the function T, so that T becomes a function of $\omega_1, \omega_2, \ldots, \omega_n, q_1, q_2, \ldots, q_n$ (we suppose for simplicity that t is not contained explicitly in T); let this form of T be denoted by $\bar{T}$.

Then we have
$$\frac{\partial T}{\partial \dot{q}_\kappa} = \sum_s \frac{\partial \bar{T}}{\partial \omega_s}\alpha_{\kappa s},$$

and therefore

$$\sum_\kappa \beta_{\kappa r} \left\{ \sum_s \alpha_{\kappa s} \frac{d}{dt}\left(\frac{\partial \bar{T}}{\partial \omega_s}\right) + \sum_s \frac{\partial \bar{T}}{\partial \omega_s} \frac{d\alpha_{\kappa s}}{dt} - \frac{\partial T}{\partial q_\kappa} \right\} = \Pi_r.$$

But $\sum_\kappa \beta_{\kappa r}\alpha_{\kappa s}$ is zero or unity according as r is different from, or equal to, s: so we have

$$\frac{d}{dt}\left(\frac{\partial \bar{T}}{\partial \omega_r}\right) + \sum_\kappa \sum_s \beta_{\kappa r} \frac{d\alpha_{\kappa s}}{dt} \frac{\partial \bar{T}}{\partial \omega_s} - \sum_\kappa \beta_{\kappa r} \frac{\partial T}{\partial q_\kappa} = \Pi_r.$$

We also have

$$\frac{\partial T}{\partial q_\kappa} = \frac{\partial \bar{T}}{\partial q_\kappa} + \sum_s \frac{\partial \bar{T}}{\partial \omega_s} \frac{\partial \omega_s}{\partial q_\kappa} = \frac{\partial \bar{T}}{\partial q_\kappa} + \sum_s \sum_m \frac{\partial \bar{T}}{\partial \omega_s} \frac{\partial \alpha_{ms}}{\partial q_\kappa} \dot{q}_m,$$

so
$$\frac{d}{dt}\left(\frac{\partial \bar{T}}{\partial \omega_r}\right) + \sum_\kappa \sum_s \sum_m \beta_{\kappa r} \frac{\partial \bar{T}}{\partial \omega_s} \dot{q}_m \left(\frac{\partial \alpha_{\kappa s}}{\partial q_m} - \frac{\partial \alpha_{ms}}{\partial q_\kappa}\right) - \sum_\kappa \beta_{\kappa r} \frac{\partial \bar{T}}{\partial q_\kappa} = \Pi_r.$$

Now $\sum_\kappa \beta_{\kappa r} \frac{\partial \bar{T}}{\partial q_\kappa}$, or $\sum_\kappa \frac{\partial \bar{T}}{\partial q_\kappa} \frac{\partial q_\kappa}{\partial \pi_r}$, would represent $\frac{\partial \bar{T}}{\partial \pi_r}$ if π_r were a true coordinate; we shall denote it by the symbol $\frac{\partial \bar{T}}{\partial \pi_r}$ whether π_r is a true coordinate or not. Also the expression

$$\sum_\kappa \sum_m \beta_{\kappa r} \beta_{ml} \left(\frac{\partial \alpha_{\kappa s}}{\partial q_m} - \frac{\partial \alpha_{ms}}{\partial q_\kappa}\right)$$

depends only on the connexion between the true coordinates and the differentials of the quasi-coordinates, and is independent of the nature or motion of the dynamical system considered: we shall denote this expression by γ_{rsl}. We have therefore

$$\frac{d}{dt}\left(\frac{\partial \bar{T}}{\partial \omega_r}\right) + \sum_s \sum_l \gamma_{rsl} \omega_l \frac{\partial \bar{T}}{\partial \omega_s} - \frac{\partial \bar{T}}{\partial \pi_r} = \Pi_r \qquad (r = 1, 2, \ldots, n).$$

These n equations are the equations of motion expressed in terms of the quasi-coordinates; when the quasi-coordinates are true coordinates, the quantities γ_{rsl} are all zero, since the conditions $\frac{\partial \alpha_{\kappa r}}{\partial q_m} = \frac{\partial \alpha_{mr}}{\partial q_\kappa}$ are satisfied, and the equations reduce to the ordinary Lagrangian equations

$$\frac{d}{dt}\left(\frac{\partial T}{\partial \dot{\pi}_r}\right) - \frac{\partial T}{\partial \pi_r} = \Pi_r \qquad (r = 1, 2, \ldots, n).$$

Example. A rigid body is free to turn about one of its points O, which is fixed, so that the coordinates of the body can be taken to be the three Eulerian angles θ, ϕ, ψ, which (§ 10) specify the position of axes $Oxyz$, fixed in the body and moving with it, with reference to axes $OXYZ$ fixed in space. Let an arbitrary displacement ($\delta\theta$, $\delta\phi$, $\delta\psi$) of the body be equivalent to the resultant of small rotations $\delta\pi_1$, $\delta\pi_2$, $\delta\pi_3$ round Ox, Oy, Oz, respectively, so that $d\pi_1$, $d\pi_2$, $d\pi_3$ can be taken as the differentials of quasi-coordinates: let ω_1, ω_2, ω_3 be the components about the axes $Oxyz$ of the angular velocity of the body at any instant, so that $d\pi_1$, $d\pi_2$, $d\pi_3$ are the differentials of quasi-coordinates corresponding respectively to the velocities ω_1, ω_2, ω_3. Shew that the equations of motion of the body are

$$\begin{cases} \dfrac{d}{dt}\left(\dfrac{\partial \bar{T}}{\partial \omega_1}\right) - \omega_3 \dfrac{\partial \bar{T}}{\partial \omega_2} + \omega_2 \dfrac{\partial \bar{T}}{\partial \omega_3} - \dfrac{\partial \bar{T}}{\partial \pi_1} = \Pi_1, \\ \dfrac{d}{dt}\left(\dfrac{\partial \bar{T}}{\partial \omega_2}\right) - \omega_1 \dfrac{\partial \bar{T}}{\partial \omega_3} + \omega_3 \dfrac{\partial \bar{T}}{\partial \omega_1} - \dfrac{\partial \bar{T}}{\partial \pi_2} = \Pi_2, \\ \dfrac{d}{dt}\left(\dfrac{\partial \bar{T}}{\partial \omega_3}\right) - \omega_2 \dfrac{\partial \bar{T}}{\partial \omega_1} + \omega_1 \dfrac{\partial \bar{T}}{\partial \omega_2} - \dfrac{\partial \bar{T}}{\partial \pi_3} = \Pi_3, \end{cases}$$

where $\bar{T}$ is the kinetic energy of the body, expressed in terms of ω_1, ω_2, ω_3, θ, ϕ, ψ; Π_1, Π_2, Π_3 are the moments about the axes Ox, Oy, Oz, respectively, of the external forces acting on the body; and $\dfrac{\partial \bar{T}}{\partial \pi_r}$ stands for $\dfrac{\partial \bar{T}}{\partial \theta}\dfrac{\partial \theta}{\partial \pi_r} + \dfrac{\partial \bar{T}}{\partial \phi}\dfrac{\partial \phi}{\partial \pi_r} + \dfrac{\partial \bar{T}}{\partial \psi}\dfrac{\partial \psi}{\partial \pi_r}$.

It will appear later that $\bar{T}$ depends only on ω_1, ω_2, ω_3, so the terms $\dfrac{\partial \bar{T}}{\partial \pi_r}$ are zero.

31. *Forces derivable from a potential-function which involves the velocities.*

In certain cases the conception of a potential-energy function can be extended to dynamical systems in which the acting forces depend not only on the position but on the velocities and accelerations of the bodies.

For consider a dynamical system whose configuration is specified by coordinates $q_1, q_2, \ldots, q_n$, and suppose that the work done by the external forces in an arbitrary displacement $(\delta q_1, \delta q_2, \ldots, \delta q_n)$ is

$$Q_1 \delta q_1 + Q_2 \delta q_2 + \ldots + Q_n \delta q_n.$$

Then if Q_r can be expressed in the form

$$Q_r = -\frac{\partial V}{\partial q_r} + \frac{d}{dt}\left(\frac{\partial V}{\partial \dot{q}_r}\right) \qquad (r = 1, 2, \ldots, n),$$

where V is a given function of $\dot{q}_1, \dot{q}_2, \ldots, \dot{q}_n, q_1, \ldots, q_n$, the Lagrangian equations of motion are

$$\frac{d}{dt}\left(\frac{\partial T}{\partial \dot{q}_r}\right) - \frac{\partial T}{\partial q_r} = -\frac{\partial V}{\partial q_r} + \frac{d}{dt}\left(\frac{\partial V}{\partial \dot{q}_r}\right) \qquad (r = 1, 2, \ldots, n),$$

and if a kinetic potential L be defined by the equation

$$L = T - V,$$

the equations take the customary form

$$\frac{d}{dt}\left(\frac{\partial L}{\partial \dot{q}_r}\right) - \frac{\partial L}{\partial q_r} = 0 \qquad (r = 1, 2, \ldots, n).$$

The function V can be regarded as a generalised potential energy function. An example of such a system is furnished by the motion of a particle subject to Weber's electrodynamic law of attraction* to a fixed point, the force per unit mass acting on the particle being

$$\frac{1}{r^2}\left(1-\frac{\dot{r}^2-2r\ddot{r}}{c^2}\right),$$

where r is the distance of the particle from the centre of force: in this case the function V is defined by the equation

$$V=\frac{1}{r}\left(1+\frac{\dot{r}^2}{c^2}\right).$$

Example. If the forces $Q_1, Q_2, \ldots, Q_n$ of a dynamical system which is specified by coordinates $q_1, q_2, \ldots, q_n$ are derivable from a generalised potential-function V, so that

$$Q_r=-\frac{\partial V}{\partial q_r}+\frac{d}{dt}\left(\frac{\partial V}{\partial \dot{q}_r}\right) \qquad (r=1, 2, \ldots, n),$$

shew that $Q_1, Q_2, \ldots, Q_n$ must be linear functions of $\ddot{q}_1, \ddot{q}_2, \ldots, \ddot{q}_n$, satisfying the $n(2n-1)$ relations

$$\frac{\partial Q_i}{\partial \ddot{q}_k}=\frac{\partial Q_k}{\partial \ddot{q}_i},$$

$$\frac{\partial Q_i}{\partial \dot{q}_k}+\frac{\partial Q_k}{\partial \dot{q}_i}=\frac{d}{dt}\left(\frac{\partial Q_i}{\partial \ddot{q}_k}+\frac{\partial Q_k}{\partial \ddot{q}_i}\right),$$

$$\frac{\partial Q_i}{\partial q_k}-\frac{\partial Q_k}{\partial q_i}=\frac{1}{2}\frac{d}{dt}\left(\frac{\partial Q_i}{\partial \dot{q}_k}-\frac{\partial Q_k}{\partial \dot{q}_i}\right).$$

On the general conditions for the existence of a kinetic potential of forces, reference may be made to

Helmholtz, *Journal für Math.*, Vol. C. (1886).
Mayer, *Leipzig. Berichte*, Vol. XLVIII. (1896).
Hirsch, *Math. Annalen*, Vol. L. (1898).

32. *Initial motions.*

The differential equations of motion of a dynamical system cannot in general be solved in a finite form in terms of known functions. It is however always possible (except in the vicinity of certain singularities which need not be considered here) to solve a set of differential equations by *power-series*, i.e. to obtain for the dependent variables $q_1, q_2, \ldots, q_n$ expressions of the type

$$\begin{cases} q_1=a_1+b_1t+c_1t^2+d_1t^3+\ldots, \\ q_2=a_2+b_2t+c_2t^2+d_2t^3+\ldots, \\ \ldots\ldots\ldots\ldots\ldots\ldots\ldots\ldots\ldots \\ q_n=a_n+b_nt+c_nt^2+d_nt^3+\ldots; \end{cases}$$

the coefficients $a, b, \ldots$ can in fact be obtained by substituting these series in the differential equations, and equating to zero the coefficients of the various

* W. Weber, *Annalen d. Phys.* LXXIII. (1848), p. 193. Cf. Whittaker's *History of the Theories of Aether and Electricity*, pp. 226—231.

powers of t; the expansions will converge in general for values of t within some definite circle of convergence in the t-plane.

It is plain that these series will give any information which may be required about the initial character of the motion (t being measured from the commencement of the motion), since a_1 is the initial value of q_1, b_1 is the initial value of $\dot{q}_1$, and so on. This method of discussing the initial motion of a system is illustrated by the following example.

Example. Consider the motion of a particle of unit mass, which is free to move in a plane and initially at rest, and which is acted on by a field of force whose components parallel to fixed rectangular axes at any point (x, y) are (X, Y); and let it be required to determine the initial radius of curvature of the path.

Let $(x+\xi, y+\eta)$ be the coordinates of any point adjacent to the initial point (x, y), so that ξ, η may be regarded as small quantities; then the equations of motion are

$$\ddot{\xi} = X(x+\xi, y+\eta)$$

$$= X(x, y) + \xi \frac{\partial X(x, y)}{\partial x} + \eta \frac{\partial X(x, y)}{\partial y} + \dots\dots,$$

$$\ddot{\eta} = Y(x, y) + \xi \frac{\partial Y(x, y)}{\partial x} + \eta \frac{\partial Y(x, y)}{\partial y} + \dots\dots.$$

If therefore we assume for ξ and η the expansions

$$\xi = at^2 + bt^3 + ct^4 + \dots,$$

$$\eta = dt^2 + et^3 + ft^4 + \dots,$$

(it is not necessary to include terms of lower order than t^2, since the quantities ξ, η, $\dot{\xi}$, $\dot{\eta}$ are initially zero), and substitute in these differential equations, we find, on comparing the coefficients of various powers of t, the relations

$$a = \tfrac{1}{2} X(x, y), \quad b = 0, \quad c = \tfrac{1}{24}\left(X\frac{\partial X}{\partial x} + Y\frac{\partial X}{\partial y}\right),$$

$$d = \tfrac{1}{2} Y(x, y), \quad e = 0, \quad f = \tfrac{1}{24}\left(X\frac{\partial Y}{\partial x} + Y\frac{\partial Y}{\partial y}\right).$$

The path of the particle near the point (x, y) is therefore given by the series

$$\xi = Xu + \tfrac{1}{6}\left(X\frac{\partial X}{\partial x} + Y\frac{\partial X}{\partial y}\right)u^2 + \dots,$$

$$\eta = Yu + \tfrac{1}{6}\left(X\frac{\partial Y}{\partial x} + Y\frac{\partial Y}{\partial y}\right)u^2 + \dots,$$

where u denotes the quantity $\tfrac{1}{2}t^2$.

Now if the coordinates ξ and η of any curve are expressed in terms of a parameter u, the radius of curvature at the point u is known to be

$$\frac{\left\{\left(\frac{d\xi}{du}\right)^2 + \left(\frac{d\eta}{du}\right)^2\right\}^{\frac{3}{2}}}{\frac{d^2\eta}{du^2}\frac{d\xi}{du} - \frac{d^2\xi}{du^2}\frac{d\eta}{du}},$$

so the radius of curvature corresponding to the zero value of u, for the curve given by the above expressions, is

$$\frac{3(X^2+Y^2)^{\frac{3}{2}}}{\left(X\frac{\partial Y}{\partial x} + Y\frac{\partial Y}{\partial y}\right)X - \left(X\frac{\partial X}{\partial x} + Y\frac{\partial X}{\partial y}\right)Y},$$

and this is the required radius of curvature of the path of the particle at the initial point.

33. *Similarity in dynamical systems*.*

If any system of connected particles and rigid bodies is given, it is possible to construct another system exactly similar to it, but on a different scale. If now the masses and forces in the two systems, which we can call the *pattern* and *model* respectively, bear certain ratios to each other, the workings of the two systems will be similar, though possibly at speeds which are not the same but bear a constant ratio to each other.

To find the relation between the various ratios involved, let the linear dimensions of the model and pattern be in the ratio $x : 1$, let the masses of corresponding particles be in the ratio $y : 1$, let the rates of working be in the ratio $z : 1$, so that the times elapsed between corresponding phases are in the ratio $1 : z$, and let the forces be in the ratio $w : 1$. Then for each particle we have an equation of motion of the form

$$m\ddot{x} = X;$$

so if m is altered in the ratio $y : 1$, $\ddot{x}$ is altered in the ratio $xz^2 : 1$, and X is altered in the ratio $w : 1$, we must have

$$w = xyz^2,$$

and this is the required relation between the numbers x, y, z, w.

Example. If the forces acting are those due to gravity, we have $w = y$, and consequently $xz^2 = 1$, so that the rates of working are inversely as the square roots of the linear dimensions.

If the forces acting are the mutual gravitations of the particles, every particle attracting every other particle with a force proportional to the product of the masses and the inverse square of the distance, we have $w = y^2/x^2$, so that the rates of working are in the ratio $y^{\frac{1}{2}} : x^{\frac{3}{2}}$.

34. *Motion with reversed forces.*

A special case of similarity is that in which the ratio w has the value -1.

We have seen that the motion of any dynamical system which is subjected to constraints independent of the time, and to forces which depend only on the positions of the particles, is expressed by the Lagrangian equations

$$\frac{d}{dt}\left(\frac{\partial T}{\partial \dot{q}_r}\right) - \frac{\partial T}{\partial q_r} = Q_r, \qquad (r = 1, 2, \ldots, n),$$

where the kinetic energy T is a homogeneous quadratic function of the velocities $\dot{q}_1, \dot{q}_2, \ldots, \dot{q}_n$, involving the coordinates $q_1, q_2, \ldots, q_n$, in any way, and Q is a function of $q_1, q_2, \ldots, q_n$ only.

Introduce a new independent variable defined by the equation

$$\tau = it, \qquad \text{where } i = \sqrt{-1},$$

and let accents denote differentiations with regard to τ. Then since

* Newton, *Principia*, Book II. Sect. 7, Prop. 32.

$\frac{d}{dt}\left(\frac{\partial T}{\partial \dot{q}_r}\right)$ and $\frac{\partial T}{\partial q_r}$ are homogeneous of degree -2 in dt, the above equations become

$$\frac{d}{d\tau}\left(\frac{\partial \mathfrak{T}}{\partial q_r'}\right) - \frac{\partial \mathfrak{T}}{\partial q_r} = -Q_r, \qquad (r = 1, 2, \ldots, n),$$

where $\mathfrak{T}$ is the same function of $q_1', q_2', \ldots, q_n', q_1, \ldots, q_n$ that T is of $\dot{q}_1, \dot{q}_2, \ldots, \dot{q}_n, q_1, q_2, \ldots, q_n$.

But if τ (instead of t) be now interpreted as denoting the time, these last equations are the equations of motion of the same system when subjected to the same forces reversed in direction. Moreover, if $\alpha_1, \alpha_2, \ldots, \alpha_n, \beta_1, \beta_2, \ldots, \beta_n$ are the initial values of $q_1, q_2, \ldots, q_n, \dot{q}_1, \dot{q}_2, \ldots, \dot{q}_n$, respectively, in any particular case of the motion of the original system, then $\alpha_1, \alpha_2, \ldots, \alpha_n, -i\beta_1, -i\beta_2, \ldots, -i\beta_n$ will be the corresponding quantities in the transformed problem. We thus have the theorem that *in any dynamical system subjected to constraints independent of the time and to forces which depend only on the position of the particles, the integrals of the equations of motion are still real if t be replaced by $\sqrt{-1}t$ and the initial velocities $\beta_1, \beta_2, \ldots, \beta_n$ by $-\sqrt{-1}\beta_1, -\sqrt{-1}\beta_2, \ldots, -\sqrt{-1}\beta_n$ respectively; and the expressions thus obtained represent the motion which the same system would have if, with the same initial conditions, it were acted on by the same forces reversed in direction.*

35. *Impulsive motion.*

In certain cases (e.g. in the collision of rigid bodies) the velocities of the particles in a dynamical system are changed so rapidly that the time occupied in the process may, for analytical purposes, be altogether neglected.

The laws which govern the impulsive motion of a system bear a close analogy to those which apply in the case of motion under finite forces: they can be formulated in the following way*.

The number which represents the mass of a particle, multiplied by the vector which represents its velocity at any instant, is a vector quantity (localised in a line through the particle) which is called the *momentum* of the particle at that instant†; the three components parallel to rectangular axes $Oxyz$ of the momentum of a particle of mass m at the point (x, y, z) are therefore $(m\dot{x}, m\dot{y}, m\dot{z})$. If any number of particles form a dynamical system, the sum of the components in any given direction of the momenta of the particles is called the *component* in that direction *of the momentum of the system.* The impulsive changes of velocity in the various particles of a connected system can be regarded as the result of sudden communications of momentum to the particles.

The effect of an agency which causes impulsive motion in the system

* They were involved in the discovery of the laws of impact in 1668 by Wallis and Wren, *Phil. Trans.* No. 43, pp. 864, 867.

† Momentum is the *quantitas motus* of Newton's *Principia*, Book I. Def. 2. The idea can be traced back to Descartes.

will be measured by the momentum which it would communicate to a single free particle. If therefore (u_0, v_0, w_0) are the components of velocity of a particle of mass m, referred to fixed axes in space, before the impulsive communication of momentum to the particle, and if (u, v, w) are the components of velocity of the particle after the impulse, then the vector quantity (localised in a line through the particle) whose components are

$$m(u-u_0), \quad m(v-v_0), \quad m(w-w_0)$$

represents *the impulse acting on the particle.*

For the discussion of the impulsive motion of a connected system of particles, it is clearly necessary to have some experimental law analogous to the law of Action and Reaction of finite forces; such a law is contained in the statement that *the total impulse acting on a particle of a connected system is equal to the resultant of the external impulse on the particle* (i.e. the impulse communicated by agencies external to the system, measured by the momentum which the particle would acquire if free) *together with impulses directed along the lines which join this particle to the other particles which constrain its motion; and the mutually induced impulses between two connected particles are equal in magnitude and opposite in sign.*

If we regard the components of an impulse as the time-integrals of the components of an ordinary finite force which is very large but acts only for a very short time, the law just stated agrees with the law of Action and Reaction for finite forces.

Change of kinetic energy due to impulses.

The change in kinetic energy of a dynamical system whose particles are acted on by a given set of impulses may be determined in the following way.

Let an impulse I, directed along a line whose direction-cosines referred to fixed axes of reference are (λ, μ, ν), be communicated to a particle of mass m, changing its velocity from v_0, in a direction whose direction-cosines are (L_0, M_0, N_0), to v, in a direction whose direction-cosines are (L, M, N). The equations of impulsive motion are

$$m(vL-v_0L_0)=I\lambda, \quad m(vM-v_0M_0)=I\mu, \quad m(vN-v_0N_0)=I\nu.$$

Multiplying these equations respectively by

$$\tfrac{1}{2}(vL+v_0L_0), \quad \tfrac{1}{2}(vM+v_0M_0), \quad \text{and} \quad \tfrac{1}{2}(vN+v_0N_0),$$

and adding, we have

$$\tfrac{1}{2}mv^2-\tfrac{1}{2}mv_0^2=\tfrac{1}{2}Iv(L\lambda+M\mu+N\nu)+\tfrac{1}{2}Iv_0(L_0\lambda+M_0\mu+N_0\nu).$$

The change in kinetic energy of the particle is therefore equal to the product of the impulse and the mean of the components, before and after the impulse, of the velocity of the particle in the direction of the impulse.

Now consider any dynamical system of connected particles and rigid bodies, to which given impulses are communicated; applying this result to each particle of the system, and summing, we see that *the change in the kinetic energy of the system is equal to the sum of the impulses applied to it, each multiplied by the mean of the components, before and after the communication of the impulse, of the velocity of its point of application in the direction of the impulse.* In this result we can clearly neglect the impulsive forces between the molecules of any rigid body of the system.

36. *The Lagrangian equations of impulsive motion.*

The equations of impulsive motion of a dynamical system can be expressed in a form* analogous to the Lagrangian equations of motion for finite forces, in the following way.

Let (X_i, Y_i, Z_i) be the components of the total impulse (external and molecular) applied to a particle m_i of the system, situated at the point (x_i, y_i, z_i). The equations of impulsive motion of the particle are

$$m_i(\dot{x}_i - \dot{x}_{i0}) = X_i, \qquad m_i(\dot{y}_i - \dot{y}_{i0}) = Y_i, \qquad m_i(\dot{z}_i - \dot{z}_{i0}) = Z_i,$$

where $(\dot{x}_{i0}, \dot{y}_{i0}, \dot{z}_{i0})$ and $(\dot{x}_i, \dot{y}_i, \dot{z}_i)$ denote the components of velocity of the particle before and after the application of the impulse.

If $q_1, q_2, \ldots, q_n$ denote the n independent coordinates in terms of which the configuration of the system can be expressed, we have therefore

$$\sum_i m_i \left\{(\dot{x}_i - \dot{x}_{i0})\frac{\partial x_i}{\partial q_r} + (\dot{y}_i - \dot{y}_{i0})\frac{\partial y_i}{\partial q_r} + (\dot{z}_i - \dot{z}_{i0})\frac{\partial z_i}{\partial q_r}\right\} = \sum_i \left(X_i \frac{\partial x_i}{\partial q_r} + Y_i \frac{\partial y_i}{\partial q_r} + Z_i \frac{\partial z_i}{\partial q_r}\right),$$

where the summation is extended over all the particles of the system.

Now in forming the summation on the right-hand side of this equation, it is seen as in § 26 that the molecular impulses between particles of the system can be omitted: the quantity

$$\sum_i \left(X_i \frac{\partial x_i}{\partial q_r} + Y_i \frac{\partial y_i}{\partial q_r} + Z_i \frac{\partial z_i}{\partial q_r}\right)$$

can therefore readily be found when the external impulses are known: we shall denote it by the symbol Q_r. We have consequently

$$\sum_i m_i \left\{(\dot{x}_i - \dot{x}_{i0})\frac{\partial x_i}{\partial q_r} + (\dot{y}_i - \dot{y}_{i0})\frac{\partial y_i}{\partial q_r} + (\dot{z}_i - \dot{z}_{i0})\frac{\partial z_i}{\partial q_r}\right\} = Q_r.$$

But as in § 26 we have

$$\frac{\partial x_i}{\partial q_r} = \frac{\partial \dot{x}_i}{\partial \dot{q}_r}, \quad \text{so} \quad \dot{x}_i \frac{\partial x_i}{\partial q_r} = \frac{\partial}{\partial \dot{q}_r}\left(\tfrac{1}{2}\dot{x}_i^2\right),$$

and similarly

$$\dot{x}_{i0}\frac{\partial x_i}{\partial q_r} = \frac{\partial}{\partial \dot{q}_{r0}}\left(\tfrac{1}{2}\dot{x}_{i0}^2\right),$$

where $\dot{q}_{r0}$ and $\dot{q}_r$ denote the velocities of the coordinate q_r before and after the impulse respectively. Thus if

$$T = \tfrac{1}{2}\sum_i m_i(\dot{x}_i^2 + \dot{y}_i^2 + \dot{z}_i^2)$$

* Due to Lagrange, *Méc. Anal.* (2e éd.), II. p. 183.

denotes the kinetic energy of the system after the impulse, the above equation can be written in the form

$$\frac{\partial T}{\partial \dot{q}_r} - \left(\frac{\partial T}{\partial \dot{q}_r}\right)_0 = Q_r,$$

where $\left(\frac{\partial T}{\partial \dot{q}_r}\right)_0$ denotes the quantity corresponding to $\frac{\partial T}{\partial \dot{q}_r}$, but relating to the instant before the impulse.

Similar equations can be found for the rest of the coordinates $q_1, q_2, \ldots, q_n$; and thus we obtain the set of n equations

$$\frac{\partial T}{\partial \dot{q}_r} - \left(\frac{\partial T}{\partial \dot{q}_r}\right)_0 = Q_r, \qquad (r = 1, 2, \ldots, n),$$

which are known as the *Lagrangian equations of impulsive motion.*

These are algebraical equations for the determination of $\dot{q}_1, \dot{q}_2, \ldots, \dot{q}_n$ in terms of $\dot{q}_{10}, \dot{q}_{20}, \ldots, \dot{q}_{n0}$; they are not differential equations like the Lagrangian equations of motion for finite forces, since the second derivates of the coordinates with respect to the time do not enter.

Miscellaneous Examples.

1. Two rigid bodies moving in space are constrained only by a taut inextensible string joining a given point of one body to a given point of the other, and one of the bodies is constrained to roll without sliding on a given fixed surface. How many degrees of freedom has the system, and how many independent coordinates are required to specify its configuration?

2. A point is referred to curvilinear coordinates a, b, c, and the square of its velocity is

$$2T \equiv A\dot{a}^2 + B\dot{b}^2 + C\dot{c}^2 + 2F\dot{b}\dot{c} + 2G\dot{c}\dot{a} + 2H\dot{a}\dot{b}.$$

Shew that p, q, r, the component accelerations in the directions of the tangents to the coordinate lines, are given by three equations of the type

$$\frac{d}{dt}\left(\frac{\partial T}{\partial \dot{a}}\right) - \frac{\partial T}{\partial a} = p\sqrt{A} + \frac{H}{\sqrt{B}}q + \frac{G}{\sqrt{C}}r. \qquad \text{(Coll. Exam.)}$$

3. A particle which is free to move in space is initially at rest at the origin, and is in a field of force whose components (X, Y, Z) at any point (x, y, z) are given by the expansions

$$X = a + bx + \text{quadratic and higher terms in } x, y, z;$$
$$Y = \quad cx + \text{quadratic and higher terms in } x, y, z;$$
$$Z = \quad dx^2 + \text{cubic and higher terms in } x, y, z.$$

Find the radii of curvature and torsion of the orbit at the origin.

CHAPTER III

PRINCIPLES AVAILABLE FOR THE INTEGRATION

37. *Problems which are soluble by quadratures.*

The determination of the motion of a holonomic dynamical system with a finite number of degrees of freedom has in the preceding chapter been shewn to depend on the solution of a set of ordinary differential equations. If n denotes the number of degrees of freedom, and $(q_1, q_2, \ldots, q_n)$ are the coordinates specifying the configuration of the system at the time t, then the set of equations consists of n differential equations, each of the second order, with $q_1, q_2, \ldots, q_n$ as dependent variables and t as independent variable. This set of equations is said to be *of order* $2n$, the *order* being defined to be the sum of the orders of the highest derivates of the dependent variables occurring in the equations. It is a well-known result of the theory of ordinary differential equations that the number of arbitrary constants of integration in the solution of a set of differential equations is equal to the order of the system; whence it follows that *there are* $2n$ *constants of integration in the general solution of a holonomic dynamical problem with* n *degrees of freedom.*

Now any given set of differential equations of order k can be reduced to the form

$$\frac{dx_r}{dt} = X_r(x_1, x_2, \ldots, x_k, t), \qquad (r = 1, 2, \ldots, k),$$

where $X_1, X_2, \ldots, X_k$ are known functions of their arguments, by taking as new variables $(x_1, x_2, \ldots, x_k)$ the original dependent variables together with their derivates up to (but not including) the highest derivates occurring in the original set of equations. Thus e.g. the set of equations

$$\frac{d^2q_1}{dt^2} = Q_1(q_1, q_2, \dot{q}_1, \dot{q}_2), \qquad \frac{d^2q_2}{dt^2} = Q_2(q_1, q_2, \dot{q}_1, \dot{q}_2),$$

(where Q_1 and Q_2 are any functions of the arguments indicated) which is of order 4, can be reduced to the set

$$\frac{dx_1}{dt} = x_3, \qquad \frac{dx_2}{dt} = x_4, \qquad \frac{dx_3}{dt} = Q_1(x_1, x_2, x_3, x_4), \qquad \frac{dx_4}{dt} = Q_2(x_1, x_2, x_3, x_4),$$

by taking

$$x_1 = q_1, \qquad x_2 = q_2, \qquad x_3 = \dot{q}_1, \qquad x_4 = \dot{q}_2.$$

The form

$$\frac{dx_r}{dt} = X_r(x_1, x_2, \ldots, x_k, t) \qquad (r = 1, 2, \ldots, k)$$

may therefore be regarded as the typical form for a set of differential equations of order k.

If a function $f(x_1, x_2, \ldots, x_k, t)$ is such that $\frac{df}{dt}$ is zero when $(x_1, x_2, \ldots, x_k)$ are any functions of t whatever which satisfy these differential equations, the equation

$$f(x_1, x_2, \ldots, x_k, t) = \text{Constant}$$

is called an *integral* of the system. The condition that a given function f may furnish an integral of the system is easily found; for the equation $df/dt = 0$ gives

$$\frac{\partial f}{\partial x_1}\dot{x}_1 + \frac{\partial f}{\partial x_2}\dot{x}_2 + \ldots + \frac{\partial f}{\partial x_k}\dot{x}_k + \frac{\partial f}{\partial t} = 0,$$

or

$$\frac{\partial f}{\partial x_1}X_1 + \frac{\partial f}{\partial x_2}X_2 + \ldots + \frac{\partial f}{\partial x_k}X_k + \frac{\partial f}{\partial t} = 0,$$

and this relation must be identically satisfied in order that the equation

$$f(x_1, x_2, \ldots, x_k, t) = \text{Constant}$$

may be an integral of the system of differential equations.

Sometimes the function f itself (as distinct from the equation f=constant) is called an integral of the system.

The complete solution of the set of differential equations of order k is furnished by k integrals

$$f_r(x_1, x_2, \ldots, x_k, t) = a_r, \qquad (r = 1, 2, \ldots, k),$$

where $a_1, a_2, \ldots, a_k$ are arbitrary constants, provided these integrals are *distinct*, i.e. no one of them is algebraically deducible from the others. For let the values of $x_1, x_2, \ldots, x_k$, obtained from these equations as functions of $t, a_1, a_2, \ldots, a_k$, be

$$x_r = \phi_r(a_1, a_2, \ldots, a_k, t), \qquad (r = 1, 2, \ldots, k);$$

then if $(x_1, x_2, \ldots, x_k)$ are any particular set of functions of t which satisfy the differential equations, it follows from what has been said above that by giving to the arbitrary constants a_r suitable constant values we can make the equations

$$f_r(x_1, x_2, \ldots, x_k, t) = a_r \qquad (r = 1, 2, \ldots, k)$$

true for this particular set of functions $(x_1, x_2, \ldots, x_k)$; and therefore this set of functions $(x_1, x_2, \ldots, x_k)$ will be included among the functions defined by the equations $x_r = \phi_r$. The solution of a dynamical problem with n degrees of freedom may therefore be regarded as equivalent to the determination of $2n$ integrals of a set of differential equations of order $2n$.

Thus the differential equation

$$\ddot{q} = -q,$$

which is of the second order, possesses the two integrals

$$\begin{cases} q^2 + \dot{q}^2 = a_1, \\ \tan^{-1}\dfrac{q}{\dot{q}} - t = a_2, \end{cases}$$

where a_1 and a_2 are arbitrary constants. On solving these equations for q and $\dot{q}$, we have

$$\begin{cases} q = a_1^{\frac{1}{2}} \sin(t + a_2), \\ \dot{q} = a_1^{\frac{1}{2}} \cos(t + a_2), \end{cases}$$

and these equations constitute the solution of the differential equation.

The more elementary division of dynamics, with which this and the immediately succeeding chapters are concerned, is occupied with the discussion of those dynamical problems which can be solved completely in terms of the known elementary functions or the indefinite integrals of such functions. These are generally referred to as *problems soluble by quadratures.* The problems of dynamics are not in general soluble by quadratures; and in those cases in which a solution by quadratures can be effected, there must always be some special reason for it,—in fact the kinetic potential of the problem must have some special character. The object of the present chapter is to discuss those peculiarities of the kinetic potential which are most frequently found in problems soluble by quadratures, and which in fact are the ultimate explanation of the solubility.

38. *Systems with ignorable coordinates.*

We have seen (§ 27) that the motion of a conservative holonomic dynamical system with n degrees of freedom, for which the coordinates are $q_1, q_2, \ldots, q_n$ and the kinetic potential is L, is determined by the differential equations

$$\frac{d}{dt}\left(\frac{\partial L}{\partial \dot{q}_r}\right) - \frac{\partial L}{\partial q_r} = 0, \qquad (r = 1, 2, \ldots, n).$$

The quantity $\dfrac{\partial L}{\partial \dot{q}_r}$ is generally called the *momentum* corresponding to the coordinate q_r.

It may happen that some of the coordinates, say $q_1, q_2, \ldots, q_k$, are not explicitly contained in L, although the corresponding velocities $\dot{q}_1, \dot{q}_2, \ldots, \dot{q}_k$ are so contained. Coordinates of this kind are said to be *ignorable* or *cyclic*; it will appear in the following chapters that the presence of ignorable coordinates is the most frequently-occurring reason for the solubility of particular problems by quadratures.

The Lagrangian equations of motion which correspond to the k ignorable coordinates are

$$\frac{d}{dt}\left(\frac{\partial L}{\partial \dot{q}_r}\right) = 0, \qquad (r = 1, 2, \ldots, k),$$

and on integration, these can be written

$$\frac{\partial L}{\partial \dot{q}_r} = \beta_r, \qquad (r = 1, 2, \ldots, k),$$

where $\beta_1, \beta_2, \ldots, \beta_k$ are constants of integration. These last equations are evidently k integrals of the system.

We shall now shew how these k integrals can be utilised to reduce the order of the set of Lagrangian differential equations of motion*.

Let R denote the function $L - \sum_{r=1}^{k} \dot{q}_r \frac{\partial L}{\partial \dot{q}_r}$. By means of the k equations

$$\frac{\partial L}{\partial \dot{q}_r} = \beta_r, \qquad (r = 1, 2, \ldots, k),$$

we can express the k quantities $\dot{q}_1, \dot{q}_2, \ldots, \dot{q}_k$, which are the velocities corresponding to the ignorable coordinates, in terms of

$$q_{k+1}, q_{k+2}, \ldots, q_n, \dot{q}_{k+1}, \dot{q}_{k+2}, \ldots, \dot{q}_n, \beta_1, \beta_2, \ldots, \beta_k;$$

we shall suppose that in this way the function R is expressed in terms of the latter set of quantities.

Now let δf denote the increment produced in any function f of the quantities $q_{k+1}, q_{k+2}, \ldots, q_n, \dot{q}_1, \dot{q}_2, \ldots, \dot{q}_n$ (or of the quantities $q_{k+1}, q_{k+2}, \ldots, q_n$, $\dot{q}_{k+1}, \ldots, \dot{q}_n, \beta_1, \beta_2, \ldots, \beta_k$) by arbitrary infinitesimal changes $\delta q_{k+1}, \delta q_{k+2}, \ldots$, $\delta q_n, \delta \dot{q}_1, \ldots, \delta \dot{q}_n$ in its arguments. Then we have

$$\delta R = \delta \left(L - \sum_{r=1}^{k} \dot{q}_r \frac{\partial L}{\partial \dot{q}_r} \right),$$

by the definition of R. But

$$\delta L = \sum_{r=k+1}^{n} \frac{\partial L}{\partial q_r} \delta q_r + \sum_{r=1}^{k} \frac{\partial L}{\partial \dot{q}_r} \delta \dot{q}_r + \sum_{r=k+1}^{n} \frac{\partial L}{\partial \dot{q}_r} \delta \dot{q}_r,$$

and

$$\delta \left(\sum_{r=1}^{k} \dot{q}_r \frac{\partial L}{\partial \dot{q}_r} \right) = \sum_{r=1}^{k} \frac{\partial L}{\partial \dot{q}_r} \delta \dot{q}_r + \sum_{r=1}^{k} \dot{q}_r \, \delta \beta_r,$$

since

$$\frac{\partial L}{\partial \dot{q}_r} = \beta_r.$$

We have therefore

$$\delta R = \sum_{r=k+1}^{n} \frac{\partial L}{\partial q_r} \delta q_r + \sum_{r=k+1}^{n} \frac{\partial L}{\partial \dot{q}_r} \delta \dot{q}_r - \sum_{r=1}^{k} \dot{q}_r \, \delta \beta_r,$$

and since the infinitesimal quantities occurring on the right-hand side of this equation are arbitrary and independent, the equation is equivalent to the

* The transformation which follows is really a case of the Hamiltonian transformation, which is discussed in Chapter X; it was however first separately given by Routh in 1876, and somewhat later by Helmholtz.

system of equations

$$\frac{\partial L}{\partial \dot{q}_r} = \frac{\partial R}{\partial \dot{q}_r}, \qquad (r = k+1, k+2, \ldots, n),$$

$$\frac{\partial L}{\partial q_r} = \frac{\partial R}{\partial q_r}, \qquad (r = k+1, k+2, \ldots, n),$$

$$\dot{q}_r = -\frac{\partial R}{\partial \beta_r}, \qquad (r = 1, 2, \ldots, k).$$

Substituting these results in the Lagrangian equations of motion, we have

$$\frac{d}{dt}\left(\frac{\partial R}{\partial \dot{q}_r}\right) - \frac{\partial R}{\partial q_r} = 0, \quad (r = k+1, k+2, \ldots, n).$$

Now R is a function only of the variables $\dot{q}_{k+1}, \dot{q}_{k+2}, \ldots, \dot{q}_n, q_{k+1}, \ldots, q_n$, and the constants $\beta_1, \beta_2, \ldots, \beta_k$: so this is a new Lagrangian system of equations, which we can regard as defining a new dynamical problem with only $(n-k)$ degrees of freedom, the new coordinates being $q_{k+1}, q_{k+2}, \ldots, q_n$, and the new kinetic potential being R. When the variables $q_{k+1}, q_{k+2}, \ldots, q_n$ have been obtained in terms of t by solving this new dynamical problem, the remainder of the original coordinates, namely $q_1, q_2, \ldots, q_k$, can be obtained from the equations

$$q_r = -\int \frac{\partial R}{\partial \beta_r} dt, \qquad (r = 1, 2, \ldots, k).$$

Hence *a dynamical problem with n degrees of freedom, which has k ignorable coordinates, can be reduced to a dynamical problem which has only* $(n-k)$ *degrees of freedom.* This process is called the *ignoration of coordinates.*

The essential basis of the ignoration of coordinates is in the theorem that when the kinetic potential does not contain one of the coordinates q_r explicitly, although it involves the corresponding velocity $\dot{q}_r$, an integral of the motion can be at once written down, namely $\frac{\partial L}{\partial \dot{q}_r} = \text{constant}$. This is a particular case of a much more general theorem which will be given later, to the effect that when a dynamical system admits a known infinitesimal contact-transformation, an integral of the system can be immediately obtained.

If the original problem relates to the motion of a conservative dynamical system in which the constraints are independent of the time, we have seen that its kinetic potential L consists of a part (the kinetic energy) which is a homogeneous quadratic function of $\dot{q}_1, \dot{q}_2, \ldots, \dot{q}_n$, and which involves $q_{k+1}, q_{k+2}, \ldots, q_n$ in any way, together with a part (the potential energy with sign reversed) which involves $q_{k+1}, q_{k+2}, \ldots, q_n$ only. But in the new dynamical system which is obtained after the ignoration of coordinates, the kinetic potential R cannot be divided into two parts in this way: in fact, R will in general contain terms linear in the velocities. And more generally when (as happens very frequently in the more advanced parts of Dynamics) the solution of one set of Lagrangian differential equations is made to depend on that of another set of Lagrangian differential equations with a smaller number of coordinates, the kinetic potential of this new system is not necessarily divisible into two groups of terms corresponding to a kinetic and a potential energy. We shall sometimes use the word *natural* to denote those systems of Lagrangian equations for which the kinetic potential contains only terms of degrees 2 and 0 in the velocities, and *non-natural* to denote those systems for which this condition is not satisfied.

As an example of the ignoration of coordinates, consider a dynamical system with two degrees of freedom, for which the kinetic energy is

$$T=\tfrac{1}{2}\frac{\dot{q}_1^2}{a+bq_2^2}+\tfrac{1}{2}\dot{q}_2^2,$$

and the potential energy is

$$V=c+dq_2^2,$$

where a, b, c, d are given constants.

It is evident that q_1 is an ignorable coordinate, since it does not appear explicitly in T or V.

The kinetic potential of the system is

$$L=\tfrac{1}{2}\frac{\dot{q}_1^2}{a+bq_2^2}+\tfrac{1}{2}\dot{q}_2^2-c-dq_2^2,$$

and the integral corresponding to the ignorable coordinate is

$$\frac{\dot{q}_1}{a+bq_2^2}=\beta,$$

where β is a constant, whose value is determined by the initial circumstances of the motion.

The kinetic potential of the new dynamical system obtained by ignoring the coordinate q_1 is

$$\begin{aligned} R &= L-\dot{q}_1\frac{\partial L}{\partial \dot{q}_1} \\ &= \tfrac{1}{2}\dot{q}_2^2-c-dq_2^2-\tfrac{1}{2}\beta^2(a+bq_2^2), \end{aligned}$$

and the problem is now reduced to the solution of the single equation

$$\frac{d}{dt}\left(\frac{\partial R}{\partial \dot{q}_2}\right)-\frac{\partial R}{\partial q_2}=0,$$

or

$$\ddot{q}_2+(2d+b\beta^2)\,q_2=0.$$

As this is a linear differential equation with constant coefficients, its solution can be immediately written down: it is

$$q_2=A\sin\{(2d+b\beta^2)^{\frac{1}{2}}\,t+\epsilon\},$$

where A and ϵ are constants of integration, to be determined by the initial circumstances of the motion. This equation gives the required expression of the coordinate q_2 in terms of the time: the value of q_1 in terms of t can then be deduced from the equation

$$q_1=\beta\int(a+bq_2^2)\,dt,$$

which gives

$$q_1=(\beta a+\tfrac{1}{2}\beta bA^2)\,t-\frac{\beta bA^2}{4\,(2d+b\beta^2)^{\frac{1}{2}}}\sin 2\{(2d+b\beta^2)^{\frac{1}{2}}\,t+\epsilon\},$$

and so completes the solution of the system.

39. *Special cases of ignoration; integrals of momentum and angular momentum.*

We shall now consider specially the two commonest types of ignorable coordinates in dynamical problems.

(i) *Systems possessing an integral of momentum.*

Let the coordinates of a conservative holonomic dynamical system with n degrees of freedom be $q_1, q_2, \ldots, q_n$; and let T be the kinetic energy of the system, and V the potential energy, so that the equations of motion of the system are

$$\frac{d}{dt}\left(\frac{\partial T}{\partial \dot{q}_r}\right) - \frac{\partial T}{\partial q_r} = -\frac{\partial V}{\partial q_r}, \qquad (r = 1, 2, \ldots, n).$$

Suppose that one of the coordinates, say q_1, is ignorable, and moreover is such that an alteration of the value of q_1 by a quantity l, the remaining coordinates $q_2, q_3, \ldots, q_n$ being unaltered, corresponds to a simple translation of the whole system through a distance l parallel to a certain fixed direction in space; we shall take this to be the direction of the x-axis in a system of fixed rectangular axes of coordinates.

Since q_1 is an ignorable coordinate, we have the integral

$$\frac{\partial T}{\partial \dot{q}_1} = \text{Constant},$$

and we shall now discuss the physical meaning of this equation.

We have

$$\frac{\partial T}{\partial \dot{q}_1} = \tfrac{1}{2}\frac{\partial}{\partial \dot{q}_1}\Sigma m_i(\dot{x}_i^2 + \dot{y}_i^2 + \dot{z}_i^2),$$

where the summation is extended over all the particles of the system,

or
$$\frac{\partial T}{\partial \dot{q}_1} = \Sigma m_i\left(\dot{x}_i\frac{\partial \dot{x}_i}{\partial \dot{q}_1} + \dot{y}_i\frac{\partial \dot{y}_i}{\partial \dot{q}_1} + \dot{z}_i\frac{\partial \dot{z}_i}{\partial \dot{q}_1}\right)$$

$$= \Sigma m_i\left(\dot{x}_i\frac{\partial x_i}{\partial q_1} + \dot{y}_i\frac{\partial y_i}{\partial q_1} + \dot{z}_i\frac{\partial z_i}{\partial q_1}\right), \quad \text{by § 26}$$

$$= \Sigma m_i\dot{x}_i, \qquad \text{since in this case } \frac{\partial x_i}{\partial q_1} = 1, \quad \frac{\partial y_i}{\partial q_1} = 0, \quad \frac{\partial z_i}{\partial q_1} = 0.$$

Now $\Sigma m_i\dot{x}_i$ represents (§ 35) the component parallel to the x-axis of the momentum of the system of particles m_i, and consequently this is the physical meaning of the quantity $\frac{\partial T}{\partial \dot{q}_1}$ in the present case.

The integral
$$\frac{\partial T}{\partial \dot{q}_1} = \text{Constant}$$

can therefore be interpreted thus: *When a dynamical system can be translated as if rigid in a given direction without violating the constraints, and the potential energy is thereby unaltered* (the way in which the kinetic energy depends on the velocities is obviously unaltered by this translation, so the corresponding coordinate is ignorable), *then the component parallel to this direction of the momentum of the system is constant.*

This result is called the *law of conservation of momentum**, and systems to which it applies are said to *possess an integral of momentum.*

(ii) *Systems possessing an integral of angular momentum.*

Again taking a system with coordinates $q_1, q_2, \ldots, q_n$ and kinetic and potential energies T and V respectively, let us now suppose that the coordinate q_1 is ignorable, and moreover is such that an alteration of q_1 by a quantity α, the other coordinates remaining unchanged, corresponds to a simple rotation of the whole system through an angle α round a given fixed line in space: we shall take this line as the axis of z in a system of fixed rectangular axes of coordinates.

Since q_1 is an ignorable coordinate, we have the integral

$$\frac{\partial T}{\partial \dot{q}_1} = \text{Constant} \quad \ldots\ldots\ldots\ldots\ldots\ldots(1),$$

and we have to determine the physical interpretation of this equation.

We have as before

$$\frac{\partial T}{\partial \dot{q}_1} = \Sigma m_i \left(\dot{x}_i \frac{\partial x_i}{\partial q_1} + \dot{y}_i \frac{\partial y_i}{\partial q_1} + \dot{z}_i \frac{\partial z_i}{\partial q_1} \right),$$

where the summation is extended over all the particles of the system. But if we write

$$x_i = r_i \cos \phi_i, \qquad y_i = r_i \sin \phi_i,$$

we have

$$d\phi_i = dq_1,$$

so

$$\frac{\partial x_i}{\partial q_1} = \frac{\partial x_i}{\partial \phi_i} = -r_i \sin \phi_i = -y_i,$$

$$\frac{\partial y_i}{\partial q_1} = \frac{\partial y_i}{\partial \phi_i} = r_i \cos \phi_i = x_i,$$

$$\frac{\partial z_i}{\partial q_1} = 0,$$

and therefore

$$\frac{\partial T}{\partial \dot{q}_1} = \Sigma m_i (-\dot{x}_i y_i + \dot{y}_i x_i) \quad \ldots\ldots\ldots\ldots\ldots\ldots(2).$$

Now if r denote the distance of any particle of mass m from a given straight line at any instant, and if ω denote the angular velocity of the particle about the line, the product $mr^2\omega$ is called the *angular momentum* of the particle about the line.

Let O be any point, and let P, P' be two consecutive positions of the moving particle, the interval of time between them being dt. Then the

* This has been evolved gradually from the observation of Newton, *Principia*, Book I. introd. to Sect. XI., that if any number of bodies are acted on only by their mutual attractions, their common centre of gravity will either be at rest, or move uniformly in a straight line.

angular momentum about any line OK through O is clearly the limiting value of the ratio

$$\frac{m}{dt} \times \text{Twice the area of the projection of the triangle } OPP' \text{ on a plane perpendicular to } OK,$$

so if (l, m, n) are the direction-cosines of OK and if (λ, μ, ν) are the direction-cosines of the normal to the triangle OPP', we see that the angular momentum about OK is equal to the product of $(l\lambda + m\mu + n\nu)$ into the angular momentum about the normal to the plane OPP'. It is evident from this that if the angular momenta of a particle about any three rectangular axes $Oxyz$ at any time are h_1, h_2, h_3 respectively, then the angular momentum about any line through O whose direction-cosines referred to these axes are (l, m, n) is $lh_1 + mh_2 + nh_3$; we may express this by saying that *angular momenta about axes through a point are compounded according to the vectorial law.*

The angular momentum of a dynamical system about a given axis is defined to be the sum of the angular momenta of the separate particles of the system about the given axis; in particular, the angular momentum of a system of particles typified by a particle of mass m, whose coordinates are (x_i, y_i, z_i), about the axis of z is $\sum_i m_i r_i^2 \dot{\phi}_i$, where

$$x_i = r_i \cos \phi_i, \qquad y_i = r_i \sin \phi_i,$$

and the summation is extended over all the particles of the system; this expression for the angular momentum of a system can be written in the form

$$\sum_i m_i (\dot{y}_i x_i - \dot{x}_i y_i),$$

and on comparing this with equation (2) we have the result that *the angular momentum of the system considered, about the axis of z, is* $\dfrac{\partial T}{\partial \dot{q}_1}$.

The equation (1) implies therefore that the angular momentum of the system about the axis of z is constant: and we have the following result: *When a dynamical system can be rotated as if rigid round a given axis without violating the constraints, and the potential energy is thereby unaltered, the angular momentum of the system about this axis is constant.*

This result is known as the *theorem of conservation of angular momentum**.

Example. A system of n free particles is in motion under the influence of their mutual forces of attraction, these forces being derived from a kinetic potential V, which contains the coordinates and components of velocity of the particles, so that the equations of motion of the particles are

$$m_r \ddot{x}_r = \frac{\partial V}{\partial x_r} - \frac{d}{dt}\left(\frac{\partial V}{\partial \dot{x}_r}\right) \text{ etc.};$$

* Kepler's law, that the radius from the sun to a planet sweeps out equal areas in equal times, was extended by Newton to all cases of motion under a central force: from this the general theorem of conservation of angular momentum has gradually developed.

shew that these equations possess the integrals

$$\sum_r \left(m_r \dot{x}_r + \frac{\partial V}{\partial \dot{x}_r} \right) = \text{Constant},$$

$$\sum_r \left(m_r \dot{y}_r + \frac{\partial V}{\partial \dot{y}_r} \right) = \text{Constant},$$

$$\sum_r \left(m_r \dot{z}_r + \frac{\partial V}{\partial \dot{z}_r} \right) = \text{Constant},$$

$$\sum_r \left\{ m_r (y_r \dot{z}_r - z_r \dot{y}_r) + y_r \frac{\partial V}{\partial \dot{z}_r} - z_r \frac{\partial V}{\partial \dot{y}_r} \right\} = \text{Constant},$$

$$\sum_r \left\{ m_r (z_r \dot{x}_r - x_r \dot{z}_r) + z_r \frac{\partial V}{\partial \dot{x}_r} - x_r \frac{\partial V}{\partial \dot{z}_r} \right\} = \text{Constant},$$

$$\sum_r \left\{ m_r (x_r \dot{y}_r - y_r \dot{x}_r) + x_r \frac{\partial V}{\partial \dot{y}_r} - y_r \frac{\partial V}{\partial \dot{x}_r} \right\} = \text{Constant},$$

which may be regarded as generalisations of the integrals of momentum and angular momentum. (Lévy.)

40. *The general theorem of angular momentum.*

The integral of angular momentum is a special case of a more general result, which may be obtained in the following way.

Consider a dynamical system formed of any number of free or connected and interacting particles: if they are subjected to any constraints other than the mutual reactions of the particles, we shall suppose the forces due to these constraints to be counted among the external forces.

Take any line fixed in space, and choose one of the coordinates which specify the configuration of the system (say q_1) to be such that a change in q_1, unaccompanied by any change in the other coordinates, implies a simple rotation of the system as if rigid round the given line, through an angle equal to the change in q_1. We suppose the constraints to be such that this is a possible displacement of the system.

The Lagrangian equation for the coordinate q_1 is

$$\frac{d}{dt}\left(\frac{\partial T}{\partial \dot{q}_1}\right) - \frac{\partial T}{\partial q_1} = Q_1,$$

and this reduces to

$$\frac{d}{dt}\left(\frac{\partial T}{\partial \dot{q}_1}\right) = Q_1,$$

since the value of q_1 (as distinguished from $\dot{q}_1$) cannot have any effect on the kinetic energy, and therefore $\frac{\partial T}{\partial q_1}$ must be zero. Now $\frac{\partial T}{\partial \dot{q}_1}$ is the angular momentum of the system about the given line; and $Q_1 \delta q_1$ is the work done on the system by the external forces in a small displacement δq_1, i.e. a small rotation of the system about the given line through an angle δq_1, from which it is easily seen that Q_1 is the moment of the external forces about the given line. We have therefore the result that *the rate of change of the angular momentum of a dynamical system about any fixed line is equal to the moment of the external forces about this line.* The law of conservation of angular momentum obviously follows from this when the moment of the external forces is zero.

Similarly we can shew that *the rate of change of the momentum of a dynamical system parallel to any fixed direction is equal to the component, parallel to this line, of the total external forces acting on the system.*

For impulsive motion it is easy to establish the following analogous results:

The impulsive increment of the component of momentum of a system in any fixed direction is equal to the component in this direction of the total external impulses applied to the system.

The impulsive increment of the angular momentum of a system round any axis is equal to the moment round that axis of the external impulses applied to the system.

41. *The Energy equation.*

We shall now introduce an integral which plays a great part in dynamical investigations, and indeed in all physical questions.

In a conservative dynamical system let $q_1, q_2, \ldots, q_n$ be the coordinates and let L be the kinetic potential: we shall suppose that the constraints are independent of the time, so that L is a given function of the variables $q_1, q_2, \ldots, q_n, \dot{q}_1, \dot{q}_2, \ldots, \dot{q}_n$ only, not involving t explicitly. We shall not, at first, restrict L by any further conditions, so that the discussion will apply to the non-natural systems obtained after ignoration of coordinates, as well as to natural systems.

We have

$$\begin{aligned}\frac{dL}{dt} &= \sum_{r=1}^{n} \ddot{q}_r \frac{\partial L}{\partial \dot{q}_r} + \sum_{r=1}^{n} \dot{q}_r \frac{\partial L}{\partial q_r} \\ &= \sum_{r=1}^{n} \ddot{q}_r \frac{\partial L}{\partial \dot{q}_r} + \sum_{r=1}^{n} \dot{q}_r \frac{d}{dt}\left(\frac{\partial L}{\partial \dot{q}_r}\right), \text{ by the Lagrangian equations} \\ &= \frac{d}{dt}\left(\sum_{r=1}^{n} \dot{q}_r \frac{\partial L}{\partial \dot{q}_r}\right).\end{aligned}$$

Integrating, we have

$$\sum_{r=1}^{n} \dot{q}_r \frac{\partial L}{\partial \dot{q}_r} - L = h,$$

where h is a constant.

This equation is an integral of the system, and is called the *integral of energy* or *law of conservation of energy**.

* Galileo was acquainted with the fact that the velocity of a particle sliding down an inclined plane from rest depends only on the vertical height through which it has descended. From this elementary particular case the principle was gradually evolved by Huygens, Newton, John and Daniel Bernoulli, and Lagrange.

We have seen that in natural systems, in which the constraints do not involve the time, the kinetic potential L can be written in the form $T-V$, where T (the kinetic energy of the system) is homogeneous and of degree 2 in the velocities, while V is a function of the coordinates only. In this case, therefore, the integral of energy becomes

$$\begin{aligned} h &= \sum_{r=1}^{n} \dot{q}_r \frac{\partial L}{\partial \dot{q}_r} - L \\ &= \sum_{r=1}^{n} \dot{q}_r \frac{\partial T}{\partial \dot{q}_r} - T + V \\ &= 2T - T + V, \text{ since } T \text{ is homogeneous of degree 2 in } \dot{q}_1, \dot{q}_2, \ldots, \dot{q}_n, \\ &= T + V. \end{aligned}$$

It follows that *in conservative natural systems, the sum of the kinetic and potential energies is constant.* This constant value h is called the *total energy of the system.*

This latter result can also be obtained directly from the elementary equations of motion. For from the equations of motion of a single particle, namely

$$m_i \ddot{x}_i = X_i, \quad m_i \ddot{y}_i = Y_i, \quad m_i \ddot{z}_i = Z_i,$$

we have

$$\sum_i m_i (\dot{x}_i \ddot{x}_i + \dot{y}_i \ddot{y}_i + \dot{z}_i \ddot{z}_i) = \sum_i (X_i \dot{x}_i + Y_i \dot{y}_i + Z_i \dot{z}_i),$$

where the summation is extended over all the particles of the system, or

$$d \, . \sum_i \tfrac{1}{2} m_i (\dot{x}_i^2 + \dot{y}_i^2 + \dot{z}_i^2) = \sum_i (X dx + Y dy + Z dz),$$

so that the increment of the kinetic energy of the system, in any infinitesimal part of its path, is equal to the work done by the forces acting on the system in this part of the path, and therefore is equal to the decrease in the potential energy of the system. The sum of the kinetic and potential energies of the system is therefore constant

The equation of energy

$$d \, . \tfrac{1}{2} m (\dot{x}^2 + \dot{y}^2 + \dot{z}^2) = X dx + Y dy + Z dz$$

(where for simplicity we suppose the system to consist of a single particle) is true not only when (x, y, z) denote coordinates referred to any fixed axes, but also when they denote coordinates referred to axes which are moving with any motion of translation in a fixed direction with constant velocity.

For let (ξ, η, ζ) denote the coordinates of the particle referred to axes fixed in space and parallel to the moving axes $Oxyz$, so that

$$x = \xi - at, \quad y = \eta - bt, \quad z = \zeta - ct,$$

where a, b, c are the constant components of velocity of the origin O of the moving axes. Then the result already proved is that

$$d \, . \tfrac{1}{2} m (\dot{\xi}^2 + \dot{\eta}^2 + \dot{\zeta}^2) = X d\xi + Y d\eta + Z d\zeta,$$

or $$d \, . \tfrac{1}{2} m \{(\dot{x}+a)^2 + (\dot{y}+b)^2 + (\dot{z}+c)^2\} = X(dx + a\,dt) + Y(dy + b\,dt) + Z(dz + c\,dt),$$

or $$d \, . \tfrac{1}{2} m (\dot{x}^2 + \dot{y}^2 + \dot{z}^2) + d \, . \, m(a\dot{x} + b\dot{y} + c\dot{z}) = X dx + Y dy + Z dz + (aX + bY + cZ)\, dt.$$

Now we have

$$d \,.\, m(a\dot{x}+b\dot{y}+c\dot{z})=m(a\ddot{x}+b\ddot{y}+c\ddot{z})\,dt$$
$$=m(a\ddot{\xi}+b\ddot{\eta}+c\ddot{\zeta})\,dt$$
$$=(aX+bY+cZ)\,dt,$$

and therefore

$$d \,.\, \tfrac{1}{2}m(\dot{x}^2+\dot{y}^2+\dot{z}^2)=Xdx+Ydy+Zdz,$$

which establishes the theorem.

It may be noted that from this result the three equations of motion of the particle can be derived, by taking $x=\xi-at$ etc., and subtracting the equation of energy in the coordinates (x, y, z) from the equation of energy in the coordinates (ξ, η, ζ).

42. *Reduction of a dynamical problem to a problem with fewer degrees of freedom, by means of the energy-equation.*

When a conservative dynamical system has only one degree of freedom, the integral of energy is alone sufficient to give the solution by quadratures. For if q be the coordinate, the integral of energy

$$\dot{q}\frac{\partial L}{\partial \dot{q}} - L = h$$

is a relation between q and $\dot{q}$; if therefore $\dot{q}$ be found explicitly in terms of q from this equation, so that it takes the form

$$\dot{q}=f(q),$$

we can integrate again and obtain the equation

$$t=\int\frac{dq}{f(q)}+\text{constant},$$

which constitutes the solution of the problem.

When the system has more than one degree of freedom, the integral of energy is not in itself sufficient for the solution; but we shall now shew that it can be used for the same purpose as the integrals corresponding to ignorable coordinates were used, namely to reduce the system to another dynamical system with a smaller number of degrees of freedom*.

In the function L, replace the quantities $\dot{q}_2, \dot{q}_3, \ldots, \dot{q}_n$ by $\dot{q}_1 q_2', \dot{q}_1 q_3', \ldots, \dot{q}_1 q_n'$, respectively, where q_r' denotes $\dfrac{dq_r}{dq_1}$: and denote the resulting function by $\Omega(\dot{q}_1, q_2', q_3', \ldots, q_n', q_1, q_2, \ldots, q_n)$. Then differentiating the equation

$$L(\dot{q}_1, \dot{q}_2, \ldots, \dot{q}_n, q_1, q_2, \ldots, q_n)=\Omega(\dot{q}_1, q_2', q_3', \ldots, q_n', q_1, q_2, \ldots, q_n),$$

we have

$$\frac{\partial L}{\partial \dot{q}_1}=\frac{\partial \Omega}{\partial \dot{q}_1}-\sum_{r=2}^{n}\frac{\dot{q}_r}{\dot{q}_1^2}\frac{\partial \Omega}{\partial q_r'} \quad \ldots\ldots\ldots\ldots(1),$$

$$\frac{\partial L}{\partial \dot{q}_r}=\frac{1}{\dot{q}_1}\frac{\partial \Omega}{\partial q_r'} \qquad (r=2, 3, \ldots, n) \quad \ldots\ldots\ldots(2),$$

$$\frac{\partial L}{\partial q_r}=\frac{\partial \Omega}{\partial q_r} \qquad (r=1, 2, 3, \ldots, n) \quad \ldots\ldots\ldots(3).$$

* Whittaker, *Mess. of Math.* xxx. (1900).

Equations (1) and (2) give

$$\frac{\partial \Omega}{\partial \dot{q}_1} = \frac{\partial L}{\partial \dot{q}_1} + \sum_{r=2}^{n} \frac{\dot{q}_r}{\dot{q}_1} \frac{\partial L}{\partial \dot{q}_r} \quad \text{..........................(4).}$$

Now in the integral of energy

$$\sum_{r=1}^{n} \dot{q}_r \frac{\partial L}{\partial \dot{q}_r} - L = h,$$

replace $\dot{q}_r$ by $\dot{q}_1 q_r'$ for all values of r from 2 to n inclusive, and then from this equation obtain $\dot{q}_1$ as a function of the quantities $(q_2', q_3', \ldots, q_n', q_1, q_2, \ldots, q_n)$; and by using this expression for $\dot{q}_1$, express the function

$$\sum_{r=1}^{n} \frac{\partial L}{\partial \dot{q}_r} \frac{\dot{q}_r}{\dot{q}_1}$$

in terms of $(q_2', q_3', \ldots, q_n', q_1, q_2, \ldots, q_n)$. Let the function thus obtained be denoted by L'; then from (4) we see that L' is the same as $\frac{\partial \Omega}{\partial \dot{q}_1}$, but differently expressed.

Differentiating the equation of energy, which by (4) may be written in the form

$$\dot{q}_1 \frac{\partial \Omega}{\partial \dot{q}_1} - \Omega = h,$$

and regarding it as a relation which implicitly determines $\dot{q}_1$ as a function of the variables $(q_2', q_3', \ldots, q_n', q_1, q_2, \ldots, q_n)$, we have

$$\dot{q}_1 \frac{\partial^2 \Omega}{\partial \dot{q}_1^2} \frac{\partial \dot{q}_1}{\partial q_r'} = \frac{\partial \Omega}{\partial q_r'} - \dot{q}_1 \frac{\partial^2 \Omega}{\partial \dot{q}_1 \partial q_r'} \quad \text{..........................(5),}$$

$$\dot{q}_1 \frac{\partial^2 \Omega}{\partial \dot{q}_1^2} \frac{\partial \dot{q}_1}{\partial q_r} = \frac{\partial \Omega}{\partial q_r} - \dot{q}_1 \frac{\partial^2 \Omega}{\partial \dot{q}_1 \partial q_r} \quad \text{..........................(6).}$$

But differentiating the equation

$$L' = \frac{\partial \Omega}{\partial \dot{q}_1},$$

regarded as an identity in the variables $(q_2', q_3', \ldots, q_n', q_1, q_2, \ldots, q_n)$, we have

$$\frac{\partial L'}{\partial q_r'} = \frac{\partial^2 \Omega}{\partial \dot{q}_1 \partial q_r'} + \frac{\partial^2 \Omega}{\partial \dot{q}_1^2} \frac{\partial \dot{q}_1}{\partial q_r'} \quad \text{..........................(7),}$$

$$\frac{\partial L'}{\partial q_r} = \frac{\partial^2 \Omega}{\partial \dot{q}_1 \partial q_r} + \frac{\partial^2 \Omega}{\partial \dot{q}_1^2} \frac{\partial \dot{q}_1}{\partial q_r} \quad \text{..........................(8).}$$

Comparing equations (5) and (7), we have

$$\frac{\partial L'}{\partial q_r'} = \frac{1}{\dot{q}_1} \frac{\partial \Omega}{\partial q_r'}, \qquad (r = 2, 3, \ldots, n),$$

and comparing equations (6) and (8), we have

$$\frac{\partial L'}{\partial q_r} = \frac{1}{\dot{q}_1} \frac{\partial \Omega}{\partial q_r}, \qquad (r = 1, 2, \ldots, n).$$

Combining these with equations (2) and (3), we have

$$\frac{\partial L'}{\partial q_r'} = \frac{\partial L}{\partial \dot{q}_r}, \text{ and } \frac{\partial L'}{\partial q_r} = \frac{1}{\dot{q}_1}\frac{\partial L}{\partial q_r}.$$

Substituting from these equations in the Lagrangian equations of motion, we obtain the system

$$\frac{d}{dt}\left(\frac{\partial L'}{\partial q_r'}\right) - \dot{q}_1 \frac{\partial L'}{\partial q_r} = 0, \qquad (r = 2, 3, \ldots, n),$$

or finally

$$\frac{d}{dq_1}\left(\frac{\partial L'}{\partial q_r'}\right) - \frac{\partial L'}{\partial q_r} = 0, \qquad (r = 2, 3, \ldots, n).$$

Now *these may be regarded as the equations of motion of a new dynamical system in which L' is the kinetic potential, $(q_2, q_3, \ldots, q_n)$ are the coordinates, and q_1 plays the part of the time as the independent variable.* The new system will, like the systems obtained by ignoration of coordinates, be in general non-natural, i.e. L' will not consist solely of terms of degrees 2 and 0 in the velocities $(q_2', q_3', \ldots, q_n')$; but on account of its possession of the Lagrangian form, most of the theorems relating to dynamical systems will be applicable to it. *The integral of energy thus enables us to reduce a given dynamical system with n degrees of freedom to another dynamical system with only $(n-1)$ degrees of freedom.*

The new dynamical system will not in general possess an integral of energy, since the independent variable q_1 occurs explicitly in the new kinetic potential L'. But if q_1 is an ignorable coordinate in the original system, then q_1 will not occur explicitly in any stage of the above process, and therefore will not occur explicitly in L'. From this it follows that the new system will also possess an integral of energy, namely

$$\sum_{r=2}^{n} q_r' \frac{\partial L'}{\partial q_r'} - L' = \text{constant},$$

and this can in its turn be used to reduce further the number of degrees of freedom of the system.

The preceding theorems shew that any conservative dynamical system with n degrees of freedom and $(n-1)$ ignorable coordinates can be completely integrated by quadratures; we can proceed either (α) by first performing the process of ignoration of the coordinates, so arriving at a system with only one degree of freedom, which possesses an integral of energy and can therefore be solved in the manner indicated at the beginning of the present article; or (β) we can first use the integral of energy to lower the number of degrees of freedom by unity, then use the integral of energy of the new system to lower the number of degrees of freedom again by unity, and so on, obtaining finally a system with one degree of freedom which again can be solved in the manner indicated.

Example. The kinetic potential of a dynamical system is

$$L = \tfrac{1}{2} f(q_2)\, \dot{q}_1^2 + \tfrac{1}{2} \dot{q}_2^2 - \psi(q_2).$$

Shew that the relation between the variables q_1 and q_2 is given by the differential equation

$$\frac{d}{dq_1}\left(\frac{\partial L'}{\partial q_2'}\right) - \frac{\partial L'}{\partial q_2} = 0,$$

where $q_2' = \dfrac{dq_2}{dq_1}$, and where L' is defined by the equation

$$L' = \{2h - 2\psi(q_2)\}^{\frac{1}{2}} \{f(q_2) + q_2'^2\}^{\frac{1}{2}}.$$

Shew that the non-natural dynamical system represented by the last differential equation possesses an integral of energy, and hence solve the system by quadratures.

43. *Separation of the variables; dynamical systems of Liouville's type.*

A class of dynamical equations which are obviously soluble by quadratures is constituted by the equations of those systems for which the kinetic energy is of the form

$$T = \tfrac{1}{2} v_1(q_1)\, \dot{q}_1^2 + \tfrac{1}{2} v_2(q_2)\, \dot{q}_2^2 + \ldots + \tfrac{1}{2} v_n(q_n)\, \dot{q}_n^2,$$

and the potential energy is of the form

$$V = w_1(q_1) + w_2(q_2) + \ldots + w_n(q_n),$$

where $v_1, v_2, \ldots, v_n, w_1, w_2, \ldots, w_n$ are arbitrary functions of their respective arguments; so that the kinetic potential breaks up into a sum of parts, each of which involves only one of the variables.

For in this case the Lagrangian equations of motion are

$$\frac{d}{dt}\{v_r(q_r) \,.\, \dot{q}_r\} - \tfrac{1}{2} v_r'(q_r)\, \dot{q}_r^2 = -w_r'(q_r), \qquad (r = 1, 2, \ldots, n),$$

or

$$v_r(q_r)\, \ddot{q}_r + \tfrac{1}{2} v_r'(q_r)\, \dot{q}_r^2 = -w_r'(q_r), \qquad (r = 1, 2, \ldots, n).$$

These equations can be immediately integrated, and give

$$\tfrac{1}{2} v_r(q_r) \,.\, \dot{q}_r^2 + w_r(q_r) = c_r, \qquad (r = 1, 2, \ldots, n),$$

where $c_1, c_2, \ldots, c_n$ are constants of integration; these equations can be further integrated, since the variables q_r and t are separable, and we thus obtain

$$t = \int \left\{\frac{v_r(q_r)}{2c_r - 2w_r(q_r)}\right\}^{\frac{1}{2}} dq_r + \gamma_r, \qquad (r = 1, 2, \ldots, n),$$

where $\gamma_1, \gamma_2, \ldots, \gamma_n$ are new constants of integration. These last equations constitute the solution of the problem.

An important extension of this class of dynamical systems was made by Liouville*, who shewed that all dynamical problems for which the kinetic and potential energies can respectively be put in the forms

$$T = \tfrac{1}{2}\{u_1(q_1) + u_2(q_2) + \ldots + u_n(q_n)\}\{v_1(q_1)\, \dot{q}_1^2 + v_2(q_2)\, \dot{q}_2^2 + \ldots + v_n(q_n)\, \dot{q}_n^2\},$$

$$V = \frac{w_1(q_1) + w_2(q_2) + \ldots + w_n(q_n)}{u_1(q_1) + u_2(q_2) + \ldots + u_n(q_n)}$$

can be solved by quadratures.

* *Journal de Math.* XIV. (1849), p. 257.

For by taking

$$\int \sqrt{v_r(q_r)}\, dq_r = q_r', \qquad (r = 1, 2, \ldots, n),$$

where $q_1', q_2', \ldots, q_n'$ are new variables, we can replace all the functions $v_1(q_1), v_2(q_2), \ldots, v_n(q_n)$ by unity; we shall suppose this done, so that the kinetic and potential energies take the form

$$T = \tfrac{1}{2} u (\dot{q}_1^2 + \dot{q}_2^2 + \ldots + \dot{q}_n^2),$$

$$V = \frac{1}{u} \{w_1(q_1) + w_2(q_2) + \ldots + w_n(q_n)\},$$

where u stands for the expression

$$u_1(q_1) + u_2(q_2) + \ldots + u_n(q_n).$$

The Lagrangian equation for the coordinate q_1 is

$$\frac{d}{dt}\left(\frac{\partial T}{\partial \dot{q}_1}\right) - \frac{\partial T}{\partial q_1} = -\frac{\partial V}{\partial q_1},$$

or

$$\frac{d}{dt}(u\dot{q}_1) - \tfrac{1}{2}\frac{\partial u}{\partial q_1}(\dot{q}_1^2 + \dot{q}_2^2 + \ldots + \dot{q}_n^2) = -\frac{\partial V}{\partial q_1}.$$

Multiplying this equation throughout by $2u\dot{q}_1$, we have

$$\frac{d}{dt}(u^2\dot{q}_1^2) - u\dot{q}_1\frac{\partial u}{\partial q_1}(\dot{q}_1^2 + \dot{q}_2^2 + \ldots + \dot{q}_n^2) = -2u\dot{q}_1\frac{\partial V}{\partial q_1}.$$

But from the integral of energy of the system, we have

$$\tfrac{1}{2} u (\dot{q}_1^2 + \dot{q}_2^2 + \ldots + \dot{q}_n^2) = h - V,$$

where h is a constant. The equation for the coordinate q_1 can therefore be written in the form

$$\frac{d}{dt}(u^2\dot{q}_1^2) = 2(h - V)\dot{q}_1\frac{\partial u}{\partial q_1} - 2u\dot{q}_1\frac{\partial V}{\partial q_1}$$

$$= 2\dot{q}_1 \frac{\partial}{\partial q_1}\{(h - V)u\}$$

$$= 2\dot{q}_1 \frac{\partial}{\partial q_1}\{hu_1(q_1) - w_1(q_1)\}$$

$$= 2\frac{d}{dt}\{hu_1(q_1) - w_1(q_1)\}.$$

Integrating, we have

$$\tfrac{1}{2}u^2\dot{q}_1^2 = hu_1(q_1) - w_1(q_1) + \gamma_1,$$

where γ_1 is a constant of integration. We obtain similar equations for each of the coordinates $(q_1, q_2, \ldots, q_n)$; the corresponding constants $(\gamma_1, \gamma_2, \ldots, \gamma_n)$ must satisfy the relation

$$\gamma_1 + \gamma_2 + \ldots + \gamma_n = 0,$$

in virtue of the integral of energy of the system.

These equations give

$$\{hu_1(q_1) - w_1(q_1) + \gamma_1\}^{-\frac{1}{2}} dq_1 = \{hu_2(q_2) - w_2(q_2) + \gamma_2\}^{-\frac{1}{2}} dq_2 = \dots$$
$$= \{hu_n(q_n) - w_n(q_n) + \gamma_n\}^{-\frac{1}{2}} dq_n,$$

and this set of equations, which can be immediately integrated since the variables are separated, furnishes the solution of the system.

For further investigations on this subject cf. Stäckel, *Math. Ann.* XLII. (1893), p. 545, Hadamard, *Bull. des Sc. Math.* XXXV. (1911), p. 106, Burgatti, *Rom. Acc. L. Rend.* (5) XX. (1911), p. 108, and Vrkljan, *Zeitschr. f. Phys.* LIX. (1930), p. 718, LXV. (1930), p. 280.

MISCELLANEOUS EXAMPLES.

1. If the components (X, Y) of the force acting on a particle of unit mass at the point (x, y) in a plane do not involve the time t, shew that by elimination of t from the differential equations the solution of the problem is made to depend on the differential equation of the third order

$$\frac{d}{dx}\left\{\frac{Y - X\dfrac{dy}{dx}}{\dfrac{d^2y}{dx^2}}\right\} - 2X = 0.$$

2. A system of free particles is in motion, and their potential energy, which depends only on their coordinates, is unaltered when the system in any configuration is translated as if rigid through any distance in any direction. What integrals of the motion can at once be written down?

3. In a dynamical system with two degrees of freedom the kinetic energy is

$$T = \frac{\dot{q}_1^2}{2(a + bq_2)} + \tfrac{1}{2} q_2^2 \dot{q}_2^2,$$

and the potential energy is

$$V = c + dq_2,$$

where a, b, c, d are constants. Shew that the value of q_2 in terms of the time is given by an equation of the form

$$(q_2 - k)(q_2 + 2k)^2 = h(t - t_0)^2$$

where h, k, and t_0 are constants.

4. The kinetic potential of a dynamical system is

$$L = \frac{\dot{q}_1^2}{aq_2 + b} + \tfrac{1}{2}\dot{q}_2^2 + 2q_2^3 + cq_2,$$

where a, b, c are given constants: shew that q_2 is given in terms of t by the equation

$$q_2 = \wp(t + \epsilon),$$

where ϵ is an arbitrary constant and $\wp$ denotes a Weierstrassian elliptic function.

5. Prove that in a system with ignorable coordinates the kinetic energy is the sum of a quadratic function T' of the velocities of the non-ignored coordinates and a quadratic function K of the cyclic momenta.

In the case where there are three coordinates x, y, ϕ and one coordinate ϕ is ignored investigate the equations of motion of the type

$$\frac{d}{dt}\left(\frac{\partial T'}{\partial \dot{x}}\right) - \frac{\partial T'}{\partial x} + \frac{\partial K}{\partial x} + \frac{\partial V}{\partial x} + k\dot{y}\left\{\frac{\partial}{\partial x}\left(\frac{\partial \dot{\phi}}{\partial \dot{y}}\right) - \frac{\partial}{\partial y}\left(\frac{\partial \dot{\phi}}{\partial \dot{x}}\right)\right\} = 0,$$

where V is the potential energy, k is the cyclic momentum, and the differential coefficients of $\dot{\phi}$ with respect to $\dot{x}$ and $\dot{y}$ are calculated from the linear equation by which k is expressed in terms of $\dot{x}$, $\dot{y}$, $\dot{\phi}$. (Camb. Math. Tripos, 1904.)

6. The kinetic potential of a dynamical system with two degrees of freedom is

$$L = \frac{\dot{q}_2^2}{4q_2} + q_2\dot{q}_1^2 + l^2\dot{q}_1^2.$$

By using the integral of energy, shew that the solution depends on the solution of the problem for which the kinetic potential is

$$L' = \left(\frac{q_2'^2}{4q_2} + q_2 + l^2\right)^{\frac{1}{2}},$$

and by using the integral of energy of this latter system, shew that the relation between q_1 and q_2 is of the form

$$cq_2 = \wp(q_1 + \epsilon) - \tfrac{1}{3}(2cl^2 - 1),$$

where c and ϵ are constants of integration, and $\wp$ denotes the Weierstrassian elliptic function.

7. The kinetic energy of a dynamical system is

$$T = \tfrac{1}{2}(q_1^2 + q_2^2)(\dot{q}_1^2 + \dot{q}_2^2),$$

and the potential energy is

$$V = \frac{1}{q_1^2 + q_2^2}.$$

Shew (by use of Liouville's theorem, or otherwise) that the relation between q_1 and q_2 is

$$a^2q_1^2 + b^2q_2^2 + 2abq_1q_2\cos\gamma = \sin^2\gamma,$$

where a, b, γ are constants of integration.

8. The kinetic energy of a particle whose rectangular coordinates are (x, y) is $\frac{1}{2}(\dot{x}^2 + \dot{y}^2)$, and its potential energy is

$$\frac{A}{x^2} + \frac{A'}{y^2} + \frac{B}{r} + \frac{B'}{r'} + C(x^2 + y^2),$$

where (A, A', B, B', C) are constants and where (r, r') are the distances of the particle from the points whose coordinates are $(c, 0)$ and $(-c, 0)$, where c is a constant. Shew that when the quantities $\frac{1}{2}(r + r')$ and $\frac{1}{2}(r - r')$ are taken as new variables, the system is of Liouville's type, and hence obtain its solution.

9. The observation that "a cat always falls on its feet" suggested the problem: A system, whose state at any instant is completely specified by the position and velocity of each element, is initially without velocity in free space *in vacuo*. Can it at a subsequent instant resume its initial configuration *but with a different orientation in space*?

Shew that if the system is not conservative, or if the forces are derived from a potential which is not one-valued, the reply is affirmative: but if the system is conservative with a one-valued potential, the reply is negative.

(Cf. Painlevé, *Comptes Rendus*, CXXXIX. (1904), p. 1170.)

CHAPTER IV

THE SOLUBLE PROBLEMS OF PARTICLE DYNAMICS

44. *The particle with one degree of freedom: the pendulum.*

As examples of the methods described in the foregoing chapters, we shall now discuss those cases of the motion of a single particle which can be solved by quadratures.

We shall consider first the motion of a particle of mass m, which is free to move in the interior of a given fixed smooth tube of small bore, under the action of forces which depend only on the position of the particle in the tube. The tube can in the most general case be supposed to have the form of a twisted curve in space.

Let s be the distance of the particle at time t from some fixed point of the tube, measured along the arc of the curve formed by the tube: and let $f(s)$ be the component of the external forces acting on the particle, in the direction of the tangent to the tube.

The kinetic energy of the particle is

$$\tfrac{1}{2} m\dot{s}^2,$$

and its potential energy is evidently

$$-\int_{s_0}^{s} f(s)\, ds,$$

where s_0 is a constant. The equation of energy is therefore

$$\tfrac{1}{2} m\dot{s}^2 = \int_{s_0}^{s} f(s)\, ds + c,$$

where c is a constant.

Integrating this equation, we have

$$t = \left(\frac{m}{2}\right)^{\frac{1}{2}} \int_{s_0}^{s} \left\{ \int_{s_0}^{s} f(s)\, ds + c \right\}^{-\frac{1}{2}} ds + l,$$

where l is another constant of integration. This equation represents the solution of the problem, since it is an integral relation between s and t, involving two constants of integration.

The two constants c and l can be physically interpreted in terms of the initial circumstances of the particle's motion; thus if the particle starts at time $t=t_0$ from the point $s=s_0$, with velocity u, then on substituting these values in the equation of energy, we have

$$c=\tfrac{1}{2}mu^2,$$

and on substituting the same values in the final equation connecting s and t, we have $l=t_0$.

The most famous problem of this type is that of the *simple pendulum*; in this case the tube is supposed to be in the form of a circle of radius a whose plane is vertical, and the only external force acting on the particle is gravity*. Using θ to denote the angle made with the downward vertical by the radius vector from the centre of the circle to the particle, we have

$$s=a\theta \quad \text{and} \quad f(s)=-mg\sin\theta;$$

so the equation of energy is

$$a\dot{\theta}^2=2g\cos\theta+\text{constant}=-4g\sin^2\tfrac{1}{2}\theta+\text{constant}.$$

Suppose that when the particle is at the lowest point of the circle, the quantity $\dfrac{a^2\dot{\theta}^2}{2g}$ has the value h. Then this last equation can be written

$$a^2\dot{\theta}^2=2gh-4ga\sin^2\tfrac{1}{2}\theta.$$

Taking $\sin\tfrac{1}{2}\theta=y$, this becomes

$$\dot{y}^2=\frac{g}{a}(1-y^2)\left(\frac{h}{2a}-y^2\right).$$

Now in the pendulum-problem there are two distinct types of motion, namely the "oscillatory," in which the particle swings to and fro about the lowest point of the circle, and the "circulating," in which the velocity of the particle is large enough to carry it over the highest point of the circle, so that it moves round and round the circle, always in the same sense. We shall consider these cases separately.

(i) In the oscillatory type of motion, since the particle comes to rest before attaining the highest point of the circle, $\dot{y}$ must be zero for some value of y less than unity, and therefore $h/2a$ must be less than unity. Writing

$$h=2ak^2,$$

where k is a new positive constant less than unity, the equation becomes

$$\dot{y}^2=\frac{gk^2}{a}\left(1-k^2\cdot\frac{y^2}{k^2}\right)\left(1-\frac{y^2}{k^2}\right);$$

* In actual pendulums, the tube is replaced by a rigid bar connecting the particle to the centre of the circle, which serves the same purpose of constraining the particle to describe the circle.

The isochronism of small oscillations of the pendulum was discovered by Galileo in 1632, and the formula for the period was given by Huygens in 1673. Oscillations of finite amplitude were first studied by Euler in 1736.

the solution of this is*

$$y = k \operatorname{sn}\left\{\left(\frac{g}{a}\right)^{\frac{1}{2}}(t - t_0),\ k\right\},$$

where t_0 is an arbitrary constant.

This equation represents the solution of the pendulum-problem in the oscillatory case: the two arbitrary constants of the solution are t_0 and k, and these must be determined from the initial conditions. From the known properties of the elliptic function sn, we see that the motion is periodic, its *period* (i.e. the interval of time between two consecutive occasions on which the pendulum is in the same configuration with the same velocity) being $4\left(\frac{a}{g}\right)^{\frac{1}{2}} K$, where

$$K = \int_0^1 (1 - t^2)^{-\frac{1}{2}}(1 - k^2t^2)^{-\frac{1}{2}}\, dt.$$

(ii) Next, suppose that the motion is of the circulating type; in this case h is greater than $2a$, so if we write $2a = hk^2$, the quantity k will be less than unity.

The differential equation now becomes

$$\dot{y}^2 = \frac{g}{ak^2}(1 - y^2)(1 - k^2y^2),$$

the solution of which is

$$y = \operatorname{sn}\left\{\sqrt{\frac{g}{a}}\frac{t - t_0}{k},\ k\right\},$$

and in this t_0 and k are the two constants which must be determined in accordance with the initial conditions.

(iii) Lastly, let h be equal to $2a$, so that the particle just reaches the vertex of the circle. The equation now becomes

$$\dot{y}^2 = \frac{g}{a}(1 - y^2)^2,$$

or

$$\dot{y} = \sqrt{\frac{g}{a}}(1 - y^2),$$

the solution of which is

$$y = \tanh\left\{\sqrt{\frac{g}{a}}(t - t_0)\right\}.$$

It was remarked by Appell† that an insight into the meaning of the imaginary period of the elliptic functions which occur in the solution of the pendulum-problem is afforded by the theorem of § 34. For we have seen that if the particle is set free with no initial velocity at a point of the circle which is at a vertical height h above the lowest point, the motion is given by

$$y = k \operatorname{sn}\left\{\sqrt{\frac{g}{a}}(t - t_0),\ k\right\}, \qquad \text{where } k^2 = \frac{h}{2a};$$

* Cf. Whittaker and Watson, *Modern Analysis*, § 22·11.

† *Comptes Rendus*, LXXXVII. (1878).

and therefore by § 34, if, with the same initial conditions, gravity were supposed to act *upwards*, the motion would be given by

$$y = k \operatorname{sn} \left\{ i \sqrt{\frac{g}{a}} (\tau - \tau_0),\ k \right\}.$$

But the period of this motion is the same as if the initial position were at a height $(2a - h)$, gravity acting downwards: and the solution of this is

$$y = k' \operatorname{sn} \left\{ \sqrt{\frac{g}{a}} (\tau - \tau_0),\ k' \right\}, \qquad \text{where } k'^2 = 1 - k^2.$$

The latter motion has a real period $4\left(\frac{a}{g}\right)^{\frac{1}{2}} K'$; and therefore the function

$$\operatorname{sn} \left\{ i \sqrt{\frac{g}{a}} (\tau - \tau_0),\ k \right\}$$

must have a period $4\left(\frac{a}{g}\right)^{\frac{1}{2}} K'$, so the function $\operatorname{sn}(u, k)$ must have a period $4iK'$ The double periodicity of the elliptic function sn is thus inferred from dynamical considerations.

Example. A particle of unit mass moves on an epicycloid, traced by a point on the circumference of a circle of radius b which rolls on a fixed circle of radius a. The particle is acted on by a repulsive force μr directed from the centre of the fixed circle, where r is the distance from this centre. Shew that the motion is periodic, its period being

$$2\pi \left\{ \frac{(a+2b)^2 - a^2}{\mu a^2} \right\}^{\frac{1}{2}}.$$

[This result is most easily obtained when the equation of the epicycloid is taken in the form

$$(a+2b)^2 - r^2 = \frac{a^2 s^2}{(a+2b)^2 - a^2},$$

s being the arc measured from the vertex of the epicycloid.]

45. *Motion in a moving tube.*

We shall now discuss some cases of the motion of a particle which is free to move in a given smooth tube, when the tube is itself constrained to move in a given manner.

(i) *Tube rotating uniformly.*

Suppose first that the tube is constrained to rotate with uniform velocity ω about a fixed axis in space. We shall suppose that the particle is of unit mass, as this involves no real loss of generality.

We shall moreover suppose that the field of external force acting on the particle is derivable from a potential-energy function which is symmetrical with respect to the fixed axis, and so can be expressed in terms of the cylindrical coordinates z and r, where z is measured parallel to the fixed axis and r is the perpendicular distance from the fixed axis; for a particle in the tube, this potential energy can therefore be expressed in terms of the arc s: we shall denote it by $V(s)$, and the equation of the tube will be written in the form

$$r = g(s).$$

By § 29, the motion of the particle is the same as if the prescribed angular velocity ω were zero, and the potential energy were to contain an additional term $-\frac{1}{2}r^2\omega^2$. Hence we can at once write down the equation of energy in the form

$$\tfrac{1}{2}\dot{s}^2 - \tfrac{1}{2}\omega^2\{g(s)\}^2 + V(s) = c,$$

where c is a constant.

Integrating again, we have

$$t = \int^s [2c + \omega^2\{g(s)\}^2 - 2V(s)]^{-\frac{1}{2}} ds + \text{constant},$$

and this relation between t and s represents the solution of the problem

Example 1. If the rotating tube is plane, and the particle can describe it with constant velocity when the fixed axis is vertical and in the plane of the tube, and the field of force is that due to gravity, shew that the tube must be in the form of a parabola with its axis vertical and vertex downwards.

Example 2. A particle moves under gravity in a circular tube of radius a which rotates uniformly about a fixed vertical axis inclined at an angle α to its plane; if θ be the angular distance of the particle from the lowest point of the circle, shew that a solution exists given by

$$\sec\theta = \frac{a\omega^2\cos\alpha}{6g} + \wp\left\{\sqrt{\frac{g\cos\alpha}{2a}}(t - t_0)\right\},$$

where the function $\wp$ is formed with the roots

$$1 - \frac{a\omega^2\cos\alpha}{6g}, \quad -1 - \frac{a\omega^2\cos\alpha}{6g}, \quad \frac{a\omega^2\cos\alpha}{3g},$$

and t_0 is a constant.

(ii) *Tube moving with constant acceleration parallel to a fixed direction.*

Consider now the motion of a particle in a straight tube, inclined at an angle α to the horizontal, which is constrained to move in its own vertical plane with constant horizontal acceleration f.

Taking the axis of x horizontal and that of y vertically upwards, with the origin at the initial position of the particle, we have for the kinetic energy

$$T = \tfrac{1}{2}(\dot{x}^2 + \dot{y}^2),$$

where
$$x = y\cot\alpha + \tfrac{1}{2}ft^2,$$

so
$$T = \tfrac{1}{2}(\dot{y}\cot\alpha + ft)^2 + \tfrac{1}{2}\dot{y}^2$$
$$= \tfrac{1}{2}\dot{y}^2\operatorname{cosec}^2\alpha + \dot{y}\cot\alpha\,.\,ft + \tfrac{1}{2}f^2t^2,$$

and the potential energy is

$$V = gy.$$

The equation of motion

$$\frac{d}{dt}\left(\frac{\partial T}{\partial \dot{y}}\right) = -\frac{\partial V}{\partial y},$$

gives therefore

$$\frac{d}{dt}(\dot{y}\operatorname{cosec}^2\alpha + ft\cot\alpha) = -g,$$

or
$$\ddot{y} = (-g - f\cot\alpha)\sin^2\alpha.$$

Integrating, we have, supposing the particle to be initially at rest,

$$y = \tfrac{1}{2}t^2(-g\sin\alpha - f\cos\alpha)\sin\alpha,$$

and therefore
$$x = \tfrac{1}{2}t^2(-g\cos\alpha + f\sin\alpha)\sin\alpha.$$

These equations constitute the solution of the problem: it will be observed that in this system the kinetic energy involves the time explicitly, so no integral of energy exists.

46. *Motion of two interacting free particles.*

We shall next consider the motion of two particles, of masses m_1 and m_2 respectively, which are free to move in space under the influence of mutual forces of attraction or repulsion, acting in the line joining the particles and dependent on their distance from each other.

The system has six degrees of freedom, since the three rectangular coordinates of either particle can have any values whatever. We shall take, as the six coordinates defining the position of the system, the coordinates (X, Y, Z) of the centre of gravity of the particles, referred to any fixed axes, and the coordinates (x, y, z) of the particle m_2 referred to moving axes whose origin is at the particle m_1 and which are parallel to the fixed axes.

The coordinates of m_1, referred to the fixed axes, are

$$\left(X - \frac{m_2 x}{m_1 + m_2},\quad Y - \frac{m_2 y}{m_1 + m_2},\quad Z - \frac{m_2 z}{m_1 + m_2}\right),$$

and those of m_2, referred to the fixed axes, are

$$\left(X + \frac{m_1 x}{m_1 + m_2},\quad Y + \frac{m_1 y}{m_1 + m_2},\quad Z + \frac{m_1 z}{m_1 + m_2}\right).$$

The kinetic energy of the system is therefore

$$T = \tfrac{1}{2}m_1\left(\dot{X} - \frac{m_2\dot{x}}{m_1 + m_2}\right)^2 + \tfrac{1}{2}m_1\left(\dot{Y} - \frac{m_2\dot{y}}{m_1 + m_2}\right)^2 + \tfrac{1}{2}m_1\left(\dot{Z} - \frac{m_2\dot{z}}{m_1 + m_2}\right)^2$$

$$+ \tfrac{1}{2}m_2\left(\dot{X} + \frac{m_1\dot{x}}{m_1 + m_2}\right)^2 + \tfrac{1}{2}m_2\left(\dot{Y} + \frac{m_1\dot{y}}{m_1 + m_2}\right)^2 + \tfrac{1}{2}m_2\left(\dot{Z} + \frac{m_1\dot{z}}{m_1 + m_2}\right)^2,$$

or
$$T = \tfrac{1}{2}(m_1 + m_2)(\dot{X}^2 + \dot{Y}^2 + \dot{Z}^2) + \tfrac{1}{2}\frac{m_1 m_2}{m_1 + m_2}(\dot{x}^2 + \dot{y}^2 + \dot{z}^2).$$

The potential energy of the system depends only on the position of the particles relative to each other, so can be expressed in terms of (x, y, z): let it be $V(x, y, z)$.

The Lagrangian equations of motion of the system are

$$\ddot{X} = 0, \qquad \ddot{Y} = 0, \qquad \ddot{Z} = 0,$$

$$\frac{m_1 m_2}{m_1 + m_2} \ddot{x} = -\frac{\partial V}{\partial x}, \qquad \frac{m_1 m_2}{m_1 + m_2} \ddot{y} = -\frac{\partial V}{\partial y}, \qquad \frac{m_1 m_2}{m_1 + m_2} \ddot{z} = -\frac{\partial V}{\partial z}.$$

The first three of these equations shew that *the centre of gravity moves in a straight line with uniform velocity*, and the other three equations shew that *the motion of m_2 relative to m_1 is the same as if m_1 were fixed and m_2 were attracted to m_1 with the force derived from the potential energy* $\frac{m_1 + m_2}{m_1} V$ *.

Example. If two free particles move in space under any law of mutual attraction, shew that the tangents to their paths meet an arbitrary fixed plane in two points, the line joining which passes through a fixed point. (Mehmke.)

47. *Central forces in general: Hamilton's theorem.*

The last article shews that the problem of two interacting free particles is reducible to the problem of the motion of a single free particle acted on by a force directed towards or from a fixed centre. This is known as the *problem of central forces.* There is clearly no loss of generality if we suppose the mass of the particle to be unity.

If the particle be projected in any way, it will always remain in the plane which passes through the centre of force and the initial direction of projection: for at no time does any force act to remove it from this plane. We can therefore define the position of the particle by polar coordinates (r, θ) in this plane, the centre of force being the origin. Let P denote the acceleration directed to the centre of force. We shall not suppose for the present that P is necessarily a function of r alone.

The kinetic energy of the particle is

$$T = \tfrac{1}{2}(\dot{r}^2 + r^2\dot{\theta}^2),$$

and the work done by the force in an arbitrary infinitesimal displacement $(\delta r, \delta\theta)$ is

$$-P\delta r.$$

The Lagrangian equations of motion of the particle are therefore

$$\begin{cases} \ddot{r} - r\dot{\theta}^2 = -P, \\ \dfrac{d}{dt}(r^2\dot{\theta}) = 0. \end{cases}$$

The latter equation gives on integration

$$r^2\dot{\theta} = h, \qquad \text{where } h \text{ is a constant;}$$

this is the integral corresponding to the ignorable coordinate θ, and can be physically interpreted as the integral of angular momentum of the particle about the centre of force.

* Newton, *Principia*, Book I. Sect. 11.

To find the differential equation of the path described (which is generally called the *orbit* or *trajectory*), we eliminate dt from the first equation by using the relation

$$\frac{d}{dt} = \frac{h}{r^2}\frac{d}{d\theta};$$

we thus obtain the equation

$$\frac{h}{r^2}\frac{d}{d\theta}\left(\frac{h}{r^2}\frac{dr}{d\theta}\right) - \frac{h^2}{r^3} = -P,$$

or, writing u for $1/r$,

$$\frac{d^2u}{d\theta^2} + u = \frac{P}{h^2u^2}.$$

This is the differential equation of the orbit*, in polar coordinates; its integration will introduce two new arbitrary constants in addition to the constant h, and a fourth arbitrary constant will occur in the determination of t by the equation

$$t = \frac{1}{h}\int r^2 d\theta + \text{constant}.$$

The differential equation of the orbit in (r, p) coordinates, (where p denotes the perpendicular from the centre of force on the tangent to the orbit), is often of use: it may be obtained directly from Siacci's theorem (§ 18), which (since h is now constant) gives at once

$$P = \frac{h^2 r}{p^3 \rho},$$

or

$$P = \frac{h^2}{p^3}\frac{dp}{dr},$$

which is the differential equation of the orbit.

Since $h = vp$, where v is the velocity in the orbit, we have from this equation

$$v^2 = P\frac{\rho p}{r},$$

which may be written in the form

$$v^2 = \tfrac{1}{2}Pq,$$

where q is the chord of curvature of the orbit through the centre of force.

We frequently require to know *the law of force which must act towards a given point in order that a given curve may be described*; this is given at once by the equation

$$P = h^2u^2\left(u + \frac{d^2u}{d\theta^2}\right),$$

if the equation of the curve is given in polar coordinates; while if the equation is given in (r, p) coordinates, the force is given by the equation

$$P = \frac{h^2}{p^3}\frac{dp}{dr}.$$

* This is substantially given in Newton's *Principia*, Book I. §§ 2 and 3, and in Clairaut's *Théorie de la Lune* (1765); and in the above form in Whewell's *Dynamics* (1823).

If the equation of the curve is given in rectangular coordinates, we proceed as follows:

Take the centre of force as origin, and let $f(x, y) = 0$ be the equation of the given curve. The equation of angular momentum is

$$x\dot{y} - y\dot{x} = h.$$

Differentiating the equation of the curve, we have

$$f_x \cdot \dot{x} + f_y \cdot \dot{y} = 0, \qquad \text{where } f_x \text{ stands for } \frac{\partial f}{\partial x}.$$

From these two equations we obtain

$$\dot{x} = \frac{-hf_y}{xf_x + yf_y}, \qquad \dot{y} = \frac{hf_x}{xf_x + yf_y}.$$

Differentiating again, we have

$$\ddot{x} = \dot{x}\frac{\partial \dot{x}}{\partial x} + \dot{y}\frac{\partial \dot{x}}{\partial y} = \frac{hf_y}{xf_x + yf_y} \cdot \frac{\partial}{\partial x}\left(\frac{hf_y}{xf_x + yf_y}\right) - \frac{hf_x}{xf_x + yf_y} \cdot \frac{\partial}{\partial y}\left(\frac{hf_y}{xf_x + yf_y}\right).$$

Performing the differentiations, this gives

$$\ddot{x} = \frac{h^2 x\,(-f_y{}^2 f_{xx} + 2f_x f_y f_{xy} - f_x{}^2 f_{yy})}{(xf_x + yf_y)^3}.$$

But the required force is P, where $\ddot{x} = -P\dfrac{x}{r}$; and therefore we have

$$P = \frac{h^2 r\,(f_y{}^2 f_{xx} - 2f_x f_y f_{xy} + f_x{}^2 f_{yy})}{(xf_x + yf_y)^3};$$

this equation gives the required central force.

The most important case of this result is that in which the curve $f(x, y) = 0$ is a conic,

$$2f(x, y) \equiv ax^2 + 2hxy + by^2 + 2gx + 2fy + c = 0.$$

In this case we find at once that the expression

$$f_{xx} f_y{}^2 - 2f_{xy} f_x f_y + f_{yy} f_x{}^2$$

has, for points on the conic, the constant value

$$-(abc + 2fgh - af^2 - bg^2 - ch^2),$$

while the quantity

$$x\frac{\partial f}{\partial x} + y\frac{\partial f}{\partial y}$$

has the value

$$-(gx + fy + c),$$

and so is a constant multiple of the perpendicular from the point (x, y) on the polar of the origin with respect to the conic. We thus obtain, for the force under which a given conic can be described, an elegant expression due to Hamilton*, namely that *the force acting on the particle in the position* (x, y)

* *Proc. Roy. Irish Acad.* 1846.

varies directly as the radius from the centre of force to the point (x, y), and inversely as the cube of the perpendicular from (x, y) on the polar of the centre of force.

The two following theorems, the proof of which is left to the student, may together be regarded as the converse of Hamilton's theorem.

(i) If a particle moves under the action of a force directed to a fixed point, varying directly as the distance from the fixed point and inversely as the cube of the distance from a given straight line, the orbit is always a conic.

(ii) If a particle moves under the action of a force directed to the origin, of magnitude

$$\mu (x^2+y^2)^{\frac{1}{2}} (ax^2+2\beta xy+\gamma y^2)^{-\frac{3}{2}},$$

where (x, y) are rectangular coordinates and μ, a, β, γ are constants, the orbits are conics which touch the lines

$$ax^2+2\beta xy+\gamma y^2=0.$$

Darboux (*Comptes Rendus*, LXXXIV. p. 936) has shewn that these two laws of force are the only laws for which the orbits are always conics, if the force depends only on the position of the particle. Suchar (*Nouv. Ann.*[4] VI. p. 532) has found other laws of force, which involve the components of velocity of the particle.

Example 1. If a conic be described under the force $\frac{\mu r}{p^3}$ given by Hamilton's theorem, shew that the periodic time is $\frac{2\pi}{\sqrt{\mu}} p_0^{\frac{3}{2}}$, where p_0 is the perpendicular from the centre of the conic on the polar of the centre of force. (Glaisher.)

Example 2. Shew that if the force be

$$\frac{\mu r}{(Ax^2+2Hxy+By^2+I)^3},$$

a particle will describe a conic having its asymptotes parallel to the lines

$$Ax^2+2Hxy+By^2=0,$$

if properly projected. (Glaisher.)

48. *The integrable cases of central forces; problems soluble in terms of circular and elliptic functions.*

The most important case of motion under central forces is that in which the magnitude of the force depends only on the distance r. Denoting the force by $f(r)$, the differential equation of the orbit is

$$\frac{d^2u}{d\theta^2}+u=\frac{f(r)}{h^2u^2}.$$

Integrating, we have

$$\left(\frac{du}{d\theta}\right)^2 = c - \frac{2}{h^2}\int^r f(r)\,dr - u^2,$$

where c is a constant: integrating this equation again, we have

$$\theta = \int^r \left\{c - \frac{2}{h^2}\int^r f(r)\,dr - \frac{1}{r^2}\right\}^{-\frac{1}{2}} \frac{dr}{r^2},$$

and this is the equation of the orbit in polar coordinates. When r has been found from this equation in terms of θ, the time is given by the integral

$$t = \frac{1}{h}\int^{\theta} r^2 d\theta + \text{constant}.$$

The problem of motion under central forces is therefore always soluble by quadratures when the force is a function of the distance only.

Example. Shew that the differential equations of motion of a point P are always integrable by a simple quadrature when the central force F is of the form

$$F = \frac{\phi(\theta)}{r^2(at+b)},$$

where ϕ is a function of θ only, while a and b are arbitrary constants. (Armellini.)

We shall now discuss the cases in which the quadrature can be effected in terms of known functions, the central force being supposed to vary as some positive or negative integral power,—say the nth,—of the distance.

Let us first find those problems for which the integration can be effected in terms of circular functions. The above integral for the determination of θ can be written in the form

$$\theta = \int (a + bu^2 + cu^{-n-1})^{-\frac{1}{2}} du,$$

where a, b, c are constants; except when $n = -1$, when a logarithm replaces the term in u^{-n-1}. If the problem is to be soluble in terms of circular functions, the polynomial under the radical in the integrand must be at most of the second degree; this gives

$$-n - 1 = 0, 1, \text{ or } 2,$$

and consequently

$$n = -1, -2, \text{ or } -3.$$

The case $n = -1$ is however excluded by what has already been said, and the case $n = 1$ is to be added, since in this case the irrationality becomes quadratic when u^2 is taken as a new variable.

Next, let us find the cases in which the integration can be effected by the aid of elliptic functions*. For this it is necessary that the irrationality to be integrated should be of the third or fourth degree† in the variable with respect to which the integration is taken. But this condition is fulfilled if

$n = 0, -4$, or -5, when u is taken as the independent variable;

$n = 3$, 5, or -7, when u^2 is taken as the independent variable.

It follows that *the problem of motion under a central force which varies as the nth power of the distance is soluble by circular or elliptic functions in the cases*

$$n = 5, 3, 1, 0, -2, -3, -4, -5, -7.$$

* These cases were first investigated by Legendre, *Théorie des Fonctions Elliptiques* (1825) and afterwards by J. F. Stader, *Crelle's Journal*, XLVI. (1853), p. 262.

† Whittaker and Watson, *Modern Analysis*, § 22·7.

Example. Shew that the problem is soluble by elliptic functions when n has the following fractional values:

$$n = -\tfrac{3}{2},\quad -\tfrac{5}{2},\quad -\tfrac{1}{3},\quad -\tfrac{5}{3},\quad -\tfrac{7}{3}.$$

The general case of fractional values of n is discussed by Nobile, *Giornale di Mat.* XLVI. (1908), p. 313.

The cases in which motion under a central force varying as a power of the distance is soluble by means of circular functions are of special interest. They correspond, as shewn above, to the values 1, − 2, − 3 of n; the case $n = -2$ will be considered in the next article: the cases $n = 1$ and $n = -3$ can be treated in the following way.

(i) $n = 1$.

In this case the attractive force is

$$f(r) = \mu r,$$

so the equation of the orbit becomes

$$\theta = -\int^u \left(c - \frac{\mu}{h^2 u^2} - u^2\right)^{-\frac{1}{2}} du$$

$$= -\tfrac{1}{2}\int^v \left(cv - \frac{\mu}{h^2} - v^2\right)^{-\frac{1}{2}} dv, \qquad \text{where } u^2 = v,$$

so
$$2\theta = -\int^v \left\{\left(\frac{c^2}{4} - \frac{\mu}{h^2}\right) - \left(v - \frac{c}{2}\right)^2\right\}^{-\frac{1}{2}} dv,$$

or
$$2(\theta - \gamma) = \arccos \frac{v - \dfrac{c}{2}}{\left(\dfrac{c^2}{4} - \dfrac{\mu}{h^2}\right)^{\frac{1}{2}}}, \qquad \text{where } \gamma \text{ is a constant of integration,}$$

or
$$\frac{1}{r^2} = \frac{c}{2} + \left(\frac{c^2}{4} - \frac{\mu}{h^2}\right)^{\frac{1}{2}} \cos(2\theta - 2\gamma).$$

This is the equation of an ellipse (when $\mu > 0$) or hyperbola (when $\mu < 0$) referred to its centre. *The orbits are therefore conics whose centre is at the centre of force**.

(ii) $n = -3$.

In this case the attractive force is

$$f(r) = \frac{\mu}{r^3},$$

so the equation of the orbit becomes

$$\theta = -\int^u \left\{c + \left(\frac{\mu}{h^2} - 1\right) u^2\right\}^{-\frac{1}{2}} du.$$

* Newton found that if a body move in an ellipse under the action of a force directed to the centre of the ellipse, the force is directly proportional to the distance: *Principia*, Book I. § 2, Prop. X.

Integrating, we have

$$\begin{cases} u = A\cos(k\theta+\epsilon), \text{ where } k^2 = 1-\dfrac{\mu}{h^2}, \text{ when } \mu < h^2, \\ u = A\cosh(k\theta+\epsilon), \text{ where } k^2 = \dfrac{\mu}{h^2}-1, \text{ when } \mu > h^2, \\ u = A\theta+\epsilon, \qquad\qquad \text{ when } \mu = h^2, \end{cases}$$

where in each case A and ϵ are constants of integration.

These curves are sometimes known as *Cotes' spirals*; the last is the reciprocal spiral*.

In connexion with forces varying as the inverse cube of the distance, it may be observed that if

$$r = f(\theta)$$

be an orbit described under a central force $P(r)$ to the origin, then the orbit

$$r = f(k\theta),$$

where k is any constant, can be described under a central force $P(r)+\frac{c}{r^3}$, where c is a constant: the intervals of time between corresponding points, i.e. points for which the radius vector has the same value, in the two orbits being the same.

For, if accented letters refer to the second orbit, we have

$$P' = h'^2u^2\left(u+\frac{d^2u}{d\theta'^2}\right)$$

$$= h'^2u^3 + \frac{h'^2}{k^2}u^2\frac{d^2u}{d\theta^2}$$

$$= h'^2u^3 + \frac{h'^2}{h^2k^2}(P-h^2u^3).$$

If therefore we choose the new constant of momentum h' so that

$$h' = hk$$

(this equation implies that the intervals of time between corresponding points in the two orbits are the same, since it can be written $\frac{dt'}{r'^2} = \frac{dt}{r^2}$), we have

$$P' = P - \frac{h^2(1-k^2)}{r^3},$$

which establishes the result. This is sometimes known as *Newton's theorem of revolving orbits.*

The types of central motion corresponding to

$$n = 5,\ 3,\ 0,\ -4,\ -5,\ -7$$

lead, as has been shewn, to elliptic integrals: on inverting the integrals, we obtain the solution in terms of elliptic functions. As an example we shall take the case of $n = -5$.

* Newton, *Principia*, Book I. § 2, Prop. IX.; R. Cotes, *Harmonia Mensurarum*, pp. 31, 98.

Let μu^5 be the force towards the centre of attraction; we shall suppose the particle initially projected with a velocity less than that which would be acquired by a fall from rest at an infinite distance to the point of projection, so that the total energy

$$\tfrac{1}{2}\dot{r}^2 + \tfrac{1}{2}r^2\dot{\theta}^2 - \frac{\mu}{4r^4}$$

is negative: call this quantity $-\tfrac{1}{2}\gamma$. Then the equation of energy

$$\dot{r}^2 + r^2\dot{\theta}^2 - \frac{\mu}{2r^4} + \gamma = 0$$

together with the equation

$$r^2\dot{\theta} = h$$

gives

$$\left(\frac{dr}{d\theta}\right)^2 = -\frac{\gamma}{h^2}r^4 - r^2 + \frac{\mu}{2h^2}.$$

Introducing in place of r a new variable ρ defined by the equation

$$r = \left(\frac{\mu}{2}\right)^{\frac{1}{4}} \frac{1}{h(\rho + \frac{1}{3})^{\frac{1}{2}}},$$

the differential equation becomes

$$\left(\frac{d\rho}{d\theta}\right)^2 = 4(\rho + \tfrac{1}{3})\left(\rho^2 - \frac{\rho}{3} - \frac{2}{9} - \frac{\mu\gamma}{2h^4}\right).$$

The roots of the quadratic

$$\rho^2 - \frac{\rho}{3} - \frac{2}{9} - \frac{\mu\gamma}{2h^4} = 0$$

are real when γ is positive; their sum is $\frac{1}{3}$, and the smaller of them is less than $-\frac{1}{3}$. Hence if the greater and less of the roots be denoted by e_1 and e_3 respectively, and if e_2 denotes $-\frac{1}{3}$, we have the relations

$$e_1 + e_2 + e_3 = 0,$$

$$e_1 > e_2 > e_3,$$

$$\left(\frac{d\rho}{d\theta}\right)^2 = 4(\rho - e_1)(\rho - e_2)(\rho - e_3),$$

so

$$\rho = \wp(\theta - \epsilon),$$

where ϵ is a constant of integration, and the function $\wp$ is formed with the roots e_1, e_2, e_3. Thus we have

$$r = \left(\frac{\mu}{2}\right)^{\frac{1}{4}} \frac{1}{h\{\wp(\theta - \epsilon) + \frac{1}{3}\}^{\frac{1}{2}}}.$$

Now r is real and positive, and, as we see from the equation of energy, cannot be greater than $\sqrt[4]{\dfrac{\mu}{2\gamma}}$. So $\wp(\theta - \epsilon) + \frac{1}{3}$ is real and positive and has a finite lower limit; but when $e_1 > e_2 > e_3$, the function $\wp(\theta - \epsilon)$ is real and has a finite lower limit for all real values of θ only when ϵ is real; so ϵ is purely real, and by measuring θ from a suitable initial line we can take ϵ to be zero. We have therefore

$$r = \left(\frac{\mu}{2}\right)^{\frac{1}{2}} \frac{1}{h\{\wp(\theta) + \frac{1}{3}\}^{\frac{1}{2}}},$$

and *this is the equation of the orbit in polar coordinates**.

The time can now be determined from the equation

$$t = \frac{1}{h}\int r^2 d\theta,$$

or

$$t = \frac{\mu}{2h^3}\int \frac{d\theta}{\wp(\theta) - e_2}.$$

Performing the integration, we have

$$t = -\frac{\mu h^{-3}}{2(e_2 - e_1)(e_2 - e_3)}\left\{\zeta(\theta) + \frac{1}{2}\frac{\wp'(\theta)}{\wp(\theta) - e_2} + e_2\theta\right\},$$

where $\zeta(\theta)$ is the Weierstrassian zeta-function†. *This equation determines t.*

Example 1. Shew that the equation of the orbit of a particle which moves under the influence of a central attractive force μ/r^5 can be written in the form

$$r = a \operatorname{sn}\left(K - \frac{\theta}{\sqrt{1+k^2}}, k\right),$$

or else in the form

$$\frac{a}{r} = k \operatorname{sn}\left(K - \frac{\theta}{\sqrt{1+k^2}}, k\right),$$

provided $h^4 > 4\mu E > 0$, where h is the angular momentum round the origin and E is the excess of the total energy over the potential energy at infinity.

(Cambridge Math. Tripos, Part I, 1894.)

Example 2. A particle is attracted to the origin with constant acceleration μ; shew that the radius vector, vectorial angle, and time, are given in terms of a real auxiliary angle u by equations of the type

$$r = \wp(iu + \omega_1) - \wp(\omega_2 + a),$$

$$\left(\frac{\mu}{2}\right)^{\frac{1}{2}} t = i\zeta(\omega_1 + iu) + u\wp(\omega_2 + a) - i\zeta(\omega_1),$$

$$e^{i\theta} = e^{-2iu\zeta(\omega_2 + a)}\frac{\sigma(\omega_1 + iu + \omega_2 + a)\,\sigma(\omega_1 - \omega_2 - a)}{\sigma(\omega_1 + iu - \omega_2 - a)\,\sigma(\omega_1 + \omega_2 + a)}.$$

(Schoute.)

Among the points of special interest on an orbit are the points at which the radius vector, after having increased for some time, begins to decrease: or after having decreased for some time, begins to increase. A point belonging to the former of these classes is called an *apocentre,* while points of the latter class are called *pericentres*; both classes are included under the

* The orbits are discussed and classified by W. D. MacMillan, *Amer. Journ. Math.* xxx. (1908), p. 282.

† Cf. Whittaker and Watson, *Modern Analysis*, § 20·4.

general term *apse*. At an apse, if the apse is not a singularity of the orbit (e.g. a cusp), we have

$$\frac{dr}{d\theta} = 0,$$

which implies that the tangent to the orbit is perpendicular to the radius vector.

The words *aphelion* and *perihelion* are generally used instead of apocentre and pericentre when the centre of force is supposed to be the Sun.

Example. A particle moves under an attraction

$$\frac{\mu}{r^2} + \frac{\nu}{r^3}$$

to a fixed centre; shew that the angle subtended at the centre of force by two consecutive apses is

$$\frac{\pi}{\sqrt{1 - \frac{\nu}{h^2}}},$$

where h is the constant of angular momentum.

49. *Motion under the Newtonian law**.

The remaining case in which motion under a central force varying as an integral power of the distance can be solved in terms of circular functions is that in which the force varies as the inverse square of the distance. This case is of great importance in Celestial Mechanics, since the mutual attractions of the heavenly bodies vary as the inverse squares of their distances apart, in accordance with the Newtonian law of universal gravitation.

(i) *The orbits.*

Consider then the motion of a particle which is acted on by a force directed to a fixed point (which we can take as the origin of coordinates), of magnitude μu^2, where u is the reciprocal of the distance from the fixed point. Let the particle be projected from the point whose polar coordinates are (c, α) with velocity v_0 in a direction making an angle γ with c; so that the angular momentum is

$$h = cv_0 \sin \gamma.$$

The differential equation of the orbit is

$$\frac{d^2u}{d\theta^2} + u = \frac{P}{h^2u^2} = \frac{\mu}{v_0^2c^2 \sin^2 \gamma};$$

this is a linear differential equation with constant coefficients, and its integral is

$$u = \frac{\mu}{v_0^2c^2 \sin^2 \gamma}\{1 + e \cos(\theta - \varpi)\},$$

* Newton, *Principia*, Book I. § 3, Props. XI., XII., XIII.

where e and ϖ are constants of integration. This is the equation, in polar coordinates, of a conic whose focus is at the origin, whose eccentricity is e, and whose semi-latus rectum l is given by the equation

$$l = \frac{v_0^2 c^2 \sin^2 \gamma}{\mu};$$

the constant ϖ determines the position of the apse-line, and is called the *perihelion-constant.*

The circumstance that the focus of the conic is at the centre of force is in accord with Hamilton's theorem; for if the centre of force is at the focus of the conic the perpendicular on the polar of the centre of force is the perpendicular on the directrix, which is proportional to r, as by Hamilton's theorem the force must be proportional to $1/r^2$.

To determine the constants e and ϖ in terms of the initial data c, α, γ, v_0, we observe that initially

$$\theta = \alpha, \quad u = \frac{1}{c}, \quad \frac{du}{d\theta} = -\frac{1}{c}\cot\gamma;$$

substituting these values in the equation of the orbit and the equation obtained by differentiating it with respect to θ, we have

$$\begin{cases} v_0^2 c \sin^2 \gamma = \mu + \mu e \cos(\alpha - \varpi), \\ v_0^2 c \sin\gamma \cos\gamma = \mu e \sin(\alpha - \varpi). \end{cases}$$

Solving these equations for e and ϖ, we obtain

$$\begin{cases} e^2 = 1 + \dfrac{v_0^4 c^2 \sin^2 \gamma}{\mu^2} - \dfrac{2 v_0^2 c \sin^2 \gamma}{\mu}, \\ \cot(\alpha - \varpi) = \dfrac{-\mu}{c v_0^2 \sin\gamma\cos\gamma} + \tan\gamma. \end{cases}$$

The semi-major axis, when the conic is an ellipse, is generally called the *mean distance* of the particle; denoting it by a, we have

$$a = \frac{l}{1 - e^2},$$

and substituting the values of l and e^2 already found, we have

$$v_0^2 = \mu\left(\frac{2}{c} - \frac{1}{a}\right);$$

this equation determines a in terms of the initial data.

The time occupied in describing the whole circumference of the ellipse, which is generally called the *periodic time,* is

$$\frac{2}{h} \times \text{area of ellipse},$$

since h represents twice the rate at which the area is swept out by the radius vector; the periodic time is therefore $\dfrac{2\pi ab}{h}$, where b is the semi-minor axis. But we have

$$h = v_0 c \sin\gamma = \sqrt{\mu l} = b\sqrt{\frac{\mu}{a}},$$

so the *periodic time is* $2\pi\sqrt{\dfrac{a^3}{\mu}}$. It is usual to denote the quantity $\mu^{\frac{1}{2}}a^{-\frac{3}{2}}$ by n; the periodic time can then be written

$$\frac{2\pi}{n};$$

n is called the *mean motion*, being the mean value of $\dot{\theta}$ for a complete period.

It has been shewn by Bertrand and Koenigs that of all laws of force which give a zero force at an infinite distance, the Newtonian law is the only one for which all the orbits are algebraic curves, and also the only one for which all the orbits are closed curves.

Example. Shew that if a centre of force repels a particle with a force varying as the inverse square of the distance, the orbit is a branch of a hyperbola, described about its outer focus.

(ii) *The velocity.*

Consider now the case in which the orbit is an ellipse; the equation

$$v_0^2 = \mu\left(\frac{2}{c} - \frac{1}{a}\right),$$

establishes a connexion between the mean distance a and the velocity v_0 and radius vector c at the initial point of the path. Since any point of the orbit can be taken as initial point, we can write this equation

$$v^2 = \mu\left(\frac{2}{r} - \frac{1}{a}\right),$$

where v is the velocity of the particle at the point whose radius vector is r.

Similarly if the orbit is a hyperbola, whose semi-major axis is a, we find

$$v^2 = \mu\left(\frac{2}{r} + \frac{1}{a}\right),$$

and if the orbit is a parabola, the relation becomes

$$v^2 = \frac{2\mu}{r}.$$

It is clear from this that *the orbit is an ellipse, parabola, or hyperbola,* according as $v_0^2 \lesseqgtr \dfrac{2\mu}{c}$, i.e. *according as the initial velocity of the particle is less than, equal to, or greater than, the velocity which the particle would acquire in falling from a position of rest at an infinite distance from the centre of force to the initial position.*

It can further be shewn that *the velocity at any point can be resolved into a component* $\frac{\mu}{h}$ *perpendicular to the radius vector and a component* $\frac{\mu e}{h}$ *perpendicular to the axis of the conic*; each of these components being constant.

For if S be the centre of force, P the position of the moving particle, G the intersection of the normal at P to the conic with the major axis, GL the perpendicular on SP from G, and SY the perpendicular on the tangent at P from S, it is known that the sides of the triangle SPG are respectively perpendicular to the velocity and to the components of the velocity in the two specified directions; and therefore we have

$$\text{Component perpendicular to the radius vector} = \frac{v \, . \, SP}{PG} = \frac{h \, . \, SP}{SY \, . \, PG} = \frac{h}{PL}$$

$$= \frac{h}{l} = \frac{\mu}{h},$$

$$\text{and Component perpendicular to the axis} = \frac{SG}{SP} \times \text{Component perpendicular to the radius vector}$$

$$= \frac{e\mu}{h},$$

which establishes the result stated.

Example 1. Shew that in elliptic motion under Newton's law, the projections, on the external bisector of two radii, of the velocities corresponding to these radii, are equal. Shew also that the sum of the projections on the inner bisector is equal to the projection of a line constant in magnitude and direction. (Cailler.)

Example 2. Shew that in elliptic motion under Newton's law, the quantity $\int T dt$, where T denotes the kinetic energy, integrated over a complete period, depends only on the mean distance and not on the eccentricity. (Grinwis.)

Example 3. At a certain point in an elliptic orbit described under a force μ/r^2, the constant μ is suddenly changed by a small amount. If the eccentricities of the former and new orbits are equal, shew that the point is an extremity of the minor axis.

(iii) *The anomalies in elliptic motion.*

If a particle is describing an ellipse under a centre of force in the focus S, the vectorial angle ASP of the point P at which the particle is situated on the ellipse, measured from the apse A which is nearer to the focus, is called the *true anomaly* of the particle and will be denoted by θ; the eccentric angle corresponding to the point P is called the *eccentric anomaly* of the particle, and will be denoted by u: and the quantity nt, where n is the mean motion and t is the time of describing the arc AP, is called the *mean anomaly* of the particle. We shall now find the connexion between the three anomalies.

The *relation between θ and u* is found thus:

We have

$$\frac{l}{r} = 1 + e\cos\theta,$$

and $r = a - ex$, where x is the rectangular coordinate of P referred to the centre of the ellipse as origin,

or $$r = a(1 - e\cos u).$$

Hence $$(1 - e\cos u)(1 + e\cos\theta) = 1 - e^2,$$

an equation which can also be written in the forms

$$\tan\frac{u}{2} = \left(\frac{1-e}{1+e}\right)^{\frac{1}{2}} \tan\frac{\theta}{2},$$

and $$\sin u = \frac{(1-e^2)^{\frac{1}{2}}\sin\theta}{1 + e\cos\theta}.$$

The *relation between u and nt* can be obtained in the following way:

We have

$$t = \frac{2}{h} \times \text{Area } ASP = \frac{2}{nab} \cdot \frac{b}{a} \times \text{Area } ASQ,$$ where Q is the point on the auxiliary circle corresponding to the point P on the ellipse

$$= \frac{2}{na^2}\{\text{Area } ACQ - \text{Area } SCQ\},$$ where C is the centre of the ellipse

$$= \frac{2}{na^2}\left\{\frac{a^2}{2}u - \frac{a^2e}{2}\sin u\right\},$$

so $$nt = u - e\sin u.$$

This is known as *Kepler's equation.*

A nomogram for the solution of this equation is described by H. Chrétien, *Assoc. Franç. Congrès*, Reims (1907), p. 83. The solution by analytical expansion has been discussed by many writers, an important recent memoir being that by Levi-Civita, *Atti della R. Acc. dei Lincei, Rendiconti*, (5) XIII. (1904), p. 260.

Lastly, the *relation between θ and nt* can be found as follows:

We have $$nt = u - e\sin u.$$

Replacing u by its value in terms of θ, this becomes

$$nt = \arcsin\left\{\frac{(1-e^2)^{\frac{1}{2}}\sin\theta}{1 + e\cos\theta}\right\} - \frac{e(1-e^2)^{\frac{1}{2}}\sin\theta}{1 + e\cos\theta},$$

which is the required relation; this equation gives the time in terms of the vectorial angle of the moving particle.

A solution of the problem of calculating the True Anomaly from the Mean Anomaly, based on a geometrical deduction, was found among the unpublished papers of Newton.

Example 1. *Shew that*

$$u = nt + 2\sum_{r=1}^{\infty} \frac{1}{r} J_r(re) \sin rnt,$$

*where the symbols J denote Bessel coefficients**.

For we have

$$\frac{1}{n}\frac{du}{dt} = \frac{1}{1 - e\cos u}$$

$$= \frac{1}{2\pi}\int_0^{2\pi} \frac{d(nt)}{1 - e\cos u} + \sum_{r=1}^{\infty} \frac{\cos rnt}{\pi}\int_0^{2\pi} \frac{\cos rnt \,.\, d(nt)}{1 - e\cos u}, \text{ by Fourier's theorem}\dagger$$

$$= \frac{1}{2\pi}\int_0^{2\pi} du + \sum_{r=1}^{\infty} \frac{\cos rnt}{\pi}\int_0^{2\pi} \cos\{r(u - e\sin u)\}\, du$$

$$= 1 + 2\sum_{r=1}^{\infty} J_r(re)\cos rnt \ddagger.$$

Integrating, we have the required result.

Example 2. Shew that

$$\theta = nt + 2e\sin nt + \tfrac{5}{4}e^2 \sin 2nt + \ldots.$$

Example 3. In hyperbolic motion under the Newtonian law, shew that

$$\mu^{-\frac{1}{2}} a^{\frac{3}{2}} t = \log\left\{\frac{(e+1)^{\frac{1}{2}}\cos\frac{1}{2}\theta - (e-1)^{\frac{1}{2}}\sin\frac{1}{2}\theta}{(e+1)^{\frac{1}{2}}\cos\frac{1}{2}\theta + (e-1)^{\frac{1}{2}}\sin\frac{1}{2}\theta}\right\} + e(e^2-1)^{\frac{1}{2}} \frac{\sin\theta}{1 + e\cos\theta},$$

and in parabolic motion, shew that

$$\left(\frac{\mu}{2p^3}\right)^{\frac{1}{2}} t = \tan\frac{\theta}{2} + \tfrac{1}{3}\tan^3\frac{\theta}{2},$$

where p is the distance from the focus to the vertex.

Example 4. In elliptic motion under Newton's law, shew that the sum of the four times (counted from perihelion) to the intersections of a circle with the ellipse is the same for all concentric circles, and remains constant when the centre of the circle moves parallel to the major axis. (Oekinghaus.)

(iv) *Lambert's theorem.*

Lambert in 1761 shewed that in elliptic motion under the Newtonian law, the time occupied in describing any arc depends only on the major axis, the sum of the distances from the centre of force to the initial and final points, and the length of the chord joining these points: so that if these three elements are given, the time is determinate, whatever be the form of the ellipse§.

* The name of Bessel is commonly connected with this expansion: but it is really due to Lagrange, *Oeuvres*, III. p. 130.

† Cf. Whittaker and Watson, *Modern Analysis*, Chapter IX.

‡ *Ibid.* Chapter XVII.

§ Lambert's original demonstration was geometrical and synthetic: the theorem was proved analytically and generalised by Lagrange in 1778 (*Oeuvres de Lagrange*, IV. p. 559).

Let u and u' be the eccentric anomalies of the points; then we have

$$n \times \text{the required time} = u' - e \sin u' - (u - e \sin u)$$

$$= (u' - u) - 2e \sin \frac{u' - u}{2} \cos \frac{u' + u}{2}.$$

Now if c be the length of the chord, and r and r' be the radii vectores, we have

$$\frac{r + r'}{a} = 1 - e \cos u + 1 - e \cos u' = 2 - 2e \cos \frac{u + u'}{2} \cos \frac{u' - u}{2},$$

and

$$c^2 = a^2 (\cos u' - \cos u)^2 + b^2 (\sin u' - \sin u)^2$$

$$= 4a^2 \sin^2 \frac{u' - u}{2} \left(1 - e^2 \cos^2 \frac{u + u'}{2}\right),$$

so

$$\frac{c}{a} = 2 \sin \frac{u' - u}{2} \left(1 - e^2 \cos^2 \frac{u + u'}{2}\right)^{\frac{1}{2}}.$$

Hence we have

$$\frac{r + r' + c}{a} = 2 - 2 \cos \left\{\frac{u' - u}{2} + \arccos \left(e \cos \frac{u + u'}{2}\right)\right\},$$

and

$$\frac{r + r' - c}{a} = 2 - 2 \cos \left\{-\frac{u' - u}{2} + \arccos \left(e \cos \frac{u + u'}{2}\right)\right\},$$

and therefore*

$$2 \arcsin \frac{1}{2} \left(\frac{r + r' + c}{a}\right)^{\frac{1}{2}} = \frac{u' - u}{2} + \arccos \left(e \cos \frac{u + u'}{2}\right),$$

and

$$2 \arcsin \frac{1}{2} \left(\frac{r + r' - c}{a}\right)^{\frac{1}{2}} = -\frac{u' - u}{2} + \arccos \left(e \cos \frac{u + u'}{2}\right).$$

Thus if quantities α and β are defined by the equations

$$\sin \frac{\alpha}{2} = \frac{1}{2} \left(\frac{r + r' + c}{a}\right)^{\frac{1}{2}}, \qquad \sin \frac{\beta}{2} = \frac{1}{2} \left(\frac{r + r' - c}{a}\right)^{\frac{1}{2}},$$

the last equations give

$$\alpha - \beta = u' - u, \text{ and } \cos \frac{\alpha + \beta}{2} = e \cos \frac{u + u'}{2}.$$

Thus finally we have

$$n \times \text{the required time} = \alpha - \beta - 2 \cos \frac{\alpha + \beta}{2} \sin \frac{\alpha - \beta}{2}$$

$$= (\alpha - \sin \alpha) - (\beta - \sin \beta).$$

This is *Lambert's theorem.*

Example 1. Examine the limiting case when the minor axis of the ellipse vanishes, so that the orbit is rectilinear.

* It will be noticed that owing to the presence of the radicals, Lambert's theorem is not free from ambiguity of sign. The reader will be able to determine without difficulty the interpretation of sign corresponding to any given position of the initial and final points.

Example 2. *To obtain the form of Lambert's theorem applicable to parabolic motion.*

If we suppose the mean distance a to become large, the angles α and β become very small, so Lambert's theorem can be written in the approximate form

$$\text{Required time} = \frac{\alpha^3 - \beta^3}{6n}$$

$$= \left(\frac{a^3}{\mu}\right)^{\frac{1}{2}} \frac{1}{6} \left\{ \left(\frac{r+r'+c}{a}\right)^{\frac{3}{2}} - \left(\frac{r+r'-c}{a}\right)^{\frac{3}{2}} \right\}$$

$$= \frac{1}{6\mu^{\frac{1}{2}}} \{(r+r'+c)^{\frac{3}{2}} - (r+r'-c)^{\frac{3}{2}}\},$$

and this is the required form*.

Example 3. Establish Lambert's theorem for parabolic motion directly from the formulae of parabolic motion.

50. *The mutual transformation of fields of central force and fields of parallel force.*

If in the general problem of central forces we suppose the centre of force to be at a very great distance from the part of the field considered, the lines of action of the force in different positions of the particle will be almost parallel to each other; and on passing to the limiting case in which the centre of force is regarded as being at an infinite distance, we arrive at the problem of the motion of a particle under the influence of a force which is always parallel to a given fixed direction.

For the discussion of this problem, take rectangular axes Ox, Oy in the plane of the motion, Ox being parallel to the direction of the force; and let $X(x)$ be the magnitude of the force, which will be supposed to be independent of the coordinate y. The equations of motion are

$$\ddot{x} = X(x), \qquad \ddot{y} = 0,$$

and the motion is therefore represented by the equations

$$t = ay + b = \int^x \{2\int X(x)\,dx + c\}^{-\frac{1}{2}}\,dx + l,$$

where a, b, c, l are the constants of integration; the values of these are determined by the circumstances of projection, i.e. by the initial values of x, y, $\dot{x}$, $\dot{y}$.

While the problem of motion in a parallel field of force is a limiting case of the problem of motion under central forces, it is not difficult to reduce the latter more general problem to the former more special one.

For if a particle is in motion under a force of magnitude P directed to

* This result was given by Newton (*Principia*, Book III, lemma x), and later by Euler in his *Determinatio Orbitae Cometae Anni* 1742 (1743), before Lambert published the general theorem.

a fixed centre (which we may take as origin of coordinates), the equations of motion are

$$\ddot{x} = -P\frac{x}{r}, \qquad \ddot{y} = -P\frac{y}{r}.$$

The angular momentum of the particle round the origin (which is constant) is $x\dot{y} - y\dot{x}$: let this be denoted by h. Introduce new coordinates X, Y, defined by the homographic transformation

$$X = \frac{x}{y}, \qquad Y = \frac{1}{y},$$

and let T be a new variable defined by the equation

$$T = \int \frac{dt}{y^2}.$$

Then we have

$$\frac{dX}{dT} = \frac{d}{dt}\left(\frac{x}{y}\right) \cdot \frac{dt}{dT} = \left(\frac{\dot{x}}{y} - \frac{\dot{y}x}{y^2}\right) y^2 = -h,$$

$$\frac{dY}{dT} = \frac{d}{dt}\left(\frac{1}{y}\right) \cdot \frac{dt}{dT} = -\frac{\dot{y}}{y^2} y^2 = -\dot{y},$$

so $$\frac{d^2X}{dT^2} = 0, \qquad \frac{d^2Y}{dT^2} = -y^2\ddot{y} = P.\frac{y^3}{r}.$$

These equations shew that a particle whose coordinates are (X, Y) would, if T were interpreted as the time, move as if acted on by a force parallel to the axis of Y and of magnitude $\frac{Py^3}{r}$. As the solution of this transformed problem will yield the solution of the original problem, it follows that *the general problem of motion under central forces is reducible to the problem of motion in a parallel field of force.*

Example 1. Shew that the path of a free particle moving under the influence of gravity alone is a parabola with its axis vertical and vertex upwards.

Example 2. Shew that the magnitude of the force parallel to the axis of x under which the curve $f(x, y) = 0$ can be described is a constant multiple of

$$\left(\frac{\partial f}{\partial x}\right)^{-3} \left\{-\frac{\partial^2 f}{\partial x^2}\left(\frac{\partial f}{\partial y}\right)^2 + 2\frac{\partial^2 f}{\partial x \partial y}\frac{\partial f}{\partial x}\frac{\partial f}{\partial y} - \frac{\partial^2 f}{\partial y^2}\left(\frac{\partial f}{\partial x}\right)^2\right\}.$$

Example 3. If a parallel field of force is such that the path described by a free particle is a conic whatever be the initial conditions, shew that the force varies as the inverse cube of the distance from some line perpendicular to the direction of the force.

51. *Bonnet's theorem.*

We now proceed to discuss the motion of a particle which is simultaneously attracted by more than one centre of force. An indefinite number of particular cases of motion of this kind can be obtained by means of a theorem due to Legendre*, but generally known as *Bonnet's theorem*, which may be stated thus:

* Legendre, *Exerc. de Calc. Int.* II. (1817), p. 383.

If a given orbit can be described in each of n given fields of force, taken separately, the velocities at any point P of the orbit being $v_1, v_2, \ldots, v_n$, *respectively, then the same orbit can be described in the field of force which is obtained by superposing all these fields, the velocity at the point P being* $(v_1^2 + v_2^2 + \ldots + v_n^2)^{\frac{1}{2}}$.

For suppose that in the field of force which is obtained by superposing the original fields, an additional normal force R is required in order to make the particle move on the curve in question; and let it be projected from a point A so that the square of its velocity at A is equal to the sum of the squares of its velocities at A in the original fields of force. Then on adding the equations of energy corresponding to the original motions, and comparing with the equation of energy for the motion in question, we see that the kinetic energy of the motion in question is the sum of the kinetic energies of the original motions, i.e. that the velocity at any point P is

$$(v_1^2 + v_2^2 + \ldots + v_n^2)^{\frac{1}{2}}.$$

Hence, resolving along the normal to the orbit, we have

$$m\frac{v_1^2 + v_2^2 + \ldots + v_n^2}{\rho} = F_1 + F_2 + \ldots + F_n + R,$$

where m is the mass of the particle, ρ the radius of curvature of the orbit, and $F_1, F_2, \ldots, F_n$ are the normal components of the original fields of force at P.

But $$\frac{mv_1^2}{\rho} = F_1, \quad \frac{mv_2^2}{\rho} = F_2, \quad \ldots, \quad \frac{mv_n^2}{\rho} = F_n,$$

and therefore R is zero; the given orbit is therefore a free path in the field of force which is obtained by superposing the original fields.

Example. *Shew that an ellipse can be described if forces*

$$\mu \cdot \frac{r^3 + 8a^3}{8a^3 r^2} \quad \textit{and} \quad \mu \cdot \frac{r'^3 + 8a^3}{8a^3 r'^2}$$

respectively act in the directions of the foci.

This result follows at once from Bonnet's theorem when it is observed that the given forces are equivalent to forces $\frac{\mu}{r^2}$ and $\frac{\mu}{r'^2}$ acting in the directions of the foci, together with a force $\frac{\mu}{4a^3}\times$distance acting in the direction of the centre of the ellipse.

52. *Determination of the most general field of force under which a given curve or family of curves can be described.*

Let $$\phi(x, y) = c$$

be the equation of a curve; on varying the constant c, this equation will represent a family of curves. We shall now find an expression for the most general field of force (the force being supposed to depend only on the position of the particle on which it acts) for which this family of curves is a family of orbits of a particle.

Let v denote the velocity of the particle, and (X, Y) the components of force per unit mass parallel to the coordinate axes. The tangential and normal components of acceleration being $\frac{1}{2}\frac{dv^2}{ds}$ and $\frac{v^2}{\rho}$ respectively, we have

$$X = -\frac{v^2}{\rho}\phi_x(\phi_x^2+\phi_y^2)^{-\frac{1}{2}} - \frac{1}{2}\frac{dv^2}{ds}\phi_y(\phi_x^2+\phi_y^2)^{-\frac{1}{2}},$$

$$Y = -\frac{v^2}{\rho}\phi_y(\phi_x^2+\phi_y^2)^{-\frac{1}{2}} + \frac{1}{2}\frac{dv^2}{ds}\phi_x(\phi_x^2+\phi_y^2)^{-\frac{1}{2}}.$$

Substituting for $\frac{1}{\rho}$ its value, namely

$$\frac{\phi_y^2\phi_{xx} - 2\phi_x\phi_y\phi_{xy} + \phi_x^2\phi_{yy}}{(\phi_x^2+\phi_y^2)^{\frac{3}{2}}},$$

we have

$$X = -\phi_x \, . \, v^2 . \frac{\phi_y^2\phi_{xx} - 2\phi_x\phi_y\phi_{xy} + \phi_x^2\phi_{yy}}{(\phi_x^2+\phi_y^2)^2} - \frac{1}{2}\frac{dv^2}{ds}\phi_y(\phi_x^2+\phi_y^2)^{-\frac{1}{2}}.$$

Writing $$v^2 = -u(\phi_x^2+\phi_y^2),$$

and replacing $\frac{d}{ds}$ by $(\phi_x^2+\phi_y^2)^{-\frac{1}{2}}\left(\phi_x\frac{\partial}{\partial y} - \phi_y\frac{\partial}{\partial x}\right)$, this equation becomes

$$X = u(\phi_x\phi_{yy} - \phi_y\phi_{xy}) + \tfrac{1}{2}\phi_y\frac{du}{ds}(\phi_x^2+\phi_y^2)^{\frac{1}{2}}.$$

Now u is arbitrary, since it depends on the velocity with which the given orbits are described; and as X and Y are to be functions of the position of the particle, we can take u to be an arbitrary function of x and y; we have therefore

$$X = u(\phi_x\phi_{yy} - \phi_y\phi_{xy}) + \tfrac{1}{2}\phi_y(\phi_x u_y - \phi_y u_x),$$

and similarly

$$Y = u(\phi_y\phi_{xx} - \phi_x\phi_{xy}) + \tfrac{1}{2}\phi_x(\phi_y u_x - \phi_x u_y),$$

where u is an arbitrary function of x and y. These expressions for the field of force under which the curves of the given family are orbits were first given by Dainelli*.

Example 1. ***Shew that a particle can describe a given curve under any arbitrary forces $P_1, P_2, \ldots$ directed to given fixed points, provided these forces satisfy the relations***

$$\sum_k \frac{1}{p_k^2}\frac{d}{ds}\left(\frac{P_k p_k^3 \rho}{r_k}\right) = 0,$$

where r_k is the radius and p_k the perpendicular on the tangent, from the kth of the given fixed points, and where ρ is the radius of curvature of the given curve.

For the tangential and normal components of force on the particle are

$$T = -\sum_k P_k\frac{dr_k}{ds} \text{ and } N = \sum_k P_k\frac{p_k}{r_k},$$

* *Giornale di Mat.* XVIII. (1880), p. 271.

so from the equation

$$2T = \frac{d(v^2)}{ds} = \frac{d}{ds}(\rho N),$$

we have

$$\sum_k \left\{ 2P_k \frac{dr_k}{ds} + \frac{d}{ds}\left(P_k \frac{\rho p_k}{r_k}\right)\right\} = 0,$$

or

$$\sum_k \frac{1}{p_k^2} \frac{d}{ds}\left(\frac{P_k p_k^3 \rho}{r_k}\right) = 0.$$

Example 2. A particle can describe a given curve under the single action of any one of the forces $\phi_1, \phi_2, \ldots,$ acting in given (variable) directions. Shew that the condition to be satisfied in order that the same curve may be described under the joint action of forces $F_1, F_2, \ldots,$ acting in the directions of $\phi_1, \phi_2, \ldots,$ respectively, is

$$\sum_k c_k \phi_k d\left(\frac{F_k}{\phi_k}\right) = 0,$$

where c_k is the chord of curvature of the curve in the direction of ϕ_k. (Curtis.)

Example 3. A point moves in a field of force in two dimensions of which the work function is V; shew that an equipotential curve is a possible path, provided V satisfy the equation

$$0 = f(V)\left\{\frac{\partial^2 V}{\partial x^2}\left(\frac{\partial V}{\partial y}\right)^2 - 2\frac{\partial^2 V}{\partial x \partial y}\frac{\partial V}{\partial x}\frac{\partial V}{\partial y} + \frac{\partial^2 V}{\partial y^2}\left(\frac{\partial V}{\partial x}\right)^2\right\} + \left\{\left(\frac{\partial V}{\partial x}\right)^2 + \left(\frac{\partial V}{\partial y}\right)^2\right\}^2. \quad \text{(Coll. Exam.)}$$

53. *The problem of two centres of gravitation.*

The equations of motion of a particle moving in a plane under arbitrary forces cannot be integrated by quadratures in the general case. The most famous of the known soluble problems of this class, other than problems of central motion, is the *problem of two centres of gravitation*, i.e. the problem of determining the motion of a free particle in a plane, attracted by two fixed Newtonian centres of force in the plane; its integrability was discovered by Euler*.

Let $2c$ denote the distance between the two centres of force; and take the point midway between them as origin, and the line joining them as axis of x, so that their coordinates can be taken to be $(c, 0)$ and $(-c, 0)$. The potential energy of the particle (whose mass is taken to be unity) is therefore

$$V = -\mu\{(x-c)^2 + y^2\}^{-\frac{1}{2}} - \mu'\{(x+c)^2 + y^2\}^{-\frac{1}{2}},$$

where μ and μ' are constants depending on the strength of the centres of attraction.

Now any ellipse or hyperbola with the two centres of force as foci is a possible orbit when either centre of force acts alone, and therefore by Bonnet's theorem it is a possible orbit when both centres of force are acting. It is therefore natural, in defining the position of the particle, to replace the rectangular coordinates (x, y) by *elliptic coordinates* (ξ, η), defined by the equations

$$x = c \cosh \xi \cos \eta, \quad y = c \sinh \xi \sin \eta.$$

* Euler, *Mém. de Berlin*, 1760, p. 228; *Nov. Comm. Petrop.* x. (1764), p. 207; xi. (1765), p. 152: Lagrange, *Mém. de Turin*, iv. (1766–9), pp. 118, 215, or *Oeuvres*, ii. p. 67.

The equations $\xi =$ Constant and $\eta =$ Constant then represent respectively ellipses and hyperbolas whose foci are at the centres of force; and these are a particular family of orbits.

The potential energy, when expressed in terms of ξ and η, becomes

$$V = -\frac{\mu}{c(\cosh\xi - \cos\eta)} - \frac{\mu'}{c(\cosh\xi + \cos\eta)},$$

and the kinetic energy T is given by the equations

$$T = \tfrac{1}{2}\dot{x}^2 + \tfrac{1}{2}\dot{y}^2$$

$$= \frac{c^2}{2}(\cosh^2\xi - \cos^2\eta)(\dot{\xi}^2 + \dot{\eta}^2).$$

This problem is evidently of Liouville's type (§ 43), and can therefore be integrated by the method applicable to this class of questions. The Lagrangian equation for the coordinate ξ is

$$c^2\frac{d}{dt}\{(\cosh^2\xi - \cos^2\eta)\dot{\xi}\} - c^2\cosh\xi\sinh\xi(\dot{\xi}^2 + \dot{\eta}^2) = -\frac{\partial V}{\partial \xi},$$

or

$$c^2\frac{d}{dt}\{(\cosh^2\xi - \cos^2\eta)^2\dot{\xi}^2\} - 2c^2\cosh\xi\sinh\xi(\cosh^2\xi - \cos^2\eta)\dot{\xi}(\dot{\xi}^2 + \dot{\eta}^2)$$
$$= -2(\cosh^2\xi - \cos^2\eta)\dot{\xi}\frac{\partial V}{\partial \xi},$$

or, using the equation of energy $T + V = h$,

$$c^2\frac{d}{dt}\{(\cosh^2\xi - \cos^2\eta)^2\dot{\xi}^2\}$$
$$= -2(\cosh^2\xi - \cos^2\eta)\dot{\xi}\frac{\partial V}{\partial \xi} + 2(h - V)\dot{\xi}\frac{\partial}{\partial \xi}(\cosh^2\xi - \cos^2\eta)$$
$$= 2\dot{\xi}\frac{\partial}{\partial \xi}\{(h - V)(\cosh^2\xi - \cos^2\eta)\}$$
$$= 2\dot{\xi}\frac{\partial}{\partial \xi}\left\{h(\cosh^2\xi - \cos^2\eta) + \frac{\mu}{c}(\cosh\xi + \cos\eta) + \frac{\mu'}{c}(\cosh\xi - \cos\eta)\right\}$$
$$= 2\frac{d}{dt}\left(h\cosh^2\xi + \frac{\mu + \mu'}{c}\cosh\xi\right).$$

Integrating, we have

$$\frac{c^2}{2}(\cosh^2\xi - \cos^2\eta)^2\dot{\xi}^2 = h\cosh^2\xi + \frac{\mu + \mu'}{c}\cosh\xi - \gamma,$$

where γ is a constant of integration.

Subtracting this from the equation of energy, which can be written

$$\frac{c^2}{2}(\cosh^2\xi - \cos^2\eta)^2(\dot{\xi}^2 + \dot{\eta}^2)$$
$$= h(\cosh^2\xi - \cos^2\eta) + \frac{\mu}{c}(\cosh\xi + \cos\eta) + \frac{\mu'}{c}(\cosh\xi - \cos\eta),$$

we have

$$\frac{c^2}{2}(\cosh^2\xi - \cos^2\eta)^2\dot{\eta}^2 = -h\cos^2\eta - \frac{\mu'-\mu}{c}\cos\eta + \gamma.$$

Eliminating dt between these equations, we have

$$\frac{(d\xi)^2}{h\cosh^2\xi + \dfrac{\mu+\mu'}{c}\cosh\xi - \gamma} = \frac{(d\eta)^2}{-h\cos^2\eta - \dfrac{\mu'-\mu}{c}\cos\eta + \gamma}$$

Introducing an auxiliary variable u, we have therefore

$$u = \int\left\{h\cosh^2\xi + \frac{\mu+\mu'}{c}\cosh\xi - \gamma\right\}^{-\frac{1}{2}}d\xi,$$

$$u = \int\left\{-h\cos^2\eta - \frac{\mu'-\mu}{c}\cos\eta + \gamma\right\}^{-\frac{1}{2}}d\eta.$$

These are elliptic integrals, and we can therefore express ξ and η as elliptic functions of the parameter u, say

$$\xi = \chi(u), \quad \eta = \phi(u).$$

These equations determine the orbit of the particle, the elliptic coordinates (ξ, η) being expressed in terms of the parameter u*.

54. *Motion on a surface* †.

We shall next proceed to consider the motion of a particle which is free to move on a smooth surface, and is acted on by any forces.

Let (X, Y, Z) be the components, parallel to fixed rectangular axes, of the external force on the particle, not including the pressure of the surface: let (x, y, z) be the coordinates and v the velocity of the particle, s the arc and ρ the radius of curvature of its path, χ the angle between the principal normal to the path and the normal to the surface, and (λ, μ, ν) the direction-cosines of the line which lies in the tangent-plane to the surface and is perpendicular to the path at time t; the mass of the particle is taken as unity.

The acceleration of the particle consists of components vdv/ds along the tangent to the path and v^2/ρ along the principal normal; the latter component can be resolved into $(v^2/\rho)\sin\chi$ along the line whose direction-cosines are (λ, μ, ν) and $(v^2/\rho)\cos\chi$ along the normal to the surface. We have therefore the equations of motion

$$\begin{cases} v\dfrac{dv}{ds} = X\dfrac{dx}{ds} + Y\dfrac{dy}{ds} + Z\dfrac{dz}{ds} \quad \text{(A)}, \\ (v^2/\rho)\sin\chi = X\lambda + Y\mu + Z\nu \quad \text{(B)}, \end{cases}$$

* Some generalisations of the problem of two fixed centres will be found in a paper by Hiltebeitel, *Amer. Journ. Math.* XXXIII. (1911), p. 337.

† The earliest investigation of motion on a surface was Galileo's study of the motion of a heavy particle on an inclined plane (*Discourses*, Third Dialogue, 1638). The motion of a heavy particle moving in a horizontal circle on a sphere was examined by Huygens (*Horolog. oscill.*, 1673).

and these, together with the equation of the surface, are sufficient to determine the motion; for the equation of the surface may be regarded as giving z in terms of x and y, and by using this value for z we can express all the quantities occurring in equations (A) and (B) in terms of $x, y, \dot{x}, \dot{y}, \ddot{x}, \ddot{y}$: equations (A) and (B) thus become a system of differential equations of the fourth order for the determination of x and y in terms of t.

If the forces are conservative, the expression

$$-Xdx - Ydy - Zdz$$

will be the differential of a potential-energy function $V(x, y, z)$; equation (A) can therefore be integrated, and gives on integration the equation of energy

$$\tfrac{1}{2}v^2 + V(x, y, z) = c,$$

where c is a constant. Substituting the value of v^2 given by this equation in (B), we have

$$2(c-V)\frac{\sin\chi}{\rho} = X\lambda + Y\mu + Z\nu.$$

This is (on eliminating z by means of the equation to the surface) a differential equation of the second order between x and y, and is in fact the differential equation of the orbits on the surface.

The differential equations of motion on a surface are not integrable by quadratures in the general case: there are however two cases in which the problem can be formulated in such a way as to utilise results obtained in other connexions.

(i) *Motion under no forces.*

When no external forces act on the particle, equation (B) gives $\chi = 0$, so *the orbit is a geodesic on the surface**; the integral of energy shews that this geodesic is described with constant velocity.

Example. A particle moves under no forces on the fixed smooth ruled surface whose line of striction is the axis of z, the direction-cosines of the generator at the point z being

$$\sin a \cos\frac{z}{m}, \quad \sin a \sin\frac{z}{m}, \quad \cos a,$$

respectively. To determine the motion.

Let v denote the distance of the point on the surface whose coordinates are (x, y, z) from the line of striction, measured along the generator, and let $(0, 0, \zeta)$ be the coordinates of the point in which this generator meets the line of striction. Then we have

$$x = v \sin a \cos\frac{\zeta}{m}, \quad y = v \sin a \sin\frac{\zeta}{m}, \quad z = \zeta + v\cos a.$$

* This theorem is due to Euler, *Mechanica* (1736), II. cap. 4.

The kinetic energy of the particle is

$$T = \tfrac{1}{2}(\dot{x}^2 + \dot{y}^2 + \dot{z}^2)$$

$$= \tfrac{1}{2}\left(\dot{v}^2 + \dot{\zeta}^2 \frac{v^2}{m^2}\sin^2 a + \dot{\zeta}^2 + 2\dot{\zeta}\dot{v}\cos a\right).$$

We can take v and ζ as the two coordinates which define the position of the particle; it is evident that the coordinate ζ is ignorable, and the corresponding integral is

$$\frac{\partial T}{\partial \dot{\zeta}} = k, \qquad \text{where } k \text{ is a constant,}$$

or

$$\left(\frac{v^2}{m^2}\sin^2 a + 1\right)\dot{\zeta} + \dot{v}\cos a = k.$$

The integral of energy is

$$T = h, \qquad \text{where } h \text{ is a constant.}$$

Eliminating $\dot{\zeta}$ between these two integrals, we have

$$\dot{v}^2(v^2 + m^2) = 2hv^2 + (2h - k^2)\, m^2 \operatorname{cosec}^2 a.$$

If $\dot{v}$ is initially sufficiently large compared with $\dot{\zeta}$, the quantity $(2h - k^2)$ is positive; we shall suppose this to be the case, and shall write

$$(2h - k^2)\, m^2 \operatorname{cosec}^2 a = 2h\lambda^2, \qquad \text{where } \lambda \text{ is a new constant;}$$

the equation thus becomes

$$\dot{v}^2 (v^2 + m^2) = 2h(v^2 + \lambda^2).$$

The integration of this equation can be effected by introducing a real auxiliary variable u, defined by the equation

$$u = \int_0^v \{(m^2 + v^2)(\lambda^2 + v^2)\}^{-\frac{1}{2}}\, dv.$$

Writing $v = \lambda m x^{-\frac{1}{2}}$, this becomes

$$u = \int_x^\infty \{4x(x + \lambda^2)(x + m^2)\}^{-\frac{1}{2}}\, dx,$$

and this is equivalent to the equation

$$x = \wp(u) - e_1,$$

where the roots e_1, e_2, e_3 of the function $\wp(u)$ are real and are defined by the equations

$$e_1 - e_2 = \lambda^2, \quad e_1 - e_3 = m^2, \quad e_1 + e_2 + e_3 = 0.$$

The connexion between the variables v and u is therefore expressed by the equation

$$v = \lambda m \{\wp(u) - e_1\}^{-\frac{1}{2}}.$$

Substituting this value of v in the equation which connects v and t, we have

$$(2h)^{\frac{1}{2}} t = \int \frac{(e_1 - e_3)\{\wp(u) - e_2\}}{\wp(u) - e_1}\, du + \text{Constant}$$

$$= \int \{-e_3 + \wp(u + \omega_1)\}\, du + \text{Constant*}$$

$$= -e_3 u - \zeta(u + \omega_1) + \text{Constant}.$$

* Cf. Whittaker and Watson, *Modern Analysis*, § 20·33.

This equation expresses the time t in terms of the auxiliary variable u, and thus in conjunction with the equation

$$v = \lambda m \{\wp(u) - e_1\}^{-\frac{1}{2}},$$

gives the connexion between v and t.

(ii) *Motion on a developable surface.*

If the surface on which the particle moves is developable, we can utilise the known theorems that the arc s and the quantity $\frac{\sin \chi}{\rho}$ are unaltered by developing the surface on a plane: these results, applied to the equations of motion given above, shew that *if in the motion of a particle on a developable surface under any forces the surface is developed on a plane, the particle will describe the plane curve thus derived from its orbit with the same velocity as before, provided the force acting in the plane-motion is the same in amount and direction relative to the curve as the component of force in the tangent-plane to the surface in the surface-motion.*

Example 1. *A smooth particle is projected along the surface of a right circular cone, whose axis is vertical and vertex upwards, with the velocity due to the depth below the vertex. Prove that the path traced out on the cone, when developed into a plane, will be of the form*

$$r^{\frac{3}{2}} \sin \tfrac{3}{2}\theta = a^{\frac{3}{2}}. \qquad \text{(Coll. Exam.)}$$

For on developing the cone, the problem becomes the same as that of motion in a plane under a constant repulsive force from the origin, and with the velocity compatible with rest at the origin. We therefore have the integrals

$$\dot{r}^2 + r^2\dot{\theta}^2 = Cr, \qquad \text{where } C \text{ is a constant,}$$

$$r^2\dot{\theta} = h, \qquad \text{where } h \text{ is a constant.}$$

These equations give

$$\left(\frac{1}{r}\frac{dr}{d\theta}\right)^2 + 1 = \frac{Cr^3}{h^2}$$

$$= \frac{r^3}{a^3} \text{ say,} \qquad \text{where } a \text{ is a new constant,}$$

so if $u = \frac{1}{r}$, we have

$$\left(\frac{du}{d\theta}\right)^2 = \frac{1 - a^3u^3}{a^3u},$$

and therefore

$$\theta = \int \frac{a^{\frac{3}{2}} u^{\frac{1}{2}} du}{(1 - a^3u^3)^{\frac{1}{2}}}$$

$$= \tfrac{2}{3}\int \frac{dv}{(1 - v^2)^{\frac{1}{2}}} \qquad \text{where } v = u^{\frac{3}{2}} a^{\frac{3}{2}},$$

$$= \tfrac{2}{3} \sin^{-1} v,$$

which is equivalent to the equation

$$r^{\frac{3}{2}} \sin \tfrac{3}{2}\theta = a^{\frac{3}{2}}.$$

Example 2. If in the motion of a point P on a developable surface the tangent IP to the edge of regression describes areas proportional to the times, shew that the component of force perpendicular to IP and in the tangent-plane is proportional to $\frac{\rho}{IP^4}$, where ρ is the radius of curvature of the edge of regression. (Hazzidakis.)

55. *Motion on a surface of revolution; cases soluble in terms of circular and elliptic functions**.

The most important case of surface-motion which is soluble by quadratures is the motion of a particle on a smooth surface of revolution, under forces derivable from a potential-energy function which is symmetrical with respect to the axis of revolution of the surface.

Let the position of a point in space be defined by cylindrical coordinates (z, r, ϕ), where z is a coordinate measured parallel to the axis of the surface, r is the perpendicular distance of the point from this axis, and ϕ is the azimuthal angle made by r with a fixed plane through the axis. The equation of the surface will be a relation between z and r, say

$$r = f(z),$$

and the potential energy will be a function of z and r (it cannot involve ϕ, since it is symmetrical with respect to the axis), which for points on the surface can, on replacing r by its value $f(z)$, be expressed as a function of z only, say $V(z)$; the mass of the particle can be taken as unity.

The kinetic energy is, by § 18,

$$T = \tfrac{1}{2}(\dot{z}^2 + \dot{r}^2 + r^2\dot{\phi}^2)$$
$$= \tfrac{1}{2}\{[\{f'(z)\}^2 + 1]\dot{z}^2 + \{f(z)\}^2\dot{\phi}^2\}.$$

The coordinate ϕ is evidently ignorable; the corresponding integral is

$$\frac{\partial T}{\partial \dot{\phi}} = k, \qquad \text{where } k \text{ is a constant,}$$

or

$$\{f(z)\}^2\dot{\phi} = k;$$

this equation can be interpreted as the integral of angular momentum about the axis of the surface.

The equation of energy is

$$T + V = h, \qquad \text{where } h \text{ is a constant,}$$

and substituting for $\dot{\phi}$ in this equation from the preceding, we have

$$[\{f'(z)\}^2 + 1]\dot{z}^2 + k^2\{f(z)\}^{-2} + 2V(z) = 2h;$$

integrating this equation, we have

$$t = \int [\{f'(z)\}^2 + 1]^{\frac{1}{2}}[2h - 2V(z) - k^2\{f(z)\}^{-2}]^{-\frac{1}{2}}dz + \text{Constant}.$$

The relation between t and z is thus given by a quadrature; the values of r and ϕ are then obtained from the equation of the surface and the equation

$$\{f(z)\}^2\dot{\phi} = k,$$

respectively.

* The motion of a particle on a surface of revolution was investigated by Newton, *Principia*, Book I. Section 10.

We shall now discuss the motion on those surfaces for which this quadrature can be effected by means of known functions, when the axis of the surface is vertical (z being measured positively upwards) and gravity is the only external force, so that

$$V(z) = gz.$$

(i) *The circular cylinder.*

When the surface is the circular cylinder $r = a$, the above integral becomes

$$t = \int \left(2h - 2gz - \frac{k^2}{a^2}\right)^{-\frac{1}{2}} dz,$$

and if the origin of coordinates is so chosen that $2ha^2 = k^2$, we have

$$t = \int (-2gz)^{-\frac{1}{2}} dz,$$

or

$$z = -\tfrac{1}{2} g (t - t_0)^2, \qquad \text{where } t_0 \text{ is a constant.}$$

The equation

$$a^2 \dot{\phi} = k$$

then gives

$$\phi - \phi_0 = \frac{k}{a^2}(t - t_0), \qquad \text{where } \phi_0 \text{ is a constant.}$$

(ii) *The sphere.*

The case in which the surface is the sphere

$$r = (l^2 - z^2)^{\frac{1}{2}}$$

is called the *problem of the spherical pendulum**, and can be realised by supposing a heavy particle attached to a fixed point by a light rigid wire capable of moving freely about the point.

In this case the quadrature for t becomes

$$t = \int \left\{\frac{z^2}{l^2 - z^2} + 1\right\}^{\frac{1}{2}} \left\{2h - 2gz - \frac{k^2}{l^2 - z^2}\right\}^{-\frac{1}{2}} dz,$$

or

$$t = l \int \{(2h - 2gz)(l^2 - z^2) - k^2\}^{-\frac{1}{2}} dz.$$

The integral on the right-hand side of this equation is an elliptic integral, which we shall now reduce to Weierstrass' canonical form. Denote by z_1, z_2, z_3 the roots of the cubic

$$2(h - gz)(l^2 - z^2) - k^2 = 0;$$

since the expression

$$2(h - gz)(l^2 - z^2) - k^2$$

* Lagrange, *Mécanique Analytique.* The complete solution in terms of (Jacobian) elliptic functions was obtained by A. Tissot, *Liouville's Journal*, (1) XVII. (1852), p. 88: Jacobi's own solution of the problem of a rotating rigid body in terms of elliptic functions had been published previously, in 1839. The analysis connected with the spherical pendulum is essentially that for Lamé's equation of order 2.

is negative for the values l and $-l$ of z, and positive for very large positive values of z and also for the values of z which occur in the problem considered (which must necessarily lie between $-l$ and $+l$, since the particle is on the sphere) we see that one of the roots (say z_1) is greater than l and the other two (say z_2 and z_3, where $z_2 > z_3$) are between l and $-l$. The values of z in the actual motion will lie between z_2 and z_3, since for them the cubic must be positive.

Write $$z = \frac{h}{3g} + \frac{2l^2}{g}\zeta, \quad \text{where } \zeta \text{ is a new variable.}$$

and $$z_r = \frac{h}{3g} + \frac{2l^2}{g} e_r, \qquad (r = 1, 2, 3)$$

so that e_1, e_2, e_3 are new constants, which satisfy the relation

$$e_1 + e_2 + e_3 = \frac{g}{2l^2}\left(z_1 + z_2 + z_3 - \frac{h}{g}\right) = 0,$$

and also satisfy the inequalities $e_1 > e_2 > e_3$.

The relation between t and z now becomes

$$t = \int \{4(\zeta - e_1)(\zeta - e_2)(\zeta - e_3)\}^{-\frac{1}{2}} d\zeta,$$

or $$\zeta = \wp(t + \epsilon),$$

where ϵ is a constant of integration and the function $\wp$ is formed with the roots e_1, e_2, e_3.

Now when e_1, e_2, e_3 are real and in descending order of magnitude, $\wp(u)$ and $\wp'(u)$ are both real when u is real, in which case $\wp(u)$ is greater than e_1, and also when u is of the form $\omega_3 +$ a real quantity, where ω_3 is the half-period corresponding to the root e_3; in this latter case, $\wp(u)$ lies between e_2 and e_3. Since in the actual motion z lies between z_2 and z_3, it follows that ζ lies between e_2 and e_3, and therefore the constant ϵ must consist of an imaginary part ω_3 and a real part depending on the instant from which time is measured: by a suitable choice of the origin of time, we can take this real part of ϵ to be zero, and we then have

$$z = \frac{h}{3g} + \frac{2l^2}{g}\wp(t + \omega_3).$$

This equation gives the connexion between z and t. We have now to determine the azimuth ϕ. For this we have the equation

$$d\phi = \frac{k}{r^2}dt = \frac{k\,dt}{l^2 - z^2},$$

so $$\phi - \phi_0 = k\int \frac{dt}{l^2 - \left\{\frac{h}{3g} + \frac{2l^2}{g}\wp(t + \omega_3)\right\}^2},$$

where ϕ_0 is a constant of integration.

To effect the integration, we take λ and μ to be the (imaginary) values of $t + \omega_3$ corresponding to the values l and $-l$ of z respectively; so that λ and μ are new constants defined by the equations

$$l - \frac{h}{3g} = \frac{2l^2}{g}\wp(\lambda) \quad \text{and} \quad -l - \frac{h}{3g} = \frac{2l^2}{g}\wp(\mu);$$

these equations give

$$\wp'(\lambda) = \wp'(\mu) = \frac{ikg}{2l^3}.$$

The integral now becomes

$$\phi - \phi_0 = -\frac{kg^2}{4l^4}\int \frac{dt}{\{\wp(t+\omega_3) - \wp(\lambda)\}\{\wp(t+\omega_3) - \wp(\mu)\}}$$

$$= -\frac{kg}{4l^3}\int\left\{\frac{dt}{\wp(t+\omega_3) - \wp(\lambda)} - \frac{dt}{\wp(t+\omega_3) - \wp(\mu)}\right\}$$

$$= \frac{i}{2}\int\left\{\frac{\wp'(\lambda)\,dt}{\wp(t+\omega_3) - \wp(\lambda)} - \frac{\wp'(\mu)\,dt}{\wp(t+\omega_3) - \wp(\mu)}\right\}.$$

But* we have

$$\frac{\wp'(\lambda)}{\wp(z) - \wp(\lambda)} = \zeta(z-\lambda) - \zeta(z+\lambda) + 2\zeta(\lambda),$$

so

$$\int \frac{\wp'(\lambda)\,dt}{\wp(t+\omega_3) - \wp(\lambda)} = \log\frac{\sigma(t+\omega_3-\lambda)}{\sigma(t+\omega_3+\lambda)} + 2\zeta(\lambda)\,t,$$

and therefore

$$e^{2i(\phi-\phi_0)} = e^{2\{\zeta(\mu)-\zeta(\lambda)\}t}\,\frac{\sigma(t+\omega_3-\mu)\,\sigma(t+\omega_3+\lambda)}{\sigma(t+\omega_3+\mu)\,\sigma(t+\omega_3-\lambda)};$$

this equation expresses the angle ϕ as a function of t, and so completes the solution of the problem.

We see that when t increases by $2\omega_1$, ϕ_1 increases by

$$-2i\omega_1\{\zeta(\mu) - \zeta(\lambda)\} - 2i\eta_1(\lambda - \mu).$$

Example. When the bob of the spherical pendulum is executing periodic oscillations between two parallels on the sphere, shew that one of the points reached on the higher parallel, and the point on the lower parallel at which the bob arrives after a half-period, have a difference of azimuth which always lies between one and two right angles.

(Puiseux and Halphen.)

The problem of the spherical pendulum has been discussed from the standpoint of periodic solutions by F. R. Moulton, *Palermo Rend.* XXXII. (1911), p. 338.

(iii) *The paraboloid.*

Consider next the problem of motion on the paraboloid, whose equation is

$$r = 2a^{\frac{1}{2}}z^{\frac{1}{2}}.$$

In this case the quadrature for t becomes

$$t = \int (a+z)^{\frac{1}{2}}\left(2hz - 2gz^2 - \frac{k^2}{4a}\right)^{-\frac{1}{2}} dz.$$

* Cf. Whittaker and Watson, *Modern Analysis*, § 20·53, Ex. 2.

To obtain the solution of the problem in terms of elliptic functions, we introduce an auxiliary quantity v, defined by the equation

$$v = \int^z (a+z)^{-\frac{1}{2}} \left(2hz - 2gz^2 - \frac{k^2}{4a}\right)^{-\frac{1}{2}} dz.$$

If α and β (where $\alpha \geqslant \beta$) denote the roots of the quadratic

$$2hz - 2gz^2 - \frac{k^2}{4a} = 0,$$

we can write this integral in the form

$$v = \left(-\frac{g}{2}\right)^{-\frac{1}{2}} \int^z \{4(z+a)(z-\beta)(z-\alpha)\}^{-\frac{1}{2}} dz.$$

Define a new variable ζ by the equation

$$z = -(a+\alpha)\zeta + \frac{-a+\alpha+\beta}{3},$$

and let e_1, e_2, e_3 be the values of ζ corresponding to the values of $-a$, β, α respectively of z; then the integrals become

$$\left\{\frac{g(a+\alpha)}{2}\right\}^{\frac{1}{2}} v = \int_\zeta \{4(\zeta-e_1)(\zeta-e_2)(\zeta-e_3)\}^{-\frac{1}{2}} d\zeta,$$

and it is easily proved that the quantities e_1, e_2, e_3 satisfy the relations

$$e_1 + e_2 + e_3 = 0, \qquad e_1 > e_2 > e_3.$$

The auxiliary quantity v can now be replaced by an auxiliary quantity u, defined by the equation

$$v = \left\{\frac{2}{g(a+\alpha)}\right\}^{\frac{1}{2}} u,$$

and then the inversion of the integral gives

$$\zeta = \wp(u+\epsilon),$$

where ϵ is a constant of integration and the function $\wp$ is formed with the roots e_1, e_2, e_3, which are given by the equations

$$e_1 = \frac{2a+\alpha+\beta}{3(a+\alpha)}, \qquad e_2 = \frac{-a+\alpha-2\beta}{3(a+\alpha)}, \qquad e_3 = \frac{-a-2\alpha+\beta}{3(a+\alpha)}.$$

As in the actual motion z evidently lies between α and β, it follows that $\wp(u+\epsilon)$ lies between e_3 and e_2, and therefore (as we wish u to be real) the imaginary part of the constant ϵ must be the half-period ω_3; the real part can be taken to be zero, since it depends merely on the lower limit of the integral for u.

We have therefore

$$z = -(a+\alpha)\wp(u+\omega_3) + \frac{h-ag}{3g}, \qquad \text{since } \alpha+\beta = \frac{h}{g}$$

The equation to determine t is

$$t = \int (a+z)\,dv$$

$$= -\left\{\frac{2(a+\alpha)}{g}\right\}^{\frac{1}{2}} \int \{\wp(u+\omega_3) - e_1\}\,du$$

$$t = -\left\{\frac{2(a+\alpha)}{g}\right\}^{\frac{1}{2}} \{-\zeta(u+\omega_3) - e_1 u\},$$

and this equation gives t in terms of the auxiliary variable u.

Lastly, we have to determine the azimuth ϕ: for this we have

$$d\phi = \frac{k\,dt}{r^2} = \frac{k\,dt}{4az}$$

$$= \frac{k}{4a}\left\{\frac{2}{g(a+\alpha)}\right\}^{\frac{1}{2}} \cdot \frac{\wp(u+\omega_3) - e_1}{\wp(u+\omega_3) - \dfrac{-a+\alpha+\beta}{3(a+\alpha)}}\,du,$$

and therefore

$$\frac{4a}{k}\left\{\frac{g(a+\alpha)}{2}\right\}^{\frac{1}{2}}(\phi - \phi_0) = u + \left\{\frac{-a+\alpha+\beta}{3(a+\alpha)} - e_1\right\} \int \frac{du}{\wp(u+\omega_3) - \dfrac{-a+\alpha+\beta}{3(a+\alpha)}}$$

$$= u - \frac{a(a+\alpha)^{\frac{1}{2}}(-2g)^{\frac{1}{2}}}{k} \int \frac{\wp'(l)\,du}{\wp(u+\omega_3) - \wp(l)},$$

where ϕ_0 is a constant of integration, and l is an auxiliary constant defined by the equation

$$\wp(l) = \frac{-a+\alpha+\beta}{3(a+\alpha)}, \text{ so } \wp'(l) = \frac{k}{(-2g)^{\frac{1}{2}}(a+\alpha)^{\frac{3}{2}}}.$$

The equation can now be written

$$\phi - \phi_0 = \frac{ku}{a\{8g(a+\alpha)\}^{\frac{1}{2}}} - \frac{i}{2}\int \frac{\wp'(l)\,du}{\wp(u+\omega_3) - \wp(l)},$$

the integral of which is found (as in the problem of the spherical pendulum) to be

$$e^{2i(\phi-\phi_0)} = e^{\left[\frac{2ik}{a\{8g(a+\alpha)\}^{\frac{1}{2}}} + 2\zeta(l)\right]u} \frac{\sigma(u+\omega_3-l)}{\sigma(u+\omega_3+l)};$$

this equation expresses ϕ in terms of the auxiliary variable u, and so completes the solution.

(iv) *The cone.*

Consider next the cone, whose equation is

$$r = z\tan\alpha,$$

where α is the semi-vertical angle.

Since this is a developable surface, we can apply the theorem of § 54, and we see that the orbit of a particle on the cone under gravity becomes, when the cone is developed on a plane, the same as the orbit of a particle of unit mass in the plane under a force of constant magnitude $g \cos \alpha$ acting towards a fixed centre of force (namely the point on the plane which corresponds to the vertex of the cone). This (§ 48) is one of the known cases in which the problem of central motion can be solved in terms of elliptic functions, and this solution furnishes at once the solution of the problem of motion on the cone.

Example 1. Shew that the motion of a particle under gravity on a surface of revolution whose axis is vertical can also be solved in terms of elliptic functions when the surface is given by any one of the following equations

$$9ar^2 = z(z-3a)^2,$$

$$2r^4 + 3a^2r^2 - 2za^3 = 0,$$

$$(r^2 - az - \tfrac{1}{2}a^2)^2 = a^3 z. \qquad \text{(Kobb and Stäckel.)}$$

Example 2. Shew that the same problem can be solved in terms of elliptic functions when the surface is

$$(x^2+y^2)^3 + 2a^6 = 8a^3 z(x^2+y^2). \qquad \text{(Salkowski.)}$$

Example 3. Shew that if an algebraic surface of revolution is such that the equations of its geodesics can be expressed in terms of elliptic functions of a parameter, the surface must be such that r^2 and z can be expressed as rational functions of a parameter, i.e. the equation of the surface regarded as an equation between r^2 and z is the equation of a unicursal curve; where z, r, ϕ are the cylindrical coordinates of a point on the surface. (Kobb.)

Example 4. Shew that in the following cases of the motion of a particle on a surface of revolution, the trajectories are all closed curves:

1°. When the surface is a sphere, and the force is directed along the tangent to the meridian and proportional to $\operatorname{cosec}^2 \theta$, where θ is the angular distance from the particle to the pole. (The trajectories are in this case sphero-conics having one focus in the pole.)

2°. When the surface is a sphere, and the force is directed along the tangent to the meridian and proportional to $\tan \theta \sec^2 \theta$. (The trajectories in this case are sphero-conics having the pole as centre*.)

56. *Joukovsky's theorem.*

We shall now shew how to determine the potential-energy function under which a given family of curves on a surface can be described as the orbits of a particle constrained to move on the surface.

The three rectangular coordinates of a point on the surface can be expressed in terms of two parameters, say u and v, so that an element of arc ds on the surface is given in terms of the increments of u and v to which it corresponds by an equation of the form

$$ds^2 = E du^2 + 2F du dv + G dv^2,$$

where E, F, G are known functions of u and v.

* Darboux has examined the possibility of other cases, in *Bull. de la Soc. Math. de France*, v. (1877).

Let the family of curves which are to be the orbits under the required system of forces be defined by an equation

$$q(u, v) = \text{Constant},$$

and let

$$p(u, v) = \text{Constant}$$

denote the family of curves which is orthogonal to these.

Then instead of u and v we can take p and q as the two parameters which define the position of a point on the surface; let the line-element in this system of parameters be expressed by the equation

$$ds^2 = E'dq^2 + G'dp^2,$$

the term in $dqdp$ being absent, because the curves $p =$ Constant and $q =$ Constant cut at right angles: E' and G' being known functions of p and q.

The kinetic energy of a particle which moves on the surface is

$$T = \tfrac{1}{2}(E'\dot{q}^2 + G'\dot{p}^2);$$

the Lagrangian equations of motion are therefore

$$\begin{cases} \dfrac{d}{dt}(E'\dot{q}) - \tfrac{1}{2}\left(\dfrac{\partial E'}{\partial q}\dot{q}^2 + \dfrac{\partial G'}{\partial q}\dot{p}^2\right) = -\dfrac{\partial V}{\partial q}, \\ \dfrac{d}{dt}(G'\dot{p}) - \tfrac{1}{2}\left(\dfrac{\partial E'}{\partial p}\dot{q}^2 + \dfrac{\partial G'}{\partial p}\dot{p}^2\right) = -\dfrac{\partial V}{\partial p}, \end{cases}$$

where V denotes the unknown potential-energy function, which it is required to determine.

These equations are to be satisfied by the value $\dot{q} = 0$; they then become

$$\begin{cases} \dfrac{1}{2}\dfrac{\partial G'}{\partial q}\dot{p}^2 = \dfrac{\partial V}{\partial q}, \\ \dfrac{1}{2}\dfrac{\partial}{\partial p}(G'\dot{p}^2) = -\dfrac{\partial V}{\partial p}. \end{cases}$$

Eliminating $\dot{p}^2$, we have

$$\frac{\partial}{\partial p}\left(\frac{G'\dfrac{\partial V}{\partial q}}{\dfrac{\partial G'}{\partial q}}\right) + \frac{\partial V}{\partial p} = 0.$$

Integrating this equation, we have

$$\frac{G'\dfrac{\partial V}{\partial q}}{\dfrac{\partial G'}{\partial q}} + V = f(q), \quad \text{where } f \text{ is an arbitrary function,}$$

or
$$\frac{\partial}{\partial q}(VG') = \frac{\partial G'}{\partial q} f(q),$$
and therefore
$$V = \frac{g(p)}{G'} + \frac{1}{G'}\int \frac{\partial G'}{\partial q} f(q)\, dq,$$
where g denotes an arbitrary function.

Now $\frac{1}{G'}$ is $\Delta_1(p)$, the differential parameter* of the first order of the function p; and thus we have a theorem enunciated by Joukovsky in 1890, that *if $q =$ Constant is the equation of a family of curves on a surface, and $p =$ Constant denotes the family of curves orthogonal to these, then the curves $q =$ Constant can be freely described by a particle under the influence of forces derived from the potential-energy function*
$$V = \Delta_1(p)\, g(p) + \Delta_1(p) \int f(q) \frac{\partial}{\partial q}\left\{\frac{1}{\Delta_1(p)}\right\} dq,$$
where f and g are arbitrary functions, and Δ_1 denotes the first differential parameter.

The above equations give
$$\tfrac{1}{2}\dot{p}^2 = \frac{\partial V}{\partial q} \Big/ \frac{\partial G'}{\partial q} = -\frac{1}{G'} V + \frac{f(q)}{G'},$$
and hence the *equation of energy* in the motion is
$$\tfrac{1}{2} G' \dot{p}^2 + V = f(q).$$

Miscellaneous Examples.

1. A particle moves under gravity on the smooth cycloid whose equation is
$$s = 4a \sin\phi,$$
where s denotes the arc and ϕ the angle made by the tangent to the curve with the horizontal: shew that the motion is periodic, the period being $4\pi\sqrt{\frac{a}{g}}$.

2. A particle moves in a smooth circular tube under the influence of a force directed to a fixed point and proportional to the distance from the point. Shew that the motion is of the same character as in the pendulum-problem.

* If the line-element on a surface is given by the equation
$$ds^2 = E\,du^2 + 2F\,du\,dv + G\,dv^2,$$
the first differential parameter of a function $\phi(u, v)$ is given by the formula
$$\Delta_1(\phi) = \frac{1}{EG - F^2}\left\{E\left(\frac{\partial\phi}{\partial v}\right)^2 - 2F\frac{\partial\phi}{\partial u}\frac{\partial\phi}{\partial v} + G\left(\frac{\partial\phi}{\partial u}\right)^2\right\}.$$

The differential parameter is a *deformation-covariant* of the surface, i.e. when a change of variables is made from (u, v) to (u', v'), the differential parameter transforms into the expression formed in the same way with the new variables (u', v') and the corresponding new coefficients (E', F', G').

3. A particle moves in a straight line under the action of two centres of repulsive force of equal strength μ, each varying as the inverse square of the distance. Shew that, if the centres of force are at a distance $2c$ apart and the particle starts from rest at a distance kc, where $k < 1$, from the middle point of the line joining them, it will perform oscillations of period

$$2\sqrt{c^3(1-k^2)/\mu}\int_0^{\frac{\pi}{2}}(1-k^2\sin^2\theta)^{\frac{1}{2}}\,d\theta.$$

(Camb. Math. Tripos, Part I, 1899.)

4. A particle under the action of gravity travels in a smooth curved tube, starting from rest at a given point O of the tube. If the particle describes every arc OP in the same time that would be taken to slide down the corresponding chord OP, shew that the tube has the form of a lemniscate.

5. A particle is projected downwards along the concave side of the curve $y^3 + ax^2 = 0$ with a velocity $\frac{2}{3}(2ag)^{\frac{1}{2}}$ from the origin, the axis of x being horizontal; shew that the vertical component of the velocity is constant. (Nicomedi.)

6. A particle moves in a smooth tube in the form of the curve $r^2 = 2a^2\cos 2\theta$, under the action of two attractive forces, varying inversely as the cube of the distance, towards the two points on the initial line which are at a distance a from the pole. Prove that if the absolute force is μ, and the velocity at the node $2\mu^{\frac{1}{2}}/a$, the time of describing one loop of the curve is $\pi a^2/2\mu^{\frac{1}{2}}$. (Camb. Math. Tripos, Part I, 1898.)

7. A particle describes a space-curve under the influence of a force whose direction always intersects a given straight line. Shew that its velocity is inversely proportional to the distance of the particle from the line and to the cosine of the angle which the plane through the particle and the line makes with the normal plane to the orbit.

(Dainelli.)

8. A heavy particle is constrained to move on a straight line, which is made to rotate with constant angular velocity ω round a fixed vertical axis at given distance from it. Shew that the motion is given by the equation

$$r = Ae^{\omega t\cos\alpha} + Be^{-\omega t\cos\alpha},$$

where r is the distance of the particle from a fixed point on the line, α is the angle made by the line with the horizontal, and A, B are constants. (H. am Ende.)

9. A heavy particle is constrained to move on a straight line, which is made to rotate with given variable angular velocity round a fixed horizontal axis. Shew that the equation of motion is

$$\ddot{r} = \mp g\sin\alpha\sin\theta - r\dot{\theta}^2\sin^2\alpha \mp a\ddot{\theta}\sin\alpha,$$

where α is the angle between the line and the axis of rotation, θ the angle made with the vertical by the shortest distance a between the lines, and r the distance of the particle from the intersection of this shortest distance with the moving line.

(Vollhering.)

10. A particle slides in a smooth straight tube which is made to rotate with uniform angular velocity ω about a vertical axis: shew that, if the particle starts from relative rest from the point where the shortest distance between the axis and the tube meets the tube, the distance through which the particle moves along the tube in time t is

$$\frac{2g}{\omega^2}\cot\alpha\operatorname{cosec}\alpha\sinh^2\left(\tfrac{1}{2}\omega t\sin\alpha\right),$$

where α is the inclination of the tube to the vertical.

(Camb. Math. Tripos, Part I, 1899.)

11. A particle is constrained to move under no external forces in a plane circular tube which is constrained to rotate uniformly about any point in its plane. Shew that the motion of the particle in the tube is similar to that in the pendulum-problem.

12. A small bead is strung upon a smooth circular wire of radius a, which is constrained to rotate with uniform angular velocity ω about a point on itself. The bead is initially at the extremity of the diameter through the centre of rotation, and is projected with velocity $2\omega b$ relative to the wire: shew that the position of the bead at time t is given by the equation

$$\sin\phi = \operatorname{sn} b\omega t/a \qquad \text{(modulus } a/b)$$

or

$$\sin\phi = (b/a)\operatorname{sn}\omega t, \qquad \text{(modulus } b/a)$$

according as $a <$ or $> b$, ϕ being the angle which the radius vector to the bead makes with the diameter of the circle through the centre of rotation.

(Camb. Math. Tripos, Part I, 1900.)

13. Shew that the force perpendicular to the asymptote under which the curve

$$x^3 + y^3 = a^3$$

can be described is proportional to

$$xy\,(x^2+y^2)^{-3}.$$

14. A particle is acted on by a force whose components (X, Y) parallel to fixed axes are conjugate functions of the coordinates (x, y). Shew that the problem of its motion is always soluble by quadratures.

15. If (C) be a closed orbit described by a particle under the action of a central force, S the centre of force, O the centre of gravity of the curve (C), G the centre of gravity of the curve (C) on the supposition that the density at each point varies inversely as the velocity, shew that the points S, O, G are collinear and that $2SG = 3SO$.

(Laisant.)

16. Shew that the motion of a particle which is constrained to move in a plane, under a constant force directed to a point out of the plane, can be expressed by means of elliptic functions.

17. Shew that the curves

$$ax + by + c = xf\left(\frac{y}{x}\right),$$

where a, b, c are arbitrary constants and f is a given function, can be described under the same law of central force to the origin.

18. Shew that when a circle is described under a central attraction directed to a point in its circumference, the law of force is the inverse fifth power of the distance.

19. A particle describes the pedal of a circle, taken with respect to any point in its plane, under the influence of a centre of force at this point. Shew that the law of force is of the form

$$\frac{A}{r^4} + \frac{B}{r^5},$$

where A and B are constants.

Shew that the law of force is also of this form when the inverse of an ellipse with respect to a focus is described under a centre of force in the focus. (Curtis.)

20. Prove that, if when projected from $r=R$, $\theta=0$ with a velocity V in a direction making an angle a with the radius vector the path of a particle be $f(r, \theta, R, V, \sin a)=0$, the path with the same initial conditions but under the action of an additional central force $\frac{\mu}{r^3}$ is

$$f(r, n\theta, R, V(n^2\sin^2 a+\cos^2 a)^{\frac{1}{2}}, \quad n\sin a\,(n^2\sin^2 a+\cos^2 a)^{-\frac{1}{2}})=0,$$

where

$$n^2=f-\frac{\mu}{V^2R^2\sin^2 a}. \qquad \text{(Coll. Exam.)}$$

21. A particle of unit mass describes an orbit under an attractive force P to the origin and a transverse force T perpendicular to the radius vector. Prove that the differential equation of the orbit is given by

$$\frac{d^2u}{d\theta^2}+u=\frac{P}{h^2u^2}-\frac{T}{h^2u^3}\frac{du}{d\theta}, \quad \frac{dh^2}{d\theta}=2Tu^{-3}.$$

If the attractive force is always zero, and the particle moves in an equiangular spiral of angle a, prove that

$$T=\mu r^{2\sec^2 a-3} \quad \text{and} \quad h=(\mu\sin a\cos a)^{\frac{1}{2}}\,r^{\sec^2 a}$$

(Camb. Math. Tripos, Part I, 1901.)

22. A particle, acted on by a central force towards a point O varying as the distance, is projected from a point P so as to pass through a point Q such that OP is equal to OQ; shew that the least possible velocity of projection is $OP(\mu\sin POQ)^{\frac{1}{2}}$, where $\mu \,.\, OP$ is the force per unit mass at P. (Camb. Math. Tripos, Part I, 1901.)

23. Find a plane curve such that the curve and its pedal, with regard to some point in the plane, can be simultaneously described by particles under central forces to that point, in such a manner that the moving particles are always at corresponding points of the curve and the pedal; and find the law of force for the pedal curve.

(Camb. Math. Tripos, Part I, 1897.)

24. If $f(x, y)$ be a homogeneous function of one dimension, then the necessary and sufficient condition that the curve $f(x, y)=1$ be capable of description under acceleration tending to the origin and varying with the distance alone, is that f be subject to a condition of the form

$$\frac{\partial^2 f}{\partial x^2}+\frac{\partial^2 f}{\partial y^2}+F(r)=0.$$

Hence shew that the only curves of this class are necessarily included in the equation

$$r(A+B\sin\theta+C\cos\theta)=1.$$

Proceed to the discussion of the case wherein $f(x, y)$ is homogeneous and of n dimensions. (Coll. Exam.)

25. An ellipse of centre C is described under the influence of a centre of force at a point O on the major axis of the ellipse; shew that

$$nt=u-e\sin u,$$

where $2\pi/n$ is the periodic time, e is the ratio of CO to the semi-major axis, and u is the eccentric angle of the point reached by the particle in time t from the vertex.

26. Two free particles μ and M move in a plane under the influence of a central force to a fixed point O. Shew that the ratio of the velocity of the particle μ at an arbitrary point m of its path, to the velocity which is possessed at m by the central projection of M on the orbit of μ, is equal to the constant ratio of the areas described in unit time by the radii $O\mu$, OM, multiplied by the square of a certain function f of the coordinates of m, which expresses the ratio of OM, Om. (Dainelli.)

27. A particle is moving freely in a parabola under an attraction to the focus. Shew that, if at every instant a point be taken on the tangent through the particle, at distance $4a \cos \frac{1}{2}\theta/(\theta + \sin\theta)$ from the particle, this point will describe a central orbit about the focus, and the rate of description of areas will be the same as in the parabola; where $4a$ is the latus rectum, and θ the vertical angle of the particle measured from the apse line. (Camb. Math. Tripos, Part I, 1896.)

28. When a periodic comet is at its greatest distance from the sun, its velocity receives a small increment δv. Shew that the comet's least distance from the sun will be increased by the quantity

$$4\delta v . \{a^3(1-e)/\mu(1+e)\}^{\frac{1}{2}}. \qquad \text{(Coll. Exam.)}$$

29. If POP' is a focal chord of an elliptic path described round the sun, shew that the time from P' to P through perihelion is equal to the time of falling towards the sun from a distance $2a$ to a distance $a(1+\cos\alpha)$, where $\alpha = 2\pi - (u'-u)$, and $u'-u$ is the difference of the eccentric anomalies of the points P, P'. (Cayley.)

30. A particle moves in a plane under attractive forces $\mu/r^3r'^2$, $\mu/r^2r'^3$ along the radii r, r' drawn to two fixed points at distance $2d$ apart. Shew that, if it is projected with the velocity due to a fall from rest at infinity, a possible path is a circle with regard to which the two fixed centres are inverse points, and that, if the radius of this circle is a, the periodic time is

$$4\pi a^2 \mu^{-\frac{1}{2}}(a^2+d^2)^{\frac{1}{2}}. \qquad \text{(Coll. Exam.)}$$

31. A heavy particle is projected horizontally with a velocity v inside a smooth sphere at an angular distance α from the vertical diameter drawn downwards: shew that it will never fall below or never rise above its initial level according as

$$v^2 > \text{ or } < ag \sin\alpha \tan\alpha. \qquad \text{(Coll. Exam.)}$$

32. A particle is projected horizontally with velocity V along the interior of a smooth sphere of radius a from a point whose angular distance from the lowest point is α. Shew that the highest point of the spherical surface attained is at an angular distance β from the lowest point, where β is the smaller of the values of ψ, χ given respectively by the equations

$$\left.\begin{array}{r}(3\cos\psi - 2\cos\alpha)\,ag + V^2 = 0\\ (\cos\chi + \cos\alpha)\,V^2 - 2ag\sin^2\chi = 0\end{array}\right\}. \qquad \text{(Coll. Exam.)}$$

33. If the motion of a spherical pendulum of length a be wholly between the levels $\frac{3}{5}a$, $\frac{4}{5}a$ below the point of support, shew that at a time t after passing a point of greatest depth, the depth of the bob is

$$\tfrac{1}{5}a\{4 - \mathrm{sn}^2 t\sqrt{(13g/14a)}\} \qquad (\text{mod. } \sqrt{(7/65)})$$

and that a horizontal coordinate referred to the point of support as origin is determined by the equation

$$\ddot{x} = (gx/a)\{-\tfrac{12}{7} + \tfrac{3}{5}\mathrm{sn}^2 t\sqrt{(13g/14a)}\},$$

which is a case of Lamé's equation. (Coll. Exam.)

34. Shew that if a conical pendulum is executing small oscillations, the horizontal projection of the bob describes an ellipse whose axes turn in the sense of the motion with the angular velocity

$$\tfrac{3}{8}\theta_1\theta_0\sqrt{\frac{l}{g}},$$

where θ_0 is the angle of greatest deviation from the vertical, θ_1 that of least deviation, l the length of the pendulum, and g gravity. (Résal.)

35. A particle is constrained to move on the surface of a sphere, and is attracted to a fixed point M on the surface of the sphere with a force that varies as $r^{-2}(d^2-r^2)^{-\frac{3}{2}}$, where d is the diameter of the sphere and r is the rectilinear distance from the particle to M. If the position of the particle on the sphere be defined by its colatitude θ and longitude ϕ, with M as pole, shew that the equations of motion furnish the differential equation

$$\frac{1}{\sin^4\theta}\left(\frac{d\theta}{d\phi}\right)^2+\frac{1}{\sin^2\theta}=a\cot\theta+b,$$

where a and b are constants; and integrate this equation, shewing that the orbit is a sphero-conic.

36. A particle of mass m moves on the inner surface of a cone of revolution whose semi-vertical angle is α, under the action of a repulsive force $m\mu/r^3$ from the axis; the angular momentum of the particle about the axis being $m\sqrt{\mu}\tan\alpha$, shew that the path is an arc of a hyperbola whose eccentricity is $\sec\alpha$.

(Camb. Math. Tripos, Part I, 1897.)

37. Shew that the necessary central force to the vertex of a circular cone in order that the path on the cone may be a plane section is

$$\frac{A}{r^2}-\frac{B}{r^3}.$$ (Coll. Exam.)

38. A particle of unit mass moves on the inner surface of a paraboloid of revolution, latus rectum $4a$, under the action of a repulsive force μr from the axis, where r is the distance from the axis; shew that, if the particle is projected along the surface in a direction perpendicular to the axis with velocity $2a\mu^{\frac{1}{2}}$, it will describe a parabola.

(Coll. Exam.)

39. A smooth surface of revolution is formed by rotating the catenary $s=c\tan\phi$ about its axis of symmetry, and a particle is projected along its surface from a point distant b from that axis with velocity $h(a^2+b^2)^{\frac{1}{2}}/b^2$. The direction of projection is such that the component velocity perpendicular to the axis is h/b and the particle moves in contact with the surface, under the influence of a force of attraction $h^2(r^2+2a^2)/r^5$ in the direction of the perpendicular r to the axis. Shew that, if gravity be neglected, the projection of the path on a plane at right angles to the axis will have a polar equation

$$c\sinh\frac{r}{c}=a\theta.$$ (Coll. Exam.)

40. A particle moves on a smooth helicoid, $z=a\phi$, under the action of a force μr per unit mass directed at each point along the generator inwards, r being the distance from the axis of z. The particle is projected along the surface perpendicularly to the generator at a point where the tangent plane makes an angle α with the plane of xy, its velocity of projection being $\mu^{\frac{1}{2}}a$. Shew that the equation of the projection of its path on the plane of xy is

$$a^2/r^2=\sec^2\alpha\cosh^2(\phi\cos\alpha)-1.$$

(Camb. Math. Tripos, Part I, 1896.)

41. Shew that the problem of the motion of a particle under no forces on a ruled surface, whose generators cut the line of striction at a constant angle, and for which the ratio of the length of the common perpendicular to two adjacent generators to the angle between these generators is constant, can be solved by quadratures. (Astor.)

42. A particle (x, y, z) whose potential energy is $(ax^2+by^2+cz^2)$ is constrained to move on the sphere $x^2+y^2+z^2=1$. Determine the motion.

(C. Neumann, *Journal für Math.* LVI. (1859), p. 46.)

CHAPTER V

THE DYNAMICAL SPECIFICATION OF BODIES

57. *Definitions.*

Before proceeding to discuss those problems in the dynamics of rigid bodies which can be solved by quadratures, it is convenient to introduce and calculate a number of constants which can be assigned to a rigid body, and which depend on its constitution: it will be found that these constants determine the dynamical behaviour of the body.

Let any rigid body be considered; and let the particles of which (from the dynamical point of view) it is constituted be typified by a particle of mass m situated at a point whose coordinates referred to fixed rectangular axes are (x, y, z).

The quantity
$$\Sigma m (y^2 + z^2),$$
where the symbol Σ denotes a summation extended over all the particles of the system, is called the *moment of inertia* of the body about the axis Ox*. Similarly the moment of inertia about any other line is defined to be the sum of the masses of the particles of the body, each multiplied by the square of its perpendicular distance from the line. These summations are evidently in the case of ordinary rigid bodies equivalent to integrations; thus $\Sigma m (y^2 + z^2)$ is equivalent to $\iiint (y^2 + z^2)\, \rho\, dx\, dy\, dz$, where ρ is the *density*, or mass per unit volume, of the body at the point (x, y, z).

The quantity
$$\Sigma mxy$$
is called the *product of inertia* of the body about the axes Ox, Oy; and similarly the quantities Σmyz and Σmzx are the products of inertia about the other pairs of axes.

For the moments and products of inertia with reference to the coordinate axes, the notation
$$A = \Sigma m (y^2 + z^2), \quad B = \Sigma m (z^2 + x^2), \quad C = \Sigma m (x^2 + y^2),$$
$$F = \Sigma myz, \quad G = \Sigma mzx, \quad H = \Sigma mxy$$
will be generally used.

Two bodies whose moments of inertia about every line in space are equal to each other are said to be *equimomental*. It will be seen later that this involves also the equality of the products of inertia of the bodies with respect to any pair of orthogonal lines.

* Moments of inertia were first introduced by Huygens in his researches on the pendulum (*Horolog. oscill.*, 1673). The name is due to Euler.

If M denotes the mass of a body and if k is a quantity such that Mk^2 is equal to the moment of inertia of the body about a given line, the quantity k is called the *radius of gyration* of the body about the line.

In the case of a plane body, the moment of inertia about a line perpendicular to its plane is often spoken of as the moment of inertia about the *point* in which this line meets the plane.

58. *The moments of inertia of some simple bodies*.*

(i) *The rectangle.*

Let it be required to find the moment of inertia of a uniform rectangular plate, whose sides are of lengths $2a$ and $2b$ respectively, about a line through its centre O parallel to the sides of length $2a$. Taking this line as axis Ox, and a line through O parallel to the other sides as axis Oy, the required moment of inertia is

$$\Sigma my^2, \quad \text{or} \quad \int_{-b}^{b}\int_{-a}^{a} \sigma y^2 \, dx\, dy,$$

where σ is the mass per unit area of the plate, or the *surface-density* as it is frequently called; evaluating the integral, we have for the required moment of inertia

$$\tfrac{4}{3}\sigma ab^3, \quad \text{or} \quad \text{Mass of rectangle} \times \tfrac{1}{3}b^2.$$

The moment of inertia of a uniform rod, about a line through its middle point perpendicular to the rod, can be deduced from this result by regarding the rod as the limiting form of a rectangle in which the length of one pair of sides is indefinitely small. It follows that the moment of inertia in question is

$$\text{Mass of rod} \times \tfrac{1}{3}b^2,$$

where $2b$ is the length of the rod.

(ii) *The rectangular block.*

Consider next a uniform rectangular block whose edges are of lengths $2a$, $2b$, $2c$; let it be required to find the moment of inertia about an axis Ox passing through the centre O and parallel to the edges of length $2a$. This moment of inertia is

$$\Sigma m(y^2+z^2), \quad \text{or} \quad \int_{-a}^{a}\int_{-b}^{b}\int_{-c}^{c} \rho\,(y^2+z^2)\, dz\, dy\, dx,$$

where ρ is the density. Evaluating the integral, we have for the moment of inertia

$$\frac{8\rho abc}{3}(b^2+c^2), \quad \text{or} \quad \text{Mass of block} \times \tfrac{1}{3}(b^2+c^2).$$

* For practical purposes the moments of inertia of a body are determined experimentally; convenient apparatus is described by W. H. Derriman, *Phil. Mag.* v. (1903), p. 648, and by W. R. Cassie, *Phys. Soc. Proc.* XXI. (1909), p. 497. See also Amsler, *Zeits. Instrument.* XLVI. (1926), pp. 16 and 19.

(iii) *The ellipse and the circle.*

Let it now be required to find the moment of inertia of a uniform elliptic plate whose equation is

$$\frac{x^2}{a^2} + \frac{y^2}{b^2} = 1,$$

about the axis of x. It is

$$\int_{-a}^{a}\int_{-\frac{b}{a}(a^2-x^2)^{\frac{1}{2}}}^{\frac{b}{a}(a^2-x^2)^{\frac{1}{2}}} \sigma y^2 dy dx, \text{ where } \sigma \text{ is the surface-density.}$$

Evaluating the integral, we have for the required moment of inertia

$$\tfrac{1}{4}\pi ab^3\sigma, \quad \text{or} \quad \text{Mass of ellipse} \times \tfrac{1}{4}b^2.$$

The moment of inertia of a circle of radius b about a diameter is therefore

$$\text{Mass of circle} \times \tfrac{1}{4}b^2.$$

(iv) *The ellipsoid and the sphere.*

The moment of inertia of a uniform solid ellipsoid of density ρ, whose equation is

$$\frac{x^2}{a^2} + \frac{y^2}{b^2} + \frac{z^2}{c^2} = 1,$$

about the axis of x is similarly

$$\iiint \rho\,(y^2 + z^2)\,dx dy dz, \text{ integrated throughout the ellipsoid.}$$

To evaluate this integral, write

$$x = a\xi, \quad y = b\eta, \quad z = c\zeta,$$

where ξ, η, ζ are new variables: the integral becomes

$$\rho ab^3c \iiint \eta^2 d\xi d\eta d\zeta + \rho abc^3 \iiint \zeta^2 d\xi d\eta d\zeta,$$

where the integration is now taken throughout a sphere whose equation is

$$\xi^2 + \eta^2 + \zeta^2 = 1.$$

Since the integrals

$$\iiint \xi^2 d\xi d\eta d\zeta, \quad \iiint \eta^2 d\xi d\eta d\zeta, \quad \text{and} \quad \iiint \zeta^2 d\xi d\eta d\zeta$$

are evidently equal, the required moment of inertia can be written in the form

$$\rho abc\,(b^2 + c^2) \iiint \xi^2 d\xi d\eta d\zeta,$$

or
$$\pi\rho abc\,(b^2 + c^2) \int_{-1}^{1} \xi^2 (1 - \xi^2)\,d\xi,$$

or
$$\tfrac{4}{15}\pi\rho abc\,(b^2 + c^2),$$

or
$$\text{Mass of ellipsoid} \times \tfrac{1}{5}(b^2 + c^2).$$

The moment of inertia of a uniform sphere of radius a about a diameter is therefore

$$\text{Mass of sphere} \times \tfrac{2}{5}a^2.$$

(v) *The triangle.*

Let it now be required to find the moment of inertia of a uniform triangular plate of surface-density σ, with respect to any line in its plane; the position of the line can be specified by the lengths α, β, γ of the perpendiculars drawn to it from the vertices of the triangle.

Taking (x, y, z) to be the areal coordinates of a point of the plate, the perpendicular distance from this point to the given line is $(\alpha x + \beta y + \gamma z)$, and the required moment of inertia is therefore

$$\sigma \iint (\alpha x + \beta y + \gamma z)^2 \, dS,$$

where dS denotes an element of area of the plate.

Now if Y denotes the length of the perpendicular from the point (x, y, z) on the side c of the triangle, and if X denotes the length intercepted on the side c between the vertex A and the foot of this perpendicular, we have

$$Y = zb \sin A$$

and $$X \sin A - Y \cos A = \text{perpendicular from } (x, y, z) \text{ on the side } b$$
$$= yc \sin A.$$

We have therefore

$$dydz = \frac{\partial (y, z)}{\partial (X, Y)} \, dX \, dY = \frac{1}{bc \sin A} \, dX \, dY = \frac{1}{2\Delta} \, dS,$$

where Δ denotes the area of the triangle. Hence the integral $\iint y^2 dS$, where the integration is extended over the area of the triangle, can be written in the form $2\Delta \iint y^2 dydz$, where the integration is extended over all positive values of y and z whose sum is less than unity: this is equal to

$$2\Delta \int_0^1 y^2 (1 - y) \, dy,$$

or $\frac{1}{6}\Delta$. By symmetry, the integrals $\iint x^2 dS$ and $\iint z^2 dS$ have the same value, and a similar calculation shews that the integrals

$$\iint yz \, dS, \quad \iint zx \, dS, \quad \iint xy \, dS$$

each have the value $\frac{1}{12}\Delta$.

Substituting these values in the integral $\sigma \iint (\alpha x + \beta y + \gamma z)^2 dS$, the moment of inertia of the triangle about the given line becomes

$$\tfrac{1}{6}\sigma\Delta\,(\alpha^2 + \beta^2 + \gamma^2 + \beta\gamma + \gamma\alpha + \alpha\beta),$$

or $$\tfrac{1}{3} \times \text{Mass of triangle} \times \left\{\left(\frac{\beta+\gamma}{2}\right)^2 + \left(\frac{\gamma+\alpha}{2}\right)^2 + \left(\frac{\alpha+\beta}{2}\right)^2\right\}.$$

But this expression evidently represents the moment of inertia about the given line of three particles situated respectively at the middle points of the sides of the triangle, the mass of each particle being one-third the mass of the triangle; the triangle is therefore equimomental to this set of three particles.

Example. Shew that a uniform solid tetrahedron of mass M is equimomental to a set of five particles, four of which are each of mass $\frac{1}{20}M$ and are situated at the vertices of the tetrahedron, while the fifth particle is at the centre of gravity of the tetrahedron and is of mass $\frac{4}{5}M$.

59. *Derivation of the moment of inertia about any axis when the moment of inertia about a parallel axis through the centre of gravity is known.*

The moments of inertia found in the preceding article were for the most part taken with respect to lines specially related to the bodies concerned: these results can however be applied to determine the moments of inertia of the same bodies with respect to other lines, by means of a theorem which will now be given.

Let $f(x, y, z, \dot{x}, \dot{y}, \dot{z}, \ddot{x}, \ddot{y}, \ddot{z})$ be any polynomial (with no constant term) of the second degree in the coordinates and the components of velocity and acceleration of a particle of mass m. Let $(\bar{x}, \bar{y}, \bar{z})$ denote the coordinates of the centre of gravity of a body which is formed of such particles, and write

$$x = \bar{x} + x_1, \qquad y = \bar{y} + y_1, \qquad z = \bar{z} + z_1.$$

If now we substitute these values for x, y, z, respectively, in the function f, we obtain the following classes of terms:

(1) Terms which do not involve x_1, y_1, z_1: these terms together evidently give

$$f(\bar{x}, \bar{y}, \bar{z}, \dot{\bar{x}}, \dot{\bar{y}}, \dot{\bar{z}}, \ddot{\bar{x}}, \ddot{\bar{y}}, \ddot{\bar{z}}).$$

(2) Terms which do not involve $\bar{x}, \bar{y}, \bar{z}$: these terms give

$$f(x_1, y_1, z_1, \dot{x}_1, \dot{y}_1, \dot{z}_1, \ddot{x}_1, \ddot{y}_1, \ddot{z}_1).$$

(3) Terms which are linear in $x_1, y_1, z_1, \dot{x}_1, \dot{y}_1, \dot{z}_1, \ddot{x}_1, \ddot{y}_1, \ddot{z}_1$; when the expression $\Sigma m f(x, y, z, \dot{x}, \dot{y}, \dot{z}, \ddot{x}, \ddot{y}, \ddot{z})$ is formed, the summation being taken over all the particles of the body, these terms will vanish in consequence of the relations

$$\Sigma m x_1 = 0, \qquad \Sigma m y_1 = 0, \qquad \Sigma m z_1 = 0.$$

We have therefore the equation

$$\Sigma mf(x, y, z, \dot{x}, \dot{y}, \dot{z}, \ddot{x}, \ddot{y}, \ddot{z}) = \Sigma mf(x_1, y_1, z_1, \dot{x}_1, \dot{y}_1, \dot{z}_1, \ddot{x}_1, \ddot{y}_1, \ddot{z}_1) + f(\bar{x}, \bar{y}, \bar{z}, \dot{\bar{x}}, \dot{\bar{y}}, \dot{\bar{z}}, \ddot{\bar{x}}, \ddot{\bar{y}}, \ddot{\bar{z}}) . \Sigma m,$$

and consequently the value of the expression Σmf, taken with respect to any system of coordinate axes, is equal to its value taken with respect to a parallel set of axes through the centre of gravity of the body, together with the mass of the body multiplied by the value of the function f at the centre of gravity, taken with respect to the original system of axes.

From this it immediately follows that *the moments and products of inertia of a body with respect to any axes are equal to the corresponding moments and products of inertia with respect to a set of parallel axes through the centre of gravity of the body, together with the corresponding moments and products of inertia, with respect to the original axes, of a particle of mass equal to that of the body and placed at the centre of gravity.*

As an example of this result, let it be required to determine the moment of inertia of a straight uniform rod of mass M and length l about a line through one extremity perpendicular to the rod. It follows from the last article that the moment of inertia about a parallel line through the centre of the rod is $\frac{1}{3}M\left(\frac{l}{2}\right)^2$; and hence, applying the above result, we see that the required moment of inertia is

$$M\left(\frac{l}{2}\right)^2 + \frac{1}{3}M\left(\frac{l}{2}\right)^2;$$

or

$$\frac{1}{3}Ml^2.$$

60. *Connexion between moments of inertia with respect to different sets of axes through the same origin.*

The result of the last article enables us to find the moments of inertia of a given body with respect to any set of axes, when the moments of inertia are already known with respect to a set of axes parallel to these. We shall now shew how the moments of inertia of a body with respect to any set of rectangular axes can be found when the moments of inertia are known with respect to another set of rectangular axes having the same origin.

Let A, B, C, F, G, H be the moments and products of inertia with respect to a set of axes $Oxyz$, and let $Ox'y'z'$ be another set of rectangular axes having the same origin O; the direction-cosines of either set of axes with respect to the other will be supposed to be given by the scheme

	x'	y'	z'
x	l_1	l_2	l_3
y	m_1	m_2	m_3
z	n_1	n_2	n_3

If the moments and products of inertia with respect to the axes $Ox'y'z'$ are denoted by A', B', C', F', G', H', we have

$A' = \Sigma m(y'^2 + z'^2)$, where the summation is extended over all the particles of the body,

$$\begin{aligned}
&= \Sigma m\{(l_2x + m_2y + n_2z)^2 + (l_3x + m_3y + n_3z)^2\}\\
&= \Sigma m\{x^2(l_2^2 + l_3^2) + y^2(m_2^2 + m_3^2) + z^2(n_2^2 + n_3^2) + 2yz(m_2n_2 + m_3n_3)\\
&\qquad + 2zx(n_2l_2 + n_3l_3) + 2xy(l_2m_2 + l_3m_3)\}\\
&= \Sigma m\{x^2(m_1^2 + n_1^2) + y^2(n_1^2 + l_1^2) + z^2(l_1^2 + m_1^2) - 2m_1n_1yz - 2n_1l_1zx - 2l_1m_1xy\}\\
&= \Sigma m\{l_1^2(y^2 + z^2) + m_1^2(z^2 + x^2) + n_1^2(x^2 + y^2) - 2m_1n_1yz - 2n_1l_1zx - 2l_1m_1xy\}\\
&= Al_1^2 + Bm_1^2 + Cn_1^2 - 2Fm_1n_1 - 2Gn_1l_1 - 2Hl_1m_1,
\end{aligned}$$

and similarly

$$B' = Al_2^2 + Bm_2^2 + Cn_2^2 - 2Fm_2n_2 - 2Gn_2l_2 - 2Hl_2m_2,$$

$$C' = Al_3^2 + Bm_3^2 + Cn_3^2 - 2Fm_3n_3 - 2Gn_3l_3 - 2Hl_3m_3.$$

We have also

$$\begin{aligned}
F' &= \Sigma my'z'\\
&= \Sigma m(l_2x + m_2y + n_2z)(l_3x + m_3y + n_3z)\\
&= l_2l_3 . \Sigma mx^2 + m_2m_3 . \Sigma my^2 + n_2n_3 . \Sigma mz^2 + (m_2n_3 + m_3n_2) . \Sigma myz\\
&\qquad + (n_2l_3 + n_3l_2) . \Sigma mzx + (l_2m_3 + l_3m_2) . \Sigma mxy\\
&= \tfrac{1}{2}l_2l_3(B + C - A) + \tfrac{1}{2}m_2m_3(C + A - B) + \tfrac{1}{2}n_2n_3(A + B - C)\\
&\qquad + (m_2n_3 + m_3n_2)F + (n_2l_3 + n_3l_2)G + (l_2m_3 + l_3m_2)H,
\end{aligned}$$

or

$$-F' = Al_2l_3 + Bm_2m_3 + Cn_2n_3 - F(m_2n_3 + m_3n_2) - G(l_3n_2 + l_2n_3) - H(l_2m_3 + l_3m_2),$$

and similarly

$$-G' = Al_3l_1 + Bm_3m_1 + Cn_3n_1 - F(m_3n_1 + m_1n_3) - G(l_1n_3 + l_3n_1) - H(l_3m_1 + l_1m_3),$$

$$-H' = Al_1l_2 + Bm_1m_2 + Cn_1n_2 - F(m_1n_2 + m_2n_1) - G(l_2n_1 + l_1n_2) - H(l_1m_2 + l_2m_1).$$

The quantities A', B', C', F', G', H' are thus determined; these results, combined with those of the last article, are sufficient to determine the moments and products of inertia of a given body with respect to any set of rectangular axes when the moments and products of inertia with respect to any other set of rectangular axes are known.

Example. If the origin of coordinates is at the centre of gravity of the body, prove that the moments and products of inertia with respect to three mutually orthogonal and intersecting lines whose coordinates are

$$(l_1, m_1, n_1, \lambda_1, \mu_1, \nu_1), \quad (l_2, m_2, n_2, \lambda_2, \mu_2, \nu_2), \quad (l_3, m_3, n_3, \lambda_3, \mu_3, \nu_3)$$

are $A' + M(\lambda_1^2 + \mu_1^2 + \nu_1^2)$ etc. and $F' - M(\lambda_3\lambda_1 + \mu_3\mu_1 + \nu_3\nu_1)^{\frac{1}{2}}(\lambda_1\lambda_2 + \mu_1\mu_2 + \nu_1\nu_2)^{\frac{1}{2}}$ etc., where A', B', C', F', G', H' have the same values as above and M is the mass of the body. (Coll. Exam.)

61. *The principal axes of inertia; Cauchy's momental ellipsoid.*

If now we consider the quadric surface whose equation is

$$Ax^2 + By^2 + Cz^2 - 2Fyz - 2Gzx - 2Hxy = 1,$$

where A, B, C, F, G, H are the moments and products of inertia of a given body with respect to the axes of reference $Oxyz$, it follows from the equation

$$A' = Al_1^2 + Bm_1^2 + Cn_1^2 - 2Fm_1n_1 - 2Gn_1l_1 - 2Hl_1m_1,$$

that the reciprocal of the square of any radius vector of the quadric is equal to the moment of inertia of the body about this radius. The quadric is therefore the same whatever be the axes of reference provided the origin is unchanged, and consequently its equation referred to any other rectangular axes $Ox'y'z'$ having the same origin is

$$A'x'^2 + B'y'^2 + C'z'^2 - 2F'y'z' - 2G'z'x' - 2H'x'y' = 1;$$

where A', B', C', F', G', H' are the moments and products of inertia with respect to these axes.

This quadric is called the *momental ellipsoid* of the body at the point O; its principal axes are called the *principal axes of inertia* of the body at O; the equation of the quadric referred to these axes contains no product-terms, and therefore the products of inertia with respect to them are zero: and the moments of inertia with respect to these axes are called the *principal moments of inertia* of the body at the point O*.

The momental ellipsoid is also called the *ellipsoid of inertia*; its polar reciprocal with regard to its centre is another ellipsoid, which is sometimes called the *ellipsoid of gyration*.

Example. The height of a solid homogeneous right circular cone is half the radius of its base. Shew that its momental ellipsoid at the vertex is a sphere.

62. *Calculation of the angular momentum of a moving rigid body.*

We shall now shew how the angular momentum of a moving rigid body about any line, at any instant of its motion, can be determined.

Let M be the mass of the body, $(\bar{x}, \bar{y}, \bar{z})$ the coordinates of its centre of gravity G, and $(\bar{u}, \bar{v}, \bar{w})$ the components of velocity of the point G, at the instant t, resolved along any (fixed or moving) rectangular axes $Oxyz$ whose origin O is fixed; and let $(\omega_1, \omega_2, \omega_3)$ be the components of the angular velocity of the body about G, resolved along axes $Gx_1y_1z_1$, parallel to the axes $Oxyz$ and passing through G. Let m denote a typical particle of the body, and let (x, y, z) be its coordinates and (u, v, w) its components of velocity at the instant t; and write

$$x = \bar{x} + x_1, \qquad y = \bar{y} + y_1, \qquad z = \bar{z} + z_1,$$
$$u = \bar{u} + u_1, \qquad v = \bar{v} + v_1, \qquad w = \bar{w} + w_1,$$

* The existence of principal axes was discovered by Euler, *Mém. de Berl.*, 1750, 1758, and by J. A. Segner, *Specimen Th. Turbinum*, 1755. The momental ellipsoid was introduced by Cauchy in 1827, *Exerc. de math.* I. p. 93.

so in virtue of the properties of the centre of gravity we have

$$\Sigma m x_1 = 0, \quad \Sigma m y_1 = 0, \quad \Sigma m z_1 = 0;$$

moreover since (§ 17) we have

$$u_1 = z_1\omega_2 - y_1\omega_3, \quad v_1 = x_1\omega_3 - z_1\omega_1, \quad w_1 = y_1\omega_1 - x_1\omega_2,$$

it follows that

$$\Sigma m u_1 = 0, \quad \Sigma m v_1 = 0, \quad \Sigma m w_1 = 0.$$

If h_3 denotes the angular momentum of the body about the axis Oz, we have therefore

$$\begin{aligned} h_3 &= \Sigma m\,(xv - yu) \\ &= \Sigma m\,\{(\bar{x} + x_1)(\bar{v} + v_1) - (\bar{y} + y_1)(\bar{u} + u_1)\} \\ &= \Sigma m\,(\bar{x}\bar{v} - \bar{y}\bar{u}) + \Sigma m\,(x_1 v_1 - y_1 u_1) \\ &= M\,(\bar{x}\bar{v} - \bar{y}\bar{u}) + \Sigma m\,(x_1^2\omega_3 - x_1 z_1\omega_1 - y_1 z_1\omega_2 + y_1^2\omega_3) \\ &= M\,(\bar{x}\bar{v} - \bar{y}\bar{u}) - G\omega_1 - F\omega_2 + C\omega_3, \end{aligned}$$

where A, B, C, F, G, H are the moments and products of inertia of the body with respect to the axes $Gx_1y_1z_1$.

Similarly the angular momenta about the axes Ox and Oy respectively are

$$h_1 = M\,(\bar{y}\bar{w} - \bar{z}\bar{v}) + A\omega_1 - H\omega_2 - G\omega_3,$$
$$h_2 = M\,(\bar{z}\bar{u} - \bar{x}\bar{w}) - H\omega_1 + B\omega_2 - F\omega_3.$$

The angular momentum about any other line through the origin can be found (§ 39) by resolving these angular momenta along the line in question.

Corollary. If the body is constrained to turn round one of its points, which is fixed in space, it is unnecessary to introduce the centre of gravity. For let $(\omega_1, \omega_2, \omega_3)$ be the components of the angular velocity of the body about the fixed point with respect to any rectangular axes (fixed or moving) which have the fixed point as origin, and let A, B, C, F, G, H denote the moments and products of inertia with respect to these axes. The components of velocity (u, v, w) with respect to these axes of a particle m whose coordinates are (x, y, z) are (§ 17)

$$u = z\omega_2 - y\omega_3, \quad v = x\omega_3 - z\omega_1, \quad w = y\omega_1 - x\omega_2,$$

and the angular momentum about the axis of z, which is $\Sigma m\,(xv - yu)$, can therefore be written in the form

$$\Sigma m\,(x^2\omega_3 - xz\omega_1 - yz\omega_2 + y^2\omega_3)$$

or

$$- G\omega_1 - F\omega_2 + C\omega_3.$$

Similarly the angular momenta of the body about the axes of x and y respectively are

$$A\omega_1 - H\omega_2 - G\omega_3$$

and

$$- H\omega_1 + B\omega_2 - F\omega_3.$$

63. *Calculation of the kinetic energy of a moving rigid body.*

The kinetic energy of a rigid body which is in motion can be calculated in the same way as the angular momenta. If the general theorem obtained in § 59 is applied to the case in which the polynomial $f(x, y, z, \dot{x}, \dot{y}, \dot{z}, \ddot{x}, \ddot{y}, \ddot{z})$ has the form $(\dot{x}^2 + \dot{y}^2 + \dot{z}^2)$, we immediately obtain the result that *the kinetic energy of a moving rigid body of mass M is equal to the kinetic energy of a particle of mass M which moves with the centre of gravity of the body, together with the kinetic energy of the motion of the body relative to its centre of gravity.*

To determine the kinetic energy of the motion of the body relative to its centre of gravity G, take any rectangular axes (whose directions may be fixed or moving) having their origin at G; let $(\omega_1, \omega_2, \omega_3)$ be the components of the angular velocity of the body about G, relative to these axes, and let (x, y, z) denote the coordinates of a typical particle m of the body referred to these axes. The components of velocity of the particle parallel to these axes, in the motion relative to G, are (§ 17)

$$z\omega_2 - y\omega_3, \quad x\omega_3 - z\omega_1, \quad y\omega_1 - x\omega_2,$$

and therefore the kinetic energy of the motion relative to the centre of gravity is

$$\tfrac{1}{2}\Sigma m \{(z\omega_2 - y\omega_3)^2 + (x\omega_3 - z\omega_1)^2 + (y\omega_1 - x\omega_2)^2\},$$

or

$$\tfrac{1}{2}(A\omega_1^2 + B\omega_2^2 + C\omega_3^2 - 2F\omega_2\omega_3 - 2G\omega_3\omega_1 - 2H\omega_1\omega_2),$$

where A, B, C, F, G, H are the moments and products of inertia relative to the axes.

This expression may (by use of § 60) be interpreted as half the square of the resultant angular velocity of the body in the motion relative to the centre of gravity, multiplied by the moment of inertia of the body about the instantaneous axis of rotation in this motion.

Corollary. If one of the points of the body is fixed in space, it is not necessary to introduce the centre of gravity. For let $(\omega_1, \omega_2, \omega_3)$ denote the components of angular velocity of the body about the fixed point O resolved along any rectangular axes (fixed or moving) $Oxyz$ which have the point O as origin, and let (x, y, z) be the coordinates of a typical particle m of the body referred to these axes. The components of velocity of the particle are (§ 17)

$$z\omega_2 - y\omega_3, \quad x\omega_3 - z\omega_1, \quad y\omega_1 - x\omega_2,$$

and so as before we see that the kinetic energy of the motion is

$$\tfrac{1}{2}(A\omega_1^2 + B\omega_2^2 + C\omega_3^2 - 2F\omega_2\omega_3 - 2G\omega_3\omega_1 - 2H\omega_1\omega_2),$$

where A, B, C, F, G, H denote the moments and products of inertia of the body with respect to the axes $Oxyz$.

From this it follows that if one of the coordinate axes—say the axis of x —is the instantaneous axis of rotation of the body, the kinetic energy is $\frac{1}{2}A\omega_1^2$; and hence, since the directions of the axes can be arbitrarily chosen, the kinetic energy of any body moving about one of its points, which is fixed, is $\frac{1}{2}I\omega^2$, where I is the moment of inertia of the body about the instantaneous axis of rotation, and ω is the angular velocity of the body about this axis.

Example. A lamina can turn freely about a horizontal axis in its own plane, and the axis turns about a fixed vertical line, which it intersects. If ϕ be the azimuth of the horizontal axis, and ψ the inclination of the plane of the lamina to the vertical, shew that the kinetic energy is

$$\tfrac{1}{2}A(\dot{\psi}^2+\dot{\phi}^2\sin^2\psi)+\tfrac{1}{2}B\dot{\phi}^2+H\dot{\psi}\dot{\phi}\cos\psi,$$

where A, B, H are the moments and product of inertia of the lamina about the horizontal axis and a perpendicular to it at the point of intersection with the vertical. (Coll. Exam.)

64. *Independence of the motion of the centre of gravity and the motion relative to it.*

The result of the last article shews that the kinetic energy of a moving body can be regarded as consisting of two parts, of which one depends on the motion of the centre of gravity and the other is the kinetic energy of the motion relative to the centre of gravity. We shall now shew that these two parts of the motion of the body can be treated quite independently of each other*.

Let a rigid body of mass M be in motion under the influence of any forces As coordinates defining its position we can take the three rectangular coordinates (x, y, z) of its centre of gravity G, relative to axes fixed in space, and the three Eulerian angles (θ, ϕ, ψ) which specify the position, relative to axes fixed in direction, of any three orthogonal lines, intersecting in G, which are fixed in the body and move with it. The kinetic energy is therefore

$$T=\tfrac{1}{2}M(\dot{x}^2+\dot{y}^2+\dot{z}^2)+f(\theta, \phi, \psi, \dot{\theta}, \dot{\phi}, \dot{\psi}),$$

where $f(\theta, \phi, \psi, \dot{\theta}, \dot{\phi}, \dot{\psi})$ denotes the kinetic energy of the motion relative to G.

Let $$X\delta x+Y\delta y+Z\delta z+\Theta\delta\theta+\Phi\delta\phi+\Psi\delta\psi$$

denote the work done on the body by the external forces in an arbitrary displacement $(\delta x, \delta y, \delta z, \delta\theta, \delta\phi, \delta\psi)$ of the body. The Lagrangian equations of motion are

$$M\ddot{x}=X,\quad M\ddot{y}=Y,\quad M\ddot{z}=Z,$$

$$\frac{d}{dt}\left(\frac{\partial f}{\partial\dot{\theta}}\right)-\frac{\partial f}{\partial\theta}=\Theta,$$

$$\frac{d}{dt}\left(\frac{\partial f}{\partial\dot{\phi}}\right)-\frac{\partial f}{\partial\phi}=\Phi,$$

$$\frac{d}{dt}\left(\frac{\partial f}{\partial\dot{\psi}}\right)-\frac{\partial f}{\partial\psi}=\Psi.$$

* Euler, *Scientia navalis*, I. (1749), § 128.

The first three of these equations shew that *the motion of the centre of gravity of the body is the same as that of a particle of mass equal to the whole mass of the body, under the influence of forces equivalent to the total external forces acting on the body, applied to the particle parallel to their actual directions*; since the work done on such a particle in an arbitrary displacement would evidently be $X\,\delta x + Y\delta y + Z\delta z$.

The second three equations shew that *the motion of the body about its centre of gravity is the same as if the centre of gravity were fixed and the body subjected to the action of the same forces*; for in the motion relative to the centre of gravity, the kinetic energy of the body is $f(\theta, \phi, \psi, \dot{\theta}, \dot{\phi}, \dot{\psi})$, and the work done by the forces in an arbitrary displacement is

$$\Theta\delta\theta + \Phi\delta\phi + \Psi\delta\psi.$$

These results are evidently true also for impulsive motion.

Corollary. If a plane rigid body (e.g. a disc of any shape) is in motion in its plane, and if (x, y) are the coordinates of its centre of gravity, M its mass, θ the angle made by a line fixed in the body with a line fixed in the plane, Mk^2 the moment of inertia of the body about its centre of gravity, and if (X, Y) are the total components parallel to the axes of the external forces acting on the body, and L the moment of the external forces about the centre of gravity, then the kinetic energy is

$$\tfrac{1}{2}M(\dot{x}^2 + \dot{y}^2 + k^2\dot{\theta}^2),$$

and the work done by the external forces in a displacement $(\delta x, \delta y, \delta\theta)$ is

$$X\,\delta x + Y\delta y + L\,\delta\theta,$$

and therefore the equations of motion of the body are

$$M\ddot{x} = X, \quad M\ddot{y} = Y, \quad Mk^2\ddot{\theta} = L.$$

Example. Obtain one of the equations of motion of a rigid body in two dimensions in the form

$$M(pf + k^2\ddot{\theta}) = L,$$

where M is the mass of the body, f is the acceleration of its centre of gravity, p is the perpendicular from the origin upon this vector, Mk^2 is the moment of inertia about the centre of gravity, θ is the angle made by a line fixed in the body with a line fixed in its plane, and L is the moment about the origin of the external forces. (Coll. Exam.)

MISCELLANEOUS EXAMPLES.

1. A homogeneous right circular cone is of mass M; its semi-vertical angle is β, and the length of a slant side is l. Shew that its moment of inertia about its axis is

$$\tfrac{3}{10}Ml^2\sin^2\beta,$$

and that its moment of inertia about a line through its vertex perpendicular to its axis is

$$\tfrac{3}{5}Ml^2\left(1-\tfrac{3}{4}\sin^2\beta\right),$$

and its moment of inertia about a generator is

$$\tfrac{3}{4}Ml^2\sin^2\beta\left(\cos^2\beta+\tfrac{1}{5}\right).$$

2. Shew that the moment of inertia of the area enclosed by the two loops of the lemniscate

$$r^2=a^2\cos 2\theta$$

about the axis of the curve is

$$\frac{(3\pi-8)\,a^2}{48}\times\text{mass of area.}$$

3. Any number of particles are in one plane; if the masses are m_1, m_2, ..., their distances apart d_{12}, ..., the relative descriptions of area $\dot{h}_{12}$, ..., and the relative velocities v_{12}, ..., prove that

$$(\Sigma m_1 m_2 d_{12}{}^2)/\Sigma m,\quad (\Sigma m_1 m_2 \dot{h}_{12})/\Sigma m,\quad (\Sigma m_1 m_2 v_{12}{}^2)/2\Sigma m$$

are respectively the moment of inertia about the centre of inertia, the angular momentum about the centre of inertia, and the kinetic energy relative to the centre of inertia.

(Coll. Exam.)

4. Prove that the moment of inertia of a hollow cubical box about an axis through the centre of gravity of the box and perpendicular to one of the faces is

$$\tfrac{10}{9}Ma^2,$$

where M is the mass of the box and $2a$ the length of an edge. The sides of the box are supposed to be thin. (Coll. Exam.)

5. Shew that the moment of inertia of an anchor-ring about its axis is

$$2\pi\rho^2a^2c\left(c^2+\tfrac{3}{4}a^2\right),$$

where a is the radius of the generating circle, c is the distance of its centre from the axis of the anchor-ring, and ρ is the density.

6. Shew how to find at what point, if any, a given straight line is a principal axis of a body, and if there is such a point find the other two principal axes through it.

A uniform square lamina is bounded by the axes of x and y and the lines $x=2c$, $y=2c$, and a corner is cut off it by the line $x/a+y/b=2$. Shew that the two principal axes at the centre of the square which are in its own plane are inclined to the axis of x at angles given by

$$\tan 2\theta=\frac{ab-2c\,(a+b)+3c^2}{(a-b)\,(a+b-2c)}.$$

(Coll. Exam.)

7. Shew that the envelope of lines in the plane of an area about which that area has a constant moment of inertia is a set of confocal ellipses and hyperbolas. Hence find the direction of the principal axes at any point. (Coll. Exam.)

8. Find the principal moments of inertia at the vertex of a parabolic lamina, latus rectum $4a$, bounded by a line perpendicular to the axis at a distance h from the vertex.

Prove that, if $15h > 28a$, two principal axes at the point on the parabola whose abscissa is $-a+(a^2-4ah/5+3h^2/7)^{\frac{1}{2}}$ are the tangent and normal. (Coll. Exam.)

9. Find how the principal axes of inertia are arranged in a plane body. Write down the conditions that particles m_i at (x_i, y_i), where $i=1, 2, \ldots$, may be equimomental to a given plate. Shew that the six quantities $m_1, m_2, x_1, x_2, y_1, y_2$ can be eliminated from these conditions.

If three equal particles are equimomental to a given plate, the area of the triangle formed by them is $3\sqrt{3}/2$ times the product of the principal radii of gyration at the centre of gravity. (Coll. Exam.)

10. A uniform lamina bounded by the ellipse $b^2x^2+a^2y^2=a^2b^2$ has an elliptic hole (semi-axes c, d) in it whose major axis lies in the line $x=y$, the centre being at a distance r from the origin; prove that if one of the principal axes at the point (x, y) makes an angle θ with the axis of x, then

$$\tan 2\theta = \frac{8abxy - cd\,[4\,(x\sqrt{2}-r)\,(y\sqrt{2}-r)-(c^2-d^2)]}{ab\,[4\,(x^2-y^2)+a^2-b^2]-cd\,[2\,(x\sqrt{2}-r)^2-2\,(y\sqrt{2}-r)^2]}.$$

(Coll. Exam.)

11. If a system of bodies or particles is moved or deformed in any way, shew that the sum of the products of the mass of each particle into the square of its displacement is equal to the product of the mass of the system into the square of the projection in any given direction of the displacement of the centre of gravity, together with the sum of the products of the masses of the particles into the squares of the distances through which they must be moved in order to bring them to their final positions after communicating to them a displacement equal to the projection in the given direction of the displacement of the centre of gravity. (Fouret.)

12. The principal moments of inertia of a body at its centre of gravity are (A, B, C); if a small mass, whose moments of inertia referred to these axes are (A', B', C'), be added to the body, shew that the moments of inertia of the compound body about its new principal axes at its new centre of gravity are

$$A+A', \quad B+B', \quad C+C',$$

accurately to the first order of small quantities. (Hoppe.)

13. Shew that the principal axes of a given material system at any point are the normals to the three quadrics which pass through the point and belong to a certain confocal system.

If $(l, m, n, \lambda, \mu, \nu)$ be the six coordinates of a principal axis and the associated Cartesian system be the principal axes at the centre of gravity, then shew that

$$Al\lambda + Bm\mu + Cn\nu = 0,$$

and therefore all principal axes of a given system belong to a quadratic complex. (Coll. Exam.)

14. A smoothly jointed framework is in the form of a parallelogram formed by attaching the ends of a pair of rods of mass m and length $2a$ to those of a pair of rods of mass m' and length $2b$. Masses M are attached to each of the four corners. Express the angular momentum of the system about the origin of coordinates, in terms of the coordinates (x, y) of the centre of gravity and the angles θ and ϕ between the two pairs of sides and the axis of x. (Coll. Exam.)

CHAPTER VI

THE SOLUBLE PROBLEMS OF RIGID DYNAMICS

65. *The motion of systems with one degree of freedom: motion round a fixed axis, etc.*

We now proceed to apply the principles which have been developed in the foregoing chapters in order to determine the motion of holonomic systems of rigid bodies in those cases which admit of solution by quadratures.

It is natural to consider first those systems which have only one degree of freedom. We have seen (§ 42) that such a system is immediately soluble by quadratures when it possesses an integral of energy: and this principle is sufficient for the integration in most cases. Sometimes, however (e.g. when we are dealing with systems in which one of the surfaces or curves of constraint is forced to move in a given manner), the problem as originally formulated does not possess an integral of energy, but can be reduced (e.g. by the theorem of § 29) to another problem for which the integral of energy holds; when this reduction has been performed, the problem can be integrated as before.

The following examples will illustrate the application of these principles.

(i) *Motion of a rigid body round a fixed axis.*

Consider the motion of a single rigid body which is free to turn about an axis, fixed in the body and in space. Let I be the moment of inertia of the body about the axis, so that its kinetic energy is $\frac{1}{2}I\dot{\theta}^2$, where θ is the angle made by a moveable plane, passing through the axis and fixed in the body, with a plane passing through the axis and fixed in space. Let Θ be the moment round the axis of all the external forces acting on the body, so that $\Theta\delta\theta$ is the work done by these forces in the infinitesimal displacement which changes θ to $\theta+\delta\theta$. The Lagrangian equation of motion

$$\frac{d}{dt}\left(\frac{\partial T}{\partial \dot{\theta}}\right)-\frac{\partial T}{\partial \theta}=\Theta$$

then gives

$$I\ddot{\theta}=\Theta,$$

which is a differential equation of the second order for the determination of θ.

If the forces are conservative, and $V(\theta)$ denotes the potential energy, this equation becomes

$$I\ddot{\theta} = -\frac{\partial V}{\partial \theta},$$

which on integration gives the equation of energy

$$\tfrac{1}{2}I\dot{\theta}^2 + V(\theta) = c, \qquad \text{where } c \text{ is a constant.}$$

Integrating again, we have

$$t = I^{\frac{1}{2}} \int \{2(c-V)\}^{-\frac{1}{2}} d\theta + \text{constant},$$

and this relation between θ and t determines the motion, the two constants of integration being determined by the initial conditions.

The most important case is that in which gravity is the only external force, and the axis is horizontal. In this case let G be the centre of gravity of the body, C the foot of the perpendicular drawn from G to the axis, and let $CG = h$. The potential energy is $-Mgh\cos\theta$, where M is the mass of the body and θ is the angle made by CG with the downward vertical: and the equation of motion is

$$\ddot{\theta} + \frac{Mgh}{I}\sin\theta = 0.$$

This is the same as the equation of motion of a simple pendulum of length I/Mh, and the motion can therefore be expressed in terms of elliptic functions as in § 44, the solution being of the form

$$\sin\frac{\theta}{2} = k\,\mathrm{sn}\left\{\left(\frac{Mgh}{I}\right)^{\frac{1}{2}}(t-t_0),\ k\right\}$$

in the oscillatory case, and of the form

$$\sin\frac{\theta}{2} = \mathrm{sn}\left\{\frac{1}{k}\left(\frac{Mgh}{I}\right)^{\frac{1}{2}}(t-t_0),\ k\right\}$$

in the circulating case. The quantity I/Mh is called the *length of the equivalent simple pendulum.*

If O be a point on the line CG such that $OC = I/Mh$, the points O and C are called respectively the *centre of oscillation* and the *centre of suspension.* A curious result in this connexion is that *the centre of oscillation and the centre of suspension are convertible,* i.e. if O is the centre of oscillation when C is the centre of suspension, then C will be the centre of oscillation when O is the centre of suspension. To prove this result, we have by § 59

$$\begin{aligned}\text{Moment of inertia of the body about } O &= \text{Moment of inertia about } G + M\,.\,GO^2\\ &= I - M\,.\,CG^2 + M\,.\,GO^2,\end{aligned}$$

and therefore we have

$$\begin{aligned}\frac{\text{Moment of inertia of body about } O}{\text{Distance of centre of gravity from } O} &= \frac{I - Mh^2 + M(I/Mh - h)^2}{I/Mh - h}\\ &= Mh + M(I/Mh - h)\\ &= I/h.\end{aligned}$$

If therefore the body were suspended from O, the equation of motion would still be

$$\ddot{\theta} + \frac{Mgh}{I}\sin\theta = 0,$$

which establishes the result. It is evident that the period of oscillation would be the same about either of the points C and O.

(ii) ***Motion of a rod on which an insect is crawling.***

We shall next study the motion of a straight uniform rod, of mass m and length $2a$, whose extremities can slide on the circumference of a smooth fixed horizontal circle of radius c; an insect of mass equal to that of the rod is supposed to crawl along the rod at a constant rate v relative to the rod.

Let θ be the angle made by the rod at time t with some fixed direction, and let x be the distance traversed by the insect from the middle point of the rod. The kinetic energy of the rod is $\frac{1}{2}m\left(c^2-\frac{2a^2}{3}\right)\dot{\theta}^2$, and the kinetic energy of the insect is due to a component of velocity $\{\dot{x}-(c^2-a^2)^{\frac{1}{2}}\dot{\theta}\}$ along the rod and a component of velocity $x\dot{\theta}$ perpendicular to the rod, so the total kinetic energy of the system is

$$T=\tfrac{1}{2}m\left(c^2-\frac{2a^2}{3}\right)\dot{\theta}^2+\tfrac{1}{2}m\{\dot{x}-(c^2-a^2)^{\frac{1}{2}}\dot{\theta}\}^2+\tfrac{1}{2}mx^2\dot{\theta}^2;$$

there is no potential energy.

Since $x=vt$, (t being measured from the epoch when x is zero), we have

$$T=\tfrac{1}{2}m(c^2-2a^2/3)\dot{\theta}^2+\tfrac{1}{2}m\{v-(c^2-a^2)^{\frac{1}{2}}\dot{\theta}\}^2+\tfrac{1}{2}mv^2t^2\dot{\theta}^2.$$

The coordinate θ, which is now the only coordinate, is ignorable, and we have therefore

$$\frac{\partial T}{\partial \dot{\theta}}=\text{constant},$$

or

$$m\left(c^2-\frac{2a^2}{3}\right)\dot{\theta}-m(c^2-a^2)^{\frac{1}{2}}\{v-(c^2-a^2)^{\frac{1}{2}}\dot{\theta}\}+mv^2t^2\dot{\theta}=\text{constant}.$$

or

$$\dot{\theta}(2c^2-\tfrac{5}{3}a^2+v^2t^2)=\text{constant}.$$

Integrating this equation, we have

$$\theta-\theta_0=k\arctan\{vt(2c^2-\tfrac{5}{3}a^2)^{-\frac{1}{2}}\},$$

where θ_0 and k are constants. This formula determines the position of the rod at any time.

(iii) *Motion of a cone on a perfectly rough inclined plane.*

Consider now the motion of a homogeneous solid right circular cone, of mass M and semi-vertical angle β, which moves on a perfectly rough plane (i.e. a plane on which only rolling without sliding can take place) inclined at an angle α to the horizon. Let l be the length of a slant side of the cone, and let θ be the angle between the generator which is in contact with the plane at time t and the line of greatest slope downwards in the plane. Then if χ be the angle made by the axis of the cone with the upward vertical, χ is one side of a spherical triangle whose vertices represent respectively the normal to the plane, the upward vertical, and the axis of the cone; the other two sides are α and $(\frac{1}{2}\pi-\beta)$, the angle included by these sides being $(\pi-\theta)$. We have therefore

$$\cos\chi=\cos\alpha\sin\beta-\sin\alpha\cos\beta\cos\theta;$$

but the vertical height of the centre of gravity of the cone above its vertex is $\frac{3}{4}l\cos\beta\cos\chi$, and the potential energy of the cone is $Mg\times$ this height; if therefore we denote by V the potential energy of the cone, we have (disregarding a constant term)

$$V=-\tfrac{3}{4}Mgl\sin\alpha\cos^2\beta\cos\theta.$$

We have next to calculate the kinetic energy of the cone; for this the moments of inertia of the cone about its axis and about a line through the vertex perpendicular to the axis are required: these are easily found (by direct integration, regarding the cone as composed of discs perpendicular to its axis) to be $\frac{3}{10}Ml^2\sin^2\beta$ and $\frac{3}{5}Ml^2(\cos^2\beta+\frac{1}{4}\sin^2\beta)$ respectively, and so the moment of inertia about a generator is, by the theorem of § 60 (since the direction-cosines of the generator can be taken to be $\sin\beta$, 0, $\cos\beta$ with respect to rectangular axes at the vertex, of which the axis of z is the axis of the cone),

$$\tfrac{3}{5}Ml^2(\cos^2\beta+\tfrac{1}{4}\sin^2\beta)\sin^2\beta+\tfrac{3}{10}Ml^2\sin^2\beta\cos^2\beta,$$

or

$$\tfrac{3}{4}Ml^2\sin^2\beta(\cos^2\beta+\tfrac{1}{5}).$$

Now all points of that generator which is in contact with the plane are instantaneously at rest, since the motion is one of pure rolling, and therefore this generator is the instantaneous axis of rotation of the cone. If ω denotes the angular velocity of the cone about this generator, the kinetic energy of the cone is therefore (§ 63, Corollary)

$$\tfrac{3}{8}Ml^2\sin^2\beta(\cos^2\beta+\tfrac{1}{5})\,\omega^2.$$

But (§ 15) we have

$$\omega=\dot\theta\cot\beta,$$

and substituting this value for ω, we have finally for the kinetic energy T of the cone the value

$$T=\tfrac{3}{8}Ml^2\cos^2\beta(\cos^2\beta+\tfrac{1}{5})\,\dot\theta^2.$$

The Lagrangian equation of motion

$$\frac{d}{dt}\left(\frac{\partial T}{\partial\dot\theta}\right)-\frac{\partial T}{\partial\theta}=-\frac{\partial V}{\partial\theta}$$

becomes therefore in this case

$$\tfrac{3}{4}Ml^2\cos^2\beta(\cos^2\beta+\tfrac{1}{5})\,\ddot\theta+\tfrac{3}{4}Mgl\sin\alpha\cos^2\beta\sin\theta=0,$$

or

$$\ddot\theta+\frac{g\sin\alpha}{l(\cos^2\beta+\frac{1}{5})}\sin\theta=0.$$

This is the same as the equation of motion of a simple pendulum of length

$$l\operatorname{cosec}\alpha(\cos^2\beta+\tfrac{1}{5});$$

the integration can therefore be effected in terms of elliptic functions, as in § 44.

(iv) *Motion of a rod on a rotating frame.*

Consider next the motion of a heavy uniform rod, whose ends are constrained to move in horizontal and vertical grooves respectively, when the framework containing the grooves is made to rotate with constant angular velocity ω about the line of the vertical groove.

Let $2a$ be the length of the rod, M its mass, and θ its inclination to the vertical. By § 29, the effect of the rotation may be allowed for by adding to the potential energy a term

$$-\tfrac{1}{2}\omega^2\rho\int x^2\sin^2\theta\,dx,$$

where ρ is the density of the rod and x denotes distance measured from the end of the rod which is in the vertical groove; integrating, this term can be written

$$-\tfrac{2}{3}M\omega^2a^2\sin^2\theta.$$

The term in the potential energy due to gravity is

$$-Mga\cos\theta,$$

and the total potential energy V is therefore given by the equation

$$V=-Mga\cos\theta-\tfrac{2}{3}M\omega^2a^2\sin^2\theta.$$

The horizontal and vertical components of velocity of the centre of gravity of the rod are $a\cos\theta\,.\,\dot\theta$ and $a\sin\theta\,.\,\dot\theta$, so the part of the kinetic energy due to the motion of the centre of gravity is $\frac{1}{2}Ma^2\dot\theta^2$; and since the moment of inertia of the rod about its centre is $\frac{1}{3}Ma^2$, the part of the kinetic energy due to the rotation of the rod about its centre is $\frac{1}{6}Ma^2\dot\theta^2$; we have therefore for the total kinetic energy T the equation

$$T=\tfrac{2}{3}Ma^2\dot\theta^2.$$

The integral of energy is therefore

$$\tfrac{2}{3}Ma^2\dot\theta^2-Mga\cos\theta-\tfrac{2}{3}M\omega^2a^2\sin^2\theta=\text{constant},$$

or, writing $\cos\theta=x$,

$$\dot x^2=(1-x^2)\left\{\epsilon^2-\left(\omega x-\frac{3g}{4a\omega}\right)^2\right\},$$

where ϵ^2 denotes a constant: this constant must evidently be positive, since $\dot x^2$ and $(1-x^2)$ are positive. We shall suppose for definiteness that ϵ is not very large and that $3g/4a\omega^2$ is less than unity, so that x oscillates between the values $3g/4a\omega^2\pm\epsilon/\omega$.

To integrate this equation, we write*

$$x=1+\frac{\frac{1}{2}\omega^2\left(1-\dfrac{3g}{4a\omega^2}-\dfrac{\epsilon}{\omega}\right)\left(1-\dfrac{3g}{4a\omega^2}+\dfrac{\epsilon}{\omega}\right)}{\xi+\dfrac{3g}{8a}-\dfrac{5}{12}\omega^2-\dfrac{3g^2}{64a^2\omega^2}+\dfrac{\epsilon^2}{12}},$$

where ξ is a new dependent variable. Substituting this value for x in the differential equation, we have

$$\dot\xi^2=4(\xi-e_1)(\xi-e_2)(\xi-e_3),$$

where the values

$$\xi=e_1,\quad \xi=e_2,\quad \xi=e_3$$

correspond respectively to the values

$$x=-1,\qquad x=\frac{3g}{4a\omega^2}-\frac{\epsilon}{\omega},\qquad x=\frac{3g}{4a\omega^2}+\frac{\epsilon}{\omega};$$

it is easily seen that $e_1+e_2+e_3$ is zero and that $e_1>e_2>e_3$.

We have therefore $\xi=\wp(t+\gamma)$, where the function $\wp$ is formed with the roots e_1, e_2, e_3, and where γ denotes a constant. Since $e_1>e_2>e_3$, and $\wp(t+\gamma)$ lies between e_2 and e_3 for real values of t (since x lies between $3g/4a\omega^2-\epsilon/\omega$ and $3g/4a\omega^2+\epsilon/\omega$), the imaginary part of the constant γ must be the half-period ω_3; the real part of γ can then be taken as zero, since it depends only on the choice of the origin of time. We have therefore

$$\xi=\wp(t+\omega_3),$$

and hence

$$\cos\theta=1+\frac{\frac{1}{2}\omega^2\left(1-\dfrac{3g}{4a\omega^2}-\dfrac{\epsilon}{\omega}\right)\left(1-\dfrac{3g}{4a\omega^2}+\dfrac{\epsilon}{\omega}\right)}{\wp(t+\omega_3)+\dfrac{3g}{8a}-\dfrac{5\omega^2}{12}-\dfrac{3g^2}{64a^2\omega^2}+\dfrac{\epsilon^2}{12}};$$

this equation determines θ in terms of t.

(v) *Motion of a disc, one of whose points is forced to move in a given manner.*

Consider next the motion of a disc of mass M resting on a perfectly smooth horizontal plane, when one of the points A of the disc is constrained to describe a circle of radius c in the horizontal plane, with uniform angular velocity ω.

* Cf. Whittaker and Watson, *A Course of Modern Analysis*, § 20·6.

Let G be the centre of gravity of the disc, and let AG be of length a. The acceleration of the point A is of magnitude $c\omega^2$, and is directed along the inward normal to the circle: if therefore we impress an acceleration $c\omega^2$, directed along the outward normal to the circle, on all the particles of the body and suppose that A is at rest, we shall obtain the motion relative to A. The resultant force acting on the body in this motion relative to A is therefore $Mc\omega^2$, acting at G in a direction parallel to the outward normal to the circle.

Let θ and ϕ be the angles made with a fixed direction in the plane by the line AG and the outward normal to the circle respectively; then the work done by this force in a small displacement $\delta\theta$ is

$$Mc\omega^2 a \sin(\phi - \theta)\,\delta\theta,$$

and the kinetic energy of the body is $\frac{1}{2}Mk^2\dot{\theta}^2$, where Mk^2 is the moment of inertia of the body about the point A. The Lagrangian equation of motion is therefore

$$Mk^2\ddot{\theta} = Mac\omega^2 \sin(\phi - \theta).$$

But since $\dot{\phi} = \omega$, we have $\ddot{\phi} = 0$; so if ψ be written for $(\theta - \phi)$, we have

$$\ddot{\psi} + \frac{ac\omega^2}{k^2}\sin\psi = 0.$$

This is the same as the equation of motion of a simple pendulum of length $k^2 g/ac\omega^2$; the integration can therefore be performed by means of elliptic functions as in § 44.

(vi) *Motion of a disc rolling on a constrained disc and linked to it.*

Consider the motion of two equal circular discs, of radius a and mass M, with edges perfectly rough, which are kept in contact in a vertical plane by means of a link (in the form of a uniform bar of mass m) which joins their centres: the centre of one disc is fixed, and this disc A is constrained to rotate with uniform angular acceleration a; it is required to determine the motion of the other disc B and the link.

Let ϕ be the angle which the link makes with the downward vertical at time t, and let θ be the angle turned through at time t by the disc A. The angular velocity of disc A is $\dot{\theta}$, and the velocities of the points of the discs which are instantaneously in contact are therefore each $a\dot{\theta}$. Since the velocity of the centre of the disc B is $2a\dot{\phi}$, it follows that the angular velocity of the disc B about its centre is $2\dot{\phi} - \dot{\theta}$. Since the moment of inertia of each disc about its centre is $\frac{1}{2}Ma^2$, the kinetic energy of the system is

$$T = \tfrac{1}{2}M\,.\,\frac{a^2}{2}\dot{\theta}^2 + \tfrac{1}{2}M\,.\,\frac{a^2}{2}(2\dot{\phi} - \dot{\theta})^2 + \tfrac{1}{2}M\,.\,(2a)^2\dot{\phi}^2 + \tfrac{1}{2}m\,.\,\frac{4a^2}{3}\dot{\phi}^2;$$

and $\dot{\theta} = at + \epsilon$, where ϵ is a constant.

The potential energy of the system is

$$V = -(2M + m)\,ag\cos\phi,$$

and the Lagrangian equation of motion is

$$\frac{d}{dt}\left(\frac{\partial T}{\partial \dot{\phi}}\right) - \frac{\partial T}{\partial \phi} = -\frac{\partial V}{\partial \phi},$$

or

$$\frac{d}{dt}\{(6M + \tfrac{4}{3}m)\,a^2\dot{\phi} - Ma^2\dot{\theta}\} = -(2M + m)\,ag\sin\phi.$$

Since $\ddot{\theta} = a$, this equation gives

$$(6M + \tfrac{4}{3}m)\,a^2\ddot{\phi} - Ma^2 a + (2M + m)\,ag\sin\phi = 0.$$

Integrating, we have

$$(3M + \tfrac{2}{3}m)\,a^2\dot{\phi}^2 - Ma^2 a\phi - (2M + m)\,ag\cos\phi = c,$$

where c is a constant depending on the initial conditions: and as the variables t and ϕ are separable, this equation can again be integrated by a quadrature: this final integral determines the motion.

Example. If the system is initially at rest with the bar vertically downwards, shew that the bar will reach the horizontal position if

$$a > \frac{4g}{\pi a}\left(1+\frac{m}{2M}\right).$$

66. *The motion of systems with two degrees of freedom.*

In the dynamics of rigid bodies, as in the dynamics of a particle, the possibility of solving by quadratures a problem with two degrees of freedom generally depends on the presence of an ignorable coordinate. The integral corresponding to the ignorable coordinate can often be interpreted physically as an integral of momentum or angular momentum. The formation and solution of the differential equations is effected by application of the principles developed in the preceding chapters: this will be shewn by the following illustrative examples.

(i) *Rod passing through ring.*

Consider, as a first example, the motion of a uniform straight rod which passes through a small fixed ring on a horizontal plane, being able to slide through the ring or turn in any way about it in the plane.

Let the distance from the ring to the middle point of the rod at time t be r, and let the rod make an angle θ with a fixed line in the plane; let $2l$ be the length of the rod, and M its mass.

The moment of inertia of the rod about its middle point is $\frac{1}{3}Ml^2$, and the kinetic energy is therefore

$$T=\tfrac{1}{2}M(\dot{r}^2+r^2\dot{\theta}^2+\tfrac{1}{3}l^2\dot{\theta}^2);$$

there is no potential energy.

The coordinate θ is ignorable and the corresponding integral is

$$\frac{\partial T}{\partial \dot{\theta}}=\text{constant},$$

or

$$(r^2+\tfrac{1}{3}l^2)\,\dot{\theta}=\text{constant}.$$

The integral of energy is

$$\dot{r}^2+r^2\dot{\theta}^2+\tfrac{1}{3}l^2\dot{\theta}^2=\text{constant}.$$

Dividing the second of these integrals by the square of the first, we have

$$\frac{\left(\frac{dr}{d\theta}\right)^2}{(r^2+\frac{1}{3}l^2)^2}+\frac{1}{r^2+\frac{1}{3}l^2}=c, \qquad \text{where } c \text{ is a constant,}$$

or

$$\theta+\text{constant}=\int\{(r^2+\tfrac{1}{3}l^2)(cr^2+\tfrac{1}{3}cl^2-1)\}^{-\frac{1}{2}}\,dr.$$

Writing $cr^2=s$, this becomes

$$\theta+\text{constant}=\int\{4s(s+\tfrac{1}{3}cl^2)(s+\tfrac{1}{3}cl^2-1)\}^{-\frac{1}{2}}\,ds.$$

If therefore $\wp$ denotes the Weierstrassian elliptic function with the roots

$$e_1=\tfrac{1}{3}(-1+\tfrac{2}{3}cl^2),\quad e_2=\tfrac{1}{3}(2-\tfrac{1}{3}cl^2),\quad e_3=\tfrac{1}{3}(-1-\tfrac{1}{3}cl^2),$$

which satisfy the relation $e_1>e_2>e_3$ if $\frac{dr}{d\theta}$ is sufficiently great initially, we have

$$s=\wp(\theta-\theta_0)-e_1,\quad \text{where } \theta_0 \text{ is a constant of integration};$$

since s is positive, we have $\wp(\theta-\theta_0)>e_1$ for real values of θ, and consequently the constant θ_0 is real.

The solution of the problem is therefore contained in the equation

$$cr^2=\wp(\theta-\theta_0)+\tfrac{1}{3}-\tfrac{2}{9}cl^2.$$

(ii) *One cylinder rolling on another under gravity.*

Let it now be required to determine the motion of a perfectly rough heavy solid homogeneous cylinder of mass m and radius r, which rolls inside a hollow cylinder of mass M and radius R, which in turn is free to turn about its axis (supposed horizontal).

Let ϕ denote the angle which the plane through the axes of the cylinders at time t makes with the downward vertical, and let θ be the angle through which the cylinder of mass M has turned since some fixed epoch. The angular velocities of the cylinders about their axes are easily seen to be $\dot\theta$ and $\{(R-r)\dot\phi-R\dot\theta\}/r$ respectively; and the moments of inertia of the cylinders about their axes are MR^2 and $\frac{1}{2}mr^2$ respectively; so the kinetic energy T of the system is given by the equation

$$T=\tfrac{1}{2}MR^2\dot\theta^2+\tfrac{1}{4}mr^2\left(\frac{R-r}{r}\dot\phi-\frac{R}{r}\dot\theta\right)^2+\tfrac{1}{2}m(R-r)^2\dot\phi^2,$$

while the potential energy is given by the equation

$$V=-mg(R-r)\cos\phi.$$

The coordinate θ is clearly ignorable; the integral corresponding to it is

$$\frac{\partial T}{\partial\dot\theta}=\text{constant},$$

or $$MR^2\dot\theta-\tfrac{1}{2}mR\{(R-r)\dot\phi-R\dot\theta\}=k,\qquad \text{where } k \text{ is a constant.}$$

The integral of energy is

$$T+V=h,\qquad \text{where } h \text{ is a constant,}$$

or $$\tfrac{1}{2}MR^2\dot\theta^2+\tfrac{1}{4}m\{(R-r)\dot\phi-R\dot\theta\}^2+\tfrac{1}{2}m(R-r)^2\dot\phi^2-mg(R-r)\cos\phi=h.$$

Eliminating $\dot\theta$ between the two integrals, we obtain the equation

$$\frac{m(3M+m)}{2(2M+m)}(R-r)^2\dot\phi^2-mg(R-r)\cos\phi=h-\frac{k^2}{(2M+m)R^2}.$$

This is the same as the equation of energy of a simple pendulum of length

$$\frac{3M+m}{2M+m}(R-r);$$

the solution can be effected by means of elliptic functions as in § 44.

(iii) *Rod moving in a free circular frame.*

We shall next consider the motion of a rod, whose ends can slide freely on a smooth vertical circular ring, the ring being free to turn about its vertical diameter, which is fixed.

Let m be the mass of the rod and $2a$ its length; let M be the mass of the ring and r its radius; let θ be the inclination of the rod to the horizontal, and ϕ the azimuth of the ring referred to some fixed vertical plane, at any time t.

The moment of inertia of the rod about an axis through the centre of the ring perpendicular to its plane is $m(r^2 - \frac{2}{3}a^2)$, and the moment of inertia of the rod about the vertical diameter of the ring is $m\{(r^2 - a^2)\sin^2\theta + \frac{1}{3}a^2\cos^2\theta\}$. The kinetic energy of the system is therefore

$$T = \tfrac{1}{2}m(r^2 - \tfrac{2}{3}a^2)\dot{\theta}^2 + \tfrac{1}{4}Mr^2\dot{\phi}^2 + \tfrac{1}{2}m\dot{\phi}^2(r^2\sin^2\theta - a^2\sin^2\theta + \tfrac{1}{3}a^2\cos^2\theta).$$

The potential energy is

$$V = -mg(r^2 - a^2)^{\frac{1}{2}}\cos\theta.$$

The coordinate ϕ is evidently ignorable; the corresponding integral is

$$\frac{\partial T}{\partial \dot{\phi}} = \text{constant},$$

or

$$\tfrac{1}{2}Mr^2\dot{\phi} + m\dot{\phi}(r^2\sin^2\theta - a^2\sin^2\theta + \tfrac{1}{3}a^2\cos^2\theta) = k,$$

where k is a constant. Substituting the value of $\dot{\phi}$ found from this equation in the integral of energy

$$T + V = h,$$

we have

$$\tfrac{1}{2}m(r^2 - \tfrac{2}{3}a^2)\dot{\theta}^2 = h + mg(r^2 - a^2)^{\frac{1}{2}}\cos\theta - \tfrac{1}{2}\frac{k^2}{\frac{1}{2}Mr^2 + m(r^2\sin^2\theta - a^2\sin^2\theta + \frac{1}{3}a^2\cos^2\theta)}.$$

In this equation the variables θ and t are separable; a further integration will therefore give θ in terms of t, and so furnish the solution of the problem.

(iv) *Hoop and ring.*

We shall next discuss the motion of a system consisting of a uniform smooth circular hoop of radius a, which lies in a smooth horizontal plane, and is so constrained that it can only move by rolling on a fixed straight line in that plane, while a small ring whose mass is $1/\lambda$ that of the hoop slides on it. The hoop is initially at rest, and the ring is projected from the point furthest from the fixed line with velocity v.

Let ϕ denote the angle turned through by the hoop after a time t from the commencement of the motion, and suppose that the diameter of the hoop which passes through the ring has then turned through an angle ψ. Taking the ring to be of unit mass, so that the mass of the hoop is λ, the moment of inertia of the hoop about its centre is λa^2, and this centre moves with velocity $a\dot{\phi}$, while the velocity of the ring is compounded of components $a\dot{\phi}$ and $a\dot{\psi}$, whose directions are inclined to each other at an angle ψ. The kinetic energy of the system is therefore

$$\begin{aligned} T &= \tfrac{1}{2}\lambda a^2\dot{\phi}^2 + \tfrac{1}{2}\lambda a^2\dot{\phi}^2 + \tfrac{1}{2}(a^2\dot{\phi}^2 + a^2\dot{\psi}^2 + 2a^2\dot{\phi}\dot{\psi}\cos\psi) \\ &= \tfrac{1}{2}(2\lambda + 1)a^2\dot{\phi}^2 + \tfrac{1}{2}a^2\dot{\psi}^2 + a^2\dot{\phi}\dot{\psi}\cos\psi, \end{aligned}$$

and the potential energy is zero.

The coordinate ϕ is evidently ignorable, and the corresponding integral is

$$\frac{\partial T}{\partial \dot{\phi}} = \text{constant},$$

or

$$\begin{aligned} (2\lambda + 1)a^2\dot{\phi} + a^2\dot{\psi}\cos\psi &= \text{the initial value of this expression} \\ &= av. \end{aligned}$$

Integrating this equation, we have

$$(2\lambda+1)\,\phi+\sin\psi-\frac{vt}{a}=\text{the initial value of this expression}$$
$$=0,$$

or
$$\phi=\frac{1}{2\lambda+1}\left(\frac{vt}{a}-\sin\psi\right).$$

This equation determines ϕ in terms of ψ.

The equation of energy is

$$T=\text{its initial value}=\tfrac{1}{2}v^2,$$

and substituting for $\dot{\phi}$ its value $(v/a-\cos\psi\,.\,\dot{\psi})/(2\lambda+1)$ in this equation, we have

$$a^2(2\lambda+\sin^2\psi)\,\dot{\psi}^2=2\lambda v^2,$$

so
$$t=\frac{a}{v\sqrt{2\lambda}}\int_0^{\psi}(2\lambda+\sin^2\psi)^{\frac{1}{2}}\,d\psi.$$

Writing $\sin\psi=x$, this becomes

$$t=\frac{a}{v\sqrt{2\lambda}}\int_0^{x}(2\lambda+x^2)^{\frac{1}{2}}(1-x^2)^{-\frac{1}{2}}\,dx.$$

In order to evaluate this integral, we introduce an auxiliary variable u, defined by the equation

$$u=\int_0^{x}(2\lambda+x^2)^{-\frac{1}{2}}(1-x^2)^{-\frac{1}{2}}\,dx.$$

Write $x^2=2\lambda/\xi$, where ξ is a new variable; the last integral becomes

$$u=\int_{\xi}^{\infty}\{4\xi(\xi+1)(\xi-2\lambda)\}^{-\frac{1}{2}}\,d\xi,$$

which is equivalent to

$$\xi=\wp(u)-\tfrac{1}{3}(1-2\lambda),$$

where the function $\wp(u)$ is formed with the roots

$$e_1=\tfrac{1}{3}(1+4\lambda),\quad e_2=\tfrac{1}{3}(1-2\lambda),\quad e_3=-\tfrac{2}{3}(1+\lambda);$$

these roots are real and satisfy the inequality $e_1>e_2>e_3$, so $\wp(u)$ is real and greater than e_1 for real values of u.

Now we have
$$dt=\frac{a}{v\sqrt{2\lambda}}(2\lambda+x^2)^{\frac{1}{2}}(1-x^2)^{-\frac{1}{2}}\,dx,$$

or
$$\frac{\sqrt{2\lambda}\,v\,dt}{a}=\left\{2\lambda+\frac{2\lambda}{\wp(u)-e_2}\right\}du.$$

Integrating, we have

$$\frac{\sqrt{2\lambda}\,vt}{a}=\tfrac{1}{3}(1+4\lambda)\,u+\zeta(u)+\tfrac{1}{2}\,\frac{\wp'(u)}{\wp(u)-\frac{1}{3}(1-2\lambda)},$$

where $\zeta(u)$ denotes the Weierstrassian Zeta-function.

Thus finally *the coordinate ψ and the time t are expressed in terms of an auxiliary variable u by the equations*

$$\sin^2\psi=\frac{2\lambda}{\wp(u)-\frac{1}{3}(1-2\lambda)},$$
$$\frac{\sqrt{2\lambda}\,vt}{a}=\tfrac{1}{3}(1+4\lambda)\,u+\zeta(u)+\tfrac{1}{2}\,\frac{\wp'(u)}{\wp(u)-\frac{1}{3}(1-2\lambda)}.$$

67. *Initial motions.*

We have already explained in § 32 the general principles used in finding the initial character of the motion of a system which starts from rest at a given time. The following examples will serve to illustrate the procedure for systems of rigid bodies.

(i) *A particle hangs by a string of length b from a point in the circumference of a disc of twice its mass and of radius a. The disc can turn about its axis, which is horizontal, and the diameter through the point of attachment of the string is initially horizontal. To find the initial path of the particle.*

Let θ denote the angle through which the disc has turned, and ϕ the inclination of the string to the vertical, at time t from the beginning of the motion: let m be the mass of the particle. The horizontal and (downward) vertical coordinates of the particle with respect to the centre of the disc are

$$a\cos\theta + b\sin\phi \quad \text{and} \quad a\sin\theta + b\cos\phi,$$

so the square of the particle's velocity is

$$a^2\dot{\theta}^2 + b^2\dot{\phi}^2 - 2ab\sin(\theta+\phi)\,\dot{\theta}\dot{\phi},$$

and the kinetic energy of the system is

$$T = ma^2\dot{\theta}^2 + \tfrac{1}{2}mb^2\dot{\phi}^2 - mab\sin(\theta+\phi)\,\dot{\theta}\dot{\phi},$$

while the potential energy is

$$V = -mg\,(a\sin\theta + b\cos\phi).$$

The Lagrangian equations of motion are

$$\begin{cases} \dfrac{d}{dt}\left(\dfrac{\partial T}{\partial \dot{\theta}}\right) - \dfrac{\partial T}{\partial \theta} = -\dfrac{\partial V}{\partial \theta}, \\ \dfrac{d}{dt}\left(\dfrac{\partial T}{\partial \dot{\phi}}\right) - \dfrac{\partial T}{\partial \phi} = -\dfrac{\partial V}{\partial \phi}, \end{cases}$$

or

$$\begin{cases} 2a^2\ddot{\theta} - ab\cos(\theta+\phi)\,\dot{\phi}^2 - ga\cos\theta - ab\sin(\theta+\phi)\,\ddot{\phi} = 0, \\ b^2\ddot{\phi} - ab\cos(\theta+\phi)\,\dot{\theta}^2 + gb\sin\phi - ab\sin(\theta+\phi)\,\ddot{\theta} = 0. \end{cases}$$

Initially the quantities θ, ϕ, $\dot{\theta}$, $\dot{\phi}$ are all zero: these equations therefore give initially $\ddot{\theta} = g/2a$ and $\ddot{\phi} = 0$, so the expansion of θ begins with a term $gt^2/4a$ and that of ϕ with a term higher than the square of t. Assuming

$$\theta = \frac{gt^2}{4a} + At^3 + Bt^4 + \ldots,$$

$$\phi = Ct^3 + Dt^4 + Et^5 + Ft^6 + \ldots,$$

substituting in the above differential equations, and equating powers of t, we can evaluate the coefficients A, B, C, ...; we thus find

$$\theta = \frac{gt^2}{4a} + 0\,.\,t^4 + \ldots,$$

$$\phi = \frac{g^2}{32ab}t^4 - \frac{g^3t^6}{1920ab^2} + \ldots.$$

Now if x and y are the coordinates of the particle referred to horizontal and (downward) vertical axes through its initial position, we have

$$x = a(1-\cos\theta) - b\sin\phi = \tfrac{1}{2}a\theta^2 - b\phi = \frac{g^3t^6}{1920ab}, \text{ approximately,}$$

and

$$y = a\sin\theta + b(\cos\phi - 1) = a\theta = \frac{gt^2}{4}, \text{ approximately.}$$

Eliminating t between these equations, we have

$$y^3 = 30abx,$$

and this is the required approximate equation of the path of the particle in the neighbourhood of its initial position.

(ii) *A ring of mass m can slide freely on a uniform rod of mass M and length 2a, which can turn about one end. Initially the rod is horizontal, with the ring at a distance r_0 from the fixed end. To find the initial curvature of the path of the ring in space.*

Let (r, θ) denote the polar coordinates of the ring at time t, referred to the fixed end of the rod and a horizontal initial line, θ being measured downwards from the initial line. For the kinetic and potential energies we have

$$T = \tfrac{1}{2} m (\dot{r}^2 + r^2\dot{\theta}^2) + \tfrac{1}{2} M \cdot \frac{4a^2}{3} \dot{\theta}^2,$$

$$V = -mrg \sin\theta - Mag \sin\theta.$$

The Lagrangian equations of motion are

$$\begin{cases} \dfrac{d}{dt}\left(\dfrac{\partial T}{\partial \dot{r}}\right) - \dfrac{\partial T}{\partial r} = -\dfrac{\partial V}{\partial r} \\ \dfrac{d}{dt}\left(\dfrac{\partial T}{\partial \dot{\theta}}\right) - \dfrac{\partial T}{\partial \theta} = -\dfrac{\partial V}{\partial \theta}, \end{cases}$$

or

$$\begin{cases} \ddot{r} - r\dot{\theta}^2 - g\sin\theta = 0, \\ \tfrac{4}{3} Ma^2\ddot{\theta} + mr^2\ddot{\theta} + 2mr\dot{r}\dot{\theta} - Mga\cos\theta - mgr\cos\theta = 0. \end{cases}$$

Since $\dot{r}$, θ, and $\dot{\theta}$ are initially zero, we can assume expansions of the form

$$\begin{cases} r = r_0 + a_2t^2 + a_3t^3 + a_4t^4 + \dots, \\ \theta = b_2t^2 + b_3t^3 + \dots ; \end{cases}$$

substituting these expansions in the differential equations, and equating coefficients of powers of t, we find

$$a_2 = 0, \quad a_3 = 0, \quad a_4 = \tfrac{1}{12} b_2 (g + 4b_2 r_0),$$

$$b_2 = \frac{3g(Ma + mr_0)}{2(4Ma^2 + 3mr_0^2)}.$$

The coordinates of the particle, referred to horizontal and vertical axes at its initial position, are

$$x = r\cos\theta - r_0 \quad \text{and} \quad y = r\sin\theta,$$

or approximately $\quad x = (a_4 - \tfrac{1}{2} r_0 b_2^2)\, t^4, \quad y = r_0 b_2 t^2.$

The curvature of the path is given by the equation

$$\frac{1}{\rho} = \text{Lt.}\ \frac{2x}{y^2} = \frac{2a_4}{b_2^2 r_0^2} - \frac{1}{r_0},$$

and on substituting the above values of b_2 and a_4, we have

$$\frac{1}{\rho} = \frac{Ma(4a - 3r_0)}{9r_0^2(Ma + mr_0)}.$$

This is the required initial curvature of the path of the ring.

Example. Two uniform rods AB, BC, of masses m_1 and m_2, and lengths a and b respectively, are freely hinged at B, and can turn round the point A, which is fixed. Initially, AB is horizontal and BC vertical. Shew that, if C be released, the equation of the initial path of the point of trisection of BC nearer to C can be put in the form

$$y^3 = 60(1 + 2m_2/m_1)\, abx.$$

(Camb. Math. Tripos, Part I, 1896.)

68. *The motion of systems with three degrees of freedom.*

The possibility of solving by quadratures the motion of a system of rigid bodies which has three degrees of freedom depends generally (as in the case of systems with two degrees of freedom) either on the occurrence of ignorable coordinates, giving rise to integrals of momentum and angular momentum, or on a disjunction of the kinetic potential into parts which depend on the coordinates separately. The following examples illustrate the procedure.

(i) *Motion of a rod in a given field of force.*

Consider the motion of a uniform rod, of mass m and length $2a$, which is free to move on a smooth table, when each element of the rod is attracted to a fixed line of the table with a force proportional to its mass and its distance from the line.

Let (x, y) be the coordinates of the middle point of the rod, and θ its inclination to the fixed line. The kinetic energy is

$$T=\tfrac{1}{2}m(\dot{x}^2+\dot{y}^2+\tfrac{1}{3}a^2\dot{\theta}^2),$$

and the potential energy is

$$V=\frac{m\mu}{4a}\int_{-a}^{a}(y+r\sin\theta)^2\,dr, \qquad \text{where } \mu \text{ is a constant,}$$

or

$$V=\mu m(\tfrac{1}{2}y^2+\tfrac{1}{6}a^2\sin^2\theta).$$

The Lagrangian equations of motion are therefore

$$\begin{cases}\ddot{x}=0,\\ \ddot{y}=-\mu y,\\ (2\ddot{\theta})+\mu\sin 2\theta=0.\end{cases}$$

The first two equations give

$$\begin{cases}x=ct+d,\\ y=f\sin(\mu^{\frac{1}{2}}t+\epsilon),\end{cases}$$

where c, d, f, ϵ are constants of integration; the centre of the rod therefore describes a sine curve in the plane. The equation for θ is of the pendulum type, and can be integrated as in § 44.

(ii) *Motion of a rod and cylinder on a plane.*

We shall next discuss the motion of a system consisting of a smooth solid homogeneous circular cylinder, of mass M and radius c, which is moveable on a smooth horizontal plane, and a heavy straight rail of mass m and length $2a$, placed with its length in contact with the cylinder, in a vertical plane perpendicular to the axis of the cylinder and passing through the centre of gravity of the cylinder, and with one extremity on the plane.

Let θ be the inclination of the rail to the vertical, and x the distance traversed on the plane by the line of contact of the cylinder and plane, at any time t. The coordinates of the centre of the rod referred to horizontal and vertical axes, the origin being the initial point of contact of the cylinder and plane, are easily seen to be

$$x-c\cot\left(\frac{\pi}{4}-\frac{\theta}{2}\right)+a\sin\theta \quad\text{and}\quad a\cos\theta.$$

Let ϕ be the angle through which the cylinder has turned at time t. The kinetic energy of the system is

$$T=\tfrac{1}{6}ma^2\dot{\theta}^2+\tfrac{1}{2}m\left\{\dot{x}-\tfrac{1}{2}c\operatorname{cosec}^2\left(\frac{\pi}{4}-\frac{\theta}{2}\right).\dot{\theta}+a\cos\theta.\dot{\theta}\right\}^2+\tfrac{1}{2}ma^2\sin^2\theta.\dot{\theta}^2+\tfrac{1}{2}M\dot{x}^2+\tfrac{1}{4}Mc^2\dot{\phi}^2$$

The potential energy is given by the equation

$$V = mga\cos\theta.$$

The coordinates x and ϕ are evidently ignorable: the corresponding integrals are

$$\frac{\partial T}{\partial \dot{x}} = \text{constant}$$

(which may be interpreted as the integral of momentum of the system parallel to the axis of x) and

$$\frac{\partial T}{\partial \dot{\phi}} = \text{constant}$$

(which may be interpreted as the integral of angular momentum of the cylinder about its axis). These integrals can be written

$$\begin{cases} m\left\{\dot{x} - \frac{1}{2}c\,\text{cosec}^2\left(\frac{\pi}{4} - \frac{\theta}{2}\right).\dot{\theta} + a\cos\theta.\dot{\theta}\right\} + M\dot{x} = \text{constant}, \\ \frac{1}{2}Mc^2\dot{\phi} = \text{constant}. \end{cases}$$

Substituting for $\dot{x}$ and $\dot{\phi}$ the values obtained from these equations in the integral of energy

$$T + V = \text{constant},$$

we have the equation

$$\dot{\theta}^2\left[\frac{1}{3}a^2 + a^2\sin^2\theta + \frac{M}{m+M}\left\{a\cos\theta - \frac{1}{2}c\,\text{cosec}^2\left(\frac{\pi}{4} - \frac{\theta}{2}\right)\right\}^2\right] = d - 2ga\cos\theta,$$

where d is a constant. This equation is again integrable, since the variables t and θ are separable; in its integrated form it gives the expression of θ in terms of t: the two integrals found above then give x and ϕ in terms of t.

69. *Motion of a body about a fixed point under no forces.*

One of the most important problems in the dynamics of systems with three degrees of freedom is that of determining the motion of a rigid body, one of whose points O is fixed, when no external forces are supposed to act*. This problem is realised (§ 64) in the motion of a rigid body relative to its centre of gravity, under the action of any forces whose resultant passes through the centre of gravity.

In this system the angular momentum of the body about every line which passes through the fixed point and is fixed in space is constant (§ 40), and consequently the line through the fixed point for which this angular momentum has its greatest value is fixed in space. Let this line, which is called the *invariable line*, be taken as axis OZ, and let OX and OY be two other axes through the fixed point which are perpendicular to OZ and to each other. The angular momenta about the axes OX and OY are zero, for if this were not the case the resultant of the angular momenta about OX, OY, OZ would give a line about which the angular momentum would be greater than the

* Euler, *Mémoires de Berlin*, Annee 1758. Elliptic functions were applied to the problem first by Rueb, *Specimen inaugurale*... (Utrecht, 1834): and the solution was completed by Jacobi, *Journal für Math.* XXXIX. (1849), p. 293.

angular momentum about OZ, which is contrary to hypothesis. It follows (§ 39) that the angular momentum about any line through O making an angle θ with OZ is $d\cos\theta$, where d denotes the angular momentum about OZ.

The position of the body at any time t is sufficiently specified by the knowledge of the positions at that time of its three principal axes of inertia at the fixed point: let these lines be taken as moving axes $Oxyz$; let (θ, ϕ, ψ) denote the three Eulerian angles which specify the position of the axes $Oxyz$ with reference to the axes $OXYZ$, let (A, B, C) be the principal moments of inertia of the body at O, supposed arranged in descending order of magnitude, and let $(\omega_1, \omega_2, \omega_3)$ be the three components of angular velocity of the system about the axes Ox, Oy, Oz respectively, so that (§§ 10, 62)

$$\begin{cases} A\omega_1 = -d\sin\theta\cos\psi, \\ B\omega_2 = d\sin\theta\sin\psi, \\ C\omega_3 = d\cos\theta, \end{cases}$$

or (§ 16)

$$\begin{cases} \dot\theta\sin\psi - \dot\phi\sin\theta\cos\psi = -\dfrac{d}{A}\sin\theta\cos\psi, \\ \dot\theta\cos\psi + \dot\phi\sin\theta\sin\psi = \dfrac{d}{B}\sin\theta\sin\psi, \\ \dot\psi + \dot\phi\cos\theta = \dfrac{d}{C}\cos\theta. \end{cases}$$

These are really three integrals of the differential equations of motion of the system (only *one* arbitrary constant however occurs, namely d, our special set of axes being such as to make the other two constants of integration zero); we can therefore take these instead of the usual Lagrangian differential equations of motion in order to determine θ, ϕ, ψ.

Solving for $\dot\theta, \dot\phi, \dot\psi$, we have

$$\begin{cases} \dot\theta = \dfrac{(A-B)\,d}{AB}\sin\theta\cos\psi\sin\psi, \\ \dot\phi = \dfrac{d}{A}\cos^2\psi + \dfrac{d}{B}\sin^2\psi, \\ \dot\psi = \left(\dfrac{d}{C} - \dfrac{d}{A}\cos^2\psi - \dfrac{d}{B}\sin^2\psi\right)\cos\theta. \end{cases}$$

The integral of energy (which is a consequence of these three equations) may be written down at once by use of § 63; it is

$$A\omega_1^2 + B\omega_2^2 + C\omega_3^2 = c,$$

where c is a constant: replacing $\omega_1, \omega_2, \omega_3$ by their values in terms of θ and ψ, this equation can be written in either of the forms

$$\frac{A-B}{AB}\sin^2\theta\cos^2\psi = -\frac{Bc-d^2}{Bd^2}+\frac{B-C}{BC}\cos^2\theta,$$

or
$$\frac{A-B}{AB}\sin^2\theta\sin^2\psi = \frac{Ac-d^2}{Ad^2}-\frac{A-C}{AC}\cos^2\theta.$$

Since $A > B > C$, the quantity $(cA-d^2)$ or $B(A-B)\omega_2{}^2 + C(A-C)\omega_3{}^2$ is positive, and $(cC-d^2)$ is negative: the quantity $(Bc-d^2)$ may be either positive or negative: for definiteness we shall suppose it to be positive.

The first of the three differential equations may, by use of the last equations, be written

$$\frac{d}{dt}(\cos\theta) = -d\,.\left\{\frac{Bc-d^2}{Bd^2}+\frac{B-C}{BC}\cos^2\theta\right\}^{\frac{1}{2}}\left\{\frac{Ac-d^2}{Ad^2}-\frac{A-C}{AC}\cos^2\theta\right\}^{\frac{1}{2}}.$$

This equation shews that $\cos\theta$ is a Jacobian elliptic function* of a linear function of t; and the two preceding equations shew that $\sin\theta\cos\psi$ and $\sin\theta\sin\psi$ are the other two Jacobian functions.

We therefore write

$$\sin\theta\cos\psi = P\,\mathrm{cn}\,u, \qquad \sin\theta\sin\psi = Q\,\mathrm{sn}\,u, \qquad \cos\theta = R\,\mathrm{dn}\,u,$$

where P, Q, R are constants and u is a linear function of t, say $\lambda t+\epsilon$; the quantities P, Q, R, λ, and the modulus k of the elliptic functions, are then to be chosen so as to make the above equations coincide with the equations

$$\begin{cases} k^2\,\mathrm{cn}^2\,u = -k'^2+\mathrm{dn}^2\,u, \\ k^2\,\mathrm{sn}^2\,u = \quad 1-\mathrm{dn}^2\,u, \\ \dfrac{d}{du}\,\mathrm{dn}\,u = -k^2\,\mathrm{sn}\,u\,\mathrm{cn}\,u. \end{cases}$$

The comparison gives

$$P^2 = \frac{A(d^2-cC)}{d^2(A-C)}, \qquad Q^2 = \frac{B(d^2-cC)}{d^2(B-C)}, \qquad R^2 = \frac{C(cA-d^2)}{d^2(A-C)},$$

$$k^2 = \frac{(A-B)(d^2-cC)}{(B-C)(Ac-d^2)}, \qquad \lambda^2 = \frac{(B-C)(cA-d^2)}{ABC}.$$

The equation for k^2 shews that k is real, and the equation

$$1-k^2 = \frac{(A-C)(Bc-d^2)}{(B-C)(Ac-d^2)}$$

shews that $(1-k^2)$ is positive, i.e. that $k<1$. The quantities P, Q, R, λ are also evidently real.

Now a real quantity a may be defined by the mutually consistent equations

$$\mathrm{sn}\,ia = i\left\{\frac{C(Ac-d^2)}{A(d^2-cC)}\right\}^{\frac{1}{2}}, \qquad \mathrm{cn}\,ia = \left\{\frac{d^2(A-C)}{A(d^2-cC)}\right\}^{\frac{1}{2}}, \qquad \mathrm{dn}\,ia = \left\{\frac{B(A-C)}{A(B-C)}\right\}^{\frac{1}{2}}$$

* The theory of elliptic functions required in this and the succeeding problems will be found in Whittaker and Watson's *Modern Analysis*, Chs. xx.—xxii.

Since
$$(k')^{-\frac{1}{2}} \operatorname{dn} ia = \frac{\vartheta_{00}(ia/2K)}{\vartheta_{01}(ia/2K)},$$
where the theta-functions are defined by the expansions
$$\vartheta_{00}(\nu) = 1 + 2q\cos 2\pi\nu + 2q^4\cos 4\pi\nu + 2q^9\cos 6\pi\nu + \ldots,$$
$$\vartheta_{01}(\nu) = 1 - 2q\cos 2\pi\nu + 2q^4\cos 4\pi\nu - 2q^9\cos 6\pi\nu + \ldots,$$
$$\vartheta_{10}(\nu) = 2q^{\frac{1}{4}}\cos \pi\nu + 2q^{\frac{9}{4}}\cos 3\pi\nu + 2q^{\frac{25}{4}}\cos 5\pi\nu + \ldots,$$
$$\vartheta_{11}(\nu) = 2q^{\frac{1}{4}}\sin \pi\nu - 2q^{\frac{9}{4}}\sin 3\pi\nu + 2q^{\frac{25}{4}}\sin 5\pi\nu + \ldots,$$
and $q = e^{-\pi K'/K}$, we have
$$\frac{1 + 2q\cosh 2\gamma + 2q^4\cosh 4\gamma + \ldots}{1 - 2q\cosh 2\gamma + 2q^4\cosh 4\gamma - \ldots} = (k')^{-\frac{1}{2}}\left\{\frac{B(A-C)}{A(B-C)}\right\}^{\frac{1}{2}}$$
where γ stands for $\pi a/2K$: from this equation γ (and consequently a) may readily be determined by successive approximation.

The Eulerian angles θ and ψ at time t are now given by the equations
$$\begin{cases} \sin\theta\cos\psi = \dfrac{\operatorname{cn}(\lambda t + \epsilon)}{\operatorname{cn} ia}, \\[2ex] \sin\theta\sin\psi = \dfrac{\operatorname{dn} ia \operatorname{sn}(\lambda t + \epsilon)}{\operatorname{cn} ia}, \\[2ex] \cos\theta = \dfrac{\operatorname{sn} ia \operatorname{dn}(\lambda t + \epsilon)}{i \operatorname{cn} ia}, \end{cases}$$
or (omitting the ϵ)
$$\sin\theta\cos\psi = \frac{\vartheta_{01}(ia/2K)\,\vartheta_{10}(\lambda t/2K)}{\vartheta_{10}(ia/2K)\,\vartheta_{01}(\lambda t/2K)},$$
$$\sin\theta\sin\psi = \frac{\vartheta_{00}(ia/2K)\,\vartheta_{11}(\lambda t/2K)}{\vartheta_{10}(ia/2K)\,\vartheta_{01}(\lambda t/2K)},$$
$$\cos\theta = \frac{\vartheta_{11}(ia/2K)\,\vartheta_{00}(\lambda t/2K)}{i\vartheta_{10}(ia/2K)\,\vartheta_{01}(\lambda t/2K)}.$$

The modulus k of the elliptic functions is known; we can therefore determine the parameter q of the theta-functions by the equation
$$q = \frac{k^2}{16} + \frac{k^4}{32} + \frac{21k^6}{1024} + \ldots,$$
or by the more rapidly convergent series
$$q = \tfrac{1}{2}\tan^2\beta + \tfrac{1}{16}\tan^{10}\beta + \tfrac{15}{512}\tan^{18}\beta + \ldots,$$
where $\cos\beta = (k')^{\frac{1}{2}}$. K may then be calculated from the series
$$(2K/\pi)^{\frac{1}{2}} = \vartheta_{00} = 1 + 2q + 2q^4 + 2q^9 + \ldots,$$
and thus the period $4K/\lambda$ of the inclinations of the axes $Oxyz$ to the line OZ is determined.

If now we write $(\pi a/2K) = \gamma$ and $(\pi\lambda/2K) = \mu$, we have

$$\sin\theta\cos\psi = \frac{(1-2q\cosh 2\gamma + 2q^4\cosh 4\gamma - \ldots)(\cos\mu t + q^2\cos 3\mu t + \ldots)}{(\cosh\gamma + q^2\cosh 3\gamma + \ldots)(1-2q\cos 2\mu t + 2q^4\cos 4\mu t + \ldots)},$$

$$\sin\theta\sin\psi = \frac{(1+2q\cosh 2\gamma + 2q^4\cosh 4\gamma + \ldots)(\sin\mu t - q^2\sin 3\mu t + \ldots)}{(\cosh\gamma + q^2\cosh 3\gamma + \ldots)(1-2q\cos 2\mu t + 2q^4\cos 4\mu t + \ldots)},$$

$$\cos\theta = \frac{(\sinh\gamma - q^2\sinh 3\gamma + \ldots)(1+2q\cos 2\mu t + 2q^4\cos 4\mu t + \ldots)}{(\cosh\gamma + q^2\cosh 3\gamma + \ldots)(1-2q\cos 2\mu t + 2q^4\cos 4\mu t + \ldots)}.$$

The quantities q, μ, γ may be regarded as the constants which specify the motion.

Example. Suppose that the body is a homogeneous ellipsoid of unit density, whose three semi-axes are

$$a=1, \quad b=2, \quad c=3.$$

The three principal moments of inertia are

$$A=\tfrac{4}{15}\pi abc\,(b^2+c^2)=20{\cdot}8\pi, \quad B=16\pi, \quad C=8\pi.$$

Suppose that the initial velocities of rotation round the principal axes are

$$\omega_1=\tfrac{1}{4}, \quad \omega_2=\tfrac{1}{2}, \quad \omega_3=1.$$

The constant of energy is

$$c=A\omega_1^2+B\omega_2^2+C\omega_3^2=13{\cdot}3\pi,$$

and the constant of angular momentum is given by the equation

$$d^2=A^2\omega_1^2+B^2\omega_2^2+C^2\omega_3^2=155{\cdot}04\pi^2,$$

so $\quad d=12{\cdot}452\pi, \quad Ac-d^2=121{\cdot}60\pi^2, \quad Bc-d^2=57{\cdot}76\pi^2, \quad d^2-cC=48{\cdot}64\pi^2.$

The modulus of the elliptic functions is given by the equation

$$k^2=\frac{(A-B)(d^2-cC)}{(B-C)(Ac-d^2)}=0{\cdot}240,$$

whence we have

$$k'^2=1-k^2=0{\cdot}760,$$

$$q=\frac{1}{2}\cdot\frac{1-k'^{\frac{1}{2}}}{1+k'^{\frac{1}{2}}}+2\left\{\frac{1}{2}\cdot\frac{1-k'^{\frac{1}{2}}}{1+k'^{\frac{1}{2}}}\right\}^5+\ldots=0{\cdot}0171,$$

$$\left(\frac{2K}{\pi}\right)^{\frac{1}{2}}=1+2q+2q^4+2q^9+\ldots=1{\cdot}0342,$$

so

$$K=1{\cdot}68013,$$

$$K'=-\frac{K}{\pi}\log_e q=2{\cdot}176.$$

We have also

$$\lambda^2=\frac{(B-C)(Ac-d^2)}{ABC}=0{\cdot}3654,$$

so

$$\lambda=0{\cdot}6045$$

and

$$\mu=\frac{\pi\lambda}{2K}=0{\cdot}5651.$$

The period of the angles θ and ψ is $\dfrac{4K}{\lambda}$ or $\dfrac{2\pi}{\mu}$, which has the value 11·118.

In order to express θ and ψ as trigonometric series in terms of t, we must determine γ. For this we have

$$\frac{B(A-C)}{A(B-C)}=1\cdot 2308,$$

so

$$\left\{\frac{B(A-C)}{A(B-C)}\right\}^{\frac{1}{2}}=1\cdot 1094,$$

and therefore if q^4 be neglected we have

$$\frac{1+2q\cosh 2\gamma}{1-2q\cosh 2\gamma}=\frac{1\cdot 1094}{0\cdot 9337},$$

giving

$$\cosh 2\gamma=2\cdot 503,$$

and hence

$$2\gamma=1\cdot 568$$

and

$$\gamma=0\cdot 784.$$

The quantity a is then given by the equation

$$a=\frac{2K}{\pi}\gamma=0\cdot 8385.$$

A limiting case of the general problem is that in which $A=B$, so that k reduces to zero and the elliptic functions become circular functions. In this case the solution may be written

$$\left.\begin{aligned}\sin\theta\cos\psi&=\frac{\cos\lambda t}{\cosh a}\\ \sin\theta\sin\psi&=\frac{\sin\lambda t}{\cosh a}\\ \cos\theta&=\tanh a\end{aligned}\right\},\quad\text{where}\quad\left\{\begin{aligned}\lambda&=\left\{\frac{(A-C)(Ac-d^2)}{A^2C}\right\}^{\frac{1}{2}}\\ \sinh a&=\left\{\frac{C(Ac-d^2)}{A(d^2-cC)}\right\}^{\frac{1}{2}}\\ \cosh a&=\left\{\frac{d^2(A-C)}{A(d^2-cC)}\right\}^{\frac{1}{2}}\end{aligned}\right.$$

so the motion is a steady precession about the invariable line OZ, the body rotating also about its own axis of symmetry Oz.

Another limiting case is that in which $d^2=cB$, so that $k^2=1$ and the elliptic functions degenerate into hyperbolic functions; this is illustrated by the following examples.

Example 1. *A rigid body is moving about a fixed point under no forces: shew that if (in the notation used above)* $d^2=Bc$, *and if* ω_2 *is zero when* t *is zero,* ω_1 *and* ω_3 *being initially positive, then the direction-cosines of the B-axis at time t, referred to the initial directions of the principal axes, are*

$$a\tanh\chi-\gamma\sin\mu\operatorname{sech}\chi,\quad \cos\mu\operatorname{sech}\chi,\quad \gamma\tanh\chi+a\sin\mu\operatorname{sech}\chi,$$

where

$$\mu=\frac{dt}{B},\quad \chi=\frac{dt}{B}\left\{\frac{(A-B)(B-C)}{AC}\right\}^{\frac{1}{2}},\quad a=\left\{\frac{A(B-C)}{B(A-C)}\right\}^{\frac{1}{2}},\quad \gamma=\left\{\frac{C(A-B)}{B(A-C)}\right\}^{\frac{1}{2}}.$$

(Camb. Math. Tripos, Part I, 1899.)

To obtain this result, we observe that when $Bc=d^2$, the differential equation for the coordinate θ becomes

$$\frac{d}{dt}(\sec\theta)=d\left(\frac{B-C}{BC}\right)^{\frac{1}{2}}\left\{\frac{Ac-d^2}{Ad^2}\sec^2\theta-\frac{A-C}{AC}\right\}^{\frac{1}{2}},$$

the integral of which is

$$\cos\theta=\gamma\operatorname{sech}\chi,$$

where γ and χ are the quantities above defined. The equation

$$\frac{A-B}{AB}\sin^2\theta\sin^2\psi=\frac{Ac-d^2}{Ad^2}-\frac{A-C}{AC}\cos^2\theta$$

then gives

$$\sin\theta\sin\psi=\tanh\chi,$$

and the equation

$$\dot{\phi}=\frac{d}{A}\cos^2\psi+\frac{d}{B}\sin^2\psi$$

gives

$$\sin(\phi-\mu)=-\gamma\sin\psi.$$

These equations shew that the direction-cosines of the B-axis referred to the axes $OXYZ$, which (§ 10) are

$$-\cos\phi\cos\theta\sin\psi-\sin\phi\cos\psi,\quad -\sin\phi\cos\theta\sin\psi+\cos\phi\cos\psi,\quad \sin\theta\sin\psi,$$

can be written

$$-\sin\mu\,\text{sech}\,\chi,\quad \cos\mu\,\text{sech}\,\chi,\quad \tanh\chi.$$

But if ω_{10}, ω_{20}, ω_{30} denote the initial directions of the principal axes, since

$$A^2\omega_1{}^2+C^2\omega_3{}^2=d^2=Bc=B(A\omega_1{}^2+C\omega_3{}^2),$$

so that $A\omega_1=ad$ and $C\omega_3=\gamma d$, we see that the direction-cosines of ω_{10}, ω_{20}, ω_{30}, referred to $OXYZ$, are given by the scheme

	X	Y	Z
ω_{10}	γ	0	a
ω_{20}	0	1	0
ω_{30}	$-a$	0	γ

and hence the direction-cosines of the B-axis, referred to ω_{10}, ω_{20}, ω_{30}, are

$$-\gamma\sin\mu\,\text{sech}\,\chi+a\tanh\chi,\quad \cos\mu\,\text{sech}\,\chi,\quad a\sin\mu\,\text{sech}\,\chi+\gamma\tanh\chi.$$

Example 2. When $d^2=cB$, shew that the axis Oy describes, on a sphere with the fixed point as centre, a rhumb line with respect to the meridians passing through the invariable line. (Coll. Exam.)

Returning now to the general case, we have to express the third Eulerian angle ϕ in terms of the time. We have

$$\dot{\phi}=\frac{d}{A}+d\left(\frac{1}{B}-\frac{1}{A}\right)\sin^2\psi.$$

Now

$$\cot\psi=\frac{\text{cn}\,\lambda t}{\text{dn}\,ia\,\text{sn}\,\lambda t},$$

whence

$$\sin^2\psi=\frac{\text{dn}^2\,ia\,\text{sn}^2\,\lambda t}{1-k^2\,\text{sn}^2\,ia\,\text{sn}^2\,\lambda t}.$$

This function of t vanishes with t, and has poles when the denominator vanishes, i.e. when

$$\text{sn}\,\lambda t=\pm\frac{1}{k\,\text{sn}\,ia}=\pm\,\text{sn}\,(ia\pm iK');$$

so in one period-parallelogram $(2K, 2iK')$ it has poles at the points

$$\lambda t=ia+iK' \text{ and } \lambda t=-ia+iK'.$$

Near the former of these points, writing $\lambda t = ia + iK + \epsilon$, and retaining only the lowest powers of ϵ, we have

$$\sin^2\psi = \frac{\mathrm{dn}^2\, ia/k^2\,\mathrm{sn}^2\, ia}{1-\{\mathrm{sn}^2\, ia/\mathrm{sn}^2\,(ia+\epsilon)\}},$$

$$= \frac{\mathrm{dn}^2\, ia}{k^2\,\mathrm{sn}^2\, ia + \epsilon k^2 . 2\,\mathrm{sn}\, ia\,\mathrm{cn}\, ia\,\mathrm{dn}\, ia - k^2\,\mathrm{sn}^2\, ia},$$

so the residue at this pole of $\sin^2\psi$, considered as a function of λt, is

$$\frac{\mathrm{dn}\, ia}{2k^2\,\mathrm{sn}\, ia\,\mathrm{cn}\, ia}, \text{ or } \frac{1}{2id\,(A-B)}\left\{\frac{(B-C)(Ac-d^2)AB}{C}\right\}^{\frac{1}{2}}.$$

Therefore the residue of $d\left(\frac{1}{B}-\frac{1}{A}\right)\sin^2\psi$ at this point (considered as a function of λt) is

$$\frac{1}{2i}\left\{\frac{(B-C)(Ac-d^2)}{ABC}\right\}^{\frac{1}{2}} \text{ or } \frac{\lambda}{2i}:$$

and the residue when $\lambda t/2K$ is regarded as the variable is consequently $-i\lambda/4K$. As we now know the zeros, poles and residues of this function, we can write down its expression as a sum of logarithmic derivates of theta-functions: in fact, since $\vartheta_{01}(\nu)$ has a simple zero at $\nu = \frac{1}{2}\omega = iK'/2K$, we have

$$\phi = \frac{d}{A} - \frac{i\lambda}{4K}\left\{\frac{\vartheta_{01}'\left(\frac{\lambda t - ia}{2K}\right)}{\vartheta_{01}\left(\frac{\lambda t - ia}{2K}\right)} - \frac{\vartheta_{01}'\left(\frac{\lambda t + ia}{2K}\right)}{\vartheta_{01}\left(\frac{\lambda t + ia}{2K}\right)} + 2\frac{\vartheta_{01}'\left(\frac{ia}{2K}\right)}{\vartheta_{01}\left(\frac{ia}{2K}\right)}\right\},$$

and therefore

$$e^{2i\phi} = \text{constant}\,.\,\frac{\vartheta_{01}\left(\frac{\lambda t - ia}{2K}\right)}{\vartheta_{01}\left(\frac{\lambda t + ia}{2K}\right)}\,.\,e^{\left\{\frac{2id}{A} + \frac{\lambda}{K}\frac{\vartheta_{01}'(ia/2K)}{\vartheta_{01}(ia/2K)}\right\}t}.$$

Now $\vartheta_{01}\left(\frac{\lambda t - ia}{2K}\right)\Big/\vartheta_{01}\left(\frac{\lambda t + ia}{2K}\right)$ is purely periodic with respect to the real period $2K/\lambda$ of t, so the exponential on the right-hand side gives the *mean motion* of ϕ, i.e. the precessional motion of the system round the invariable line. We have

$$\vartheta_{01}(\nu) = 1 - 2q\cos 2\pi\nu + 2q^4\cos 4\pi\nu - \ldots,$$

$$\vartheta_{01}'(\nu) = 4\pi q\sin 2\pi\nu - 8\pi q^4\sin 4\pi\nu + \ldots,$$

so the coefficient of t in ϕ, i.e. the constant part of $\dot\phi$, or the precession, which is

$$\frac{d}{A} + \frac{\lambda}{2iK}\frac{\vartheta_{01}'(ia/2K)}{\vartheta_{01}(ia/2K)},$$

may be written

$$\frac{d}{A} + 4\mu\,.\,\frac{q\sinh 2\gamma - 2q^4\sinh 4\gamma + \ldots}{1 - 2q\cosh 2\gamma + 2q^4\cosh 4\gamma - \ldots},$$

in which form it may be calculated readily.

Example 1. In the case previously discussed, of an ellipsoid whose semi-axes are $a=1$, $b=2$, $c=3$, we have

$$2\gamma=1\cdot568,\quad \sinh 2\gamma=2\cdot294,\quad \cosh 2\gamma=2\cdot503,$$

$$d=12\cdot452\pi,\quad A=20\cdot8\pi,\quad \mu=0\cdot5651,\quad q=0\cdot0171,$$

so the mean motion of ϕ, which when q^4 is neglected may be written

$$\frac{d}{A}+4\mu\,.\,\frac{q\sinh 2\gamma}{1-2q\cosh 2\gamma},$$

is

$$0\cdot5986+0\cdot0970,$$

or

$$0\cdot6956.$$

Example 2. A uniform circular disc has its centre O fixed, and moves under the action of no external forces. The disc is given initial angular velocities Ω about a diameter coinciding with $O\xi$ in space, and n about its axis coinciding with $O\zeta$ in space. Shew that at any subsequent time

$$\chi=2\arcsin\left[\frac{\Omega}{(\Omega^2+4n^2)^{\frac{1}{2}}}\sin\{(\Omega^2+4n^2)^{\frac{1}{2}}\,.\,\tfrac{1}{2}t\}\right],$$

$$\omega=\operatorname{arccot}\left[\frac{\Omega}{(\Omega^2+4n^2)^{\frac{1}{2}}}\tan\{(\Omega^2+4n^2)^{\frac{1}{2}}\,.\,\tfrac{1}{2}t\}\right],$$

where χ is the angle between $O\zeta$ and the axis of the disc Oz and ω is the angle between the planes $\zeta O\xi$ and ζOz. (Coll. Exam.)

For let OZ denote as usual the invariable line, and consider the spherical triangle $Z\zeta z$, whose vertices are the intersections of the lines OZ, $O\zeta$, Oz respectively with a sphere of centre O. In this spherical triangle we have $Zz=\theta$, $\zeta\hat{Z}z=\phi$. Moreover we have for the disc $C=2B=2A$, so

$$d^2=A^2\Omega^2+C^2n^2=A^2(\Omega^2+4n^2)$$

and

$$\frac{d}{A}=(\Omega^2+4n^2)^{\frac{1}{2}}.$$

The equations of motion for θ and ϕ therefore become

$$\dot\theta=0,\qquad \dot\phi=d/A=(\Omega^2+4n^2)^{\frac{1}{2}},$$

so

$$\theta=Z\zeta=\arccos\frac{2n}{(\Omega^2+4n^2)^{\frac{1}{2}}},\quad \phi=(\Omega^2+4n^2)^{\frac{1}{2}}t.$$

In the spherical triangle $Z\zeta z$, we have therefore

$$Z\zeta=Zz=\arccos\frac{2n}{(\Omega^2+4n^2)^{\frac{1}{2}}},\quad \zeta\hat{Z}z=(\Omega^2+4n^2)^{\frac{1}{2}}t,\quad Z\hat{\zeta}z=\omega,\quad \zeta z=\chi,$$

and hence

$$\sin\frac{\chi}{2}=\sin Z\zeta\sin\tfrac{1}{2}\zeta Zz=\frac{\Omega}{(\Omega^2+4n^2)^{\frac{1}{2}}}\sin\{(\Omega^2+4n^2)^{\frac{1}{2}}\,.\,\tfrac{1}{2}t\}$$

and

$$\cot\omega=\cos Z\zeta\tan\tfrac{1}{2}\zeta Zz=\frac{2n}{(\Omega^2+4n^2)^{\frac{1}{2}}}\tan\{(\Omega^2+4n^2)^{\frac{1}{2}}\,.\,\tfrac{1}{2}t\},$$

which are the required equations.

70. *Poinsot's kinematical representation of the motion; the polhode and herpolhode.*

An elegant method of representing kinematically the motion of a body about a fixed point under no forces is the following, which is due to Poinsot*.

* Poinsot, *Théorie nouvelle de la rotation des corps*, Paris, 1834.

The equation of the momental ellipsoid of the body at the fixed point, referred to the moving axes $Oxyz$, is

$$Ax^2 + By^2 + Cz^2 = 1.$$

Consider the tangent-plane to the ellipsoid which is perpendicular to the invariable line. If p denotes the perpendicular on this tangent-plane from the origin, we have (since the direction-cosines of p are $A\omega_1/d$, $B\omega_2/d$, $C\omega_3/d$)

$$p^2 = \frac{A\omega_1^2 + B\omega_2^2 + C\omega_3^2}{A^2\omega_1^2 + B^2\omega_2^2 + C^2\omega_3^2}$$

$$= \frac{c}{d^2}, \text{ which is constant.}$$

Since the perpendicular on the plane is constant in magnitude and direction, the plane is fixed in space: so the momental ellipsoid always touches a fixed plane.

Moreover, if (x', y', z') are the coordinates of the point of contact of the ellipsoid and the plane, we have on identifying the equations

$$Axx' + Byy' + Czz' = 1 \quad \text{and} \quad A\omega_1 x + B\omega_2 y + C\omega_3 z = pd$$

the values $$x' = \frac{\omega_1}{pd} = \frac{\omega_1}{\sqrt{c}}, \quad y' = \frac{\omega_2}{pd} = \frac{\omega_2}{\sqrt{c}}, \quad z' = \frac{\omega_3}{pd} = \frac{\omega_3}{\sqrt{c}},$$

and hence the radius vector to the point (x', y', z') is the instantaneous axis of rotation of the body. It follows that *the body moves as if it were rigidly connected to its momental ellipsoid, and the latter body were to roll about the fixed point on a fixed plane perpendicular to the invariable line, without sliding; the angular velocity being proportional to the radius to the point of contact, so that the component of angular velocity about the invariable line is constant.*

Example 1. If a body which is moveable about a fixed point is initially at rest and then is acted on continually by a couple of constant magnitude and orientation, shew that Poinsot's construction still holds good, but that the component angular velocity about the invariable line is no longer constant but varies directly as the time. (Coll. Exam.)

For in any interval of time dt the addition of angular momentum to the body is Ndt about the fixed axis OZ of the couple; so that the resultant angular momentum of the system at time t is Nt about OZ. Now the components of angular momentum about the principal axes of inertia $Oxyz$ are $A\omega_1$, $B\omega_2$, $C\omega_3$, where A, B, C are the principal moments of inertia and $(\omega_1, \omega_2, \omega_3)$ are the components of angular velocity: hence we have

$$A\omega_1 = -Nt \sin\theta \cos\psi, \quad B\omega_2 = Nt \sin\theta \sin\psi, \quad C\omega_3 = Nt \cos\theta,$$

where θ, ϕ, ψ are the Eulerian angles which fix the position of the axes $Oxyz$ with reference to fixed axes $OXYZ$. But these equations differ from those which occur in the motion of a body under no forces only in the substitution of $t\,dt$ for dt; so the motion will be the same as in the problem of motion under no forces, except that the velocities are multiplied by t; whence the result follows.

Example 2. In the motion of a body, one of whose points is fixed, under no forces, let a hyperboloid be rigidly connected with the body, so as to have the principal axes of inertia of the body at the point as axes, and to have the squares of its axes respectively proportional to $d^2 - Ac$, $d^2 - Bc$, $d^2 - Cc$, where A, B, C are the moments of inertia of the body at the fixed point, c is twice its kinetic energy, and d is the resultant angular momentum. Shew that the motion of this hyperboloid can be represented by causing it to roll without sliding on a circular cylinder, whose axis passes through the fixed point and is parallel to the axis of resultant angular momentum. (Siacci.)

The curve which in Poinsot's construction is traced on the momental ellipsoid by the point of contact with the fixed plane is called the *polhode*. Its equations, referred to the principal moments of inertia, are clearly the equation of the ellipsoid together with the equation $p =$ constant, i.e. they are

$$Ax^2 + By^2 + Cz^2 = 1,$$

$$A^2x^2 + B^2y^2 + C^2z^2 = d^2/c.$$

Example 1. Shew that when $A = B$, the polhode is a circle.

Example 2. Taking $A \geqslant B \geqslant C$, shew that there are two kinds of polhodes, one kind consisting of curves which surround the axis Oz of the momental ellipsoid, and correspond to $cB > d^2 > cC$, while the other kind consists of curves which surround the axis Ox, and correspond to $cA > d^2 > cB$; and that the limiting case between these two kinds of polhodes is a singular polhode which corresponds to $cB - d^2 = 0$, and consists of two ellipses which pass through the extremities of the mean axis.

The curve which is traced on the fixed plane by the point of contact with the moving ellipsoid is called the *herpolhode*.

To find the equation of the herpolhode, let ρ, χ be the polar coordinates of the point of contact, when the foot of the perpendicular from the fixed point on the fixed plane is taken as pole. If (x', y', z') denote the coordinates of the same point referred to the moving axes $Oxyz$, we have

$$x'^2 + y'^2 + z'^2 = \text{square of radius from point of suspension to point of contact}$$
$$= \rho^2 + \frac{c}{d^2}.$$

Substituting for x', y', z' their values as given by the equations

$$\begin{cases} x' = \omega_1/\sqrt{c} = -\,d\sin\theta\cos\psi/A\sqrt{c}, \\ y' = \omega_2/\sqrt{c} = \quad d\sin\theta\sin\psi/B\sqrt{c}, \\ z' = \omega_3/\sqrt{c} = \quad d\cos\theta/C\sqrt{c}, \end{cases}$$

we have

$$\rho^2 = -\frac{c}{d^2} + \frac{d^2}{A^2c}\sin^2\theta\cos^2\psi + \frac{d^2}{B^2c}\sin^2\theta\sin^2\psi + \frac{d^2}{C^2c}\cos^2\theta.$$

Replacing θ and ψ by their values in terms of t, this becomes

$$\rho^2 = \frac{(cA - d^2)(d^2 - cC)}{cd^2A^2B^2C^2}\left\{ACB^2 - \frac{(B-C)(A-B)\,d^2}{\wp(t) - e_3}\right\}$$

$$= \frac{(cA - d^2)(d^2 - cC)}{cd^2AC}\,\frac{\wp(t) - \wp(l+\omega)}{\wp(t) - e_3}.$$

where ω denotes the half-period corresponding to the root e_1; this equation expresses the radius vector of the herpolhode in terms of the time.

We have next to find the vectorial angle χ in terms of t. For this we observe that $\sqrt{c}\,\rho^2\dot{\chi}/d$ is six times the volume of the tetrahedron whose vertices are the fixed point, the foot of the perpendicular from the fixed point on the fixed plane, and two consecutive positions of the point of contact, divided by the interval of time elapsed between these positions, and that this quantity can also be expressed in the form

$$\begin{vmatrix} x', & y', & z' \\ Acx'/d^2, & Bcy'/d^2, & Ccz'/d^2 \\ \dot{x}', & \dot{y}', & \dot{z}' \end{vmatrix} \text{ or } \frac{c}{d^2}x'y'z' \begin{vmatrix} 1, & 1, & 1 \\ A, & B, & C \\ \dot{x}'/x', & \dot{y}'/y', & \dot{z}'/z' \end{vmatrix}$$

All the quantities involved, except $\dot{\chi}$, are known functions of t: on substituting their values in terms of t, and reducing, we have

$$\dot{\chi} = \frac{d}{B\{\wp(t) - \wp(l+\omega)\}}\left\{\wp(t) - \frac{(B-C)e_2 + (A-B)e_1}{A-C}\right\},$$

which can be written in the form

$$\dot{\chi} = \frac{d}{B} + \frac{i}{2}\frac{\wp'(l+\omega)}{\wp(t) - \wp(l+\omega)}.$$

This equation can be integrated in the same way as the equation for the Eulerian angle ϕ, and gives

$$e^{2i(\chi-\chi_0)} = e^{\{2id/B - 2\zeta(l+\omega)\}t}\frac{\sigma(t+l+\omega)}{\sigma(t-l-\omega)},$$

where χ_0 is a constant of integration. The current coordinates (ρ, χ) of the herpolhode are thus expressed as functions of t.

Example 1. A particle moves in such a way that its angular momentum round the origin is a linear function of the square of the radius vector, while the square of its velocity is a quadratic function of the square of the radius vector, the coefficient of the highest power being negative; shew that the path is the herpolhode of a Poinsot motion, in which however A, B, C are not restricted to be positive.

Example 2. Discuss the cases in which the polhode consists of (α) two ellipses intersecting on the mean axis of the momental ellipsoid, (β) two parallel circles, (γ) two points; shewing that in these cases the herpolhode becomes respectively a spiral curve (whose equation can be expressed in terms of elementary functions), a circle, or a point.

71. *Motion of a top on a perfectly rough plane; determination of the Eulerian angle θ.*

A *top* is defined to be a material body which is symmetrical about an axis and terminates in a sharp point (called the *apex* or *vertex*) at one end of the axis.

We shall now study the motion of a top when spinning with its apex O placed on a perfectly rough plane, so that O is practically a fixed point. The problem is essentially that of determining the motion of a solid of revolution under the influence of gravity, when a point on its axis is fixed in space*.

Let (A, A, C) denote the moments of inertia of the top about rectangular axes $Oxyz$, fixed relative to the top and moving with it, the origin being the apex and the axis Oz being the axis of symmetry of the top; let (θ, ϕ, ψ) be the Eulerian angles defining the position of these axes with reference to fixed rectangular axes $OXYZ$, of which OZ is directed vertically upwards.

The kinetic energy is (§ 63)

$$T = \tfrac{1}{2}(A\omega_1^2 + A\omega_2^2 + C\omega_3^2),$$

where $\omega_1, \omega_2, \omega_3$ denote the components relative to the moving axes of the angular velocity of the top, so that (§ 16) we have

$$\begin{cases} \omega_1 = \dot\theta \sin\psi - \dot\phi \sin\theta \cos\psi, \\ \omega_2 = \dot\theta \cos\psi + \dot\phi \sin\theta \sin\psi, \\ \omega_3 = \dot\psi + \dot\phi \cos\theta; \end{cases}$$

the kinetic energy is therefore

$$T = \tfrac{1}{2}A\dot\theta^2 + \tfrac{1}{2}A\dot\phi^2 \sin^2\theta + \tfrac{1}{2}C(\dot\psi + \dot\phi\cos\theta)^2,$$

and the potential energy is $V = Mgh\cos\theta$, where M is the mass of the top and h is the distance of its centre of gravity from the apex.

The kinetic potential is therefore

$$L = T - V = \tfrac{1}{2}A\dot\theta^2 + \tfrac{1}{2}A\dot\phi^2 \sin^2\theta + \tfrac{1}{2}C(\dot\psi + \dot\phi\cos\theta)^2 - Mgh\cos\theta.$$

The coordinates ϕ and ψ are evidently ignorable; the corresponding integrals are

$$\frac{\partial T}{\partial \dot\phi} = \text{constant, and } \frac{\partial T}{\partial \dot\psi} = \text{constant,}$$

or

$$A\dot\phi \sin^2\theta + C(\dot\psi + \dot\phi\cos\theta)\cos\theta = a,$$
$$C(\dot\psi + \dot\phi\cos\theta) = b,$$

where a and b are constants: the former of these may be interpreted as the integral of angular momentum about the axis OZ, and so is obvious *à priori* from general dynamical principles.

The modified kinetic potential (§ 38) is

$$R = L - a\dot\phi - b\dot\psi$$
$$= \tfrac{1}{2}A\dot\theta^2 - \frac{(a - b\cos\theta)^2}{2A\sin^2\theta} - \frac{b^2}{2C} - Mgh\cos\theta.$$

The term $-b^2/2C$ can be neglected, as it is merely a constant; the equation of motion is

$$\frac{d}{dt}\left(\frac{\partial R}{\partial \dot\theta}\right) - \frac{\partial R}{\partial \theta} = 0,$$

* Lagrange, *Méc. Anal.* (*Oeuvres*, XII. p. 251).

so the variation of θ is the same as in a dynamical system with one degree of freedom for which the kinetic energy is $\frac{1}{2}A\dot{\theta}^2$ and the potential energy is

$$\frac{(a - b\cos\theta)^2}{2A\sin^2\theta} + Mgh\cos\theta.$$

The connexion between θ and t is therefore given by the integral of energy of this reduced system, namely

$$\tfrac{1}{2}A\dot{\theta}^2 = -\frac{(a - b\cos\theta)^2}{2A\sin^2\theta} - Mgh\cos\theta + c,$$

where c is a constant.

Writing $\cos\theta = x$, this equation becomes

$$A^2\dot{x}^2 = -(a - bx)^2 - 2AMgh(x - x^3) + 2Ac(1 - x^2).$$

The right-hand side of this equation is a cubic polynomial in x; now when $x = -1$, the cubic is negative; for some real values of θ, i.e. for some values of x between -1 and 1, the cubic must be positive, since the left-hand side of the equation is positive; when $x = 1$, the cubic is again negative; and when $x = +\infty$, the cubic is positive. The cubic has therefore two real roots which lie between -1 and 1, and the remaining root is also real and is greater than unity. Let these roots be denoted by

$$\cos\alpha, \quad \cos\beta, \quad \cosh\gamma,$$

where $\cos\beta > \cos\alpha$, so that $\alpha > \beta$.

The differential equation now becomes

$$(Mgh/2A)^{\frac{1}{2}}\,dt = \{4(x - \cos\alpha)(x - \cos\beta)(x - \cosh\gamma)\}^{-\frac{1}{2}}\,dx.$$

If we write

$$x = \frac{2A}{Mgh}z + \tfrac{1}{3}(\cos\alpha + \cos\beta + \cosh\gamma) = \frac{2A}{Mgh}z + \frac{2Ac + b^2}{6AMgh},$$

we have therefore $t + \text{constant} = \int \{4(z - e_1)(z - e_2)(z - e_3)\}^{-\frac{1}{2}}dz$,

where the constants e_1, e_2, e_3 are given by the equations

$$\begin{cases} e_1 = \dfrac{Mgh}{2A}\cosh\gamma - \dfrac{2Ac + b^2}{12A^2}, \\[2ex] e_2 = \dfrac{Mgh}{2A}\cos\beta - \dfrac{2Ac + b^2}{12A^2}, \\[2ex] e_3 = \dfrac{Mgh}{2A}\cos\alpha - \dfrac{2Ac + b^2}{12A^2}, \end{cases}$$

so that e_1, e_2, e_3 are all real and satisfy the relations

$$e_1 + e_2 + e_3 = 0, \quad e_1 > e_2 > e_3.$$

The connexion between z and t is therefore

$$z = \wp(t + \epsilon),$$

where ϵ is a constant of integration, and the function $\wp$ is formed with the roots e_1, e_2, e_3; and hence we have

$$x = \frac{2A}{Mgh}\wp(t+\epsilon) + \frac{2Ac+b^2}{6AMgh}.$$

Now in order that $\dot{x}$ may be real for real values of t, it is evident that x must lie between $\cos\alpha$ and $\cos\beta$, i.e. $\wp(t+\epsilon)$ must lie between e_2 and e_3 for real values of t: and therefore the imaginary part of the constant ϵ must be the half-period ω_3 corresponding to the root e_3. The real part of ϵ depends on the epoch from which the time is measured, and so can be taken to be zero by suitably choosing this epoch. We have therefore finally

$$\cos\theta = \frac{2A}{Mgh}\wp(t+\omega_3) + \frac{2Ac+b^2}{6AMgh},$$

and this is the equation which expresses the Eulerian angle θ in terms of the time*.

Example 1. If the circumstances of projection of the top are such that initially

$$\theta = 60^\circ, \quad \dot\theta = 0, \quad \dot\phi = 2(Mgh/3A)^{\frac{1}{2}}, \quad \dot\psi = (3A-C)(Mgh/3AC^2)^{\frac{1}{2}},$$

shew that the value of θ at any time t is given by the equation

$$\sec\theta = 1 + \operatorname{sech}\left(\sqrt{\frac{Mgh}{A}}\,t\right),$$

so that the axis of the top continually approaches the vertical.

For in this case we readily find for the constants a, b, c the values

$$a = b = (3MghA)^{\frac{1}{2}}, \quad c = Mgh,$$

so the differential equation to determine x is

$$\left(\frac{dx}{dt}\right)^2 = \frac{Mgh}{A}(1-x)^2(2x-1),$$

whence the result follows.

Example 2. A solid of revolution can turn freely about a fixed point in its axis of symmetry, and is acted on by forces derived from a potential-energy function $\mu\cot^2\theta$, where θ is the angle between this axis and a fixed line; shew that the equations of motion can be integrated in terms of elementary functions.

For proceeding as in the problem of the top on the perfectly rough plane, we find for the integral of energy of the reduced problem the equation

$$\tfrac{1}{2}A\dot\theta^2 = -\frac{(a-b\cos\theta)^2}{2A\sin^2\theta} - \mu\frac{\cos^2\theta}{\sin^2\theta} + c.$$

Writing $\cos\theta = x$, this becomes

$$A^2\dot{x}^2 = -(a-bx)^2 - 2A\mu x^2 + 2Ac(1-x^2).$$

The quadratic on the right-hand side is negative when $x=1$ and $x=-1$, but is positive for some values of x between -1 and $+1$, since the left-hand side is positive for some real

* It may be remarked that the present problem reduces to that of the spherical pendulum (§ 55) when the quantities M, C, A, h, a, b, c, $\cos\theta$, ϕ, l, k are replaced respectively by 1, 0, l^2, l, k, 0, h, z/l, ϕ, λ, μ.

values of θ: the quadratic has therefore two real roots between -1 and $+1$. Calling these $\cos\alpha$ and $\cos\beta$, the equation is of the form

$$\lambda^2\dot{x}^2 = (\cos\alpha - x)(x - \cos\beta),$$

the solution of which is

$$x = \cos\alpha \sin^2(t/2\lambda) + \cos\beta\cos^2(t/2\lambda).$$

72. *Determination of the remaining Eulerian angles, and of the Cayley-Klein parameters; the spherical top.*

When the Eulerian angle θ has been obtained in terms of the time, as in the last article, it remains to determine the other Eulerian angles ϕ and ψ. For this purpose we use the two integrals corresponding to the ignorable coordinates: these, when solved for $\dot\phi$ and $\dot\psi$, give

$$\begin{cases} \dot\phi = \dfrac{a - b\cos\theta}{A\sin^2\theta}, \\ \dot\psi = \dfrac{b}{C} - \dfrac{(a - b\cos\theta)\cos\theta}{A\sin^2\theta}. \end{cases}$$

If we regard the motion as specified by the constants of the body (M, A, C, h) and the constants of integration (a, b, c), it is evident from these equations and the equation for $\dot\theta$ that C does not occur except in the constant term of the expression for $\dot\psi$; and therefore an auxiliary top whose moments of inertia are (A, A, A) can be projected in such a way that its axis of symmetry always occupies the same position as the axis of symmetry of the top considered, the only difference in the motion of the two tops being that the auxiliary top has throughout the motion a constant extra spin $b(C-A)/AC$ about its axis of symmetry. A top such as this auxiliary top, whose moments of inertia are all equal, is called a *spherical top*. It follows therefore that the motion of any top can be simply expressed in terms of the motion of a spherical top, and that there is no real loss of generality in supposing any top under consideration to be spherical.

If then we take $C = A$, the equations to determine ϕ and ψ become

$$\begin{cases} \dot\phi = \dfrac{a - b\cos\theta}{A\sin^2\theta} = \dfrac{a+b}{2A(\cos\theta+1)} - \dfrac{a-b}{2A(\cos\theta-1)}, \\ \dot\psi = \dfrac{b - a\cos\theta}{A\sin^2\theta} = \dfrac{a+b}{2A(\cos\theta+1)} + \dfrac{a-b}{2A(\cos\theta-1)}. \end{cases}$$

Substituting for $\cos\theta$ its value from the equation

$$\cos\theta = \frac{2A}{Mgh}\wp(t+\omega_3) + \frac{2Ac+b^2}{6AMgh},$$

and writing

$$\wp(l) = \frac{Mgh}{2A} - \frac{2Ac+b^2}{12A^2},$$

$$\wp(k) = -\frac{Mgh}{2A} - \frac{2Ac+b^2}{12A^2},$$

so that l and k are known imaginary constants (being in fact the values of $t+\omega_3$ corresponding to the values 0 and π of θ), the differential equations become

$$\begin{cases} \dot{\phi} = \dfrac{Mgh\,(a+b)}{4A^2}\cdot\dfrac{1}{\wp(t+\omega_3)-\wp(k)} - \dfrac{Mgh\,(a-b)}{4A^2}\cdot\dfrac{1}{\wp(t+\omega_3)-\wp(l)}, \\ \dot{\psi} = \dfrac{Mgh\,(a+b)}{4A^2}\cdot\dfrac{1}{\wp(t+\omega_3)-\wp(k)} + \dfrac{Mgh\,(a-b)}{4A^2}\cdot\dfrac{1}{\wp(t+\omega_3)-\wp(l)}. \end{cases}$$

Now the connexion between the function $\wp$ and its derivate $\wp'$ can be at once written down by substituting for x from the equation

$$x = \frac{2A}{Mgh}\,\wp(t+\omega_3) + \frac{2Ac+b^2}{6AMgh}$$

in the equation

$$A^2\left(\frac{dx}{dt}\right)^2 = -(a-bx)^2 - 2AMgh\,(x-x^3) + 2Ac\,(1-x^2);$$

if the argument of the $\wp$-function is k, it follows from the definition of k that the corresponding value of x is -1; and so the last equation gives

$$A^2.\{2A\wp'(k)/Mgh\}^2 = -(a+b)^2,$$

or

$$\wp'(k) = iMgh\,(a+b)/2A^2.$$

Similarly we have

$$\wp'(l) = iMgh\,(a-b)/2A^2,$$

and therefore the equations for ϕ and ψ can be written in the form

$$\begin{cases} 2i\dot{\phi} = \dfrac{\wp'(k)}{\wp(t+\omega_3)-\wp(k)} - \dfrac{\wp'(l)}{\wp(t+\omega_3)-\wp(l)}, \\ 2i\dot{\psi} = \dfrac{\wp'(k)}{\wp(t+\omega_3)-\wp(k)} + \dfrac{\wp'(l)}{\wp(t+\omega_3)-\wp(l)}. \end{cases}$$

Now the function

$$\frac{\wp'(k)}{\wp(t+\omega_3)-\wp(k)}$$

is an elliptic function, whose poles in any period-parallelogram are congruent with $t+\omega_3=k$ and $t+\omega_3=-k$, the corresponding residues being 1 and -1; and the function is zero when $t+\omega_3=0$. Hence we have

$$\frac{\wp'(k)}{\wp(t+\omega_3)-\wp(k)} = \zeta(t+\omega_3-k) - \zeta(t+\omega_3+k) + 2\zeta(k),$$

and therefore

$$\int\frac{\wp'(k)\,dt}{\wp(t+\omega_3)-\wp(k)} = \log\frac{\sigma(t+\omega_3-k)}{\sigma(t+\omega_3+k)} + 2\zeta(k)\,t + \text{constant}.$$

The integrals of the equations for ϕ and ψ can therefore be written in the form

$$\begin{cases} e^{2i(\phi-\phi_0)} = e^{2\{\zeta(k)-\zeta(l)\}t} \cdot \dfrac{\sigma(t+\omega_3-k)\,\sigma(t+\omega_3+l)}{\sigma(t+\omega_3+k)\,\sigma(t+\omega_3-l)}, \\[2ex] e^{2i(\psi-\psi_0)} = e^{2\{\zeta(k)+\zeta(l)\}t} \cdot \dfrac{\sigma(t+\omega_3-k)\,\sigma(t+\omega_3-l)}{\sigma(t+\omega_3+k)\,\sigma(t+\omega_3+l)}, \end{cases}$$

where ϕ_0 and ψ_0 are constants of integration.

These equations lead to simple expressions for the Cayley-Klein parameters α, β, γ, δ (§ 12), which define the position of the moving axes $Oxyz$ with reference to the fixed axes $OXYZ$: for by definition we have

$$\alpha = \cos\tfrac{1}{2}\theta \,.\, e^{\frac{1}{2}i(\phi+\psi)}, \qquad \beta = i\sin\tfrac{1}{2}\theta \,.\, e^{\frac{1}{2}i(\phi-\psi)},$$

$$\gamma = i\sin\tfrac{1}{2}\theta \,.\, e^{\frac{1}{2}i(\psi-\phi)}, \qquad \delta = \cos\tfrac{1}{2}\theta \,.\, e^{-\frac{1}{2}i(\phi+\psi)}.$$

But we have

$$\begin{aligned} 2\cos^2\tfrac{1}{2}\theta &= 1+\cos\theta \\ &= 1 + \frac{2A}{Mgh}\wp(t+\omega_3) + \frac{2Ac+b^2}{6AMgh} \\ &= \frac{2A}{Mgh}\{\wp(t+\omega_3)-\wp(k)\} \\ &= -\frac{2A}{Mgh} \cdot \frac{\sigma(t+\omega_3+k)\,\sigma(t+\omega_3-k)}{\sigma^2(k)\,\sigma^2(t+\omega_3)}, \end{aligned}$$

or $$\cos\tfrac{1}{2}\theta = \left(\frac{-A}{Mgh}\right)^{\frac{1}{2}} \frac{\{\sigma(t+\omega_3+k)\,\sigma(t+\omega_3-k)\}^{\frac{1}{2}}}{\sigma(k)\,\sigma(t+\omega_3)}.$$

Similarly we find

$$\sin\tfrac{1}{2}\theta = \left(\frac{A}{Mgh}\right)^{\frac{1}{2}} \frac{\{\sigma(t+\omega_3+l)\,\sigma(t+\omega_3-l)\}^{\frac{1}{2}}}{\sigma(l)\,\sigma(t+\omega_3)},$$

and on combining these with the expressions for $e^{2i\phi}$ and $e^{2i\psi}$ already found, we have

$$\begin{cases} \alpha = \left(\dfrac{-A}{Mgh}\right)^{\frac{1}{2}} \cdot \dfrac{e^{\frac{1}{2}i(\phi_0+\psi_0)}}{\sigma(k)} \cdot \dfrac{\sigma(t+\omega_3-k)}{\sigma(t+\omega_3)} \cdot e^{t\zeta(k)}, \\[2ex] \beta = \left(\dfrac{-A}{Mgh}\right)^{\frac{1}{2}} \cdot \dfrac{e^{\frac{1}{2}i(\phi_0-\psi_0)}}{\sigma(l)} \cdot \dfrac{\sigma(t+\omega_3+l)}{\sigma(t+\omega_3)} \cdot e^{-t\zeta(l)}, \\[2ex] \gamma = \left(\dfrac{-A}{Mgh}\right)^{\frac{1}{2}} \cdot \dfrac{e^{\frac{1}{2}i(\psi_0-\phi_0)}}{\sigma(l)} \cdot \dfrac{\sigma(t+\omega_3-l)}{\sigma(t+\omega_3)} \cdot e^{t\zeta(l)}, \\[2ex] \delta = \left(\dfrac{-A}{Mgh}\right)^{\frac{1}{2}} \cdot \dfrac{e^{-\frac{1}{2}i(\phi_0+\psi_0)}}{\sigma(k)} \cdot \dfrac{\sigma(t+\omega_3+k)}{\sigma(t+\omega_3)} \cdot e^{-t\zeta(k)}. \end{cases}$$

These equations express the parameters α, β, γ, δ as functions of the time.

Example 1. *A gyrostat of mass M moves about a fixed point in its axis of symmetry: the moments of inertia about the axis of figure and a perpendicular to it through the fixed point are C and A respectively, and the centre of gravity is at a distance h from the fixed point. The gyrostat is held so that its axis makes an angle* arccos $1/\sqrt{3}$ *with the downward vertical, and is given an angular velocity* $\sqrt{AMgh\sqrt{3}/C}$ *about its axis. If the axis be now left free to move about the fixed point, shew that it will describe the cone*

$$\sin^2\theta \sin 2\phi = (-\cos\theta - 1/\sqrt{3})^{\frac{3}{2}} (-\cos\theta + \sqrt{3})^{\frac{1}{2}},$$

or

$$\sin^2\theta \cos 2\phi = \frac{2\sqrt{2}}{\sqrt{3}\sqrt[4]{3}} (\sqrt{3}/2 + \cos\theta)^{\frac{1}{2}},$$

where ϕ *is the azimuthal angle and* θ *the inclination of the axis to the upward vertical.*

(Camb. Math. Tripos, Part I, 1894.)

For in this problem we have initially

$$\cos\theta = -1/\sqrt{3}, \quad \phi = 0, \quad \dot{\theta} = 0, \quad \dot{\phi} = 0, \quad \dot{\psi} = \sqrt{AMgh\sqrt{3}/C},$$

and these initial values give

$$a = -\sqrt{MAgh}/\sqrt[4]{3}, \quad b = \sqrt[4]{3}\sqrt{MAgh}, \quad c = -Mgh/\sqrt{3}.$$

Substituting in the general differential equation for θ, namely

$$\tfrac{1}{2}A\dot{\theta}^2 = -\frac{(a - b\cos\theta)^2}{2A\sin^2\theta} - Mgh\cos\theta + c.$$

we have

$$A\dot{\theta}^2 \sin^2\theta = -Mgh(\cos\theta + 1/\sqrt{3})(\sqrt{3} + 2\cos\theta)(-\cos\theta + \sqrt{3}),$$

while the equation

$$\dot{\phi} = \frac{a - b\cos\theta}{A\sin^2\theta}$$

gives

$$\dot{\phi} = -\sqrt{\frac{Mgh\sqrt{3}}{A}} \cdot \frac{\cos\theta + 1/\sqrt{3}}{\sin^2\theta}.$$

Dividing this equation by the square root of the preceding equation, we have

$$\phi = 3^{\frac{1}{4}} \int (-\cos\theta - 1/\sqrt{3})^{\frac{1}{2}} (\sqrt{3} + 2\cos\theta)^{-\frac{1}{2}} (-\cos\theta + \sqrt{3})^{-\frac{1}{2}} \operatorname{cosec}\theta \, d\theta,$$

$$\phi = 3^{\frac{1}{4}} \int (x - 1/\sqrt{3})^{\frac{1}{2}} (\sqrt{3} - 2x)^{-\frac{1}{2}} (x + \sqrt{3})^{-\frac{1}{2}} (1 - x^2)^{-1} dx, \qquad \text{where } x = -\cos\theta.$$

Now if we write

$$u = (x - 1/\sqrt{3})^{\frac{3}{2}} (x + \sqrt{3})^{\frac{1}{2}} (\sqrt{3}/2 - x)^{-\frac{1}{2}},$$

we have by differentiation

$$\frac{du}{dx} = \tfrac{3}{2}(1 - x^2)(x - 1/\sqrt{3})^{\frac{1}{2}} (x + \sqrt{3})^{-\frac{1}{2}} (\sqrt{3}/2 - x)^{-\frac{3}{2}}$$

and

$$1 + \frac{3^{\frac{3}{2}}}{8} u^2 = \frac{3^{\frac{3}{2}}(1 - x^2)^2}{8(\sqrt{3}/2 - x)}.$$

We have therefore

$$\phi = \frac{3^{\frac{3}{4}}}{4\sqrt{2}} \int \frac{du}{1 + 3^{\frac{3}{2}} u^2/8},$$

or

$$2\phi = \arctan(3^{\frac{3}{4}} 2^{-\frac{3}{2}} u),$$

or

$$\tan 2\phi = 3^{\frac{3}{4}} 2^{-\frac{3}{2}} (-\cos\theta - 1/\sqrt{3})^{\frac{3}{2}} (-\cos\theta + \sqrt{3})^{\frac{1}{2}} (\sqrt{3}/2 + \cos\theta)^{-\frac{1}{2}},$$

which is equivalent to the result given above.

Example 2. Shew that the logarithms of the Cayley-Klein parameters, considered as functions of $\cos\theta$, are elliptic integrals of the third kind.

Example 3. Obtain the expressions found above for the Cayley-Klein parameters as functions of the time t by shewing that they satisfy differential equations typified by

$$\frac{d^2y}{dt^2} + Yy = 0,$$

where Y denotes a doubly-periodic function of t, these equations being of the Hermite-Lame type which is soluble by doubly-periodic functions of the second kind.

A simple type of motion of the top is that in which the axis of symmetry maintains a constant inclination to the vertical; in this case, which is generally known as the *steady motion* of the top, $\dot{\theta}$ and $\ddot{\theta}$ are permanently zero; since we have

$$\tfrac{1}{2}A\dot{\theta}^2 = -\frac{(a - b\cos\theta)^2}{2A\sin^2\theta} - Mgh\cos\theta + c,$$

it follows that

$$0 = \frac{d}{d\theta}\left\{\frac{(a - b\cos\theta)^2}{2A\sin^2\theta} + Mgh\cos\theta\right\}.$$

Performing the differentiation, and substituting for $(a - b\cos\theta)$ its value $A\dot{\phi}\sin^2\theta$, we have

$$0 = -b\dot{\phi} + A\dot{\phi}^2\cos\theta + Mgh.$$

This equation gives the relation between the constants $\dot{\phi}$, θ, and b (which depends on the rate of spinning of the top on its axis) in steady motion.

73. *Motion of a top on a perfectly smooth plane.*

We shall now consider the motion of a top which is spinning with its apex in contact with a smooth horizontal plane*. The reaction of the plane is now vertical, so the horizontal component of the velocity of the centre of gravity, G, of the top is constant; we can therefore without loss of generality suppose that this component is zero, so that the point G moves vertically in a fixed line, which we shall take as axis of Z; two horizontal lines fixed in space and perpendicular to each other will be taken as axes of X and Y.

Let $Gxyz$ be the principal axes of inertia of the top at G, and (A, A, C) the moments of inertia about them, Gz being the axis of symmetry: and let (θ, ϕ, ψ) be the Eulerian angles defining their position with reference to the axes of X, Y, Z.

The height of G above the plane is $h\cos\theta$, where h denotes the distance of G from the apex of the top; the part of the kinetic energy due to the motion of G is therefore $\frac{1}{2}Mh^2\sin^2\theta \,.\, \dot{\theta}^2$, where M is the mass of the top; and so, as in § 71, the total kinetic energy is

$$T = \tfrac{1}{2}Mh^2\sin^2\theta \,.\, \dot{\theta}^2 + \tfrac{1}{2}A\dot{\theta}^2 + \tfrac{1}{2}A\dot{\phi}^2\sin^2\theta + \tfrac{1}{2}C(\dot{\psi} + \dot{\phi}\cos\theta)^2,$$

and the potential energy is

$$V = Mgh\cos\theta.$$

* Poisson, *Traité de Mécanique* (1811), II. p. 198.

Proceeding now exactly as in § 71, we have two integrals corresponding to the ignorable coordinates ϕ and ψ, namely

$$\begin{cases} A\dot{\phi}\sin^2\theta + C(\dot{\psi} + \dot{\phi}\cos\theta)\cos\theta = a, \\ \qquad\qquad C(\dot{\psi} + \dot{\phi}\cos\theta) = b, \end{cases}$$

where a and b are constants; and on performing the process of ignoration of coordinates we obtain for the modified kinetic potential the expression

$$\tfrac{1}{2}(A + Mh^2\sin^2\theta)\dot{\theta}^2 - \frac{(a - b\cos\theta)^2}{2A\sin^2\theta} - Mgh\cos\theta,$$

so the variation of θ is the same as in the system with one degree of freedom for which the kinetic energy is

$$\tfrac{1}{2}(A + Mh^2\sin^2\theta)\dot{\theta}^2,$$

and the potential energy is

$$\frac{(a - b\cos\theta)^2}{2A\sin^2\theta} + Mgh\cos\theta.$$

The connexion between θ and t is given by the integral of energy of this latter system, namely

$$\tfrac{1}{2}(A + Mh^2\sin^2\theta)\dot{\theta}^2 = -\frac{(a - b\cos\theta)^2}{2A\sin^2\theta} - Mgh\cos\theta + c,$$

where c is a constant. Writing $\cos\theta = x$, this becomes

$$A(A + Mh^2 - Mh^2x^2)\dot{x}^2 = -(a - bx)^2 - 2AMgh(x - x^3) + 2Ac(1 - x^2).$$

The variables x and t are separated in this equation, so the solution can be expressed as a quadrature; but the evaluation of the integral involved will require in general hyperelliptic functions, or automorphic functions of genus two.

74. *Kowalevski's top.*

The problem of the motion under gravity of a body one of whose points is fixed is not in general soluble by quadratures: and the cases considered in § 69 (in which the fixed point is the centre of gravity of the body, so that gravity does not influence the motion), and in § 71 (in which the fixed point and the centre of gravity lie on an axis of symmetry of the body), were for long the only ones known to be integrable. In 1888 however Mme. S. Kowalevski* shewed that the problem is also soluble when two of the principal moments of inertia at the fixed point are equal and double the third, so that $A = B = 2C$, and when further the centre of gravity is situated in the plane of the equal moments of inertia.

Let the line through the fixed point O and the centre of gravity be taken as the axis Ox, and let the centre of gravity be at a distance a from the fixed

* *Acta Math.* XII. (1888), p. 177.

point; let (θ, ϕ, ψ) be the Eulerian angles which define the position of the principal axes of inertia $Oxyz$ with reference to fixed rectangular axes $OXYZ$, of which the axis OZ is vertical; let $(\omega_1, \omega_2, \omega_3)$ be the components along the axes $Oxyz$ of the angular velocity of the body, and let M be its mass. The kinetic and potential energies are given by the equations

$$\begin{aligned} T &= \tfrac{1}{2}(A\omega_1^2 + A\omega_2^2 + C\omega_3^2) \\ &= C\{\dot\theta^2 + \dot\phi^2 \sin^2\theta + \tfrac{1}{2}(\dot\psi + \dot\phi\cos\theta)^2\}, \\ V &= -Mga \sin\theta \cos\psi. \end{aligned}$$

The coordinate ϕ is evidently ignorable, giving an integral

$$\frac{\partial T}{\partial \dot\phi} = \text{constant},$$

or

$$2\dot\phi \sin^2\theta + (\dot\psi + \dot\phi \cos\theta)\cos\theta = k,$$

where k is a constant: and the integral of energy is

$$T + V = \text{constant},$$

or

$$\dot\theta^2 + \dot\phi^2 \sin^2\theta + \tfrac{1}{2}(\dot\psi + \dot\phi\cos\theta)^2 - \frac{Mga}{C}\sin\theta\cos\psi = h.$$

Mme. Kowalevski shewed that another algebraic integral exists, which can be found in the following way.

The kinetic potential is

$$L = C\dot\theta^2 + C\dot\phi^2 \sin^2\theta + \tfrac{1}{2}C(\dot\psi + \dot\phi\cos\theta)^2 + Mga \sin\theta\cos\psi,$$

and the equations of motion are

$$\frac{d}{dt}\left(\frac{\partial L}{\partial \dot\theta}\right) - \frac{\partial L}{\partial \theta} = 0,$$

$$\frac{d}{dt}\left(\frac{\partial L}{\partial \dot\psi}\right) - \frac{\partial L}{\partial \psi} = 0,$$

$$\frac{d}{dt}\left(\frac{\partial L}{\partial \dot\phi}\right) = 0;$$

the first of these is

$$2\ddot\theta = (\dot\phi\cos\theta - \dot\psi)\,\dot\phi\sin\theta + \frac{Mga}{C}\cos\theta\cos\psi,$$

and on eliminating $\ddot\psi$ between the second and third, we obtain

$$2\frac{d}{dt}(\dot\phi\sin\theta) = -(\dot\phi\cos\theta - \dot\psi)\,\dot\theta + \frac{Mga}{C}\cos\theta\sin\psi.$$

Adding the first of these equations multiplied by i to the second, we have

$$2\frac{d}{dt}(\dot\phi\sin\theta + i\dot\theta) = i(\dot\phi\cos\theta - \dot\psi)(\dot\phi\sin\theta + i\dot\theta) + i\,.\,\frac{Mga}{C}\cos\theta e^{-i\psi},$$

an equation which can be written in the form

$$\frac{d}{dt}\left\{(\dot\phi\sin\theta+i\dot\theta)^2+\frac{Mag}{C}\sin\theta e^{-i\psi}\right\}$$
$$=i(\dot\phi\cos\theta-\dot\psi)\left\{(\dot\phi\sin\theta+i\dot\theta)^2+\frac{Mag}{C}\sin\theta e^{-i\psi}\right\},$$

or
$$\frac{1}{U}\frac{dU}{dt}=i(\dot\phi\cos\theta-\dot\psi),$$

where
$$U=(\dot\phi\sin\theta+i\dot\theta)^2+\frac{Mag}{C}\sin\theta e^{-i\psi}.$$

Similarly, if
$$V=(\dot\phi\sin\theta-i\dot\theta)^2+\frac{Mag}{C}\sin\theta e^{i\psi},$$

we have
$$\frac{1}{V}\frac{dV}{dt}=-i(\dot\phi\cos\theta-\dot\psi).$$

It follows that
$$\frac{1}{U}\frac{dU}{dt}+\frac{1}{V}\frac{dV}{dt}=0,$$

or
$$UV=\text{constant}.$$

We have therefore the equation

$$\left\{(\dot\phi\sin\theta+i\dot\theta)^2+\frac{Mag}{C}\sin\theta e^{-i\psi}\right\}\left\{(\dot\phi\sin\theta-i\dot\theta)^2+\frac{Mag}{C}\sin\theta e^{i\psi}\right\}=\text{constant},$$

or

$$(\dot\theta^2+\dot\phi^2\sin^2\theta)^2+\left(\frac{Mag}{C}\right)^2\sin^2\theta+\frac{Mag}{C}\sin\theta\{e^{i\psi}(\dot\phi\sin\theta+i\dot\theta)^2+e^{-i\psi}(\dot\phi\sin\theta-i\dot\theta)^2\}$$
$$=\text{constant},$$

and *this is the required third algebraic integral of the system.*

The first integrals which have been found constitute a system of three differential equations, each of the first order, for the determination of $\dot\theta$, $\dot\phi$, $\dot\psi$, and they can be regarded as replacing the original differential equations of motion. The variable ϕ does not occur explicitly in them and we can therefore use one of the three equations in order to eliminate $\dot\phi$ from the other two: we shall then have a system of two differential equations, each of the first order, to determine θ and ψ. It has been shewn by Mme. Kowalevski that these equations can be solved by means of hyperelliptic functions: for this solution reference may be made to the memoir already referred to*.

* Cf. also Kötter, *Acta Math.* XVII. (1893), p. 209; Stekloff, Gorjatscheff, and Tchapligine, *Trav. Soc. Imp. Nat. Moscou*, X. (1899) and XII. (1904); G. Dumas, *Nouv. Ann.* (4) IV. (1904), p. 355; Husson, *Toulouse Ann.* (2) VIII. (1906), p. 73; Husson, *Acta Math.* XXXI. (1907), p. 71; N. Kowalevski, *Math. Ann.* LXV. (1908), p. 528; P. Stäckel, *Math. Ann.* LXV. (1908), p. 538; O. Olsson, *Arkiv för Mat.* IV. Nr. 7 (1908); R. Marcolongo, *Rom. Acc. Rend.* (5) XVII. (1908), p. 698; F. de Brun, *Arkiv för Mat.* VI. Nr. 9 (1910); P. Burgatti, *Palermo Rend.* XXIX. (1910), p. 396; O. Lazzarino, *Rend. d. Soc. reale di Napoli*, (3ª) XVII. (1911), p. 68.

Example. Let γ_1, γ_2, γ_3 denote the direction-cosines of Ox, Oy, Oz referred to OZ, and let variables x, y, τ be defined by the equations

$$\omega_2^2 x = \left(\omega_1^2 - \omega_2^2 + \frac{Mga\gamma_1}{C}\right)\left\{\left(\omega_3\omega_1 + \frac{Mga\gamma_3}{C}\right)^2 - \omega_3^2\omega_2^2\right\}$$
$$+ 2\omega_3\omega_2\left(2\omega_1\omega_2 + \frac{Mga\gamma_2}{C}\right)\left(\omega_3\omega_1 + \frac{Mga\gamma_3}{C}\right),$$

$$\omega_2^2 y = \left(2\omega_1\omega_2 + \frac{Mga\gamma_3}{C}\right)\left\{\left(\omega_3\omega_1 + \frac{Mga\gamma_3}{C}\right)^2 - \omega_3^2\omega_2^2\right\}$$
$$- 2\omega_3\omega_2\left(\omega_3\omega_1 + \frac{Mga\gamma_3}{C}\right)\left(\omega_1^2 - \omega_2^2 + \frac{Mga\gamma_1}{C}\right),$$

$$\omega_2^2 d\tau = \left\{\left(\omega_3\omega_1 + \frac{Mga\gamma_3}{C}\right)^2 + \omega_3^2\omega_2^2\right\} dt.$$

Shew by use of Kowalevski's integral (without using the integrals of energy or angular momentum) that the equations of motion can be written in the form

$$\frac{d^2x}{d\tau^2} = -\frac{\partial V}{\partial x}, \quad \frac{d^2y}{d\tau^2} = -\frac{\partial V}{\partial y},$$

where V is a function of x and y only, so that the problem is transformed into that of the motion of a particle in a plane conservative field of force. (Kolosoff.)

R. Liouville* has stated that the only other general case in which the motion under gravity of a rigid body with one point fixed has a third algebraic integral is that in which

1°. The momental ellipsoid of the point of suspension is an ellipsoid of revolution.

2°. The centre of gravity of the body is in the equatorial plane of the momental ellipsoid.

3°. If (A, A, C) are the principal moments of inertia at the point of suspension, the ratio $2C/A$ is an integer: this integer can be arbitrarily chosen.

On this, cf. the memoirs cited in the footnote on the preceding page.

Example. A heavy body rotates about a fixed point O, the principal moments of inertia at which satisfy the relation $A = B = 4C$: and the centre of gravity of the body lies in the equatorial plane of the momental ellipsoid, at a distance h from O. Shew that if the constant of angular momentum about the vertical through O vanishes, there exists an integral

$$\omega_3(\omega_1^2 + \omega_2^2) + gh\omega_1 \cos\theta = \text{constant},$$

where ω_1, ω_2, ω_3 are the components of angular velocity about the principal axes $Oxyz$, Ox being the line from O to the centre of gravity; and hence that the problem can be solved by quadratures, leading to hyperelliptic integrals. (Tchapligine.)

75. *Impulsive motion.*

As has been observed in § 36, the solution of problems in impulsive motion does not depend on the integration of differential equations, and can generally be effected by simple algebraic methods. The following examples illustrate various types of impulsive systems.

Example 1. Two uniform rods AB, BC, each of length 2a, are smoothly jointed at B and rest on a horizontal table with their directions at right angles. An impulse is applied to the middle point of AB, and the rods start moving as a rigid body: determine the direction of the impulse that this may be the case, and prove that the velocities of A, C will be in the ratio $\sqrt{13} : 1$. (Coll. Exam.)

We can without loss of generality suppose the mass of each rod to be unity. Let $(\dot{x}, \dot{y})$ be the component velocities of B referred to fixed axes Ox, Oy parallel to the undisturbed

* *Acta Math.* xx. (1897), p. 239

position BA, BC of the rods, and let $\dot{\theta}$, $\dot{\phi}$ be the angular velocities of BA and BC. The components of velocity of the middle point of AB are $(\dot{x}, \dot{y}+a\dot{\theta})$, and the components of velocity of the middle point of BC are $(\dot{x}-a\dot{\phi}, \dot{y})$, so the kinetic energy of the system is given by the equation

$$T=\tfrac{1}{2}\dot{x}^2+\tfrac{1}{2}(\dot{y}+a\dot{\theta})^2+\tfrac{1}{6}a^2\dot{\theta}^2+\tfrac{1}{2}(\dot{x}-a\dot{\phi})^2+\tfrac{1}{2}\dot{y}^2+\tfrac{1}{6}a^2\dot{\phi}^2.$$

Let the components parallel to the axes of the impulse be I, J. The components of the displacement of the point of application of the impulse in a small displacement of the system are $(\delta x, \delta y+a\delta\theta)$; and hence the equations of § 36 become

$$\frac{\partial T}{\partial \dot{x}}=I, \quad \frac{\partial T}{\partial \dot{y}}=J, \quad \frac{\partial T}{\partial \dot{\theta}}=Ja, \quad \frac{\partial T}{\partial \dot{\phi}}=0,$$

or

$$\begin{cases} I=2\dot{x}-a\dot{\phi}, \\ J=2\dot{y}+a\dot{\theta}, \\ Ja=a\dot{y}+\tfrac{4}{3}a^2\dot{\theta}, \\ 0=-a\dot{x}+\tfrac{4}{3}a^2\dot{\phi}, \end{cases}$$

while the condition that the system moves as if rigid is $\dot{\theta}=\dot{\phi}$. These equations give

$$\tfrac{1}{4}\dot{x}=\dot{y}=\tfrac{1}{3}a\dot{\theta}=\tfrac{1}{3}a\dot{\phi}=\tfrac{1}{5}I=\tfrac{1}{5}J.$$

Hence $I=J$, which shews that the direction of the impulse makes an angle of 45° with BA; and as the components of velocity of A are $(\dot{x}, \dot{y}+2a\dot{\theta})$, and the components of velocity of C are $(\dot{x}-2a\dot{\phi}, \dot{y})$, we have for the velocities of A and of C the values $\sqrt{65}\dot{y}$ and $\sqrt{5}\dot{y}$ respectively, so the velocity of A is $\sqrt{13}\times$ the velocity of C: which is the required result.

Example 2. A framework in the form of a parallelogram is made by smoothly jointing the ends of two pairs of uniform bars of lengths 2a, 2b, masses m, m', and radii of gyration k, k'. The parallelogram is moving without any rotation of its sides, and with velocity V, in the direction of one of its diagonals; it impinges on a smooth fixed wall with which the sides make angles θ, φ and the direction of the velocity V a right angle, the vertex which impinges being brought to rest by the impact. Shew that the impulse on the wall is

$$2V\{(m+m')^{-1}+(mk^2+m'a^2)^{-1}a^2\cos^2\theta+(mb^2+m'k'^2)^{-1}b^2\cos^2\phi\}^{-1}.$$

(Coll. Exam.)

Let x and y be the coordinates of the centre of the parallelogram, x being measured at right angles to the wall and towards it. The kinetic energy is

$$T=(m+m')(\dot{x}^2+\dot{y}^2)+(mk^2+m'a^2)\dot{\theta}^2+(mb^2+m'k'^2)\dot{\phi}^2.$$

The x-coordinate of the point of contact is $x+a\sin\theta+b\sin\phi$, so the displacement of the point of contact parallel to the axis of x corresponding to an arbitrary displacement $(\delta x, \delta y, \delta\theta, \delta\phi)$ is $\delta x+a\cos\theta\,\delta\theta+b\cos\phi\,\delta\phi$. The equations of motion, denoting the impulse by I, are therefore

$$\begin{cases} \dfrac{\partial T}{\partial \dot{x}}-\left(\dfrac{\partial T}{\partial \dot{x}}\right)_0=-I, \\ \dfrac{\partial T}{\partial \dot{\theta}}-\left(\dfrac{\partial T}{\partial \dot{\theta}}\right)_0=-Ia\cos\theta, \\ \dfrac{\partial T}{\partial \dot{\phi}}-\left(\dfrac{\partial T}{\partial \dot{\phi}}\right)_0=-Ib\cos\phi, \end{cases}$$

or

$$\begin{cases} 2(m+m')(\dot{x}-V)=-I, \\ 2(mk^2+m'a^2)\dot{\theta} \quad =-Ia\cos\theta, \\ 2(mb^2+m'k'^2)\dot{\phi} \quad =-Ib\cos\phi. \end{cases}$$

Moreover since the final velocity of the point of contact is zero, we have

$$\dot{x}+a\cos\theta\,.\,\dot{\theta}+b\cos\phi\,.\,\dot{\phi}=0.$$

Eliminating $\dot{x}$, $\dot{\theta}$, $\dot{\phi}$ from these equations, we have

$$V=I\left\{\frac{1}{2(m+m')}+\frac{a^2\cos^2\theta}{2(mk^2+m'a^2)}+\frac{b^2\cos^2\phi}{2(mb^2+m'k'^2)}\right\},$$

which is the result stated.

The next example relates to a case of *sudden fixture*; if one point (or line) of a freely-moving rigid body is suddenly seized and compelled to move in a given manner, there will be an impulsive change in the motion of the body, which can be determined from the condition that the angular momentum of the body about any line through the point seized (or about the line seized) is unchanged by the seizure; this follows from the fact that the impulse of seizure has no moment about the point (or line).

Example 3. *A uniform circular disc is spinning with an angular velocity* Ω *about a diameter when a point P on its rim is suddenly fixed. Prove that the subsequent velocity of the centre is equal to* $\frac{1}{5}$ *of the velocity of the point P immediately before the impact.*

(Coll. Exam.)

Let m be the mass of the disc, and let a be the angle between the radius to P and the diameter about which the disc was originally spinning. The original velocity of P is $\Omega c\sin a$, where c is the radius of the disc. The original angular momentum about P is about an axis through P parallel to the original axis of rotation, and of magnitude $\frac{1}{4}mc^2\Omega$; and this is unchanged by the fixing of P, so when P has been fixed, the angular momentum about the tangent at P is $\frac{1}{4}mc^2\Omega\sin a$. But the moment of inertia of the disc about its tangent at P is $\frac{5}{4}mc^2$, and so the angular velocity about the tangent at P is $\frac{1}{5}\Omega\sin a$. The velocity of the centre of the disc is therefore $\frac{1}{5}\Omega c\sin a$, which is $\frac{1}{5}$ of the original velocity of P.

Example 4. A lamina in the form of a parallelogram whose mass is m has a smooth pivot at each of the middle points of two parallel sides. It is struck at an angular point by a particle of mass m which adheres to it after the blow. Shew that the impulsive reaction at one of the pivots is zero. (Coll. Exam.)

Miscellaneous Examples.

1. Prove that for a disc free to turn about a horizontal axis perpendicular to its plane the locus on the disc of the centres of suspension for which the simple equivalent pendulum has a given length L consists of two circles; and that, if A and B are two points, one on each circle, and L' is the length of the simple equivalent pendulum when the centre of suspension is the middle point of AB, the radius of gyration k of the disc about its centre of inertia is given by the equation

$$k^2L'^2=(\tfrac{1}{2}L^2-c^2)(L'^2-\tfrac{1}{2}L^2+c^2),$$

where $2c$ is the length of AB. (Coll. Exam.)

2. A heavy rigid body can turn about a fixed horizontal axis. If one point in the body is given through which the horizontal axis has to pass, discuss the problem of choosing the direction of the axis in the body in such a way that the simple equivalent pendulum shall have a given length; shewing that the axes which satisfy this condition are the generators of a quartic cone. (Coll. Exam.)

3. A sphere of radius b rolls without slipping down the cycloid

$$x=a(\theta+\sin\theta), \quad y=a(1-\cos\theta).$$

It starts from rest with its centre on the horizontal line $y=2a$. Prove that the velocity V of its centre when at the lowest point is given by

$$V^2=\tfrac{10}{7}g(2a-b). \qquad \text{(Coll. Exam.)}$$

4. A uniform smooth cube of edge $2a$ and mass M rests symmetrically on two shelves each of breadth b and mass m and attached to walls at a distance $2c$ apart. Shew that, if one of the shelves gives way and begins to turn about the edge where it is attached to the wall, the initial angular acceleration of the cube will be

$$\frac{Mg(c-a)^2(c-b)+\frac{1}{2}mgb(c-a)(c-b+a)}{M(c-a)^2\{k^2+(c-b)^2\}+I(c-b+a)^2},$$

where Mk^2 and I are respectively the moments of inertia of the cube about its centre and of the shelf about its edge. (Camb. Math. Tripos, Part I, 1899.)

5. A homogeneous rod of mass M and length $2a$ moves on a horizontal plane, one end being constrained to slide without friction in a fixed straight line. The rod is initially perpendicular to the line, and is struck at the free end by a blow I parallel to the line. Shew that after time t the perpendicular distance y of the middle point of the rod from the line is given by the equation

$$\int_{y/a}^{1}(1-\tfrac{3}{4}x^2)^{\frac{1}{2}}(1-x^2)^{-\frac{1}{2}}dx=3It/2Ma. \qquad \text{(Coll. Exam.)}$$

6. Four equal uniform rods, of length $2a$, are smoothly jointed so as to form a rhombus $ABCD$. The joint A is fixed, whilst C is free to move on a smooth vertical rod through A. Initially C coincides with A and the whole system is rotating about the vertical with angular velocity ω. Prove that, if in the subsequent motion $2a$ is the least angle between the upper rods,

$$a\omega^2\cos a=3g\sin^2 a.$$

(Camb. Math. Tripos, Part I, 1900.)

7. A disc of mass M rests on a smooth horizontal table, and a smooth circular groove of radius a is cut in it, passing through the centre of gravity of the disc. A particle of mass $\frac{1}{3}M$ is started in the groove from the centre of gravity of the disc. Investigate the motion. Prove that if $a\phi$ is the arc traversed by the particle and θ the angle turned round by the disc, then

$$\tan\tfrac{1}{2}\phi=\frac{k}{(a^2+k^2)^{\frac{1}{2}}}\tan\frac{(a^2+k^2)^{\frac{1}{2}}}{k}(\tfrac{1}{2}\phi-\theta),$$

Mk^2 being the moment of inertia of the disc about a vertical line through its centre of gravity. (Coll. Exam.)

8. A rigid body is moving freely under the action of gravity and rotating with angular velocity ω about an axis through its centre of gravity perpendicular to the plane of its motion. Shew that the axis of instantaneous rotation describes a parabolic cylinder of latus rectum $(\sqrt{4a}+\sqrt{2g}/\omega)^2$, whose vertex is at a distance $\sqrt{2ga}/\omega$ above that of the path of the centre of gravity of the body; where $4a$ is the latus rectum of the parabola described by the centre of gravity. (Coll. Exam.)

9. A particle of mass m is placed in a smooth uniform tube which can rotate in a vertical plane about its middle point. The system starts from rest when the tube is horizontal. If θ is the angle the tube makes with the vertical when its angular velocity is a maximum and equal to ω, prove that

$$4(mr^2+Mk^2)\omega^4-8mgr\omega^2\cos\theta+mg^2\sin^2\theta=0,$$

where Mk^2 is the moment of inertia of the tube about its centre and r the distance of the particle from the centre of the tube. (Coll. Exam.)

10. Four uniform rods, smoothly jointed at their ends, form a parallelogram which can move smoothly on a horizontal surface, one of the angular points being fixed. Initially the configuration is rectangular and the framework is set in motion in such a manner that the angular velocity of one pair of opposite sides is Ω, that of the other pair being zero. Shew that when the angle between the rods is a maximum or minimum, the angular velocity of the system is Ω. (Coll. Exam.)

11. Two homogeneous rough spheres of equal radii a and of masses m, m' rest on a smooth horizontal plane with m' at the highest point of m. If the system is disturbed, shew that the inclination of their common normal θ to the vertical is given by the equation

$$a\dot{\theta}^2 (7m + 5m' \sin^2 \theta) = 5g (m + m') (1 - \cos \theta). \qquad \text{(Coll. Exam.)}$$

12. A uniform rod AB is of length $2a$ and is attached at one end to a light inextensible string of length c. The other end of this string is fixed at O to a point in a smooth horizontal plane on which the rod moves. Initially OAB is a straight line and the rod is projected without rotation with velocity V in the direction perpendicular to its length. Prove that the cosine of the greatest subsequent angle between the rod and string is $1 - a/6c$. (Coll. Exam.)

13. To a fixed point are smoothly jointed two uniform rods of length $2a$, and upon them slides, by means of a smooth ring at each end, a third rod similar in all respects. Initially the three rods are in a horizontal line with the ends of the third rod at the middle points of the other two and, on the application of an impulse, the rods begin to rotate with angular velocity Ω in a horizontal plane. Shew that the third rod will slide right off the other two unless

$$\Omega^2 > 2g/a\,\sqrt{3}. \qquad \text{(Coll. Exam.)}$$

14. A hollow thin cylinder of radius a and mass M is maintained at rest in a horizontal position on a rough plane whose inclination is α, and contains an insect of mass m at rest on the line of contact with the plane. The cylinder is released as the insect starts off with velocity V: if this relative velocity be maintained and the cylinder roll up hill, shew that it will come to instantaneous rest when the radius through the insect makes an angle θ with the vertical given by

$$V^2 \{1 - \cos(\theta - \alpha)\} + ag (\cos \alpha - \cos \theta) = (1 + M/m)\, ag\, (\theta - \alpha) \sin \alpha.$$

(Coll. Exam.)

15. A uniform smooth plane tube can turn smoothly about a fixed axis of rotation lying in its plane and intersecting it: the moment of inertia of the tube about the axis is I. Initially the tube is rotating with angular velocity Ω about the axis, and a particle of mass m is projected with velocity V within the tube from the point of intersection of the tube with the axis. The system then moves under no external forces. Prove that, when the particle is at a distance r from the axis, the square of its velocity relative to the tube is

$$V^2 + \frac{Ir^2}{I + mr^2}\,\Omega^2. \qquad \text{(Coll. Exam.)}$$

16. A uniform straight rod of mass M is laid across two smooth horizontal pegs so that each of its ends projects beyond the corresponding peg. A second uniform rod of mass m and length $2l$ is fastened to the first at some point between the pegs by a

universal joint. This rod is initially held horizontal and in contact with the first rod; and then let go, so as to oscillate in the vertical plane through the first rod. Prove that if θ be the angle which the second rod makes with the vertical at any instant, and x the distance through which the first rod has moved from rest,

$$(M+m)\,x+ml\sin\theta=ml,$$

and $$\left(\tfrac{4}{3}-\frac{m}{m+M}\cos^2\theta\right)l\dot{\theta}^2=2g\cos\theta.$$ (Coll. Exam.)

17. A plane body is free to rotate in its plane about a fixed point, and a second plane body is free to slide along a smooth straight groove in the first body, its motion being in the same plane; shew that the relation between the relative advance x along the groove and the angle of rotation θ (no external forces being supposed to act on the system) is of the form

$$\left(\frac{dx}{d\theta}\right)^2+P\frac{dx}{d\theta}+Q=0,$$

where P and Q are respectively linear and quadratic functions of x^2. (Coll. Exam.)

18. A pendulum is formed of a straight rod and a hollow circular bob, and fitting inside the bob is a smooth vertical lamina in the shape of a segment of a circle, the distances of the centre (C) of the bob from the point of suspension (O) and from the centre of gravity (G) of the lamina being l and c respectively. Prove that if M, m are the masses of the pendulum and lamina, k and k' their respective radii of gyration about O and G, θ and ϕ the angles which OC and CG make with the vertical, then twice the work done by gravity on the system during its motion from rest is equal to

$$(Mk^2+ml^2)\,\dot{\theta}^2+m\,(k'^2+c^2)\,\dot{\phi}^2+2mcl\cos(\theta-\phi)\,\dot{\theta}\dot{\phi}.$$ (Coll. Exam.)

19. A particle of mass m is attached to the end of a fine string which passes round the circumference of a wheel of mass M, the other end of the string being attached to a point in that circumference, a length l of the string being straight initially, and the wheel (radius a and radius of gyration k) being free to move about a fixed vertical axis through its centre; the particle, which lies on a smooth horizontal plane, is projected at right angles to the string, so that the string begins to wrap round the wheel; prove that, if the string eventually unwinds from the wheel, the shortest length of the straight portion is

$$(l^2-a^2-Mk^2/m)^{\frac{1}{2}}.$$ (Coll. Exam.)

20. A carriage is placed on an inclined plane making an angle a with the horizon and rolls straight down without any slipping between the wheels and the plane. The floor of the carriage is parallel to the plane and a perfectly rough ball is placed freely on it. Shew that the acceleration of the carriage down the plane is

$$\frac{14M+4M'+14m}{14M+4M'+21m}\,g\sin a,$$

where M is the mass of the carriage excluding the wheels, m the sum of the masses of the wheels, which are uniform discs, and M' that of the ball. The friction between the wheels and the axes is neglected. (Coll. Exam.)

21. A uniform rod of mass m_1 and length $2a$ is capable of rotating freely about its fixed upper extremity and is initially inclined at an angle of $\pi/6$ to the vertical. A second rod, of mass m_2 and length $2a$, is smoothly attached to the lower end of the first and rests initially at an angle of $2\pi/3$ with it and in a horizontal position. Shew that if the centre of the lower rod commence to move in a direction making an angle $\pi/6$ with the vertical, then $3m_1=14m_2$. (Coll. Exam.)

22. A uniform circular disc is symmetrically suspended by two elastic strings of natural length c inclined at an angle a to the vertical, and attached to the highest point of the disc. If one of the strings is cut, prove that the initial curvature of the path of the centre of the disc is

$$(c \sin 4a - b \sin 2a)/b\,(b-c),$$

where b is the equilibrium length of each string. (Coll. Exam.)

23. Two rods AC, CB of equal length $2a$ are freely jointed at C, the rod AC being freely moveable about a fixed point A, and the end B of the rod CB is attached to A by an inextensible string of length $4a/\sqrt{3}$. The system being in equilibrium, the string is cut; shew that the radius of curvature of the initial path of B at B is

$$\frac{4}{181}\sqrt{\frac{41^3}{3}}\,.\,a.$$

(Camb. Math. Tripos, Part I, 1897.)

24. A rod of length $2a$ is supported in a horizontal position by two light strings which pass over two smooth pegs in a horizontal line at a distance $2a$ apart and have at their other extremities weights each equal to one half that of the rod. One of the strings is cut; prove that the initial curvature of the path of that end of the rod to which the cut string was attached is $27/25a$. (Coll. Exam.)

25. A heavy plank, straight and very rough, is free to turn in a vertical plane about a horizontal axis from which the distance of its centre of gravity is c. A rough heavy sphere is placed on this plank at a distance b from the axis, on the side remote from the centre of gravity; the plank being held horizontal. The system is now left free to move. Prove that the initial radius of curvature of the path of the centre of the sphere is $21\,b\theta/(5-11\theta)$, where $\theta=(mb-Mc)/(mb+Ma)$, m and M are the masses of the sphere and the plank, and Mab is the moment of inertia of the plank about the axis.

(Coll. Exam.)

26. A light stiff rod of length $2c$ carries two equal particles of mass m at distances k from the centre on each side of it; to each end of the rod is tied an end of an inextensible string of length $2a$ on which is a ring of mass m'. Initially the string and rod are in one straight line on a smooth horizontal table with the string taut and the ring at the loop; the ring is then projected at right angles to the rod, shew that the relative motion will be oscillatory if

$$c^2/k^2 > 1 + 2m/m'.$$

(Coll. Exam.)

27. Three equal uniform rods, each of length c, are firmly joined to form an equilateral triangle ABC of weight W; a uniform bar of length $2b$ and weight W' is freely jointed to the triangle at C. This system rests in equilibrium in contact with the surface of a fixed smooth sphere of radius a, AB being horizontal and in contact with the sphere, and the bar being in the vertical plane through the centre of the triangle; the bar, and the centre of the triangle, are on opposite sides of the vertical line through C. Prove that the inclination of the plane of the triangle to the horizon is the angle whose tangent is

$$[ab\mu+2c\lambda^2] \div [n\mu\,(a^2+\tfrac{1}{4}c^2)+\lambda^2\mu-2abc],$$

where $\lambda^2=a^2+\tfrac{1}{4}c^2-\tfrac{1}{2}bc$, $\mu^2=12a^2-c^2$, and $n=W/W'$.

(Camb. Math. Tripos, Part I, 1896.)

28. A body, under the action of no forces, moves so that the resolved part of its angular velocity about one of the principal axes at the centre of gravity is constant; shew that the angular velocity of the body must be constant, and find its resolved parts about the other two principal axes when the moments of inertia about these axes are equal.

(Coll Exam.)

29. Shew that a herpolhode cannot have a point of inflexion. (Hess.)

(A simple proof of this result is given by Lecornu, *Bull. de la Soc. Math. de France*, XXXIV. (1906), p. 40.)

30. In the motion under no forces of a body one of whose points is fixed, shew that the motion of every quadric homocyclic with the momental ellipsoid relative to the fixed point, and rigidly connected with the body, is the same as if it were made to roll without sliding on a fixed quadric of revolution, which has its centre at the fixed point, and whose axis is the invariable line. (Gebbia.)

31. In the motion of a body under no forces round a fixed point, shew that the three diameters of the momental ellipsoid at the fixed point and the diameter of the ellipsoid reciprocal to the momental ellipsoid, determined respectively by the intersection of the invariable plane with the three principal planes and with the plane perpendicular to the instantaneous axis, describe areas proportional to the times, so that the accelerations of their extremities are directed to the centre. (Siacci.)

32. When a body moveable about a fixed point is acted on by forces whose moment round the instantaneous axis is always zero, shew that the velocity of rotation is proportional to that radius vector of the momental ellipsoid which is in the direction of this axis.

Shew that this theorem is still true if the body is moveable about a fixed point and also constrained to slide on a fixed surface. (Flye St Marie.)

33. A plane lamina is initially moving with equal angular velocities Ω about the principal axes of greatest and least moment of inertia at its centre of mass, and has no angular velocity about the third principal axis; express the angular velocities about these axes as elliptic functions of the time, supposing no forces to act on the lamina.

If θ be the angle between the plane of the lamina and any fixed plane, shew that

$$\frac{d^2\theta}{dt^2}+2\Omega\left\{\Omega^2-\left(\frac{d\theta}{dt}\right)^2\right\}^{\frac{1}{2}}\operatorname{dn}(\Omega t)=\left\{\Omega^2-\left(\frac{d\theta}{dt}\right)^2\right\}\cot\theta.$$

(Camb. Math. Tripos, Part I, 1896.)

34. A rigid body is kinetically symmetrical about an axis which passes through a fixed point above its centre of gravity and is set in motion in any manner; shew that in the subsequent motion, except in one case, the centre of gravity can never be vertically over the fixed point; and find the greatest height it attains. (Coll. Exam.)

35. In the motion of the top on the rough plane, shew that there exists an auxiliary set of axes $O\xi\eta\zeta$ whose motion with respect to the fixed axes $OXYZ$ and also with respect to the moving axes $Oxyz$ is a Poinsot motion; the invariable planes being the horizontal plane in the former case, and the plane perpendicular to the axis of the body in the second case. (Jacobi.)

36. A uniform solid of revolution moves about a point, so that its motion may be represented by the uniform rolling of a cone of semi-vertical angle a fixed in the body upon an equal cone fixed in space, the axis of the former being the axis of revolution. Shew that the couple necessary to maintain the motion is of magnitude

$$\tfrac{1}{2}\Omega^2\tan a\,\{C+(C-A)\cos 2a\},$$

where Ω is the resultant angular velocity and A and C the principal moments of inertia at the point, and that the couple lies in the plane of the axes of the cones. (Coll. Exam.)

37. A vertical plane is made to rotate with uniform angular velocity about a vertical axis in itself, and a perfectly rough cone of revolution has its vertex fixed at a point of that axis. Shew that, if the line of contact make an angle θ with the vertical, and β and γ be the extreme values of θ, and a be the semi-vertical angle of the cone,

$$k^2\left(\frac{d\theta}{dt}\right)^2=2gh\,\frac{\sin^2 a}{\cos a}\,\frac{(\cos\theta-\cos\beta)(\cos\gamma-\cos\theta)}{\cos\beta+\cos\gamma},$$

where h is the distance of the centre of gravity of the cone from its vertex, and k its radius of gyration about a generator. (Camb. Math. Tripos, Part I, 1896.)

38. A body can rotate freely about a fixed vertical axis for which its moment of inertia is I: the body carries a second body in the form of a disc which can rotate about a horizontal axis, fixed in the first body and intersecting the vertical axis. In the position of equilibrium the moments and product of inertia of the disc with regard to the vertical and horizontal axes respectively are A, B, F. Prove that if the system start from rest with the plane of the disc inclined at an angle a to the vertical, the first body will oscillate through an angle

$$\frac{2F}{\{B(A+I)\}^{\frac{1}{2}}}\arctan\left\{\frac{B^{\frac{1}{2}}\sin a}{(A+I)^{\frac{1}{2}}}\right\}. \qquad \text{(Coll. Exam.)}$$

39. A gyrostat consists of a heavy symmetrical flywheel freely mounted in a heavy spherical case and is suspended from a fixed point by a string of length l fixed to a point in the case. The centres of gravity of the flywheel and case are coincident. Shew that, if the whole revolve in steady motion round the vertical with angular velocity Ω, the string and the axis of the gyrostat inclined at angles a, β to the vertical, then

$$\Omega^2(l\sin a+a\sin\beta+b\cos\beta)=g\tan a,$$

and

$$I\Omega\sin\beta-A\Omega^2\sin\beta\cos\beta=Mg\sec a\,\{a\sin(\beta-a)+b\cos(\beta-a)\},$$

where M is the mass of the gyrostat, a and b the coordinates of the point of attachment of the string with reference to axes coinciding with, and at right angles to, the axis of the flywheel, I the angular momentum of the flywheel about its axis and A the moment of inertia about a line perpendicular to its axis. (Camb. Math. Tripos, Part I, 1900.)

40. A system consisting of any number of equal uniform rods loosely jointed and initially in the same straight line is struck at any point by a blow perpendicular to the rods. Shew that if u, v, w be the initial velocities of the middle points of any three consecutive rods, $u+4v+w=0$. (Coll. Exam.)

41. Any number of uniform rods of masses A, B, C, ..., Z are smoothly jointed to each other in succession and laid in a straight line on a smooth table. If the end Z be free and the end A moved with velocity V in a direction perpendicular to the line of the rods, then the initial velocities of the joints (AB), (BC), ... and the end Z are a, b, ..., z, where

$$0=A(V+2a)+B(2a+b),\quad 0=B(a+2b)+C(2b+c),\ \ldots,\quad 0=Y(x+2y)+Z(2y+z),$$

and

$$y+2z=0. \qquad \text{(Coll. Exam.)}$$

42. Six equal uniform rods form a regular hexagon loosely jointed at the angular points: a blow is given at right angles to one of them at its middle point, shew that the opposite rod begins to move with $\frac{1}{10}$ of the velocity of the rod struck.

(Camb. Math. Tripos, 1882.)

43. A body at rest, with one point O fixed, is struck: shew that the initial axis of rotation of the body is the diametral line, with respect to the momental ellipsoid at O, of the plane of the impulsive couple acting on the body.

44. The positive octant of the ellipsoid $x^2/a^2 + y^2/b^2 + z^2/c^2 = 1$ has the origin fixed. Shew that if an impulsive couple in the plane

$$\frac{x}{b} + \frac{y}{a} = \frac{\pi}{2}\left(\frac{a}{b} + \frac{b}{a}\right)\frac{z}{c}$$

act upon the octant, it will begin to revolve about the axis of z. (Coll. Exam.)

45. An ellipsoid is rotating about its centre with angular velocity $(\omega_1, \omega_2, \omega_3)$ referred to its principal axes; the centre is free and a point (x, y, z) on the surface is suddenly brought to rest. Find the impulsive reaction at that point. (Coll. Exam.)

46. Two equal rods AB, BC inclined at an angle a are smoothly jointed at B; A is made to move parallel to the external bisector of the angle ABC: prove that the initial angular velocities of AB, BC are in the ratio

$$3\left(1 + 2\sin^2\frac{a}{2}\right) : 1 - 18\sin^2\frac{a}{2}.$$

(Coll. Exam.)

47. A uniform cone is rotating with angular velocity ω about a generator when suddenly this generator is loosed and the diameter of the base which intersects the generator is fixed. Prove that the new angular velocity is

$$(1 - h^2/4k^2)\,\omega \sin a,$$

where h is the altitude, a the semi-vertical angle, and k the radius of gyration about a diameter of the base. (Coll. Exam.)

48. A rough disc can turn about an axis perpendicular to its plane, and a rough circular cone rests on the disc with its vertex just at the axis. If the disc be made to turn with angular velocity Ω, shew that the cone takes an amount of kinetic energy equal to

$$\tfrac{1}{2}\Omega^2/\{\cos^2 a/A + \sin^2 a/C\}.$$

(Coll. Exam.)

49. One end of an inelastic string is attached to a fixed point and the other to a point in the surface of a body of mass M. The body is allowed to fall freely under gravity without rotation. Shew that just after the string becomes tight the loss of kinetic energy due to the impact is

$$\tfrac{1}{2}V^2 \Big/ \left(\frac{1}{M} + \frac{\lambda^2}{A} + \frac{\mu^2}{B} + \frac{\nu^2}{C}\right),$$

where V is the resolved velocity of the body in the direction of the string just before impact, the string only touching the body at the point of attachment, $(l, m, n, \lambda, \mu, \nu)$ are the coordinates of the string at the instant it becomes tight, and A, B, C are the principal moments of inertia of the body with respect to its principal axes at its centre of inertia. (Coll. Exam.)

CHAPTER VII

THEORY OF VIBRATIONS

76. *Vibrations about equilibrium.*

In Dynamics we frequently have to deal with systems for which there exists an *equilibrium-configuration*, i.e. a configuration in which the system can remain permanently at rest: thus in the case of the spherical pendulum, the configurations in which the bob is vertically over or vertically under the point of support are of this character. If $(q_1, q_2, \ldots, q_n)$ are the coordinates of a system and L its kinetic potential, and if $(\alpha_1, \alpha_2, \ldots, \alpha_n)$ are the values of the coordinates in an equilibrium-configuration, the equations of motion

$$\frac{d}{dt}\left(\frac{\partial L}{\partial \dot{q}_r}\right) - \frac{\partial L}{\partial q_r} = 0 \qquad (r = 1, 2, \ldots, n)$$

must be satisfied by the set of values

$$\ddot{q}_1 = 0,\ \ddot{q}_2 = 0,\ \ldots,\ \ddot{q}_n = 0,\ \dot{q}_1 = 0,\ \dot{q}_2 = 0,\ \ldots,\ \dot{q}_n = 0,\ q_1 = \alpha_1,\ q_2 = \alpha_2,\ \ldots,\ q_n = \alpha_n.$$

The values of the coordinates in the various possible equilibrium-configurations of a system are therefore obtained by solving for $q_1, q_2, \ldots, q_n$ the equations

$$\frac{\partial L}{\partial q_r} = 0 \qquad (q = 1, 2, \ldots, n),$$

in which $\dot{q}_1, \dot{q}_2, \ldots, \dot{q}_n$ are to be replaced by zero.

In many cases, if the system is initially placed near an equilibrium-configuration, its particles having very small initial velocities, the divergence from the equilibrium-configuration will never become very marked, the particles always remaining in the vicinity of their original positions and never acquiring large velocities. We shall now study motions of this type*; they are called *vibrations about an equilibrium-configuration*†.

* More strictly speaking, we study in this chapter the limiting form to which this type of motion approximates when the initial divergence from a state of rest in the equilibrium-configuration tends to zero; the study of the motions which differ by a finite, though not large, amount from a state of rest in the equilibrium-configuration is given later in Chapter XVI: the discussion of the present chapter may be regarded as a first approximation to that of Chapter XVI.

† The theory of vibrations has developed from Galileo's study of the small oscillations of a pendulum. In the first half of the eighteenth century the vibrations of a stretched cord were investigated by Brook Taylor, D'Alembert, Euler, and Daniel Bernoulli, the last-named of whom in 1753 enunciated the principle of the resolution of all compound types of vibration into independent simple modes. The general theory of the vibrations of a dynamical system with a finite number of degrees of freedom was given by Lagrange in 1762–5 (*Oeuvres*, I. p. 520).

In the present work we are of course concerned only with the vibrations of systems which have a finite number of degrees of freedom; the study of the vibrations of systems which have an infinite number of degrees of freedom, which is here excluded, will be found in treatises on the Analytical Theory of Sound.

We shall suppose that the system is defined by its kinetic energy T and its potential energy V, and that the position of the system is specified by the coordinates $(q_1, q_2, \ldots, q_n)$ independently of the time, so that T does not involve t explicitly: we shall also suppose that no coordinates have been ignored; the kinetic energy T is therefore a homogeneous quadratic function of $\dot{q}_1, \dot{q}_2, \ldots, \dot{q}_n$, with coefficients involving $q_1, q_2, \ldots, q_n$ in any way. There is evidently no loss of generality in assuming that the equilibrium-configuration corresponds to zero values of the coordinates $q_1, q_2, \ldots, q_n$; so that $q_1, q_2, \ldots, q_n, \dot{q}_1, \dot{q}_2, \ldots, \dot{q}_n$ are very small throughout the motion considered.

The coefficients of the squares and products of $\dot{q}_1, \dot{q}_2, \ldots, \dot{q}_n$ in T are functions of $q_1, q_2, \ldots, q_n$: as however all the coordinates and velocities are small, we can in approximating to the motion retain only the terms of lowest order in T, and so can replace all these coefficients by the constant values which they assume when $q_1, q_2, \ldots, q_n$ are replaced by zero. The kinetic energy is therefore for our purposes a homogeneous quadratic function of $\dot{q}_1, \dot{q}_2, \ldots, \dot{q}_n$ with constant coefficients.

Moreover, if we expand the function V by Taylor's theorem in ascending powers of $q_1, q_2, \ldots, q_n$ the term independent of $q_1, q_2, \ldots, q_n$ can be omitted, since it exercises no influence on the equations of motion; and there are no terms linear in $q_1, q_2, \ldots, q_n$, since if such terms existed the quantities $\dfrac{\partial V}{\partial q_r}$ would not be zero in the equilibrium position, as they must be. The terms of lowest order in V are therefore the terms quadratic in $q_1, q_2, \ldots, q_n$. Neglecting the higher terms of the expansion in comparison with these, we have therefore V expressed as a homogeneous quadratic form in the variables $q_1, q_2, \ldots, q_n$ with constant coefficients.

Thus *the problem of vibratory motions about a configuration of equilibrium depends on the solution of Lagrangian equations of motion in which the kinetic and potential energies are homogeneous quadratic forms in the velocities and coordinates respectively, with constant coefficients.*

77. *Normal coordinates.*

In order to solve the equations of motion of a vibrating system, we write the expressions for the kinetic and potential energies in the form

$$T = \tfrac{1}{2}(a_{11}\dot{q}_1^2 + a_{22}\dot{q}_2^2 + \ldots + a_{nn}\dot{q}_n^2 + 2a_{12}\dot{q}_1\dot{q}_2 + 2a_{13}\dot{q}_1\dot{q}_3 + \ldots + 2a_{n-1,n}\dot{q}_{n-1}\dot{q}_n),$$

$$V = \tfrac{1}{2}(b_{11}q_1^2 + b_{22}q_2^2 + \ldots + b_{nn}q_n^2 + 2b_{12}q_1q_2 + 2b_{13}q_1q_3 + \ldots + 2b_{n-1,n}q_{n-1}q_n);$$

of these T is (§ 26) a positive definite form; and the determinant formed of the quantities a_{rs} is not zero (since if this condition is not satisfied, T will depend on less than n independent velocities). The equations of motion are

$$\frac{d}{dt}\left(\frac{\partial T}{\partial \dot{q}_r}\right) = -\frac{\partial V}{\partial q_r} \qquad (r = 1, 2, \ldots, n);$$

if a change of variables is made, such that the new variables $(q_1', q_2', \ldots, q_n')$ are linear functions of $(q_1, q_2, \ldots, q_n)$, the new equations of motion will be

$$\frac{d}{dt}\left(\frac{\partial T}{\partial \dot{q}_r'}\right) = -\frac{\partial V}{\partial q_r'} \qquad (r = 1, 2, \ldots, n),$$

and these equations are clearly linear combinations of the original equations.

Suppose then* that the original equations of motion are multiplied respectively by undetermined constants $m_1, m_2, \ldots, m_n$, and added together. The resulting equation will be of the form

$$\frac{d^2Q}{dt^2} + \lambda Q = 0,$$

where
$$Q = h_1 q_1 + h_2 q_2 + \ldots + h_n q_n,$$

provided the constants $m_1, m_2, \ldots, m_n, h_1, h_2, \ldots, h_n, \lambda$ satisfy the equations

$$\begin{aligned}
b_{11} m_1 + b_{12} m_2 + \ldots + b_{1n} m_n &= \lambda (a_{11} m_1 + a_{12} m_2 + \ldots + a_{1n} m_n) = \lambda h_1, \\
b_{21} m_1 + b_{22} m_2 + \ldots + b_{2n} m_n &= \lambda (a_{21} m_1 + a_{22} m_2 + \ldots + a_{2n} m_n) = \lambda h_2, \\
&\cdots\cdots\cdots\cdots\cdots\cdots \\
&\cdots\cdots\cdots\cdots\cdots\cdots \\
b_{n1} m_1 + b_{n2} m_2 + \ldots + b_{nn} m_n &= \lambda (a_{n1} m_1 + a_{n2} m_2 + \ldots + a_{nn} m_n) = \lambda h_n.
\end{aligned}$$

These equations can coexist only if λ is a root of the determinantal equation

$$\begin{vmatrix}
a_{11}\lambda - b_{11}, & a_{12}\lambda - b_{12}, & \ldots, & a_{1n}\lambda - b_{1n} \\
a_{21}\lambda - b_{21}, & a_{22}\lambda - b_{22}, & \ldots, & a_{2n}\lambda - b_{2n} \\
\cdots & \cdots & \cdots & \cdots \\
\cdots & \cdots & \cdots & \cdots \\
a_{n1}\lambda - b_{n1}, & \cdots & \cdots & a_{nn}\lambda - b_{nn}
\end{vmatrix} = 0.$$

Moreover, if λ_1 is any root of this equation, we can determine from the preceding equations a possible set of ratios for $m_1, m_2, \ldots, m_n, h_1, h_2, \ldots, h_n$; these ratios may, in certain cases, be partly indeterminate, but in all cases at least one function Q can be obtained in this way, satisfying the equation

$$\frac{d^2Q}{dt^2} + \lambda_1 Q = 0.$$

Now let a linear change of variables be effected so that the quantity Q so determined is one of the new variables: there will be no ambiguity in

* This method of proof is due to Jordan, *Comptes Rendus*, LXXIV. (1872) p. 1395.

denoting the new variables by $q_1, q_2, \ldots, q_n$; we shall take q_1 to be identical with Q, so that the above equations are satisfied by the values $h_1=1$, $h_2=0$, $\ldots$, $h_n=0$. Since the form T is a positive definite form, the coefficients $a_{22}, a_{33}, \ldots, a_{nn}$ of the squares of $\dot{q}_2, \dot{q}_3, \ldots, \dot{q}_n$ will not be zero: so instead of $q_2, q_3, \ldots, q_n$ we can again take new variables $q_1', q_2', \ldots, q_n'$, where

$$q_1 = q_1', \quad q_2 = q_2' + \frac{A_{12}}{A_{11}} q_1', \quad q_3 = q_3' + \frac{A_{13}}{A_{11}} q_1', \ldots,$$

and A_{rs} is the co-factor of a_{rs} in the discriminant of T. It is found without difficulty that by this change of variables the terms in $\dot{q}_1\dot{q}_2, \dot{q}_1\dot{q}_3, \ldots, \dot{q}_1\dot{q}_n$ are removed from T: so we can assume that $a_{21}, a_{31}, \ldots, a_{n1}$ are zero.

Now introducing the conditions $h_1=1$, $h_2=0$, $h_3=0$, $\ldots$, $h_n=0$, $a_{21}=0$, $\ldots$, $a_{n1}=0$ in the equations which determine $m_1, m_2, \ldots, m_n, h_1, h_2, \ldots, h_n, \lambda$, we obtain the values

$$m_1 = 1/a_{11}, \quad m_2 = 0, \quad m_3 = 0, \ldots, m_n = 0,$$
$$b_{11} = \lambda_1 a_{11}, \quad b_{21} = 0, \quad b_{31} = 0, \ldots, b_{n1} = 0.$$

It follows that the equation

$$\frac{d}{dt}\left(\frac{\partial T}{\partial \dot{q}_1}\right) = -\frac{\partial V}{\partial q_1}$$

has the form

$$\frac{d^2 q_1}{dt^2} + \lambda_1 q_1 = 0,$$

while the equations

$$\frac{d}{dt}\left(\frac{\partial T}{\partial \dot{q}_r}\right) = -\frac{\partial V}{\partial q_r} \qquad (r = 2, 3, \ldots, n)$$

have the form

$$\frac{d}{dt}\left(\frac{\partial T'}{\partial \dot{q}_r}\right) = -\frac{\partial V'}{\partial q_r} \qquad (r = 2, 3, \ldots, n),$$

where

$$T' = T - \tfrac{1}{2} a_{11}\dot{q}_1^2, \quad V' = V - \tfrac{1}{2}\lambda_1 a_{11} q_1^2,$$

so that T' and V' do not involve $\dot{q}_1$ and q_1.

This last system of equations may be regarded as the system of equations corresponding to a vibrational problem with $(n-1)$ degrees of freedom. Treating them in the same manner, we can isolate another coordinate q_2 such that if

$$T'' = T' - \tfrac{1}{2} a_{22}\dot{q}_2^2, \quad V'' = V' - \tfrac{1}{2}\lambda_2 a_{22} q_2^2$$

(where λ_2 and a_{22} are certain constants), then T'' and V'' do not involve $\dot{q}_2$ or q_2, and the coordinates $q_3, q_4, \ldots, q_n$ are determined by the equations of a vibrational problem with $(n-2)$ degrees of freedom, in which the kinetic and potential energies are respectively T'' and V''

Proceeding in this way, we shall finally have the variables chosen so that the kinetic and potential energies of the original system can be written in terms of the new variables in the form

$$T = \tfrac{1}{2}(\alpha_{11}\dot{q}_1^2 + \alpha_{22}\dot{q}_2^2 + \ldots + \alpha_{nn}\dot{q}_n^2),$$
$$V = \tfrac{1}{2}(\beta_{11} q_1^2 + \beta_{22} q_2^2 + \ldots + \beta_{nn} q_n^2),$$

where $\alpha_{11}, \alpha_{22}, \ldots, \alpha_{nn}, \beta_{11}, \beta_{22}, \ldots, \beta_{nn}$ are constants.

If finally we take as variables the quantities $\sqrt{\alpha_{11}}q_1, \sqrt{\alpha_{22}}q_2, \ldots, \sqrt{\alpha_{nn}}q_n$, instead of $q_1, q_2, \ldots, q_n$, the kinetic and potential energies take the form

$$T = \tfrac{1}{2}(\dot{q}_1^2 + \dot{q}_2^2 + \ldots + \dot{q}_n^2),$$
$$V = \tfrac{1}{2}(\mu_1 q_1^2 + \mu_2 q_2^2 + \ldots + \mu_n q_n^2),$$

where μ_k stands for β_{kk}/α_{kk}.

In this reduction it is immaterial whether the determinantal equation has its roots all distinct or has groups of repeated roots. The final result can be expressed by the statement that *if the kinetic and potential energies of a vibrating system are given in the form*

$$T = \tfrac{1}{2}(a_{11}\dot{q}_1^2 + a_{22}\dot{q}_2^2 + \ldots + a_{nn}\dot{q}_n^2 + 2a_{12}\dot{q}_1\dot{q}_2 + \ldots + 2a_{n-1,n}\dot{q}_{n-1}\dot{q}_n),$$
$$V = \tfrac{1}{2}(b_{11}q_1^2 + b_{22}q_2^2 + \ldots + b_{nn}q_n^2 + 2b_{12}q_1q_2 + \ldots + 2b_{n-1,n}q_{n-1}q_n),$$

it is always possible to find a linear transformation of the coordinates such that the kinetic and potential energies, when expressed in terms of the new coordinates, have the form

$$T = \tfrac{1}{2}(\dot{q}_1^2 + \dot{q}_2^2 + \ldots + \dot{q}_n^2),$$
$$V = \tfrac{1}{2}(\mu_1 q_1^2 + \mu_2 q_2^2 + \ldots + \mu_n q_n^2),$$

where the quantities $\mu_1, \mu_2, \ldots, \mu_n$ are constants. These new coordinates are called the *normal coordinates* or *principal coordinates* of the vibrating system.

Now it is a well-known algebraical theorem that the roots of the determinantal equation

$$\begin{vmatrix} a_{11}\lambda - b_{11}, & a_{12}\lambda - b_{12}, & \ldots, & a_{1n}\lambda - b_{1n} \\ a_{21}\lambda - b_{21}, & a_{22}\lambda - b_{22}, & \ldots, & a_{2n}\lambda - b_{2n} \\ \cdots & \cdots & \cdots & \cdots \\ \cdots & \cdots & \cdots & \cdots \\ a_{n1}\lambda - b_{n1} & \cdots & \cdots & a_{nn}\lambda - b_{nn} \end{vmatrix} = 0$$

are the values of λ for which the expression

$$(a_{11}\lambda - b_{11})q_1^2 + (a_{22}\lambda - b_{22})q_2^2 + \ldots + (a_{nn}\lambda - b_{nn})q_n^2 + 2(a_{12}\lambda - b_{12})q_1q_2 + \ldots$$
$$+ 2(a_{n-1,n}\lambda - b_{n-1,n})q_{n-1}q_n$$

can be made to depend on less than n independent variables (which will be linear functions of $q_1, q_2, \ldots, q_n$). Since this is a property which persists through any linear change of variables, we see that the determinantal equation is *invariantive*, i.e. if $q_1', q_2', \ldots, q_n'$ are any n independent linear functions of $q_1, q_2, \ldots, q_n$, and if T and V when expressed in terms of $q_1', q_2', \ldots, q_n'$ take the form

$$T = \tfrac{1}{2}(a_{11}'\dot{q}_1'^2 + a_{22}'\dot{q}_2'^2 + \ldots + 2a_{12}'\dot{q}_1'\dot{q}_2' + \ldots),$$
$$V = \tfrac{1}{2}(b_{11}'q_1'^2 + b_{22}'q_2'^2 + \ldots + 2b_{12}'q_1'q_2' + \ldots),$$

then the roots of the new determinantal equation $\| a_{rs}'\lambda - b_{rs}' \| = 0$ are the same as the roots of the original determinantal equation $\| a_{rs}\lambda - b_{rs} \| = 0$.

But when the kinetic and potential energies have been brought by the introduction of normal coordinates to the form

$$T = \tfrac{1}{2}(\dot{q}_1^2 + \dot{q}_2^2 + \ldots + \dot{q}_n^2),$$

$$V = \tfrac{1}{2}(\mu_1 q_1^2 + \mu_2 q_2^2 + \ldots + \mu_n q_n^2),$$

the determinantal equation is

$$\begin{vmatrix} \lambda - \mu_1 & 0 & 0 & 0 \ldots & 0 \\ 0 & \lambda - \mu_2 & 0 & 0 \ldots & 0 \\ 0 & 0 & \lambda - \mu_3 & 0 \ldots & 0 \\ \cdots & \cdots & \cdots & \cdots & \cdots \\ \cdots & \cdots & \cdots & \cdots & \cdots \\ 0 & 0 & 0 & 0 \ldots & \lambda - \mu_n \end{vmatrix} = 0,$$

so its roots are $\mu_1, \mu_2, \ldots, \mu_n$. It follows that *the constants* $\mu_1, \mu_2, \ldots, \mu_n$, *which occur as the coefficients of the squares of the normal coordinates in the potential energy, are the n roots (distinct or repeated) of the determinantal equation* $\| a_{rs}\lambda - b_{rs} \| = 0$, *where* $a_{11}, a_{12}, \ldots, b_{11}, b_{12}, \ldots$ *are the coefficients in the original expressions for the kinetic and potential energies.*

It will be seen that the problem of reducing the kinetic and potential energies to their expressions in terms of normal coordinates is essentially the problem of simultaneously reducing each of two given homogeneous quadratic expressions in n variables to a sum of squares of n new variables: the fact that T is a function of the velocities while V is a function of the coordinates does not affect the question, since the formulae of transformation for the velocities $\dot{q}_1, \dot{q}_2, \ldots, \dot{q}_n$ are the same as the formulae of transformation for the coordinates $q_1, q_2, \ldots, q_n$.

It might be supposed from the foregoing that it is always possible to transform simultaneously each of two given homogeneous quadratic expressions in n variables to a sum of squares of n new variables; but this is not the case; for example, it is not possible to transform the two quadratic expressions

$$ax^2 + bxy + az^2 \text{ and } cx^2 + dxy + cz^2$$

to the forms

$$\xi^2 + \eta^2 + \zeta^2 \text{ and } \alpha\xi^2 + \beta\eta^2 + \gamma\zeta^2,$$

where ξ, η, ζ are linear functions of x, y, z.

The conditions which must be satisfied in order that two given quadratic expressions

$$a_{11}x_1^2 + a_{22}x_2^2 + \ldots + 2a_{12}x_1x_2 + \ldots,$$

$$b_{11}x_1^2 + b_{22}x_2^2 + \ldots + 2b_{12}x_1x_2 + \ldots,$$

may be simultaneously reducible to the form

$$\alpha_{11}\xi_1^2 + \alpha_{22}\xi_2^2 + \ldots + \alpha_{nn}\xi_n^2,$$

$$\beta_{11}\xi_1^2 + \beta_{22}\xi_2^2 + \ldots + \beta_{nn}\xi_n^2,$$

are, in fact, that the elementary divisors (*Elementartheiler*) of the determinant $\| a_{rs}\lambda - b_{rs} \|$ shall be linear*. If however one of the two given forms is a *definite* form (as we saw was the case with the kinetic energy in the dynamical problem), the elementary divisors are always linear, and the simultaneous reduction to sums of squares is therefore possible; this explains the circumstance that the reduction can always be effected in the dynamical problem of vibrations.

The universal possibility of the reduction to normal coordinates for dynamical systems was established by Weierstrass in 1858†; previous writers (following Lagrange) had supposed that in cases where the determinantal equation had repeated roots a set of normal coordinates would not exist, and that terms involving the time otherwise than in trigonometric and exponential functions would occur in the final solution of the equations of motion.

78. *Sylvester's theorem on the reality of the roots of the determinantal equation.*

We have seen in the preceding article that by introducing new variables which are linear functions of the original variables, it is always possible to reduce the kinetic and potential energies of a vibrating system to the form

$$T = \tfrac{1}{2}(\dot{q}_1^2 + \dot{q}_2^2 + \dots + \dot{q}_n^2),$$

$$V = \tfrac{1}{2}(\lambda_1 q_1^2 + \lambda_2 q_2^2 + \dots + \lambda_n q_n^2).$$

The question arises as to whether this transformation is *real*, i.e. whether the coefficients $m_1, m_2, \dots, m_n, h_1, h_2, \dots, h_n$ which occur in the transformation are real or complex. Since these coefficients are given by linear equations whose coefficients, with the possible exception of the roots $\lambda_1, \lambda_2, \dots, \lambda_n$ of the determinantal equation, are certainly real, the question reduces to an investigation of the reality or otherwise of the roots of the equation

$$\begin{vmatrix} a_{11}\lambda - b_{11} & a_{12}\lambda - b_{12} & \dots\dots & a_{1n}\lambda - b_{1n} \\ a_{21}\lambda - b_{21} & a_{22}\lambda - b_{22} & \dots\dots & a_{2n}\lambda - b_{2n} \\ \dots & \dots & \dots & \dots \\ \dots & \dots & \dots & \dots \\ a_{n1}\lambda - b_{n1} & a_{n2}\lambda - b_{n2} & \dots\dots & a_{nn}\lambda - b_{nn} \end{vmatrix} = 0;$$

it being known that the quantities a_{rs} and b_{rs} are all real, and that

$$a_{11}\dot{q}_1^2 + a_{22}\dot{q}_2^2 + \dots + a_{nn}\dot{q}_n^2 + 2a_{12}\dot{q}_1\dot{q}_2 + \dots + 2a_{n-1,n}\dot{q}_{n-1}\dot{q}_n$$

is a positive definite form.

Let Δ denote‡ the determinant $\|a_{rs}\lambda - b_{rs}\|$, and let Δ_1 denote the determinant obtained from it by striking out the first row and first column; let Δ_2 denote the determinant obtained from Δ by striking out the first two

* Cf. Muth's treatise on *Elementartheiler* (Leipzig, 1899); or Bôcher's *Introduction to Higher Algebra* (New York, 1907).

† Cf. Weierstrass' *Collected Works*, Vol. I. p. 233.

‡ The following proof is due to Nanson, *Mess. of Math.* XXVI. (1896), p. 59.

rows and first two columns, and so on. Then in any symmetrical determinant, say

$$D = \begin{vmatrix} a_{11} & a_{12} & \dots\dots & a_{1n} \\ a_{21} & a_{22} & \dots\dots & a_{2n} \\ \dots & \dots & \dots & \dots \\ \dots & \dots & \dots & \dots \\ a_{n1} & a_{n2} & \dots\dots & a_{nn} \end{vmatrix}, \text{ where } a_{rs} = a_{sr}$$

it is known that $\dfrac{\partial D}{\partial a_{11}} \dfrac{\partial D}{\partial a_{22}} - \left(\dfrac{\partial D}{\partial a_{12}}\right)^2 = D \dfrac{\partial^2 D}{\partial a_{11} \partial a_{22}}$,

and hence if $\dfrac{\partial D}{\partial a_{11}}$ vanishes the quantities D and $\dfrac{\partial^2 D}{\partial a_{11} \partial a_{22}}$ must have opposite signs; thus we have the result that in the series of quantities

$$\Delta, \Delta_1, \Delta_2, \dots, \Delta_n \qquad (\text{where } \Delta_n = 1),$$

if any one member of the series vanishes for a given value of λ, the two adjacent members must have opposite signs for that value of λ.

Let $\bar{\Delta}_r$ denote the determinant formed from Δ_r by replacing λ by unity and each of the quantities b_{rs} by zero, so that $\bar{\Delta}_r$ is the coefficient of the highest power of λ in Δ_r. Since

$$a_{11}\dot{q}_1^2 + a_{22}\dot{q}_2^2 + \dots + a_{nn}\dot{q}_n^2 + 2a_{12}\dot{q}_1\dot{q}_2 + \dots + 2a_{n-1,n}\dot{q}_{n-1}\dot{q}_n$$

is a positive definite form, $\bar{\Delta}_r$ is positive for all values of r from 0 to n. Thus the coefficients of the highest powers of λ in the functions $\Delta, \Delta_1, \dots, \Delta_n$ are all of the same sign; and therefore as λ increases from $-\infty$ to $+\infty$, these functions lose n changes of sign.

Now since Δ_n is not zero and Δ_{r-1}, Δ_{r+1} have opposite signs when Δ_r vanishes, it follows that the functions $\Delta, \Delta_1, \Delta_2, \dots, \Delta_n$ cannot lose or gain a change of sign except when λ passes through a root of Δ. But as λ passes from $-\infty$ to $+\infty$, the functions lose n changes of sign; and hence *the n roots of the determinant* Δ *are all real. The transformation to normal coordinates is therefore always a real transformation**.

Moreover, since a change of sign is lost in the pair Δ, Δ_1, every time that λ passes through a root of Δ, it is evident that Δ_1 must change sign when λ increases from one root of Δ to the consecutive root, and hence that the n roots of Δ are separated by the $(n-1)$ roots of Δ_1: similarly the roots of each of the functions Δ_r are separated by the roots of the function Δ_{r+1}. Now Δ_n has no roots: and if Δ_{n-1} has the same sign at $\lambda = 0$ as at $\lambda = -\infty$, the root of the function Δ_{n-1} will not be negative. If moreover Δ_{n-2} has the same sign at $\lambda = 0$ as at $\lambda = -\infty$, neither of the roots of Δ_{n-2} will be negative: for if this condition is satisfied, Δ_{n-2} must have either two negative roots or no negative roots, and there cannot be two negative roots since there is no negative root of Δ_{n-1} to separate them. Similarly in

* Sylvester, *Phil. Mag.* (4) IV. (1852), p. 138: *Coll. Papers*, I. p. 378.

general the condition that none of the functions in the series $\Delta, \Delta_1, \Delta_2, \ldots, \Delta_n$ shall have a negative root is that each of the functions must have the same sign at $\lambda = 0$ as at $\lambda = -\infty$. Hence the condition to be satisfied in order that all the roots of Δ may be positive is that each quantity Δ_r shall have at $\lambda = 0$ the same sign as $(-1)^{n-r}$, i.e. that each of the determinants

$$\begin{vmatrix} b_{11} & b_{12} & \ldots & b_{1n} \\ b_{21} & b_{22} & \ldots & b_{2n} \\ \ldots & \ldots & \ldots & \ldots \\ \ldots & \ldots & \ldots & \ldots \\ b_{n1} & b_{n2} & \ldots & b_{nn} \end{vmatrix}, \quad \begin{vmatrix} b_{22} & b_{23} & \ldots & b_{2n} \\ b_{32} & b_{33} & \ldots & b_{3n} \\ \ldots & \ldots & \ldots & \ldots \\ \ldots & \ldots & \ldots & \ldots \\ b_{n2} & \ldots & \ldots & b_{nn} \end{vmatrix}, \quad \begin{vmatrix} b_{33} & b_{34} & \ldots & b_{3n} \\ b_{43} & b_{44} & \ldots & \ldots \\ \ldots & \ldots & \ldots & \ldots \\ \ldots & \ldots & \ldots & \ldots \\ b_{n3} & \ldots & \ldots & b_{nn} \end{vmatrix}, \ldots b_{nn}$$

shall be positive. But these are the well-known conditions that the quadratic form

$$b_{11}q_1^2 + b_{22}q_2^2 + \ldots + b_{nn}q_n^2 + 2b_{12}q_1q_2 + \ldots + 2b_{n-1,n}q_{n-1}q_n$$

shall be a positive definite form. Hence finally *the condition that the determinantal equation* $\| a_{rs}\lambda - b_{rs} \| = 0$ *shall have all its roots positive is that the quadratic form*

$$b_{11}q_1^2 + b_{22}q_2^2 + \ldots + b_{nn}q_n^2 + 2b_{12}q_1q_2 + \ldots + 2b_{n-1,n}q_{n-1}q_n$$

shall be a positive definite form, i.e. that the potential energy in the vibratory motion shall be essentially positive.

79. *Solution of the differential equations; the periods; stability.*

In order to express the configuration of any vibrating system in terms of the time, we first determine the normal coordinates of the system, and express the kinetic and potential energies in terms of them, so that these take the form

$$T = \tfrac{1}{2}(\dot{q}_1^2 + \dot{q}_2^2 + \ldots + \dot{q}_n^2),$$
$$V = \tfrac{1}{2}(\lambda_1 q_1^2 + \lambda_2 q_2^2 + \ldots + \lambda_n q_n^2),$$

where $(q_1, q_2, \ldots, q_n)$ are the normal coordinates, and $(\lambda_1, \lambda_2, \ldots, \lambda_n)$ are the roots of the determinantal equation $\| a_{rs}\lambda - b_{rs} \| = 0$; these quantities $(\lambda_1, \lambda_2, \ldots, \lambda_n)$ have been shewn in the last article to be all real.

The Lagrangian equation of motion for any coordinate q_r, namely

$$\frac{d}{dt}\left(\frac{\partial T}{\partial \dot{q}_r}\right) - \frac{\partial T}{\partial q_r} = -\frac{\partial V}{\partial q_r},$$

is therefore

$$\ddot{q}_r + \lambda_r q_r = 0.$$

The solution of this equation is

$$\begin{cases} q_r = A_r \cos(\sqrt{\lambda_r}\, t + B_r), & \text{if } \lambda_r \text{ is positive,} \\ q_r = A_r t + B_r, & \text{if } \lambda_r \text{ is zero,} \\ q_r = A_r e^{\sqrt{-\lambda_r}\, t} + B_r e^{-\sqrt{-\lambda_r}\, t}, & \text{if } \lambda_r \text{ is negative,} \end{cases}$$

where in each case A_r and B_r denote constants of integration.

It appears from these equations that if all the normal coordinates and velocities except one, say q_r, are initially zero, and if the constant λ_r corresponding to the non-zero coordinate is positive, then the coordinates $(q_1, q_2, \ldots, q_{r-1}, q_{r+1}, \ldots, q_n)$ will be permanently zero, and the system will perform vibrations in which the coordinate q_r is alone affected. Moreover the configuration of the system will repeat itself after an interval of time $2\pi/\sqrt{\lambda_r}$. This is usually expressed by saying that *each of the normal coordinates corresponds to an independent mode of vibration of the system, provided the corresponding constant* λ_r *is positive; and the period of this vibration is* $2\pi/\sqrt{\lambda_r}$.

Moreover, if the system be referred to any other set of coordinates which are not normal coordinates, these coordinates are linear functions of the normal coordinates; and the normal coordinates perform their vibrations quite independently of each other; thus every conceivable vibration of the system may be regarded as the superposition of n independent normal vibrations. This is generally known as *Daniel Bernoulli's principle of the superposition of vibrations**.

If the quantities $(\lambda_1, \lambda_2, \ldots, \lambda_n)$ are not all positive, it appears from the above solution that those normal coordinates q_r which correspond to the non-positive roots λ_r will not oscillate about a zero value when the system is slightly disturbed from a state of rest in its equilibrium position, but will increase so as to invalidate the assumption made at the outset of the work, namely that the higher powers of the coordinates can be neglected. In this case therefore, there will not be a vibration at all, and the equilibrium configuration is said to be *unstable.* If however the initial disturbance is such that these normal coordinates which correspond to non-positive roots λ_r are not affected, the system will perform vibrations in which the rest of the normal coordinates oscillate about zero values.

The normal modes of vibration, which correspond to those normal coordinates for which the corresponding root λ_r is positive, are said to be *stable.* If the constants λ_r are all positive, the equilibrium-configuration as a whole is said to be stable. The *condition for stability* of the equilibrium-configuration is therefore, by the theorem of the last article, that *the potential energy of the vibrating system shall be a positive definite form*

This result might have been expected from a consideration of the integral of energy; for this integral is

$$T + V = h,$$

where T and V are the quadratic forms which represent the kinetic and potential energies, and where h is a constant. This constant h will be small if the initial divergence from the equilibrium state is small. But T is a positive definite form; and if V is also a positive definite form, we must have T and V each less than h, so T and V will remain small throughout the motion: the motion will therefore never differ greatly from the equilibrium-configuration, i.e. it will be stable.

* *Histoire de l'Académie de Berlin*, année 1753, p. 147.

80. *Examples of vibrations about equilibrium.*

We shall now discuss a number of illustrative cases of vibration about equilibrium.

(i) *To find the vibration-period of a cylinder of any cross-section which can roll on the outside of a perfectly rough fixed cylinder.*

Let s be the arc described on the fixed cylinder by the point of contact, s being measured from the equilibrium position; let ρ and ρ' be the radii of curvature of the cross-sections of the fixed and moving cylinders respectively at the points which are in contact in the equilibrium position; ρ and ρ' being supposed positive when the cylinders are convex to each other: let M be the mass of the moving cylinder, Mk^2 its moment of inertia about its centre of gravity, and c the distance of the centre of gravity from the initial position of the point of contact in the moving cylinder.

If a denotes the initial angle between the common normal to the cylinders and the vertical, then $a+s/\rho$ is the angle between the common normal at time t and the vertical, $a+s/\rho+s/\rho'$ is the angle made with the vertical by the line joining the centre of curvature of the moving cylinder with the original point of contact in the moving cylinder, and $s/\rho+s/\rho'$ is the angle made with the vertical by the line joining the last-named point to the centre of gravity of the moving cylinder. The angular velocity of the moving cylinder is therefore

$$\dot{s}\left(\frac{1}{\rho}+\frac{1}{\rho'}\right),$$

so its kinetic energy is

$$T=\tfrac{1}{2}M(k^2+c^2)\left(\frac{1}{\rho}+\frac{1}{\rho'}\right)^2\dot{s}^2.$$

The potential energy is

$V=Mg\times$ height of the centre of gravity of the moving cylinder above some fixed position

$$=Mg\left\{(\rho+\rho')\cos\left(a+\frac{s}{\rho}\right)-\rho'\cos\left(a+\frac{s}{\rho}+\frac{s}{\rho'}\right)+c\cos\left(\frac{s}{\rho}+\frac{s}{\rho'}\right)\right\}.$$

Neglecting s^3 and constant terms, this gives

$$V=\tfrac{1}{2}Mg\left\{\frac{\rho+\rho'}{\rho\rho'}\cos a-c\left(\frac{\rho+\rho'}{\rho\rho'}\right)^2\right\}s^2.$$

The Lagrangian equation of motion,

$$\frac{d}{dt}\left(\frac{\partial T}{\partial \dot{s}}\right)-\frac{\partial T}{\partial s}=-\frac{\partial V}{\partial s},$$

gives
$$M(k^2+c^2)\left(\frac{1}{\rho}+\frac{1}{\rho'}\right)^2\ddot{s}+Mg\left\{\frac{\rho+\rho'}{\rho\rho'}\cos a-c\left(\frac{\rho+\rho'}{\rho\rho'}\right)^2\right\}s=0,$$

and the vibrations are therefore given by the equation

$$s=A\cos(\lambda t+\epsilon),$$

where A and ϵ are constants of integration to be determined by the initial conditions, and λ is given by the equation

$$\lambda^2=\frac{g}{k^2+c^2}\left\{\frac{\rho\rho'}{\rho+\rho'}\cos a-c\right\}.$$

The vibration-period is $2\pi/\lambda$.

(ii) *To find the periods of the normal modes of vibration about an equilibrium-configuration of a particle moving on a fixed smooth surface under gravity.*

The tangent-plane to the surface at the point occupied by the particle in the equilibrium-configuration is evidently horizontal: take as axes of x and y the tangents to the lines of curvature of the surface at this point, and as axis of z a line drawn vertically upwards: so that the equation to the surface is approximately

$$z=\frac{x^2}{2\rho_1}+\frac{y^2}{2\rho_2},$$

where ρ_1 and ρ_2 denote the principal radii of curvature, measured positively upwards. The kinetic energy and potential energy are approximately

$$T=\tfrac{1}{2}m(\dot{x}^2+\dot{y}^2) \qquad \text{(where } m \text{ is the mass)},$$

and

$$V=mgz$$

$$=mg\left(\frac{x^2}{2\rho_1}+\frac{y^2}{2\rho_2}\right).$$

It is evident from these expressions that x and y are the normal coordinates: the equations of motion are

$$\ddot{x}+\frac{g}{\rho_1}x=0 \text{ and } \ddot{y}+\frac{g}{\rho_2}y=0,$$

and the periods of the normal modes of vibration are therefore

$$2\pi\left(\frac{\rho_1}{g}\right)^{\frac{1}{2}} \text{ and } 2\pi\left(\frac{\rho_2}{g}\right)^{\frac{1}{2}}.$$

(iii) *To find the normal modes of vibration of a rigid body, one of whose points is fixed, and which is vibrating about a position of stable equilibrium under the action of any system of conservative forces.*

Take as fixed axes of reference $OXYZ$ the equilibrium positions of the principal axes of inertia of the body at the fixed point; the moving axes will be taken as usual to be these principal axes of inertia. We shall suppose the position of the body at any instant defined by the symmetrical parameters (ξ, η, ζ, χ) of § 9; we shall regard ξ, η, ζ as the independent coordinates of the system, χ being defined in terms of them by the equation

$$\xi^2+\eta^2+\zeta^2+\chi^2=1.$$

The components of angular velocity of the body about the moving axes are (§ 16)

$$\begin{cases}\omega_1=2(\chi\dot{\xi}+\zeta\dot{\eta}-\eta\dot{\zeta}-\xi\dot{\chi}),\\ \omega_2=2(-\zeta\dot{\xi}+\chi\dot{\eta}+\xi\dot{\zeta}-\eta\dot{\chi}),\\ \omega_3=2(\eta\dot{\xi}-\xi\dot{\eta}+\chi\dot{\zeta}-\zeta\dot{\chi}).\end{cases}$$

On account of the smallness of the vibration, we regard ξ, η, ζ as small quantities of the first order; χ therefore differs from unity by a small quantity of the second order, and so we have, correctly to the first order of small quantities,

$$\omega_1=2\dot{\xi}, \quad \omega_2=2\dot{\eta}, \quad \omega_3=2\dot{\zeta},$$

and the kinetic energy of the body, which is given by the equation

$$2T=A\omega_1^2+B\omega_2^2+C\omega_3^2,$$

where A, B, C are the principal moments of inertia at the point of suspension, can be written

$$T=2(A\dot{\xi}^2+B\dot{\eta}^2+C\dot{\zeta}^2).$$

The potential energy is some function of the position of the body, and therefore of the parameters (ξ, η, ζ); let it be denoted by $V(\xi, \eta, \zeta)$.

Since zero values of (ξ, η, ζ) correspond to the equilibrium position, there will be no terms linear in (ξ, η, ζ) when V is expanded in ascending powers of (ξ, η, ζ): the lowest terms are therefore of the second order; neglecting terms of higher order, we can therefore write

$$V = a\xi^2 + b\eta^2 + c\zeta^2 + 2f\eta\zeta + 2g\zeta\xi + 2h\xi\eta,$$

where a, b, c, f, g, h are constants.

The problem of determining the normal coordinates is therefore the same as that of reducing the two quadratic expressions

$$\begin{cases} A\xi^2 + B\eta^2 + C\zeta^2, \\ a\xi^2 + b\eta^2 + c\zeta^2 + 2f\eta\zeta + 2g\zeta\xi + 2h\xi\eta, \end{cases}$$

to the form

$$\begin{cases} A_1 x^2 + B_1 y^2 + C_1 z^2, \\ a_1 x^2 + b_1 y^2 + c_1 z^2, \end{cases}$$

where (x, y, z) are linear functions of (ξ, η, ζ).

Now the equation, referred to the fixed axes, of the momental ellipsoid in its equilibrium position is

$$AX^2 + BY^2 + CZ^2 = 1;$$

consider in connexion with this the quadric whose equation is

$$aX^2 + bY^2 + cZ^2 + 2fYZ + 2gZX + 2hXY = 1,$$

which we shall call the "ellipsoid of equal potential energy"; and determine the common set of conjugate diameters of these quadrics. Let (X', Y', Z') be the coordinates, referred to these conjugate diameters, of a point whose coordinates referred to the fixed axes are (X, Y, Z), and let the equations connecting (X', Y', Z') and (X, Y, Z) be

$$\begin{cases} X = l_1 X' + m_1 Y' + n_1 Z', \\ Y = l_2 X' + m_2 Y' + n_2 Z', \\ Z = l_3 X' + m_3 Y' + n_3 Z'. \end{cases}$$

By this transformation the equations of the quadrics are reduced to the form

$$\begin{cases} A_1 X'^2 + B_1 Y'^2 + C_1 Z'^2 = 1, \\ a_1 X'^2 + b_1 Y'^2 + c_1 Z'^2 = 1, \end{cases}$$

and therefore the transformation which gives the normal coordinates in the dynamical problem is

$$\begin{cases} \xi = l_1 x + m_1 y + n_1 z, \\ \eta = l_2 x + m_2 y + n_2 z, \\ \zeta = l_3 x + m_3 y + n_3 z. \end{cases}$$

It follows that in a normal mode of vibration, say that in which x alone varies, the quantities (ξ, η, ζ) will be permanently in the ratio

$$\xi : \eta : \zeta = l_1 : l_2 : l_3.$$

But from the definitions of § 9 it is evident that ξ, η, ζ are, to the first order of small quantities, proportional to the direction-cosines of the line about which the rotation of the rigid body takes place, and consequently the normal mode of vibration of the rigid body consists of a small oscillation about a line whose equation is

$$X : Y : Z = l_1 : l_2 : l_3,$$

i.e. about the line

$$Y' = 0, \quad Z' = 0,$$

which is one of the common conjugate diameters of the two quadrics.

Hence finally we have the result that *the normal vibrations of the body are small oscillations about the common conjugate diameters of the momental ellipsoid and the ellipsoid of equal potential energy.*

(iv) *To find the normal coordinates and the periods of normal vibration in the system of three degrees of freedom for which*

$$T=\tfrac{1}{2}(\dot{x}^2+\dot{y}^2+\dot{z}^2),$$
$$V=\tfrac{1}{2}\{p^2(x^2+y^2)+2az(x+y)+q^2z^2\},$$

where a is small in comparison with p and q; and to shew that if such a system be let go from rest with y and z initially zero, the vibration in x will have temporarily ceased after a time $\pi p(q^2-p^2)/a^2$, and that there will then be a vibration of the same amplitude in y as the original one was in x. (Coll. Exam.)

The form of the kinetic and potential energies suggests the transformation

$$x+y=2\xi, \quad x-y=2\eta,$$

which gives

$$\begin{cases} T=\dot{\xi}^2+\dot{\eta}^2+\tfrac{1}{2}\dot{z}^2, \\ V=p^2\xi^2+p^2\eta^2+2az\xi+\tfrac{1}{2}q^2z^2. \end{cases}$$

The variable η is therefore a normal coordinate: to reduce the remaining terms in the kinetic and potential energies to sums of squares, we write

$$z=\zeta-\frac{2a}{q^2-p^2}\phi, \quad \xi=\phi+\frac{a}{q^2-p^2}\zeta,$$

and then we have

$$\begin{cases} T=\dot{\eta}^2+\left\{1+\dfrac{2a^2}{(q^2-p^2)^2}\right\}\dot{\phi}^2+\tfrac{1}{2}\left\{1+\dfrac{2a^2}{(q^2-p^2)^2}\right\}\dot{\zeta}^2, \\ V=p^2\eta^2+\left\{p^2-\dfrac{a^2(2q^2-4p^2)}{(q^2-p^2)^2}\right\}\phi^2+\tfrac{1}{2}\left\{q^2+\dfrac{(4q^2-2p^2)a^2}{(q^2-p^2)^2}\right\}\zeta^2. \end{cases}$$

The variables η, ϕ, ζ are therefore the normal coordinates.

Suppose that initially we have

$$x=k, \quad y=0, \quad z=0,$$
$$\dot{x}=0, \quad \dot{y}=0, \quad \dot{z}=0,$$

and suppose that k is so small that its product with other small quantities can be neglected. Then to this degree of approximation we have initially

$$\eta=\tfrac{1}{2}k, \quad \phi=\tfrac{1}{2}k, \quad \zeta=0.$$

The vibrations of the normal coordinates η and ϕ are therefore given by the equations

$$\eta=\tfrac{1}{2}k\cos pt,$$

$$\phi=\tfrac{1}{2}k\cos\left[t\left\{\frac{p^2-\dfrac{a^2(2q^2-4p^2)}{(q^2-p^2)^2}}{1+\dfrac{2a^2}{(q^2-p^2)^2}}\right\}^{\frac{1}{2}}\right].$$

The last equation can be written

$$\phi=\tfrac{1}{2}k\cos\left[pt\left\{1-\frac{a^2}{p^2(q^2-p^2)}\right\}\right],$$

or

$$\phi=\tfrac{1}{2}k\cos pt\cos\frac{a^2t}{p(q^2-p^2)}+\tfrac{1}{2}k\sin pt\sin\frac{a^2t}{p(q^2-p^2)}.$$

The motion can therefore be approximately represented initially by

$$\eta = \tfrac{1}{2}k\cos pt, \quad \phi = \tfrac{1}{2}k\cos pt,$$

or

$$x = k\cos pt, \quad y = 0.$$

After an interval of time $\pi p(q^2 - p^2)/a^2$, the motion is approximately represented by

$$\eta = \tfrac{1}{2}k\cos pt, \quad \phi = -\tfrac{1}{2}k\cos pt,$$

or

$$x = 0, \qquad y = -k\cos pt;$$

which establishes the result stated.

81. *Effect of a new constraint on the periods of a vibrating system.*

We shall now consider the effect produced on the periods of normal vibration of a dynamical system about a configuration of stable equilibrium when the number of degrees of freedom of the system is diminished by the introduction of an additional constraint.

Suppose that the original system is specified in terms of its normal coordinates $(q_1, q_2, \ldots, q_n)$, so that the kinetic and potential energies have the form

$$T = \tfrac{1}{2}(\dot{q}_1^2 + \dot{q}_2^2 + \ldots + \dot{q}_n^2),$$

$$V = \tfrac{1}{2}(\lambda_1^2 q_1^2 + \lambda_2^2 q_2^2 + \ldots + \lambda_n^2 q_n^2);$$

and let the additional constraint be expressed by the equation

$$f(q_1, q_2, \ldots, q_n) = 0.$$

Since $q_1, q_2, \ldots, q_n$ are small, we can expand the function f in ascending powers of $q_1, q_2, \ldots, q_n$, and retain only the first terms of the expansion: we can thus express the constraint by the equation

$$A_1 q_1 + A_2 q_2 + \ldots + A_n q_n = 0,$$

where $A_1, \ldots, A_n$ are constants. As the equilibrium-configuration is supposed to be compatible with the constraint, there will be no constant term. By means of this equation we can eliminate q_n: we thus have

$$T = \tfrac{1}{2}\left\{\dot{q}_1^2 + \dot{q}_2^2 + \ldots + \dot{q}_{n-1}^2 + \frac{1}{A_n^2}(A_1\dot{q}_1 + \ldots + A_{n-1}\dot{q}_{n-1})^2\right\},$$

$$V = \tfrac{1}{2}\left\{\lambda_1^2 q_1^2 + \ldots + \lambda_{n-1}^2 q_{n-1}^2 + \frac{\lambda_n^2}{A_n^2}(A_1 q_1 + \ldots + A_{n-1}q_{n-1})^2\right\}.$$

The Lagrangian equations of motion of the constrained system are therefore the $(n-1)$ equations

$$\ddot{q}_r + \lambda_r^2 q_r + A_r\left\{\frac{1}{A_n^2}(A_1\ddot{q}_1 + \ldots + A_{n-1}\ddot{q}_{n-1}) + \frac{\lambda_n^2}{A_n^2}(A_1 q_1 + \ldots + A_{n-1}q_{n-1})\right\} = 0$$

$$(r = 1, 2, \ldots, n-1),$$

or

$$\ddot{q}_r + \lambda_r^2 q_r + \mu A_r = 0 \qquad (r = 1, 2, \ldots, n-1)$$

where

$$\mu = \frac{1}{A_n^2}(A_1\ddot{q}_1 + \ldots + A_{n-1}\ddot{q}_{n-1}) + \frac{\lambda_n^2}{A_n^2}(A_1 q_1 + \ldots + A_{n-1}q_{n-1})$$

$$= -\frac{\ddot{q}_n}{A_n} - \frac{\lambda_n^2 q_n}{A_n},$$

so the equations of motion of the constrained system can be written in the form of the n equations

$$\ddot{q}_r + \lambda_r^2 q_r + \mu A_r = 0 \qquad (r = 1, 2, \ldots, n),$$

where μ is undetermined.

Now consider a normal mode of vibration of the modified system, defined by equations

$$q_1 = \alpha_1 \cos \lambda t, \quad q_2 = \alpha_2 \cos \lambda t, \ldots, q_n = \alpha_n \cos \lambda t, \quad \mu = \nu \cos \lambda t.$$

Substituting in the equations of motion, we have

$$\alpha_r(\lambda_r^2 - \lambda^2) + \nu A_r = 0 \qquad (r = 1, 2, \ldots, n).$$

Substituting the values of $\alpha_1, \alpha_2, \ldots, \alpha_n$ given by these equations in the equation

$$A_1\alpha_1 + A_2\alpha_2 + \ldots + A_n\alpha_n = 0,$$

we have

$$\frac{A_1^2}{\lambda_1^2 - \lambda^2} + \frac{A_2^2}{\lambda_2^2 - \lambda^2} + \ldots + \frac{A_n^2}{\lambda_n^2 - \lambda^2} = 0.$$

This equation in λ^2 has $(n-1)$ roots, which from the form of the equation are evidently interspaced between the quantities $\lambda_1^2, \lambda_2^2, \ldots, \lambda_n^2$: the quantities $2\pi/\lambda$ corresponding to these roots are the periods of the normal modes of vibration of the constrained system, and it therefore follows that *the $(n-1)$ periods of normal vibration of the constrained system are spaced between the n periods of the original system.*

82. *The stationary character of normal vibrations.*

We shall next consider the effect of adding constraints to a dynamical system to such an extent that only one degree of freedom is left to the system. Let $(q_1, q_2, \ldots, q_n)$ be the normal coordinates of the original system; the constraints may, as in the last article, be represented by linear equations between these coordinates, and can therefore be expressed in the form

$$q_1 = \mu_1 q, \quad q_2 = \mu_2 q, \ldots q_n = \mu_n q,$$

where $\mu_1, \mu_2, \ldots, \mu_n$ are constants and q is a new variable which may be taken as defining the configuration of the constrained system at time t.

Let the kinetic and potential energies of the original system be

$$T = \tfrac{1}{2}(\dot{q}_1^2 + \dot{q}_2^2 + \ldots + \dot{q}_n^2),$$

$$V = \tfrac{1}{2}(\lambda_1^2 q_1^2 + \lambda_2^2 q_2^2 + \ldots + \lambda_n^2 q_n^2),$$

so $2\pi/\lambda_1, 2\pi/\lambda_2, \ldots, 2\pi/\lambda_n$ are its periods of normal vibration: the kinetic and potential energies of the constrained system are then

$$T = \tfrac{1}{2}(\mu_1^2 + \mu_2^2 + \ldots + \mu_n^2)\dot{q}^2,$$
$$V = \tfrac{1}{2}(\lambda_1^2\mu_1^2 + \lambda_2^2\mu_2^2 + \ldots + \lambda_n^2\mu_n^2)q^2.$$

The period of a vibration of the constrained system is therefore $2\pi/\lambda$, where λ is given by the equation

$$\lambda^2 = \frac{\lambda_1^2\mu_1^2 + \lambda_2^2\mu_2^2 + \ldots + \lambda_n^2\mu_n^2}{\mu_1^2 + \mu_2^2 + \ldots + \mu_n^2}.$$

If the constraints are varied, this expression has a stationary value when $(n-1)$ of the quantities $\mu_1, \mu_2, \ldots, \mu_n$ are zero: this stationary value is one of the quantities $\lambda_1^2, \lambda_2^2, \ldots, \lambda_n^2$: and thus we have the theorem that *when constraints are put on the system so as to reduce its number of degrees of freedom to unity, the period of the constrained system has a stationary value for those constraints which make the vibration to be a normal vibration of the unconstrained system.*

83. *Vibrations about steady motion.*

A type of motion which presents many analogies with the equilibrium-configuration is that known as the *steady motion* of systems which possess ignorable coordinates: this is defined to be a motion in which the non-ignorable coordinates of the system have constant values, while the velocities corresponding to the ignorable coordinates have also constant values.

One example of a steady motion is that of the top, discussed in § 72; as another example we may take the case of a particle which is free to move in a plane and is attracted by a fixed centre of force, the potential energy depending only on the distance from the centre of force; for such a particle, a circular orbit described with constant velocity is always a possible orbit, and this is a form of steady motion, since the radius vector is constant and the angular velocity $\dot{\theta}$ corresponding to the ignorable coordinate θ is also constant.

In many cases, if a system is initially in a state of motion differing only slightly from a given form of steady motion, the divergence from this form of motion will never subsequently become very marked; we shall now consider motions of this kind, which are called *vibrations about steady motion.*

The steady motion is said to be *stable** if the vibratory motion tends to a certain limiting form, namely the steady motion, when the initial disturbance from steady motion tends to zero.

Let $(p_1, p_2, \ldots, p_k)$ be the ignorable and $(q_1, q_2, \ldots, q_n)$ be the non-ignorable coordinates of the system. Corresponding to the ignorable coordinates, there will be k integrals

$$\frac{\partial L}{\partial \dot{p}_r} = \beta_r \qquad (r = 1, 2, \ldots, k),$$

* This definition is due to Klein and Sommerfeld.

where $\beta_1, \beta_2, \ldots, \beta_k$ are constants. We shall suppose that these constants have the same value in the vibratory motion as in the undisturbed steady motion of which it is regarded as the disturbed form; this of course only amounts to coordinating each vibratory motion to some particular steady motion.

We suppose the system conservative, with constraints independent of the time; let its kinetic energy be

$$T = \tfrac{1}{2} \sum_{i=1}^{n} \sum_{j=1}^{n} a_{ij} \dot{q}_i \dot{q}_j + \sum_{i=1}^{n} \sum_{j=1}^{k} b_{ij} \dot{q}_i \dot{p}_j + \tfrac{1}{2} \sum_{i=1}^{k} \sum_{j=1}^{k} c_{ij} \dot{p}_i \dot{p}_j,$$

where the coefficients a_{ij}, b_{ij}, c_{ij} are functions of $q_1, q_2, \ldots, q_n$.

The integrals corresponding to the ignorable coordinates are

$$\sum_i c_{ij} \dot{p}_i + \sum_i b_{ij} \dot{q}_i = \beta_j \qquad (j = 1, 2, \ldots, k).$$

Let C_{ij} be the minor of c_{ij} in the determinant formed of the coefficients c_{ij}, divided by this determinant; then solving the last equations for the quantities $\dot{p}_r$, we have

$$\dot{p}_r = \sum_s C_{rs} (\beta_s - \sum_l b_{ls} \dot{q}_l).$$

Substituting for $\dot{p}_1, \dot{p}_2, \ldots, \dot{p}_k$ in the above expression for T, and utilising the properties of minors of determinants, we have

$$T = \tfrac{1}{2} \sum_{i,j} (a_{ij} - \sum_{l,s} C_{ls} b_{il} b_{js}) \dot{q}_i \dot{q}_j + \tfrac{1}{2} \sum_{l,s} C_{ls} \beta_l \beta_s.$$

Now perform the process of ignoration of coordinates. Let R be the modified kinetic potential, so

$$\begin{aligned} R &= T - V - \sum_{r=1}^{k} \dot{p}_r \beta_r \\ &= \tfrac{1}{2} \sum_{i,j} (a_{ij} - \sum_{l,s} C_{ls} b_{il} b_{js}) \dot{q}_i \dot{q}_j + \sum_{l,r,s} C_{rs} \beta_r b_{ls} \dot{q}_l - \tfrac{1}{2} \sum_{l,s} C_{ls} \beta_l \beta_s - V. \end{aligned}$$

We can without loss of generality suppose that the values of $q_1, q_2, \ldots, q_n$ in the steady motion are all zero. If then the coefficients in R are expanded in ascending powers of $q_1, q_2, \ldots, q_n$ by Taylor's theorem, and all terms in the expression of R thus obtained which are above the second degree in the variables $\dot{q}_1, \dot{q}_2, \ldots, \dot{q}_n, q_1, q_2, \ldots, q_n$ are neglected in comparison with the terms of the second degree, we obtain for R an expression consisting of terms linear and quadratic in $\dot{q}_1, \dot{q}_2, \ldots, \dot{q}_n, q_1, q_2, \ldots, q_n$. Now the terms which are linear in $\dot{q}_1, \dot{q}_2, \ldots, \dot{q}_n$ and independent of $q_1, q_2, \ldots, q_n$ disappear automatically from the equations of motion

$$\frac{d}{dt}\left(\frac{\partial R}{\partial \dot{q}_r}\right) - \frac{\partial R}{\partial q_r} = 0 \qquad (r = 1, 2, \ldots, n),$$

and these terms can therefore be omitted. Moreover, since the equations are satisfied by permanent zero values of $q_1, q_2, \ldots, q_n$, it is evident that no terms linear in $q_1, q_2, \ldots, q_n$ and independent of $\dot{q}_1, \dot{q}_2, \ldots, \dot{q}_n$ can be present in R. It follows that *the problem of vibrations about steady motion depends on the solution of Lagrangian equations of motion in which the kinetic potential is a homogeneous quadratic function of the velocities and coordinates, with constant coefficients.*

The difference between vibrations about equilibrium and vibrations about steady motion consists in the possible presence in the latter case of terms of the type $q_r \dot{q}_s$ (i.e. products of a coordinate and a velocity) in the kinetic potential. These are called *gyroscopic* terms. The vibrations about steady motion of a system are in fact the same thing as the vibrations about equilibrium of the reduced or non-natural (§ 38) system to which the problem is brought by ignoration of coordinates.

The equations of motion for the vibrating system are therefore

$$\frac{d}{dt}\left(\frac{\partial R}{\partial \dot{q}_r}\right) - \frac{\partial R}{\partial q_r} = 0 \qquad (r = 1, 2, \ldots, n),$$

where R can be written in the form

$$R = \tfrac{1}{2} \sum_{r,s} \alpha_{rs} \dot{q}_r \dot{q}_s + \tfrac{1}{2} \sum_{r,s} \beta_{rs} q_r q_s + \sum_{r,s} \gamma_{rs} q_r \dot{q}_s \qquad (r, s = 1, 2, \ldots, n),$$

and where $\alpha_{rs} = \alpha_{sr}, \quad \beta_{rs} = \beta_{sr},$

but where γ_{rs} is not in general equal to γ_{sr}. The equations of motion in the expanded form are

$$\begin{cases} \alpha_{11}\ddot{q}_1 - \beta_{11} q_1 + \alpha_{12}\ddot{q}_2 + (\gamma_{21} - \gamma_{12})\,\dot{q}_2 - \beta_{12} q_2 + \alpha_{13}\ddot{q}_3 + (\gamma_{31} - \gamma_{13})\,\dot{q}_3 - \beta_{13} q_3 + \ldots = 0, \\ \alpha_{21}\ddot{q}_1 + (\gamma_{12} - \gamma_{21})\,\dot{q}_1 - \beta_{21} q_1 + \alpha_{22}\ddot{q}_2 - \beta_{22} q_2 + \alpha_{23}\ddot{q}_3 + (\gamma_{32} - \gamma_{23})\,\dot{q}_3 - \beta_{23} q_3 + \ldots = 0, \\ \text{etc.} \end{cases}$$

These are linear equations with constant coefficients, which are of the same general character as the corresponding equations in the case of vibrations about equilibrium; they differ only in the presence of the gyroscopic terms, which involve the coefficients $(\gamma_{sr} - \gamma_{rs})$. The presence of these terms makes it impossible to transform the system to normal coordinates*; but as we shall next see, the main characteristic of vibrations about equilibrium is retained, namely that any vibration can be regarded as a superposition of n purely periodic vibrations, which we shall call (as before) the *normal* modes of vibration of the system.

84. *The integration of the equations.*

We shall now shew how the nature of the vibrations can be determined, by integration of the equations of motion.

* That is to say, impossible to transform the system to normal coordinates by a *point-transformation*: it is possible to effect the transformation to normal coordinates by a contact-transformation, and this is actually done in Chapter XVI.

It will be convenient first to transform them into a system of equations each of the first order. Let R denote the modified kinetic potential of the system, so that in the vibratory problem R is a homogeneous quadratic function of $q_1, q_2, \ldots, q_n, \dot{q}_1, \dot{q}_2, \ldots, \dot{q}_n$. Write

$$\frac{\partial R}{\partial \dot{q}_r} = q_{n+r} \qquad (r = 1, 2, \ldots, n),$$

so that $q_{n+1}, q_{n+2}, \ldots, q_{2n}$ are linear functions of $\dot{q}_1, \dot{q}_2, \ldots \dot{q}_n$ and *vice versa*; the equations of motion can be written

$$\dot{q}_{n+r} = \frac{\partial R}{\partial q_r} \qquad (r = 1, 2, \ldots, n).$$

Now if δ denote an increment of a function of the variables $q_1, q_2, \ldots, q_n, q_{n+1}, \ldots, q_{2n}$, due to small changes in these variables, we have

$$\delta R = \sum_{r=1}^{n} \left(\frac{\partial R}{\partial q_r} \delta q_r + \frac{\partial R}{\partial \dot{q}_r} \delta \dot{q}_r \right)$$

$$= \sum_{r=1}^{n} (\dot{q}_{n+r} \delta q_r + q_{n+r} \delta \dot{q}_r)$$

$$= \delta \left(\sum_{r=1}^{n} q_{n+r} \dot{q}_r \right) + \sum_{r=1}^{n} (\dot{q}_{n+r} \delta q_r - \dot{q}_r \delta q_{n+r}).$$

Let the quantity $$\sum_{r=1}^{n} q_{n+r} \dot{q}_r - R,$$

when expressed as a function of $q_1, q_2, \ldots, q_{2n}$, be denoted by H, so that H is a known homogeneous quadratic function of the variables $q_1, q_2, \ldots, q_{2n}$; the last equation can be written

$$\delta H = \sum_{r=1}^{n} (\dot{q}_r \delta q_{n+r} - \dot{q}_{n+r} \delta q_r),$$

and therefore* *the equations of motion, which consisted originally of n equations each of the second order, can be replaced by a system of $2n$ equations, each of the first order, namely*

$$\dot{q}_r = \frac{\partial H}{\partial q_{n+r}}, \quad \dot{q}_{n+r} = -\frac{\partial H}{\partial q_r} \qquad (r = 1, 2, \ldots, n)$$

the independent variables being $q_1, q_2, \ldots, q_{2n}$.

We shall now shew that the function H, which has replaced R as the determining function of the equations, represents the sum of the kinetic and potential energies of the dynamical system considered.

For R contains terms of degrees 2, 1, and 0 in the velocities, and

$$\sum_{r=1}^{n} \dot{q}_r \frac{\partial R}{\partial \dot{q}_r}$$

* This transformation is really a case of the Hamiltonian transformation given later in Chapter X.

is equivalent to twice the terms of degree two together with the terms of degree one, by Euler's theorem; it follows that H, being defined as

$$\sum_{r=1}^{n} \dot{q}_r \frac{\partial R}{\partial \dot{q}_r} - R,$$

will be equal to the terms of degree two in the velocities in R, together with the terms of zero degree in R with their signs changed: on comparing the expressions for T and R given on page 194, it follows that

$$H = T + V,$$

so *H is the total energy of the dynamical system, expressed in terms of the variables* $q_1, q_2, \ldots, q_{2n}$.

In the case of vibrations about an equilibrium-configuration, we have seen that the condition for stability is that the potential as well as the kinetic energy shall be a positive definite form; we shall now make a similar assumption for the case of vibrations about steady motion, namely that *the total energy H is a positive definite form in the variables* $q_1, q_2, \ldots, q_{2n}$; on this assumption we shall shew that the steady motion is stable, and in fact that the equations of motion

$$\frac{dq_r}{dt} = \frac{\partial H}{\partial q_{n+r}}, \qquad \frac{dq_{n+r}}{dt} = -\frac{\partial H}{\partial q_r} \qquad (r = 1, 2, \ldots, n)$$

can be integrated in the following way*.

Consider the set of linear equations in the variables $q_1, q_2, \ldots, q_{2n}$,

$$\left.\begin{aligned} sq_{n+r} + \frac{\partial H(q_1, q_2, \ldots, q_{2n})}{\partial q_r} &= y_r \\ -sq_r + \frac{\partial H(q_1, q_2, \ldots, q_{2n})}{\partial q_{n+r}} &= y_{n+r} \end{aligned}\right\} \quad (r = 1, 2, \ldots, n);$$

if we denote the determinant of the system by $f(s)$, and the minor of the element in the λth row and μth column by

$$f(s)_{\lambda\mu} \qquad (\lambda, \mu = 1, 2, \ldots, 2n),$$

the expression of $q_1, q_2, \ldots, q_{2n}$ in terms of $y_1, y_2, \ldots, y_{2n}$ is given by the equations

$$q_\mu = \sum_{\lambda=1}^{2n} \frac{f(s)_{\lambda\mu}}{f(s)} y_\lambda \qquad (\mu = 1, 2, \ldots, 2n),$$

and the degree of $f(s)$ in s is $2n$, while the degree of $f(s)_{\lambda\mu}$ is not greater than $(2n-1)$.

In order to solve the equations of motion, consider expressions for $q_1, q_2, \ldots, q_{2n}$ of the form

$$q_\mu = \int_C \frac{p_\mu(s)}{f(s)} e^{s(t-t_0)} ds \qquad (\mu = 1, 2, \ldots, 2n),$$

* The method of integration which follows is due to Weierstrass, *Berlin. Monatsberichte*, 1879.

where the integration is taken round a large circle C which encloses all the roots of the equation $f(s)=0$. These values of $q_1, q_2, \ldots, q_{2n}$ will satisfy the equations of motion, provided the equations

$$\left.\begin{aligned} \int_C e^{s(t-t_0)}\left\{sp_{n+r}+\frac{\partial H(p_1, p_2, \ldots, p_{2n})}{\partial p_r}\right\}\frac{ds}{f(s)}=0 \\ \int_C e^{s(t-t_0)}\left\{-sp_r+\frac{\partial H(p_1, p_2, \ldots, p_{2n})}{\partial p_{n+r}}\right\}\frac{ds}{f(s)}=0 \end{aligned}\right\} \quad (r=1, 2, \ldots, n)$$

are satisfied. If therefore $p_1, p_2, \ldots, p_{2n}$ are polynomials in s so chosen that the expressions in brackets under the integral sign vanish when s is equal to one of the roots of the equation $f(s)=0$, these equations will be satisfied, since the integrands will then have no singularities within the contour C*. It follows that $p_1, p_2, \ldots, p_{2n}$ must be a set of solutions of the equations

$$\left.\begin{aligned} sp_{n+r}+\frac{\partial H(p_1, p_2, \ldots, p_{2n})}{\partial p_r}=0 \\ -sp_r+\frac{\partial H(p_1, p_2, \ldots, p_{2n})}{\partial p_{n+r}}=0 \end{aligned}\right\} \quad (r=1, 2, \ldots, n),$$

when s is a root of the equation $f(s)=0$; this condition is satisfied by the expressions

$$p_\mu(s)=a_1 f(s)_{1\mu}+a_2 f(s)_{2\mu}+\ldots+a_{2n} f(s)_{2n,\mu} \quad (\mu=1, 2, \ldots, 2n),$$

where $a_1, a_2, \ldots, a_{2n}$ are arbitrary constants.

The equations of motion are therefore satisfied by the values

$q_\mu=$ coefficient of $1/s$ in the Laurent expansion† in positive and negative powers of s of the expression

$$\{a_1 f(s)_{1\mu}+a_2 f(s)_{2\mu}+\ldots+a_{2n} f(s)_{2n,\mu}\}\frac{e^{s(t-t_0)}}{f(s)} \quad (\mu=1, 2, \ldots, 2n).$$

Now on inspection of the determinant $f(s)$ we see that minors of the types

$$f(s)_{n+\mu,\mu} \text{ and } f(s)_{\mu,n+\mu} \quad (\mu=1, 2, \ldots, n)$$

are of degree $(2n-1)$ in s, and the other minors are of degree $(2n-2)$ in s; so the coefficient of $1/s$ in the Laurent expansion of $f(s)_{\lambda\mu}/f(s)$ is zero unless $\lambda=n+\mu$ or $\mu=n+\lambda$; in the former case it is -1, and in the latter case it is 1. Hence on taking $t=t_0$, we see that the quantities

$$a_1, a_2, \ldots, a_{2n}$$

are respectively the values of

$$q_{n+1}, q_{n+2}, \ldots, q_{2n}, -q_1, -q_2, \ldots, -q_n$$

at the time t_0.

* Whittaker and Watson, *A Course of Modern Analysis*, § 5·2.

† *Ibid.* § 5·6.

If therefore we write

$$\phi(t)_{\lambda\mu} = \text{coefficient of } 1/s \text{ in the Laurent expansion of } \frac{f(s)_{\lambda\mu}}{f(s)} e^{s(t-t_0)},$$

and if $\mathring{q}_1, \mathring{q}_2, \ldots, \mathring{q}_{2n}$ are the values of $q_1, q_2, \ldots, q_{2n}$ respectively corresponding to any definite value t_0 of t, we have

$$q_\mu = \sum_{\alpha=1}^{n} \{\mathring{q}_{n+\alpha}\phi(t)_{\alpha,\mu} - \mathring{q}_\alpha \phi(t)_{n+\alpha,\mu}\} \quad (\mu = 1, 2, \ldots, 2n).$$

In order to evaluate the quantities $\phi(t)_{\lambda\mu}$, it is necessary to discuss the nature of the roots of the determinantal equation $f(s) = 0$; let $ki + l$, where k and l are real and i denotes $\sqrt{-1}$, be any root of this equation; then the $2n$ equations

$$\left.\begin{aligned} (ki+l)\,q_{n+\alpha} + \frac{\partial H(q_1, q_2, \ldots, q_{2n})}{\partial q_\alpha} &= 0 \\ -(ki+l)\,q_\alpha + \frac{\partial H(q_1, q_2, \ldots, q_{2n})}{\partial q_{n+\alpha}} &= 0 \end{aligned}\right\} \quad (\alpha = 1, 2, \ldots, n)$$

can be satisfied by values of $q_1, q_2, \ldots, q_{2n}$ which are not all zero. Let a system of such values be

$$\xi_1 + i\eta_1, \quad \xi_2 + i\eta_2, \ldots, \xi_{2n} + i\eta_{2n},$$

where $\xi_1, \xi_2, \ldots, \xi_{2n}, \eta_1, \eta_2, \ldots, \eta_{2n}$ are real quantities. Then if we write

$$\frac{\partial H(q_1, q_2, \ldots, q_{2n})}{\partial q_\mu} = H(q_1, q_2, \ldots, q_{2n})_\mu,$$

we have, on separating the last equations into their real and imaginary parts,

$$\left.\begin{aligned} H(\xi_1, \xi_2, \ldots, \xi_{2n})_\alpha + l\xi_{n+\alpha} - k\eta_{n+\alpha} &= 0 \\ H(\xi_1, \xi_2, \ldots, \xi_{2n})_{n+\alpha} - l\xi_\alpha + k\eta_\alpha \quad &= 0 \\ H(\eta_1, \eta_2, \ldots, \eta_{2n})_\alpha + l\eta_{n+\alpha} + k\xi_{n+\alpha} &= 0 \\ H(\eta_1, \eta_2, \ldots, \eta_{2n})_{n+\alpha} - l\eta_\alpha - k\xi_\alpha \quad &= 0 \end{aligned}\right\} \quad (\alpha = 1, 2, \ldots, n).$$

But since H is homogeneous and of degree two in its arguments, we have

$$2H(\xi_1, \xi_2, \ldots, \xi_{2n}) = \sum_{\lambda=1}^{2n} \xi_\lambda H(\xi_1, \xi_2, \ldots, \xi_{2n})_\lambda,$$

and using the first two of the preceding equations this gives

$$\left.\begin{aligned} 2H(\xi_1, \xi_2, \ldots, \xi_{2n}) &= k \sum_{\alpha=1}^{n} (\xi_\alpha \eta_{n+\alpha} - \eta_\alpha \xi_{n+\alpha}). \\ \text{Similarly} \qquad 2H(\eta_1, \eta_2, \ldots, \eta_{2n}) &= k \sum_{\alpha=1}^{n} (\xi_\alpha \eta_{n+\alpha} - \eta_\alpha \xi_{n+\alpha}). \end{aligned}\right\} \quad \ldots\ldots\ldots\text{(A)}$$

Moreover on multiplying the first of the preceding equations by η_α and the second by $\eta_{n+\alpha}$, adding, and summing for values of α from 1 to n, we have

$$\sum_{\lambda=1}^{2n} \eta_\lambda H(\xi_1, \xi_2, \ldots, \xi_{2n})_\lambda = l \sum_{\alpha=1}^{n} (\xi_\alpha \eta_{n+\alpha} - \eta_\alpha \xi_{n+\alpha}),$$

and similarly

$$\sum_{\lambda=1}^{2n} \xi_\lambda H(\eta_1, \eta_2, \ldots, \eta_{2n})_\lambda = -l \sum_{\alpha=1}^{n} (\xi_\alpha \eta_{n+\alpha} - \eta_\alpha \xi_{n+\alpha}).$$

Since the left-hand sides of these equations are equal, we must have

$$l \sum_{\alpha=1}^{n} (\xi_\alpha \eta_{n+\alpha} - \eta_\alpha \xi_{n+\alpha}) = 0.$$

But from equations (A) we see that, as H is a positive definite form, neither k nor $\sum_{\alpha=1}^{n} (\xi_\alpha \eta_{n+\alpha} - \eta_\alpha \xi_{n+\alpha})$ can be zero; we must therefore have l zero; and so *the equation $f(s)=0$ has each of its roots of the form ik, where k is a real quantity different from zero.*

We shall next shew that in the case in which the equation $f(s)=0$ has a j-tuple root s', each of the functions $f(s)_{\lambda\mu}$ is divisible by $(s-s')^{j-1}$.

For let $c_1, c_2, \ldots, c_{2n}$ be a set of definite real quantities; define quantities $q_1, q_2, \ldots, q_{2n}$ by the equations

$$\left.\begin{aligned} sq_{n+\alpha} + H(q_1, q_2, \ldots, q_{2n})_\alpha &= c_\alpha \\ -sq_\alpha + H(q_1, q_2, \ldots, q_{2n})_{n+\alpha} &= c_{n+\alpha} \end{aligned}\right\} \quad (\alpha = 1, 2, \ldots, n) \ \ldots\text{(B)},$$

so that we have

$$q_\mu = \sum_{\lambda=1}^{2n} \frac{f(s)_{\lambda\mu}}{f(s)} c_\lambda \qquad (\mu = 1, 2, \ldots, 2n).$$

Let $s_1 i$ be any root of the equation $f(s)=0$, and let m be the smallest positive integer for which all the functions

$$(s - s_1 i)^m \cdot \frac{f(s)_{\lambda\mu}}{f(s)}$$

are finite for the value $s_1 i$ of s. When s is taken sufficiently near $s_1 i$, we can expand q_μ in a series of the form

$$(g_\mu + h_\mu i)(s - s_1 i)^{-m} + (g_\mu' + h_\mu' i)(s - s_1 i)^{-m+1} + \ldots,$$

where $g_\mu, h_\mu, g_\mu', h_\mu', \ldots$ denote real constants; and we can suppose the quantities $c_1, c_2, \ldots, c_{2n}$ so chosen that the quantities g_μ and h_μ are not zero. Substituting this value of q_μ in equations (B), and equating the coefficients of $(s-s_1 i)^{-m}$, we have

$$\left.\begin{aligned} H(g_1, g_2, \ldots, g_{2n})_\alpha - s_1 h_{n+\alpha} &= 0 \\ H(g_1, g_2, \ldots, g_{2n})_{n+\alpha} + s_1 h_\alpha &= 0 \\ H(h_1, h_2, \ldots, h_{2n})_\alpha + s_1 g_{n+\alpha} &= 0 \\ H(h_1, h_2, \ldots, h_{2n})_{n+\alpha} - s_1 g_\alpha &= 0 \end{aligned}\right\} \quad (\alpha = 1, 2, \ldots, n) \ \text{(C)},$$

and on equating the coefficients of $(s-s_1 i)^{-m+1}$, we have

$$\left.\begin{aligned} H(g_1', g_2', \ldots, g_{2n}')_\alpha - s_1 h'_{n+\alpha} + g_{n+\alpha} &= \begin{cases} 0 & \text{when } m > 1 \\ c_\alpha & \text{when } m = 1 \end{cases} \\ H(g_1', g_2', \ldots, g_{2n}')_{n+\alpha} + s_1 h_\alpha' - g_\alpha &= \begin{cases} 0 & \text{when } m > 1 \\ c_{n+\alpha} & \text{when } m = 1 \end{cases} \\ H(h_1', h_2', \ldots, h_{2n}')_\alpha + s_1 g'_{n+\alpha} + h_{n+\alpha} &= 0 \\ H(h_1', h_2', \ldots, h_{2n}')_{n+\alpha} - s_1 g_\alpha' - h_\alpha &= 0 \end{aligned}\right\} \quad (\alpha = 1, 2, \ldots, n) \ \text{(D)}.$$

Now by Euler's theorem on homogeneous functions we have

$$2H(g_1, g_2, \ldots, g_{2n}) = \sum_{\lambda=1}^{2n} g_\lambda H(g_1, g_2, \ldots, g_{2n})_\lambda,$$

or by (C),

$$2H(g_1, g_2, \ldots, g_{2n}) = s_1 \sum_{a=1}^{n} (g_a h_{n+a} - h_a g_{n+a}),$$

and similarly

$$2H(h_1, h_2, \ldots, h_{2n}) = s_1 \sum_{a=1}^{n} (g_a h_{n+a} - h_a g_{n+a}),$$

from which it is evident that $\sum_{a=1}^{n} (g_a h_{n+a} - h_a g_{n+a})$ is not zero.

Moreover, the first two of equations (C) give

$$\sum_{\lambda=1}^{2n} h_\lambda' H(g_1, g_2, \ldots, g_{2n})_\lambda + s_1 \sum_{a=1}^{n} (h_a h'_{n+a} - h_a' h_{n+a}) = 0 \quad \text{.........(E)},$$

and the last two of equations (C) give

$$\sum_{\lambda=1}^{2n} g_\lambda' H(h_1, h_2, \ldots, h_{2n})_\lambda - s_1 \sum_{a=1}^{n} (g_a g'_{n+a} - g_a' g_{n+a}) = 0 \quad \text{.........(F)}.$$

But from the first two of equations (D), when $m > 1$, we have

$$\sum_{\lambda=1}^{2n} h_\lambda H(g_1', g_2', \ldots, g_{2n}')_\lambda - s_1 \sum_{a=1}^{n} (h_a h'_{n+a} - h_a' h_{n+a}) - \sum_{a=1}^{n} (g_a h_{n+a} - h_a g_{n+a}) = 0 \quad \text{.........(G)},$$

and from the last two of equations (D) we have

$$\sum_{\lambda=1}^{2n} g_\lambda H(h_1', h_2', \ldots, h_{2n}')_\lambda + s_1 \sum_{a=1}^{n} (g_a g'_{n+a} - g_a' g_{n+a}) + \sum_{a=1}^{n} (g_a h_{n+a} - h_a g_{n+a}) = 0 \quad \text{.........(H)}.$$

Also since H is homogeneous of the second degree in its arguments, we have the identities

$$\sum_{\lambda=1}^{2n} h_\lambda' H(g_1, g_2, \ldots, g_{2n})_\lambda = \sum_{\lambda=1}^{2n} g_\lambda H(h_1', h_2', \ldots, h_{2n}')_\lambda \quad \text{.........(K)}$$

and

$$\sum_{\lambda=1}^{2n} g_\lambda' H(h_1, h_2, \ldots, h_{2n})_\lambda = \sum_{\lambda=1}^{2n} h_\lambda H(g_1', g_2', \ldots, g_{2n}')_\lambda \quad \text{.........(L)}.$$

From equations (E), (H), (K) we have

$$\sum_{a=1}^{n} (g_a h_{n+a} - h_a g_{n+a}) = s_1 \sum_{a=1}^{n} (h_a h'_{n+a} - h_a' h_{n+a}) - s_1 \sum_{a=1}^{n} (g_a g'_{n+a} - g_a' g_{n+a}),$$

and from equations (F), (G), (L) we have

$$\sum_{a=1}^{n} (g_a h_{n+a} - h_a g_{n+a}) = - s_1 \sum_{a=1}^{n} (h_a h'_{n+a} - h_a' h_{n+a}) + s_1 \sum_{a=1}^{n} (g_a g'_{n+a} - g_a' g_{n+a}).$$

Comparing these equations, we have

$$\sum_{a=1}^{n} (g_a h_{n+a} - h_a g_{n+a}) = 0,$$

which is contrary to what has already been proved. The assumption that $m > 1$, which was used in obtaining equation (G), must therefore be false; m must therefore be unity, and consequently *when* $f(s)$ *is divisible by* $(s - s_1 i)^k$, *each of the functions* $f(s)_{\lambda\mu}$ *is divisible by* $(s - s_1 i)^{k-1}$.

Now let $s_1, s_2, \ldots, s_r$ be the moduli of the *distinct* roots of the equation $f(s) = 0$, so that the functions $f(s)_{\lambda\mu}/f(s)$ are infinite only for $s = \pm s_1 i, \pm s_2 i, \ldots, \pm s_r i$; then denoting the coefficient of $(s - s_\rho i)^{-1}$ in the Laurent expansion of $f(s)_{\lambda\mu}/f(s)$ in powers of $(s - s_\rho i)$ by

$$(\lambda, \mu)_\rho + i(\lambda, \mu)_\rho',$$

where $(\lambda, \mu)_\rho$ and $(\lambda, \mu)_\rho'$ are real, and observing that the only poles of the function $f(s)_{\lambda\mu}/f(s)$ are the points $s = \pm s_\rho i$, and that these are simple poles, we have

$$\frac{f(s)_{\lambda\mu}}{f(s)} = \sum_{\rho=1}^{r} \left\{ \frac{(\lambda, \mu)_\rho + i(\lambda, \mu)_\rho'}{s - s_\rho i} + \frac{(\lambda, \mu)_\rho - i(\lambda, \mu)_\rho'}{s + s_\rho i} \right\},$$

and therefore $\phi(t)_{\lambda\mu}$ is the coefficient of $1/s$ in the Laurent expansion of

$$e^{s(t-t_0)} \sum_{\rho=1}^{r} \left\{ \frac{(\lambda, \mu)_\rho + i(\lambda, \mu)_\rho'}{s - s_\rho i} + \frac{(\lambda, \mu)_\rho - i(\lambda, \mu)_\rho'}{s + s_\rho i} \right\}$$

in powers of s.

But the coefficient of $1/s$ in the Laurent expansion of $e^{s(t-t_0)}/(s - s_\rho i)$ is $e^{s_\rho(t-t_0)i}$, and the coefficient of $1/s$ in the Laurent expansion of $e^{s(t-t_0)}/(s + s_\rho i)$ is $e^{-s_\rho(t-t_0)i}$; we have therefore

$$\phi(t)_{\lambda\mu} = 2 \sum_{\rho=1}^{r} \{(\lambda, \mu)_\rho \cos s_\rho(t - t_0) - (\lambda, \mu)_\rho' \sin s_\rho(t - t_0)\},$$

and so finally

$$q_\mu = 2 \sum_{\alpha=1}^{n} \sum_{\rho=1}^{r} [\mathring{q}_{n+\alpha} \{(\alpha, \mu)_\rho \cos s_\rho(t - t_0) - (\alpha, \mu)_\rho' \sin s_\rho(t - t_0)\}$$
$$- \mathring{q}_\alpha \{(n + \alpha, \mu)_\rho \cos s_\rho(t - t_0) - (n + \alpha, \mu)_\rho' \sin s_\rho(t - t_0)\}] \quad (\mu = 1, 2, \ldots, 2n).$$

This formula constitutes the general solution of the differential equations of motion. Hence finally we see that *when the total energy of a system vibrating about a state of steady motion is a positive definite form, the vibratory motion can be expressed in terms of circular functions of* t, *and the steady motion is stable; the periods of the normal vibrations are* $2\pi/s_1, 2\pi/s_2, \ldots$, *where* $\pm is_1$, $\pm is_2, \ldots$ *are the roots of the determinantal equation* $f(s) = 0$, *whose order in* s^2 *is equal to the number of non-ignorable coordinates of the system.*

The above investigation is valid whether the determinantal equation has repeated roots or not.

Between the coefficients $(\lambda, \mu)_\rho$, $(\lambda, \mu)_\rho'$, there exist the relations

$$(\lambda, \mu)_\rho = -(\mu, \lambda)_\rho, \quad (\lambda, \mu)_\rho' = (\mu, \lambda)_\rho',$$

and so in particular $$(\lambda, \lambda)_\rho = 0.$$

These relations follow from equations which (in virtue of their definitions) are true for $f(s)$, $f(s)_{\lambda\mu}$, namely

$$f(s) = f(-s),$$
$$f(s)_{\lambda\mu} = f(-s)_{\mu\lambda}.$$

Example. If the number of degrees of freedom of the system, after ignoration of the ignorable coordinates, is even, shew that when the ignorable velocities are large (e.g. if the ignorable coordinates are the angles through which certain fly-wheels have rotated, this would imply that the fly-wheels are rotating very rapidly), half the periods of vibration are very long and the other half are very short, the one set being proportional to the ignorable velocities and the other set being inversely proportional to these velocities.

It was pointed out by Poincaré* that the discussion of stability by the method of small oscillations does not take account of some features which are likely to be present in actual problems. Thus†, consider a particle moveable on the inner surface of a spherical bowl which rotates with constant angular velocity about its vertical diameter. If the bowl be perfectly smooth, the equilibrium of the particle in the lowest position is certainly stable, the rotation of the bowl having no effect on it. But if there be the slightest friction between the particle and the bowl, and if the angular velocity of the bowl exceeds a certain value, the particle will work its way outwards in a spiral path towards the position in which it rotates with the bowl like the bob of a conical pendulum.

85. *Examples of vibrations about steady motion.*

A number of illustrative cases of vibration about a state of steady motion will now be considered.

(i) *A particle is describing the circle $r = a$, $z = b$, in the cylindrical field of force in which the potential energy is $V = \phi(r, z)$, where $r^2 = x^2 + y^2$, it being given that $\partial V/\partial z$ is zero when $r = a$, $z = b$. To find the conditions for stability of the motion.*

If we write $$x = r\cos\theta, \quad y = r\sin\theta,$$

we have for the kinetic and potential energies of the particle, whose mass will be denoted by m,

$$T = \tfrac{1}{2}m(\dot{r}^2 + r^2\dot{\theta}^2 + \dot{z}^2),$$
$$V = \phi(r, z).$$

The integral corresponding to the ignorable coordinate θ is $mr^2\dot{\theta} = k$, where k is a constant. The modified kinetic potential after ignoration of θ is therefore

$$R = T - V - k\dot{\theta}$$
$$= \tfrac{1}{2}m\dot{r}^2 + \tfrac{1}{2}m\dot{z}^2 - \phi(r, z) - \frac{k^2}{2mr^2}.$$

* *Acta Math.* VII. (1885), p. 259.

† This illustration is due to Lamb, *Proc. Roy. Soc.* LXXX. (1908), p. 168.

For the steady motion we must have

$$\frac{\partial R}{\partial r}=0, \quad \frac{\partial R}{\partial z}=0;$$

the latter condition is satisfied by hypothesis, and the former gives $k^2=ma^3\partial\phi/\partial a$. We have therefore

$$R=\tfrac{1}{2}m\dot{r}^2+\tfrac{1}{2}m\dot{z}^2-\phi(r,z)-\frac{a^3}{2r^2}\frac{\partial\phi}{\partial a}.$$

Writing $\qquad r=a+\rho, \quad z=b+\zeta,$

and neglecting terms above the second degree in ρ and ζ, we have

$$R=\tfrac{1}{2}m\dot{\rho}^2+\tfrac{1}{2}m\dot{\zeta}^2-\tfrac{1}{2}\rho^2\left(\phi_{aa}+\frac{3}{a}\phi_a\right)-\rho\zeta\phi_{ab}-\tfrac{1}{2}\zeta^2\phi_{bb}.$$

As no terms linear in $\dot{\rho}$ or $\dot{\zeta}$ occur, this is essentially the same as a problem of vibrations about equilibrium, and the condition for stability is (§ 79) that

$$\rho^2\left(\phi_{aa}+\frac{3}{a}\phi_a\right)+2\rho\zeta\phi_{ab}+\zeta^2\phi_{bb}$$

shall be a positive definite form, i.e. that

$$\left(\phi_{aa}+\frac{3}{a}\phi_a\right)\phi_{bb}-\phi^2_{ab} \text{ and } \phi_{bb}$$

shall both be positive. These are the required conditions for stability of the steady motion.

Corollary. If a particle of unit mass is describing a circular orbit of radius a in a plane about a centre of force at the centre of the circle, the potential energy being $\phi(r)$ where r is the distance from the centre, the modified kinetic potential is

$$\tfrac{1}{2}\dot{\rho}^2-\tfrac{1}{2}\rho^2\left(\phi_{aa}+\frac{3}{a}\phi_a\right),$$

where $r=a+\rho$, so the condition for stability is

$$\phi_{aa}+\frac{3}{a}\phi_a>0,$$

and the period of a vibration about the circular motion is

$$2\pi\left\{\phi_{aa}+\frac{3}{a}\phi_a\right\}^{-\frac{1}{2}}.$$

(ii) *To find the period of the vibrations about steady circular motion of a particle moving under gravity on a surface of revolution whose axis is vertical.*

Let $z=f(r)$ be the equation of the surface, where (z, r, θ) are cylindrical coordinates with the axis of the surface as axis of z. If the particle is projected along the horizontal tangent to the surface at any point with a suitable velocity, it will describe a horizontal circle on the surface with constant velocity. Let a be the radius of the circle; we shall take the mass of the particle to be unity, as this involves no loss of generality

The kinetic potential is

$$L=\tfrac{1}{2}(\dot{r}^2+r^2\dot{\theta}^2+\dot{z}^2)-gz$$
$$=\tfrac{1}{2}\dot{r}^2\{1+f'^2(r)\}+\tfrac{1}{2}r^2\dot{\theta}^2-gf(r).$$

The integral corresponding to the ignorable coordinate θ is $r^2\dot{\theta}=k$, and the modified kinetic potential of the system after ignoration of θ is therefore

$$R=\tfrac{1}{2}\dot{r}^2\{1+f'^2(r)\}-gf(r)-k^2/2r^2.$$

The problem is thus reduced to that of finding the vibrations about equilibrium of the system with one degree of freedom for which R is the kinetic potential. The condition for equilibrium is

$$\left(\frac{\partial R}{\partial r}\right)_{r=a}=0, \quad \text{or} \quad k^2=ga^3f'(a),$$

and this gives

$$R=\tfrac{1}{2}\dot{r}^2\{1+f'^2(r)\}-gf(r)-ga^3f'(a)/2r^2.$$

Writing $r=a+\rho$, where ρ is small, and expanding in powers of ρ, we have

$$R=\tfrac{1}{2}\dot{\rho}^2\{1+f'^2(a)\}-\tfrac{1}{2}g\rho^2\left\{f''(a)+\frac{3}{a}f'(a)\right\}.$$

The equation of motion

$$\frac{d}{dt}\left(\frac{\partial R}{\partial \dot{\rho}}\right)-\frac{\partial R}{\partial \rho}=0$$

is therefore

$$\ddot{\rho}\{1+f'^2(a)\}+g\rho\left\{f''(a)+\frac{3}{a}f'(a)\right\}=0,$$

and the condition for stability is

$$f''(a)+\frac{3}{a}f'(a)>0,$$

the period of a vibration being

$$\frac{2\pi}{\sqrt{g}}\left\{\frac{1+f'^2(a)}{f''(a)+3f'(a)/a}\right\}^{\frac{1}{2}}.$$

Example. If the surface is a paraboloid of revolution whose axis is vertical and vertex downwards, shew that the vibration-period is

$$\pi\left(\frac{l^2+a^2}{gl}\right)^{\frac{1}{2}},$$

where l is the semi-latus rectum of the paraboloid.

(iii) *To determine the vibrations about steady motion of a top on a perfectly rough plane.*

Let A denote the moment of inertia of the top about a line through its apex perpendicular to its axis of symmetry, and let θ denote the angle made by the axis with the vertical, M the mass of the top, and h the distance of its centre of gravity from its apex: then we have seen (§ 71) that after ignoring the Eulerian angles ϕ and ψ, the angle θ is determined by solving the dynamical system defined by the kinetic potential

$$R=\tfrac{1}{2}A\dot{\theta}^2-\frac{(a-b\cos\theta)^2}{2A\sin^2\theta}-Mgh\cos\theta,$$

where a and b are constants depending on the initial circumstances of the motion.

Let α, n be the values of θ and $\dot{\phi}$ respectively in the steady motion, so (§ 72) we have

$$An^2\cos\alpha+Mgh=bn,$$

$$An\sin^2\alpha=a-b\cos\alpha.$$

To discuss the vibratory motion of the top about this form of steady motion, we write $\theta=\alpha+x$ where x is a small quantity, and expand R in ascending powers of x, neglecting powers of x above the second and eliminating a and b by use of the last two equations; we thus obtain for R the value

$$R=\tfrac{1}{2}A\dot{x}^2-\tfrac{1}{2}Ax^2\{n^2\sin^2\alpha+(n\cos\alpha-Mgh/An)^2\}.$$

The equation of motion for x is therefore

$$\ddot{x}+\{n^2\sin^2 a+(n\cos a-Mgh/An)^2\}x=0.$$

As the coefficient of x is positive, *the state of steady motion is stable*; and *the period of a vibration is*

$$2\pi\{n^2-2Mgh\cos a/A+M^2g^2h^2/A^2n^2\}^{-\frac{1}{2}}.$$

(iv) *The sleeping top.*

If we consider that form of steady motion of the top in which a is zero, so that the axis of the top is permanently directed vertically upwards, the top rotating about this axis with a given angular velocity, the method of the preceding example must be modified, since now the form of steady motion in which a is a small constant is to be regarded as a vibration about the type of motion in which a is zero: so that we may now expect to have *two* independent periods of normal vibration, the analogues of which in the previous example are the period of the steady motion and the period of vibration about it.

As in § 71, the kinetic and potential energies of the top are

$$T=\tfrac{1}{2}A\dot{\theta}^2+\tfrac{1}{2}A\dot{\phi}^2\sin^2\theta+\tfrac{1}{2}C(\dot{\psi}+\dot{\phi}\cos\theta)^2,$$

$$V=Mgh\cos\theta.$$

The integral corresponding to the ignorable coordinate ψ is

$$b=C(\dot{\psi}+\dot{\phi}\cos\theta),$$

and hence after ignoration of ψ we obtain for the kinetic potential of the system the value

$$R=\tfrac{1}{2}A\dot{\theta}^2+\tfrac{1}{2}A\dot{\phi}^2\sin^2\theta+b\dot{\phi}\cos\theta-Mgh\cos\theta.$$

In the two last terms we can replace $\cos\theta$ by $(\cos\theta-1)$, since the terms $-b\dot{\phi}$ and Mgh thus added disappear from the equations of motion.

As ϕ is not a small quantity throughout the motion, we take as coordinates in place of θ and ϕ the quantities ξ and η, where

$$\xi=\sin\theta\cos\phi,\quad \eta=\sin\theta\sin\phi.$$

From these equations, neglecting terms above the second degree in ξ, η, $\dot{\xi}$, $\dot{\eta}$, we have

$$\dot{\theta}^2+\dot{\phi}^2\sin^2\theta=\dot{\xi}^2+\dot{\eta}^2,$$

$$\dot{\phi}\sin^2\theta=\xi\dot{\eta}-\eta\dot{\xi},$$

$$1-\cos\theta=\tfrac{1}{2}(\xi^2+\eta^2),$$

and so we have

$$R=\tfrac{1}{2}A\dot{\xi}^2+\tfrac{1}{2}A\dot{\eta}^2-\tfrac{1}{2}b(\xi\dot{\eta}-\eta\dot{\xi})+\tfrac{1}{2}Mgh(\xi^2+\eta^2).$$

The equations of motion are

$$\frac{d}{dt}\left(\frac{\partial R}{\partial\dot{\xi}}\right)-\frac{\partial R}{\partial\xi}=0,\quad \frac{d}{dt}\left(\frac{\partial R}{\partial\dot{\eta}}\right)-\frac{\partial R}{\partial\eta}=0,$$

or

$$\begin{cases}A\ddot{\xi}+b\dot{\eta}-Mgh\xi=0,\\ A\ddot{\eta}-b\dot{\xi}-Mgh\eta=0.\end{cases}$$

If $2\pi/\lambda$ is the period of a normal vibration, on substituting $\xi=Je^{i\lambda t}$, $\eta=Ke^{i\lambda t}$ in these differential equations and eliminating J and K we obtain the equation

$$\begin{vmatrix}-\lambda^2A-Mgh & ib\lambda\\ -ib\lambda & -\lambda^2A-Mgh\end{vmatrix}=0,$$

or

$$(\lambda^2A+Mgh)^2-b^2\lambda^2=0.$$

The two roots of this quadratic in λ^2 give the values of λ corresponding to the two normal vibrations: we have therefore to determine the nature of these roots.

The solution of the quadratic is

$$\lambda^2 = \frac{1}{2A^2}\{b^2 - 2AMgh \pm b(b^2 - 4AMgh)^{\frac{1}{2}}\},$$

so

$$\pm\lambda = \frac{1}{2A}\{b \pm (b^2 - 4AMgh)^{\frac{1}{2}}\}.$$

The values of λ are therefore real or not according as b^2 is greater or less than $4AMgh$. In the former case the steady spinning motion round the vertical is stable: in the latter case, unstable.

It must not be supposed, however, that in the unstable case the axis of the top necessarily departs very far from the vertical: all that is meant by the term "unstable" is that when $b^2 < 4AMgh$ the disturbed motion does not, as the disturbance is indefinitely diminished, tend to a limiting form coincident with the undisturbed motion.

As a matter of fact, if $b^2 - 4AMgh$, though negative, is very small, it is possible for the axis of the top in its "unstable" motion to remain permanently close to the vertical: but in this case the maximum divergence from the vertical cannot be made indefinitely small (for a given value of b) by making the initial disturbance indefinitely small*.

86. *Vibrations of systems involving moving constraints.*

If a dynamical system involves a constraint which varies with the time (e.g. if one of the particles of the system is moveable on a smooth wire or surface which is made to rotate uniformly about a given axis), the kinetic potential of the system is no longer necessarily composed of terms of degrees 2 and 0 in the velocities; terms which are linear in the velocities may also occur. The equations which determine the vibrations of such a system will therefore in general include gyroscopic terms, even when the vibration is about relative equilibrium: the solution can be effected by the methods above developed for the problem of vibrations about steady motion. The following example will illustrate this.

Example. To find the periods of the normal vibrations of a heavy particle about its position of equilibrium at the lowest point of a surface which is rotating with constant angular velocity ω about a vertical axis through the point.

Let (x, y, z) be the coordinates of the particle, referred to axes which revolve with the surface, the axes of x and y being the tangents to the lines of curvature at the lowest point, and the axis of z being vertical. Let the equation of the surface be

$$z = \frac{x^2}{2\rho_1} + \frac{y^2}{2\rho_2} + \text{terms of higher order}.$$

The kinetic and potential energies of the particle are

$$T = \tfrac{1}{2}m\{(\dot{x} - y\omega)^2 + (\dot{y} + x\omega)^2 + \dot{z}^2\},$$
$$V = mgz.$$

The kinetic potential of the vibration-problem is therefore

$$L = \tfrac{1}{2}m\{\dot{x}^2 + \dot{y}^2 + 2\omega(x\dot{y} - y\dot{x}) + \omega^2(x^2 + y^2)\} - mg\left(\frac{x^2}{2\rho_1} + \frac{y^2}{2\rho_2}\right).$$

* A discussion of the stability of the sleeping top is given by Klein, *Bull. Amer. Math. Soc.* III. (1897), pp. 129—132, 292.

The equations of motion are

$$\frac{d}{dt}\left(\frac{\partial L}{\partial \dot{x}}\right)-\frac{\partial L}{\partial x}=0,\quad \frac{d}{dt}\left(\frac{\partial L}{\partial \dot{y}}\right)-\frac{\partial L}{\partial y}=0,$$

or

$$\ddot{x}-2\omega\dot{y}+x\left(\frac{g}{\rho_1}-\omega^2\right)=0,$$

$$\ddot{y}+2\omega\dot{x}+y\left(\frac{g}{\rho_2}-\omega^2\right)=0.$$

If the period of a normal vibration is $2\pi/\lambda$, we have (substituting $x=Ae^{i\lambda t}$, $y=Be^{i\lambda t}$ in the differential equations, and eliminating A and B)

$$\begin{vmatrix} -\lambda^2-\omega^2+g/\rho_1 & -2\omega i\lambda \\ 2\omega i\lambda & -\lambda^2-\omega^2+g/\rho_2 \end{vmatrix}=0,$$

or

$$(\lambda^2+\omega^2-g/\rho_1)(\lambda^2+\omega^2-g/\rho_2)-4\lambda^2\omega^2=0.$$

The roots of this quadratic in λ^2 determine the periods of the normal vibrations.

MISCELLANEOUS EXAMPLES.

1. A particle moves on a curve which rotates uniformly about a fixed axis, the potential energy $V(s)$ of the particle depending only on its position as defined by the arc s. Shew that the period of a vibration about a position of relative rest on the curve is

$$2\pi\left\{-\frac{dV}{ds}\frac{d}{ds}\log\left(\frac{-r\,dr/ds}{dV/ds}\right)\right\}^{-\frac{1}{2}},$$

where r is the distance of the particle from the axis.

2. Determine the vibrations of a solid horizontal circular cylinder rolling inside a hollow horizontal circular cylinder whose axis is fixed, shewing that the length of the simple equivalent pendulum is $(b-a)(3M+m)/(2M+m)$; where b is the radius and M the mass, of the outer cylinder, and a is the radius and m the mass, of the inner cylinder. (Coll. Exam.)

3. A thin hemispherical bowl of mass M and radius a is on a perfectly rough horizontal plane, and a particle of mass m is in contact with the inner surface of the bowl, supposed smooth. Shew that when the system performs small oscillations, the motion of the particle and the centre of gravity of the bowl being in one plane, the periods of the normal vibrations are $2\pi/\sqrt{\lambda_1}$ and $2\pi/\sqrt{\lambda_2}$, where λ_1 and λ_2 are the roots of the equation

$$ma\lambda g-(g-a\lambda)(\tfrac{1}{2}g-\tfrac{2}{3}a\lambda)M=0. \qquad \text{(Coll. Exam.)}$$

4. A string of length $4a$ is loaded at equal intervals with three weights m, M and m respectively, and is suspended from two points A and B symmetrically. Shew that if M perform small vertical vibrations, the length of the simple equivalent pendulum is

$$\frac{a\cos\alpha\cos\beta\sin(\alpha-\beta)\cos(\alpha-\beta)}{\sin\alpha\cos^2\alpha+\sin\beta\cos^2\beta},$$

where α and β are the inclinations of the parts of the string to the vertical. (Coll. Exam.)

5. A uniform bar whose length is $2a$ is suspended by a short string whose length is l; prove that the time of vibration is greater than if the bar were swinging about one extremity in the ratio $1+9l/32a : 1$ nearly. (Coll. Exam.)

6. An elliptic cylinder with plane ends at right angles to its axis rests upon two fixed smooth perpendicular planes which are each inclined at 45° to the horizon. Shew that there are two stable configurations and one unstable, and that in the former case the length of the equivalent pendulum is

$$ab\,(a^2+b^2)/2\sqrt{2}\,(a-b)^2\,(a+b),$$

a and b being the lengths of the semi-axes. (Coll. Exam.)

7. A rough circular cylinder of radius a and mass m is loaded so that its centre of gravity is at a distance h from the axis, and is placed on a board of equal mass which can move on a smooth horizontal plane. If the system is disturbed slightly when in a position of stable equilibrium, shew that the length of the simple equivalent pendulum is $k^2/h+\frac{1}{2}(a-h)^2/h$, where mk^2 is the moment of inertia of the cylinder about a horizontal axis through its centre of gravity. (Coll. Exam.)

8. One end of a uniform rod of length b and mass m is freely jointed to a point in a smooth vertical wall; the other end is freely jointed to a point in the surface of a uniform sphere of mass M and radius a which rests against the wall. Shew that the period of the vibrations about the position of equilibrium is $2\pi/p$, where

$$p^2\{\sin\beta\sin^2(\alpha-\beta)+\tfrac{2}{3}\cos\alpha\sin(\alpha-\beta)+\tfrac{2}{5}\sin\beta\cos^2\beta\}=\frac{g}{ab\cos\alpha}(a\sin\alpha\cos^2\alpha+b\sin\beta\cos^2\beta),$$

α and β being given by the equations

$$a\sin\alpha+b\sin\beta-a=0,$$

$$(\tfrac{1}{2}m+M)\tan\beta-M\tan\alpha=0.$$ (Coll. Exam.)

9. A thin circular cylinder of mass M and radius b rests on a perfectly rough horizontal plane, and inside it is placed a perfectly rough sphere of mass m and radius a. If the system be disturbed in a plane perpendicular to the generators of the cylinder, find the equations of finite motion, and deduce two first integrals of them; and if the motion be small, shew that the length of the simple equivalent pendulum is

$$14M(b-a)/(10M+7m).$$

(Camb. Math. Tripos, Part I, 1899.)

10. A sphere of radius c is placed upon a horizontal perfectly rough wire in the form of an ellipse of axes $2a$, $2b$. Prove that the time of a vibration under gravity about the position of stable equilibrium is that of a simple pendulum of length l given by $b^2dl=(a^2-b^2)(d^2+k^2)$, where $k^2=2c^2/5$ and $d^2=c^2-b^2$. (Coll. Exam.)

11. A rhombus of four equal uniform rods of length a freely jointed together is laid on a smooth horizontal plane with one angle equal to 2α. The opposite corners are connected by similar elastic strings of natural lengths $2a\cos\alpha$, $2a\sin\alpha$. Prove that if one string be slightly extended and the rhombus left free, the periods during which the strings are extended in the subsequent motion are in the ratio

$$(\cos\alpha)^{\frac{3}{2}} : (\sin\alpha)^{\frac{3}{2}}.$$ (Coll. Exam.)

12. A particle of mass m is attached by n equal elastic strings of natural length a to the fixed angular points of a regular polygon of n sides, the radius of whose circumscribing circle is c. Shew that if the particle be slightly displaced from its equilibrium position in the plane of the polygon, it will execute harmonic vibrations in a straight line, the length of the simple equivalent pendulum being $2mgac/n\lambda(2c-a)$, and that for vibrations perpendicular to the plane of the polygon, the corresponding length will be $mgac/n\lambda(c-a)$, λ being the modulus of each string. (Camb. Math. Tripos, Part I, 1900.)

13. The energy-equation of a particle is

$$f(x)\,\dot{x}^2 = 2\phi(x) + \text{constant},$$

and a is a value of x for which $\phi'(x)$ is zero. If $\phi^{(2p)}(x)$ is the first derivative of $\phi(x)$ which does not vanish for $x=a$, shew that the period of a vibration about the position a is

$$\frac{4}{h^{p-1}}\,\frac{\Gamma(1/2p)}{\Gamma(1/2p+\frac{1}{2})}\left\{-\frac{\Gamma(2p)\,f(a)\,\pi}{4p\phi^{(2p)}(a)}\right\}^{\frac{1}{2}},$$

where h is the value of $(x-a)$ corresponding to the extreme displacement. (Elliott.)

14. A cone has its centre of gravity at a distance c from its axis, there being in other respects the usual kinetic symmetry at the vertex. If the cone oscillates on a horizontal plane and the plane be perfectly rough, shew that the length of the simple equivalent pendulum is

$$(\cos\alpha/Mc)\,(A\sin^2\alpha + C\cos^2\alpha),$$

whereas if this plane be perfectly smooth, the length is

$$(\cos\alpha/Mc)\,(\sin^2\alpha/A + \cos^2\alpha/C).$$ (Coll. Exam.)

15. A number of equal uniform rods each of length $2a$ are freely jointed at a common extremity and arranged at equal angular intervals like the ribs of an umbrella. This cone of rods is put over a smooth fixed sphere of radius b, each rod being in contact with the sphere, and rests in equilibrium. Shew that, if the system be slightly disturbed so that the hinge performs vertical vibrations about the position of equilibrium, their period is

$$2\pi\left(\frac{a}{3g}\,\frac{1+3\sin^2\alpha}{1+2\sin^2\alpha}\right)^{\frac{1}{2}}\sin^{\frac{1}{2}}\alpha,$$

where $\sec^3\alpha\sin\alpha = a/b$. (Camb. Math. Tripos, Part I, 1896.)

16. A heavy rectangular board is symmetrically suspended in a horizontal position by four light elastic strings attached to the corners of the board and to a fixed point vertically above its centre. Shew that the period of the vertical vibrations is

$$2\pi\left(\frac{g}{c}+\frac{4c^2\lambda}{k^3M}\right)^{-\frac{1}{2}},$$

where c is the equilibrium distance of the board below the fixed point, a is the length of a semi-diagonal, $k=(a^2+c^2)^{\frac{1}{2}}$, and λ is the modulus. (Coll. Exam.)

17. A heavy lamina hangs in equilibrium in a horizontal position suspended by three vertical inextensible strings of unequal lengths. Shew that the normal vibrations are (1) a rotation about either of two vertical lines in a plane through the centroid, and (2) a horizontal swing parallel to this plane. (Coll. Exam.)

18. A uniform rod of length $2a$ is freely hinged at one end, at the other end a string of length b is attached which is fastened at its further end to a point on the surface of a homogeneous sphere of radius c. If the masses of the rod and sphere are equal, find the motion of the system when slightly disturbed from the vertical, and shew that the equation to determine the periods is

$$8abc\mu^3 - g\mu^2(18bc+52ca+20ab) + g^2\mu(80a+45b+63c) - 45g^3 = 0.$$

(Coll. Exam.)

19. A uniform wire, in the shape of an ellipse of semi-axes a, b, rests upon a rough horizontal plane with its minor axis vertical and a particle of equal mass is suspended by a fine string of length l attached to the highest point. If vibrations in a vertical plane be performed, prove that their periods will be those of pendulums whose lengths are the value of x given by the equation

$$\{x(3b-2a^2/b)+5b^2+k^2\}(x-l)+4b^2l=0,$$

where k is the radius of gyration about the centre of gravity. (Coll. Exam.)

20. A fine inextensible string has its ends tied to two fixed pegs in a horizontal line whose distance apart is three-quarters of the length of the string. The string also passes through two small smooth rings which are fixed to the ends of a uniform straight rod whose length is half that of the string. The rod hangs in equilibrium in a horizontal position and receives a small disturbance in the vertical plane of the string. Shew that initially its normal coordinates in terms of the time are $L\cos(pt+\alpha)$ and $M\cosh(qt+\beta)$, where p^2 and $-q^2$ are the roots of the equation

$$x^4-\frac{\sqrt{3}}{4}\frac{g}{a}x^2-\frac{3}{4}\frac{g^2}{a^2}=0. \qquad \text{(Coll. Exam.)}$$

21. A heavy uniform rod of length $2a$, suspended from a fixed point by a string of length b, is slightly disturbed from its vertical position. Shew that the periods of the normal vibrations are $2\pi/p_1$ and $2\pi/p_2$, where $p_1{}^2$ and $p_2{}^2$ are the roots of the equation

$$abp^4-(4a+3b)\,gp^2+3g^2=0.$$

22. A circular disc, mass M, is attached by a string from its centre C to a fixed point O. A particle of mass m is fixed to the disc at a point P on the rim. Find the equations of motion on a vertical plane in terms of the angles θ and ϕ which OC and CP make with the vertical, and prove that if the system vibrates about the position of equilibrium the periods in these coordinates are given by the equation

$$(M+m)(p^2a-g)\{(M+2m)\,cp^2-2mg\}=2m^2cap^4,$$

where a is the length of the string OC and c the radius of the disc. (Coll. Exam.)

23. A hemispherical bowl of radius $2b$ rests on a smooth table with the plane of its rim horizontal; within it and in equilibrium lies a perfectly rough sphere of radius b, and mass one-quarter of that of the bowl. A slight displacement in a vertical plane containing the centres of the sphere and the bowl is given: prove that the periods of the consequent vibrations are $2\pi/p_1$ and $2\pi/p_2$, where $p_1{}^2$ and $p_2{}^2$ are the roots of the equation

$$156b^2x^2-260bxg+75g^2=0. \qquad \text{(Coll. Exam.)}$$

24. A uniform circular disc of mass m and radius a is held in equilibrium on a smooth horizontal plane by three equal elastic strings of modulus λ, natural length l_0 and stretched length l. The strings are attached to the disc at the extremities of three radii equally inclined to one another and their other ends are attached to points of the plane lying on the radii produced. Shew that the periods of vibration of the disc are

$$2\pi\,\{\mu/(2l-l_0)\}^{\frac{1}{2}} \text{ and } 2\pi\,\{\mu a/4\,(a+l)\,(l-l_0)\}^{\frac{1}{2}},$$

where $\mu=2mll_0/3\lambda$. (Camb. Math. Tripos, Part I, 1898.)

25. A particle is describing a circle under the influence of a force to the centre varying as the nth power of the distance. Shew that this state of motion is unstable if n be less than -3.

Shew that, if the force vary as $e^{-\frac{r}{a}}/r^2$, the motion is stable or unstable according as the radius of the circle is less or greater than a. (Coll. Exam.)

26. A particle moves in free space under the action of a centre of force which varies as the inverse square of the distance and a field of constant force: shew that a circle described uniformly is a possible state of steady motion, but this will be stable only provided the circle as viewed from the centre of force appears to lie on a right circular cone whose semi-vertical angle is greater than $\arccos\frac{1}{3}$. (Coll. Exam.)

27. A particle describes a circle uniformly under the influence of two centres of force which attract inversely as the square of the distance. Prove that the motion is stable if $3\cos\theta\cos\phi<1$, where θ and ϕ are the angles which a radius of the circle subtends at the centres of force. (Camb. Math. Tripos, Part I, 1889.)

28. A heavy particle is projected horizontally on the interior of a smooth cone with its axis vertical and apex downwards; the initial distance from the apex is c and the semi-vertical angle of the cone is a. Find the condition that a horizontal circle should be described; and shew that the time of a vibration about this steady motion is that of a simple pendulum of length $\frac{1}{3}c\sec a$. (Coll. Exam.)

29. A circular disc has a thin rod pushed through its centre perpendicular to its plane, the length of the rod being equal to the radius of the disc; prove that the system cannot spin with the rod vertical unless the velocity of a point on the circumference of the disc is greater than the velocity acquired by a body after falling from rest vertically through ten times the radius of the disc. (Coll. Exam.)

30. Prove that for a symmetrical top spinning upright with sufficient angular velocity for stability, the two types of motion, differing slightly from the steady motion in the upright position, which are determined by simple harmonic functions of the time, are the limits of steady motions with the axis slightly inclined to the vertical, and that the period of the vibrations is the limiting value of that which corresponds to steady motion in an inclined position when the inclination is indefinitely diminished.

(Coll. Exam.)

31. One end of a uniform rod of length $2a$ whose radius of gyration about one end is k is compelled to describe a horizontal circle of radius c with uniform angular velocity ω. Prove that when the motion is steady the rod lies in the vertical plane through the centre of the circle and makes an angle a with the vertical given by

$$\omega^2(k^2+ac\operatorname{cosec}a)=ag\sec a.$$

Shew that the periods of the normal vibrations are $2\pi/\lambda_1$, $2\pi/\lambda_2$, where λ_1, λ_2 are the roots of

$$(k^2\lambda^2\sin a-\omega^2ac)(k^2\lambda^2\sin a-\omega^2ac-\omega^2k^2\sin^3 a)=4\omega^2k^4\lambda^2\sin^2 a\cos^2 a.$$

(Camb. Math. Tripos, Part I, 1889.)

32. Investigate the motion of a conical pendulum when disturbed from its state of steady motion by a small vertical harmonic oscillation of the point of support. Can the steady motion be rendered unstable by such a disturbance? (Coll. Exam.)

33. The middle point of one side of a uniform rectangle is fixed and the line joining it to the middle point of the opposite side is constrained to describe a circular cone of semi-angle a with uniform angular velocity. The rectangle being otherwise free, find the positions of steady motion and prove that the time of a vibration about the position of stable steady motion is equal to the period of revolution divided by $\sin a$.

(Coll. Exam.)

34. A solid of revolution, symmetrical about a plane through its centre of gravity perpendicular to its axis, is suspended from a fixed point by a string of length b which is attached to one end of the axis of the solid, this axis being of length $2a$. The mass of the solid is M, and its principal moments of inertia at its centre of gravity are (A, A, C). If the solid is slightly disturbed from the state of steady motion in which the string and axis are vertical, and the body is spinning on its axis with angular velocity n, shew that the periods of the normal vibrations are $2\pi/p_1$ and $2\pi/p_2$, where p_1^2 and p_2^2 are the roots of the equation

$$Ma^2gp^2=(g-bp^2)(Mag+Cnp-Ap^2).$$

35. A symmetrical top spins with its axis vertical, the tip of the peg resting in a fixed socket. A second top, also spinning, is placed on the summit of the first, the tip of the peg resting in a small socket. Shew that the arrangement is stable provided the equation

$$(Mcgx^2 + C\Omega x + A)\{(M'c' + Mh)gx^2 + C'\Omega' x + (A' + Mh^2)\} = M^2h^2c^2)$$

has all its roots real; Ω, Ω' being the spins of the upper and lower tops respectively, M, M' their masses, C, C' their moments of inertia about the axis of figure, A, A' about perpendiculars through the pegs, c, c' the distances of the centroids from the pegs, and h the distance between the pegs. (Camb. Math. Tripos, Part I, 1898.)

36. A homogeneous body spins on a smooth horizontal plane in stable steady motion, with angular velocity ω about the vertical through the point of contact and the centre of gravity. The body is symmetrical about each of two perpendicular planes through the vertical. The principal radii of curvature at the vertex on which it rests are ρ_1, ρ_2; the moments of inertia about the principal axes through the centre of gravity (parallel to the lines of curvature) are respectively A and B, and that about the vertical is C. The height of the centre of gravity about the vertex is $a = a_1 + \rho_1 = a_2 + \rho_2$; and $\lambda\omega^2$ is the weight of the body.

Shew that the following conditions must be satisfied:

(i) $(\lambda a_1 + A - C)(\lambda a_2 + B - C) > 0$,

(ii) $\lambda(a_1 A + a_2 B) < AB + (A - C)(B - C)$,

(iii) The value of λ must not lie between the two values

$$(A + B - C)[\sqrt{B}\{a_1 A + a_2(A - C)\}^{\frac{1}{2}} \pm \sqrt{A}\{a_2 B + a_1(B - C)^2\}^{\frac{1}{2}}]^2/(a_1 A - a_2 B)^2,$$

if the two radicals in the expression are both real.

(Camb. Math. Tripos, Part I, 1897.)

CHAPTER VIII

NON-HOLONOMIC SYSTEMS. DISSIPATIVE SYSTEMS

87. *Lagrange's equations with undetermined multipliers.*

We now proceed to the consideration of *non-holonomic* dynamical systems. In these systems, as was seen in § 25, the number of independent coordinates $(q_1, q_2, \ldots, q_n)$ required in order to specify the configuration of the system at any time is greater than the number of degrees of freedom of the system, owing to the fact that the system is subject to constraints which will be supposed to do no work, and which are expressed by a number of non-integrable* kinematical relations of the form

$$A_{1k}dq_1 + A_{2k}dq_2 + \ldots + A_{nk}dq_n + T_k dt = 0 \qquad (k = 1, 2, \ldots, m),$$

where $A_{11}, A_{12}, \ldots, A_{nm}, T_1, T_2, \ldots, T_m$ are given functions of $q_1, q_2, \ldots, q_n, t$.

The most familiar example of such a system is that of a body which is constrained to roll without sliding on a given fixed surface: the condition that no sliding takes place is expressed by two relations of the type given above. A still simpler example is that of a vertical wheel with a sharp edge which rolls on a horizontal sheet of paper, as in the integraph of Abdank-Abakanowicz and the integrator of Pascal: the wheel moves only in its own instantaneous plane, the friction at the sharp edge preventing it from slipping sideways. If (x, y) are the rectangular coordinates of its point of contact with the paper, and ϕ the azimuth of its plane, we have in this case the non-holonomic equation of condition

$$dy - \tan\phi \, . \, dx = 0.$$

The number of kinematical relations being m, the system will have $(n-m)$ degrees of freedom; it is not possible to apply Lagrange's equations directly to such a system, but an extension of the Lagrangian equations will now be given which will enable us to discuss the motion of non-holonomic systems in a way analogous to that previously developed for holonomic systems.

Consider then a non-holonomic system, whose configuration at any instant is completely specified by n coordinates $q_1, q_2, \ldots, q_n$; let the kinetic energy be T, and let the kinematical conditions due to the non-holonomic constraints be expressed by the relations

$$A_{1k}dq_1 + A_{2k}dq_2 + \ldots + A_{nk}dq_n + T_k dt = 0 \qquad (k = 1, 2, \ldots, m).$$

* If these relations were integrable, it would be possible to express some of the coordinates $(q_1, q_2, \ldots, q_n)$ in terms of the others, and the n coordinates would therefore not be independent: which is contrary to our assumption.

Now it is open to us either simply to regard the system as subject to these kinematical conditions, or in place of these to regard the system as acted on by certain additional external forces, namely the forces which have to be exerted by the constraints in order to compel the system to fulfil the kinematical conditions; we shall for the present take the latter point of view. Let

$$Q_1'\delta q_1 + Q_2'\delta q_2 + \ldots + Q_n'\delta q_n$$

be the work done on the system by these additional forces in an arbitrary displacement ($\delta q_1, \delta q_2, \ldots, \delta q_n$) (which is now not restricted to satisfy the kinematical conditions), and let

$$Q_1\delta q_1 + Q_2\delta q_2 + \ldots + Q_n\delta q_n$$

be the work done on the system by the original external forces in this displacement. Since the substitution of additional forces for the kinematical relations has made the system holonomic, we can apply the Lagrangian equations; we have therefore

$$\frac{d}{dt}\left(\frac{\partial T}{\partial \dot{q}_r}\right) - \frac{\partial T}{\partial q_r} = Q_r + Q_r' \qquad (r = 1, 2, \ldots, n)$$

as the equations of motion of the system.

The forces $Q_1', Q_2', \ldots, Q_n'$ are unknown: but they are such that, in any displacement consistent with the instantaneous constraints, they do no work. It follows that the quantity

$$Q_1'dq_1 + Q_2'dq_2 + \ldots + Q_n'dq_n$$

is zero for all values of the ratios $dq_1 : dq_2 : \ldots : dq_n$ which satisfy the equations

$$A_{1k}dq_1 + A_{2k}dq_2 + \ldots + A_{nk}dq_n = 0;$$

hence we must have

$$Q_r' = \lambda_1 A_{r1} + \lambda_2 A_{r2} + \ldots + \lambda_m A_{rm} \qquad (r = 1, 2, \ldots, n),$$

where the quantities $\lambda_1, \lambda_2, \ldots, \lambda_m$ are independent of r. *We thus have altogether the* $(n + m)$ *equations*

$$\frac{d}{dt}\left(\frac{\partial T}{\partial \dot{q}_r}\right) - \frac{\partial T}{\partial q_r} = Q_r + \lambda_1 A_{r1} + \lambda_2 A_{r2} + \ldots + \lambda_m A_{rm} \qquad (r = 1, 2, \ldots, n),$$

$$A_{1k}\dot{q}_1 + A_{2k}\dot{q}_2 + \ldots + A_{nk}\dot{q}_n + T_k = 0 \qquad (k = 1, 2, \ldots, m),$$

and these are sufficient to determine the $(n + m)$ *unknown quantities* $q_1, q_2, \ldots, q_n, \lambda_1, \lambda_2, \ldots, \lambda_m$. The problem is thus reduced to the solution of this set of equations*

* The extension of Lagrange's equations to non-holonomic systems is due to Ferrers, *Quart. Journ. Math.* XII. (1871), p. 1: C. Neumann, *Leipzig Berichte*, XL. (1888), p. 22: and Vierkandt, *Monatshefte für Math. u. Phys.* III. (1892), p. 31. See also Gugino, *Lincei Rend.* XII. (1930), p. 307.

88. *Equations of motion referred to axes moving in any manner.*

The method given in the preceding article depends essentially on the reduction of the non-holonomic system to a holonomic system by introducing the forces due to the non-holonomic constraints. In practice, this is often most conveniently done by forming separately the equations of motion of each of the bodies of the system. It is moreover frequently advantageous to use axes of reference which are not fixed either in space or in the body, and we shall now find the equations of motion of a rigid body referred to axes which have their origin at the centre of gravity of the body, and are turning about it in any manner*.

Let G be the centre of gravity of the body, and let $Gxyz$ be the moving axes. Let (u, v, w) be the components of velocity of the centre of gravity resolved parallel to these axes, and let $(\theta_1, \theta_2, \theta_3)$ be the components of angular velocity of the system of axes $Gxyz$ resolved along the axes themselves; further let $(\omega_1, \omega_2, \omega_3)$ be the components of angular velocity of the body, resolved along the same axes. Then (§ 64) the motion of G is the same as that of a particle of mass M, equal to that of the body, acted on by forces equal to the external forces which act on the body (including all forces of constraint, except the molecular reactions between the constituent particles of the body); let (X, Y, Z) be the components parallel to the axes $Gxyz$ of these external forces.

The component of velocity of G parallel to Gx is u, and consequently (§ 17) the component of its acceleration in this direction is $\dot{u} - v\theta_3 + w\theta_2$; we have therefore the equation

$$M(\dot{u} - v\theta_3 + w\theta_2) = X,$$

which can be written

$$\frac{d}{dt}\left(\frac{\partial T}{\partial u}\right) - \theta_3 \frac{\partial T}{\partial v} + \theta_2 \frac{\partial T}{\partial w} = X,$$

where T denotes the kinetic energy of the body, expressed in terms of $(u, v, w, \omega_1, \omega_2, \omega_3)$; and similar equations can be obtained for the motion of G parallel to the axes Gy and Gz.

Consider next the motion of the body relative to G, which (§ 64) is independent of the motion of G; from §§ 62, 63, we see that the angular momentum of the body about the axis Gx is $\partial T/\partial \omega_1$, so that the rate of increase of angular momentum about an axis fixed in space and instantaneously coinciding with Gx is

$$\frac{d}{dt}\left(\frac{\partial T}{\partial \omega_1}\right) - \theta_3 \frac{\partial T}{\partial \omega_2} + \theta_2 \frac{\partial T}{\partial \omega_3}.$$

* In the applications of this method, the axes are usually chosen subject to the condition that the moments and products of inertia of the body with respect to them do not vary: but this condition is not essential.

If L, M, N denote the moments of the external forces about the axes $Gxyz$, we have therefore (§ 40)

$$\frac{d}{dt}\left(\frac{\partial T}{\partial \omega_1}\right) - \theta_3 \frac{\partial T}{\partial \omega_2} + \theta_2 \frac{\partial T}{\partial \omega_3} = L,$$

and two similar equations.

Hence finally *the motion of the body is determined by the six equations*

$$\begin{cases} \dfrac{d}{dt}\left(\dfrac{\partial T}{\partial u}\right) - \theta_3 \dfrac{\partial T}{\partial v} + \theta_2 \dfrac{\partial T}{\partial w} = X, & \dfrac{d}{dt}\left(\dfrac{\partial T}{\partial \omega_1}\right) - \theta_3 \dfrac{\partial T}{\partial \omega_2} + \theta_2 \dfrac{\partial T}{\partial \omega_3} = L, \\ \dfrac{d}{dt}\left(\dfrac{\partial T}{\partial v}\right) - \theta_1 \dfrac{\partial T}{\partial w} + \theta_3 \dfrac{\partial T}{\partial u} = Y, & \dfrac{d}{dt}\left(\dfrac{\partial T}{\partial \omega_2}\right) - \theta_1 \dfrac{\partial T}{\partial \omega_3} + \theta_3 \dfrac{\partial T}{\partial \omega_1} = M, \\ \dfrac{d}{dt}\left(\dfrac{\partial T}{\partial w}\right) - \theta_2 \dfrac{\partial T}{\partial u} + \theta_1 \dfrac{\partial T}{\partial v} = Z, & \dfrac{d}{dt}\left(\dfrac{\partial T}{\partial \omega_3}\right) - \theta_2 \dfrac{\partial T}{\partial \omega_1} + \theta_1 \dfrac{\partial T}{\partial \omega_2} = N. \end{cases}$$

It will be observed that these are really Lagrangian equations of motion in terms of quasi-coordinates, and could have been derived by use of the theorem of § 30.

Example. If the origin of the moving axes is not fixed in the body, let (u_1, u_2, u_3) be the components of velocity of the origin of coordinates, resolved parallel to the instantaneous position of the axes; let $(\theta_1, \theta_2, \theta_3)$ be the components of angular velocity of the system of axes, resolved along themselves; let (v_1, v_2, v_3) be the components of velocity of that point of the body which is instantaneously situated at the origin of coordinates; and let $(\omega_1, \omega_2, \omega_3)$ be the components of angular velocity of the body, also referred to the moving axes. Shew that the equations of motion can be written in the form

$$\begin{cases} \dfrac{d}{dt}\left(\dfrac{\partial T}{\partial v_1}\right) - \theta_3 \dfrac{\partial T}{\partial v_2} + \theta_2 \dfrac{\partial T}{\partial v_3} = X, \\ \dfrac{d}{dt}\left(\dfrac{\partial T}{\partial v_2}\right) - \theta_1 \dfrac{\partial T}{\partial v_3} + \theta_3 \dfrac{\partial T}{\partial v_1} = Y, \\ \dfrac{d}{dt}\left(\dfrac{\partial T}{\partial v_3}\right) - \theta_2 \dfrac{\partial T}{\partial v_1} + \theta_1 \dfrac{\partial T}{\partial v_2} = Z, \\ \dfrac{d}{dt}\left(\dfrac{\partial T}{\partial \omega_1}\right) - u_3 \dfrac{\partial T}{\partial v_2} + u_2 \dfrac{\partial T}{\partial v_3} - \theta_3 \dfrac{\partial T}{\partial \omega_2} + \theta_2 \dfrac{\partial T}{\partial \omega_3} = L, \\ \dfrac{d}{dt}\left(\dfrac{\partial T}{\partial \omega_2}\right) - u_1 \dfrac{\partial T}{\partial v_3} + u_3 \dfrac{\partial T}{\partial v_1} - \theta_1 \dfrac{\partial T}{\partial \omega_3} + \theta_3 \dfrac{\partial T}{\partial \omega_1} = M, \\ \dfrac{d}{dt}\left(\dfrac{\partial T}{\partial \omega_3}\right) - u_2 \dfrac{\partial T}{\partial v_1} + u_1 \dfrac{\partial T}{\partial v_2} - \theta_2 \dfrac{\partial T}{\partial \omega_1} + \theta_1 \dfrac{\partial T}{\partial \omega_2} = N, \end{cases}$$

where (X, Y, Z, L, M, N) are the components and moments of the external forces with reference to the moving axes.

89. *Application to special non-holonomic problems.*

We shall now consider some examples illustrative of the theory of non-holonomic systems.

Example 1. *Sphere rolling on a fixed sphere.*

Let it be required to determine the motion of a perfectly rough sphere of radius a and mass m which rolls on a fixed sphere of radius b, the only external force being gravity.

Let (b, θ, ϕ) be the polar coordinates of the point of contact, referred to the centre of the fixed sphere, the polar axis being vertical. We take moving axes $GABC$, where G is the centre of the moving sphere, GC is the prolongation of the line joining the centres of the spheres, GA is horizontal and perpendicular to GC, and GB is perpendicular to GA and GC, in the direction of θ increasing.

With these axes we have, in the notation of the last article,

$$\theta_1 = -\dot\theta, \qquad \theta_2 = -\dot\phi \sin\theta, \qquad \theta_3 = \dot\phi\cos\theta,$$

$$u = -(a+b)\,\dot\phi\sin\theta, \qquad v = (a+b)\,\dot\theta, \qquad w = 0,$$

$$T = \tfrac{1}{2}m\left\{u^2+v^2+w^2+\frac{2a^2}{5}(\omega_1{}^2+\omega_2{}^2+\omega_3{}^2)\right\},$$

and if F, F' denote the components of the force at the point of contact parallel to GA and GB respectively, we have

$$X = F, \qquad Y = mg\sin\theta + F',$$

$$L = F'a, \qquad M = -Fa, \qquad N = 0.$$

The equations of motion of the last article become therefore

$$m(\dot u - v\theta_3) = F = -\tfrac{2}{5}am(\dot\omega_2 - \theta_1\omega_3 + \theta_3\omega_1),$$

$$m(\dot v + u\theta_3) - mg\sin\theta = F' = \tfrac{2}{5}am(\dot\omega_1 - \theta_3\omega_2 + \theta_2\omega_3),$$

$$\dot\omega_3 - \theta_2\omega_1 + \theta_1\omega_2 = 0.$$

Moreover, the components parallel to the axes GA, GB of the velocity of the point of contact are $u - a\omega_2$ and $v + a\omega_1$, and consequently the kinematical equations which express the condition of no sliding at the point of contact are

$$u - a\omega_2 = 0, \quad v + a\omega_1 = 0.$$

Eliminating F, F', ω_1, ω_2, we have

$$\begin{cases} \dot u - v\theta_3 - \tfrac{2}{7}a\theta_1\omega_3 = 0, \\ \dot v + u\theta_3 - \tfrac{2}{7}a\theta_2\omega_3 - \tfrac{5}{7}g\sin\theta = 0, \\ \dot\omega_3 = 0. \end{cases}$$

The last equation gives $\omega_3 = n$, where n is a constant; while substituting for $u, v, \theta_1, \theta_2, \theta_3$, in the first two equations their values in terms of $\theta, \dot\theta, \dot\phi$, we have

$$\begin{cases} (a+b)\dfrac{d}{dt}(\dot\phi\sin\theta) + (a+b)\,\dot\theta\dot\phi\cos\theta - \tfrac{2}{7}an\dot\theta = 0, \\ (a+b)\,\ddot\theta - (a+b)\,\dot\phi^2\cos\theta\sin\theta + \tfrac{2}{7}an\,\dot\phi\sin\theta - \tfrac{5}{7}g\sin\theta = 0. \end{cases}$$

The former of these equations can be integrated at once after multiplying throughout by $\sin\theta$, and gives

$$(a+b)\,\dot\phi\sin^2\theta + \tfrac{2}{7}an\cos\theta = k,$$

where k is a constant. Moreover, multiplying the second equation throughout by $\dot\theta$ and the first equation by $\dot\phi\sin\theta$, and adding, we obtain an equation which can be at once integrated, giving

$$\dot\theta^2 + \dot\phi^2\sin^2\theta + \frac{10}{7}\frac{g}{a+b}\cos$$

where h is a constant; this is really the equation of energy of the system.

Eliminating $\dot\phi$ between these two integral equations, we have

$$(a+b)^2\sin^2\theta\,.\,\dot\theta^2 = -(k - \tfrac{2}{7}an\cos\theta)^2 - \tfrac{10}{7}g(a+b)\sin^2\theta\cos\theta + h(a+b)^2\sin^2\theta;$$

and on writing $\cos\theta = x$, this equation becomes

$$(a+b)^2\dot x^2 = h(a+b)^2(1-x^2) - (k - \tfrac{2}{7}anx)^2 - \tfrac{10}{7}g(a+b)\,x(1-x^2).$$

The cubic polynomial in x on the right-hand side of this equation is positive when $x=+\infty$, negative when $x=1$, positive for some real values of θ, i.e. for some values of x between -1 and 1, and negative when $x=-1$; it has therefore one root greater than unity, and two roots between 1 and -1; we shall denote these roots by

$$\cosh\gamma, \quad \cos\beta, \quad \cos\alpha,$$

where $\cos\beta > \cos\alpha$; and we then have

$$\left(\frac{10}{7}\frac{g}{a+b}\right)^{\frac{1}{2}}(t+\epsilon)=\int\{(x-\cosh\gamma)(x-\cos\beta)(x-\cos\alpha)\}^{-\frac{1}{2}}dx,$$

where ϵ is a constant of integration.

Writing

$$x=\frac{14}{5}\frac{a+b}{g}z+\tfrac{1}{3}(\cosh\gamma+\cos\beta+\cos\alpha)=\frac{14}{5}\frac{a+b}{g}z+\frac{7h(a+b)^2+\frac{4}{7}a^2n^2}{30g(a+b)},$$

the equation becomes

$$t+\epsilon=\int\{4(z-e_1)(z-e_2)(z-e_3)\}^{-\frac{1}{2}}dz,$$

or

$$z=\wp(t+\epsilon),$$

where the function $\wp$ is formed with the roots

$$\begin{cases} e_1=\dfrac{5g}{14(a+b)}\left\{\cosh\gamma-\dfrac{7h(a+b)^2+\frac{4}{7}a^2n^2}{30g(a+b)}\right\}, \\ e_2=\dfrac{5g}{14(a+b)}\left\{\cos\beta-\dfrac{7h(a+b)^2+\frac{4}{7}a^2n^2}{30g(a+b)}\right\}, \\ e_3=\dfrac{5g}{14(a+b)}\left\{\cos\alpha-\dfrac{7h(a+b)^2+\frac{4}{7}a^2n^2}{30g(a+b)}\right\}; \end{cases}$$

these quantities e_1, e_2, e_3 are all real, and satisfy the relations

$$e_1+e_2+e_3=0, \quad e_1>e_2>e_3.$$

Now x is real for real values of t and (since $\dot{x}$ is real) lies between $\cos\alpha$ and $\cos\beta$; so z is real and lies between e_2 and e_3; hence the imaginary part of the constant ϵ in the argument of the $\wp$-function is the half-period corresponding to the root e_3, which we shall denote by ω; the real part of ϵ may be taken to be zero by suitably choosing the origin of time: and therefore we have finally

$$\cos\theta=\frac{14}{5}\frac{a+b}{g}\wp(t+\omega)+\frac{7h(a+b)^2+\frac{4}{7}a^2n^2}{30g(a+b)}.$$

This equation gives the variable θ in terms of the time: the other coordinate ϕ of the centre of the moving sphere is then obtained by integrating the equation

$$\dot{\phi}=\frac{k-\frac{2}{7}an\cos\theta}{(a+b)\sin^2\theta};$$

this integration can be effected by a procedure similar to that used (§ 72) to obtain the Eulerian angles which define the position of a top spinning on a perfectly rough plane.

Example 2. A rough sphere rolls in contact with the outside of a fixed rough sphere, under gravity; if z_2, z_3 be the greatest and least heights of its centre, during the motion, and z be the height at a time t from an instant when z was equal to z_2, prove that

$$(z_2-z)[\wp(t)-e_2]=(z_2-z_3)(e_1-e_2),$$

where e_1, e_2, e_3 $(=-e_1-e_2)$ are real quantities in descending order of magnitude.

(Coll. Exam.)

Example 3. Sphere rolling on a moving sphere.

Consider now the motion of a rough sphere of radius a and mass m which rolls under gravity on another sphere, of radius b and mass M, the latter sphere being free to turn about its centre O, which is fixed.

Let (θ, ϕ) be the polar coordinates of the point of contact referred to axes fixed in space with the fixed centre as origin, the axis from which θ is measured being vertical.

To obtain the equations of motion of the sphere m, we take (as in the last example) moving axes $GABC$ of which GC is the prolongation of the line OG joining the centres of the spheres, and GA is horizontal. Let $(\theta_1, \theta_2, \theta_3)$ denote the components of angular velocity of the coordinate-system resolved along its own axes, and let $(\omega_1, \omega_2, \omega_3)$ denote the components of angular velocity of the sphere m along the same axes. Then, as in the last example, we have

$$\theta_1 = -\dot{\theta}, \qquad \theta_2 = -\dot{\phi}\sin\theta, \qquad \theta_3 = \dot{\phi}\cos\theta.$$

$$u = -(a+b)\,\dot{\phi}\sin\theta, \qquad v = (a+b)\,\dot{\theta}, \qquad w = 0,$$

$$T = \tfrac{1}{2}m\left\{u^2 + v^2 + w^2 + \frac{2a^2}{5}(\omega_1^2 + \omega_2^2 + \omega_3^2)\right\},$$

and if F, F' be the components of the force acting on the sphere m at the point of contact parallel to GA and GB respectively, we have

$$X = F, \qquad Y = mg\sin\theta + F',$$

$$L = F'a, \qquad M = -Fa, \qquad N = 0,$$

so the equations of motion become

$$\begin{cases} m(\dot{u} - v\theta_3) = F = -\tfrac{2}{5}am(\dot{\omega}_2 - \theta_1\omega_3 + \theta_3\omega_1) \quad\ldots\ldots(1), \\ m(\dot{v} + u\theta_3) - mg\sin\theta = F' = \tfrac{2}{5}am(\dot{\omega}_1 - \theta_3\omega_2 + \theta_2\omega_3) \quad\ldots\ldots(2), \\ \dot{\omega}_3 - \theta_2\omega_1 + \theta_1\omega_2 = 0 \quad\ldots\ldots(3). \end{cases}$$

To determine the motion of the sphere M, we take moving axes parallel to the axes $GABC$, but with their origin at O; let $(\Omega_1, \Omega_2, \Omega_3)$ denote the components of angular velocity of the sphere resolved along these axes. Then for the sphere M we have

$$T = \tfrac{1}{2}M \cdot \tfrac{2}{5}b^2(\Omega_1^2 + \Omega_2^2 + \Omega_3^2),$$

and its equations of motion are

$$\begin{cases} -\tfrac{2}{5}bM(\dot{\Omega}_2 - \theta_1\Omega_3 + \theta_3\Omega_1) = F \quad\ldots\ldots(4), \\ \tfrac{2}{5}bM(\dot{\Omega}_1 - \theta_3\Omega_2 + \theta_2\Omega_3) = F' \quad\ldots\ldots(5), \\ \dot{\Omega}_3 - \theta_2\Omega_1 + \theta_1\Omega_2 = 0 \quad\ldots\ldots(6). \end{cases}$$

The conditions of no sliding at the point of contact are

$$u - a\omega_2 = b\Omega_2, \qquad v + a\omega_1 = -b\Omega_1 \quad\ldots\ldots(7).$$

In order to solve this set of equations we multiply equations (3) and (6) by a and b respectively, and add; thus, using (7), we have

$$a\dot{\omega}_3 + b\dot{\Omega}_3 + u\theta_1 + v\theta_2 = 0,$$

or

$$a\dot{\omega}_3 + b\dot{\Omega}_3 = 0.$$

Integrating, we have

$$a\omega_3 + b\Omega_3 = an, \qquad \text{where } n \text{ is a constant.}$$

Moreover, from equations (4) and (7) we have

$$-\tfrac{2}{5}M(\dot{u}-a\dot{\omega}_2-b\theta_1\Omega_3-\theta_3 v-\theta_3 a\omega_1)=F.$$

Eliminating F and $\dot{\omega}_2+\theta_3\omega_1$ between this and equations (1), we have

$$\frac{7M+5m}{2M}(\dot{u}-\theta_3 v)=an\theta_1,$$

or

$$\frac{d}{dt}(\dot{\phi}\sin\theta)+\dot{\theta}\dot{\phi}\cos\theta-\frac{2Man\dot{\theta}}{(7M+5m)(a+b)}=0 \quad \text{..................(A).}$$

Similarly from equations (5) and (7), we have

$$\tfrac{2}{5}M(-\dot{v}-a\dot{\omega}_1-u\theta_3+a\theta_3\omega_2+b\theta_2\Omega_3)=F'.$$

Eliminating F' and $\dot{\omega}_1-\theta_3\omega_2$ between this and equations (2), we have

$$\frac{5m+7M}{2M}(\dot{v}+u\theta_3)=an\theta_2+\frac{5(M+m)}{2M}g\sin\theta,$$

or

$$\ddot{\theta}-\dot{\phi}^2\sin\theta\cos\theta-\frac{5(M+m)g\sin\theta}{(5m+7M)(a+b)}=-\frac{an}{a+b}\cdot\frac{2M}{5m+7M}\cdot\dot{\phi}\sin\theta \quad \text{........(B).}$$

Now the equations (A) and (B), from which θ and ϕ are to be determined in terms of t, are of essentially the same character as the equations found for the determination of θ and ϕ in the previous example: the former equations being in fact derivable from the present ones by making M very large compared with m. The integration therefore proceeds exactly as in the former case.

Example 4. A uniform sphere rolls on a perfectly rough horizontal plane, under forces whose resultant passes through its centre. Shew that the motion of its centre is the same as that of a particle acted on by the same forces reduced in the ratio 5 : 7.

Example 5. Form the equations of motion of a perfectly rough sphere rolling under gravity inside a fixed right circular cylinder, the axis of which is inclined to the vertical at an angle α; and shew that, if the sphere be such that $k^2=\tfrac{1}{3}a^2$, a being its radius and k the radius of gyration about any diameter, and if it be placed at rest with the axial plane through its centre making an angle β with the vertical axial plane, the velocity of the centre parallel to the axis, when this angle is θ, is

$$\tfrac{1}{2}\left(\frac{3gb\cos^2\alpha}{\sin\alpha}\right)^{\frac{1}{2}}\{\sin\tfrac{1}{2}\theta\,\mathrm{arccosh}\,(\cos\tfrac{1}{2}\theta\sec\tfrac{1}{2}\beta)+\cos\tfrac{1}{2}\theta\arccos(\sin\tfrac{1}{2}\theta\,\mathrm{cosec}\,\tfrac{1}{2}\beta)\},$$

where $b+a$ is the radius of the cylinder. (Camb. Math. Tripos, Part I, 1895.)

For other examples cf. Woronetz, *Math. Ann.* LXX. (1911), p. 410.

90. *Vibrations of non-holonomic systems.*

We shall next consider the small vibratory motions of a non-holonomic system: it will appear that so far as vibrations about equilibrium are concerned, the difference between holonomic and non-holonomic systems is unimportant.

For consider the vibrations about equilibrium of a non-holonomic system with n independent coordinates and $(n-m)$ degrees of freedom, in which the constraints are independent of the time. Let T be the kinetic and V the potential energy, so that for the vibrational problem T will be supposed to be a homogeneous quadratic function of $(\dot{q}_1, \dot{q}_2, \ldots, \dot{q}_n)$, and V to be a homogeneous quadratic function of $(q_1, q_2, \ldots, q_n)$, the coefficients in both cases being constants. There are m equations of the type

$$A_{1k}\dot{q}_1 + A_{2k}\dot{q}_2 + \ldots + A_{nk}\dot{q}_n = 0 \qquad (k = 1, 2, \ldots, m),$$

which express the non-holonomic constraints; and the equations of motion are (§ 87)

$$\frac{d}{dt}\left(\frac{\partial T}{\partial \dot{q}_r}\right) = -\frac{\partial V}{\partial q_r} + \lambda_1 A_{r1} + \lambda_2 A_{r2} + \ldots + \lambda_m A_{rm} \qquad (r = 1, 2, \ldots, n).$$

From these equations it is evident that $\lambda_1, \lambda_2, \ldots, \lambda_m$ are in general small quantities of the order of the coordinates; and therefore for the vibrational problem only the constant parts of $A_{11}, A_{12}, \ldots, A_{nm}$ need be considered. The vibrational motion is therefore the same as if the coefficients $A_{11}, A_{12}, \ldots, A_{nm}$ were constants independent of the coordinates; but in this case the equations

$$A_{1k}\dot{q}_1 + A_{2k}\dot{q}_2 + \ldots + A_{nk}\dot{q}_n = 0 \qquad (k = 1, 2, \ldots, m)$$

can be integrated; in fact, they give

$$A_{1k}q_1 + A_{2k}q_2 + \ldots + A_{nk}q_n = 0 \qquad (k = 1, 2, \ldots, m),$$

the constants of integration being zero since the values

$$q_1 = 0, \quad q_2 = 0, \ldots, \; q_n = 0$$

represent a possible position of the system.

It follows that the vibratory motion of the given non-holonomic system is the same as that of the holonomic system for which the equations of constraint are expressible in the integrated form

$$A_{1k}q_1 + A_{2k}q_2 + \ldots + A_{nk}q_n = 0 \qquad (k = 1, 2, \ldots, m);$$

we can therefore determine the vibrations by using these equations to eliminate m of the coordinates $(q_1, q_2, \ldots, q_n)$ from T and V; we shall then have a holonomic system with $(n - m)$ degrees of freedom, the kinetic and potential energies being expressed in terms of $(n - m)$ coordinates and the corresponding velocities: the vibrations of this system can be determined by the usual method described in the preceding chapter.

As an example, we shall consider the following problem*.

A heavy homogeneous hemisphere is resting in equilibrium on a perfectly rough horizontal plane with its spherical surface downwards. A second heavy homogeneous hemisphere is resting in the same way on a perfectly rough plane face of the first, the point of contact being in the centre of the face. The equilibrium being slightly disturbed, it is required to find the vibrations of the system.

Take as axes of reference

(1) A rectangular set of axes Z_2xyz fixed in the upper hemisphere, the origin being its centre of gravity Z_2.

* Due to Madame Kerkhoven-Wythoff, *Nieuw Archief voor Wiskunde*, Deel IV. (1899).

(2) A rectangular set of axes $Z_1\xi\eta\zeta$, fixed in the lower hemisphere, the origin being its centre of gravity Z_1.

(3) A rectangular set of axes $Rlmn$ fixed in space, the origin R being the equilibrium position of the point of contact of the lower hemisphere and the plane.

We further define these axes by supposing that in the equilibrium position the axes Z_2z, $Z_1\zeta$, and Rn are vertical and therefore coincident, while the axes Z_2x, $Z_1\xi$, Rl are parallel, the axes Z_2y, $Z_1\eta$, and Rm being therefore also parallel.

Suppose that at time t the coordinates of a point referred to these different sets of axes are connected by the equations

$$\begin{aligned}
\xi &= \alpha + \alpha_1 x + \alpha_2 y + \alpha_3 z,\\
\eta &= \beta + \beta_1 x + \beta_2 y + \beta_3 z,\\
\zeta &= \gamma + \gamma_1 x + \gamma_2 y + \gamma_3 z,\\
l &= a + a_1 \xi + a_2 \eta + a_3 \zeta,\\
m &= b + b_1 \xi + b_2 \eta + b_3 \zeta,\\
n &= c + c_1 \xi + c_2 \eta + c_3 \zeta,
\end{aligned}$$

The 24 coefficients in these transformation formulae completely specify the position of the system at any instant. As however the system has only six degrees of freedom, there must be 18 equations connecting these coefficients or their differentials. Of these, 12 are the ordinary conditions of the types

$$\alpha_1^2+\alpha_2^2+\alpha_3^2=1, \quad \alpha_1\beta_1+\alpha_2\beta_2+\alpha_3\beta_3=0,$$
$$a_1^2+a_2^2+a_3^2=1, \quad a_1b_1+a_2b_2+a_3b_3=0,$$

which express the orthogonal character of the axes; the remaining 6 are the conditions of contact and rolling, which we shall now find.

Let R_1, R_2 be the radii of the lower and upper hemispheres respectively, and l_1, l_2 the distances of the centres of gravity from their plane faces, so $l_1=\frac{3}{8}R_1$, $l_2=\frac{3}{8}R_2$. The coordinates of the point of contact of the upper hemisphere with the lower are

$$x_2=-R_2\gamma_1, \quad y_2=-R_2\gamma_2, \quad z_2=l_2-R_2\gamma_3;$$

the conditions that this point shall be at rest relative to the lower hemisphere are

$$\begin{aligned}
\dot\alpha+\dot\alpha_1x_2+\dot\alpha_2y_2+\dot\alpha_3z_2&=0,\\
\dot\beta+\dot\beta_1x_2+\dot\beta_2y_2+\dot\beta_3z_2&=0,\\
\dot\gamma+\dot\gamma_1x_2+\dot\gamma_2y_2+\dot\gamma_3z_2&=0.
\end{aligned}$$

The last of these equations gives $\dot\gamma+l_2\dot\gamma_3=0$, which is the differentiated form of the equation

$$l_1-\gamma-\gamma_3l_2=-R_2,$$

an equation which expresses the condition of contact of the two hemispheres: while the first two of the equations give

$$\begin{aligned}
\dot\alpha-\dot\alpha_1R_2\gamma_1-\dot\alpha_2R_2\gamma_2+\dot\alpha_3(l_2-R_2\gamma_3)&=0,\\
\dot\beta-\dot\beta_1R_2\gamma_1-\dot\beta_2R_2\gamma_2+\dot\beta_3(l_2-R_2\gamma_3)&=0,
\end{aligned}$$

and these express the condition of rolling of the upper on the lower hemisphere. These equations give as a first approximation

$$\dot\alpha=\dot\alpha_3(R_2-l_2), \qquad \dot\beta=\dot\beta_3(R_2-l_2),$$

and therefore on integration

$$\alpha=\alpha_3(R_2-l_2), \quad \beta=\beta_3(R_2-l_2).$$

Similarly the condition of contact of the lower hemisphere and the horizontal plane is

$$c+c_3l_1=R_1,$$

and the conditions of rolling are

$$a=a_3(R_1-l_1), \quad b=b_3(R_1-l_1).$$

We have thus now obtained the 18 equations connecting the 24 coefficients: taking α_2, β_3, γ_1, a_2, b_3, c_1 as the 6 independent coordinates of the system, and solving for the other 18 coefficients in terms of these, we find with the necessary approximation

$$\begin{cases}\alpha=\gamma_1(l_2-R_2),\\ \alpha_1=1-\frac{1}{2}(\alpha_2{}^2+\gamma_1{}^2),\\ \alpha_3=-\gamma_1.\end{cases} \qquad \begin{cases}a=c_1(l_1-R_1),\\ a_1=1-\frac{1}{2}(a_2{}^2+c_1{}^2),\\ a_3=-c_1.\end{cases}$$

$$\begin{cases}\beta=\beta_3(R_2-l_2),\\ \beta_1=-\alpha_2,\\ \beta_2=1-\frac{1}{2}(\alpha_2{}^2+\beta_3{}^2).\end{cases} \qquad \begin{cases}b=b_3(R_1-l_1),\\ b_1=-a_2,\\ b_2=1-\frac{1}{2}(a_2{}^2+b_3{}^2).\end{cases}$$

$$\begin{cases}\gamma=R_2+l_1-l_2\{1-\frac{1}{2}(\gamma_1{}^2+\beta_3{}^2)\},\\ \gamma_2=-\beta_3,\\ \gamma_3=1-\frac{1}{2}(\gamma_1{}^2+\beta_3{}^2).\end{cases} \qquad \begin{cases}c=R_1-l_1\{1-\frac{1}{2}(c_1{}^2+b_3{}^2)\},\\ c_2=-b_3,\\ c_3=1-\frac{1}{2}(c_1{}^2+b_3{}^2).\end{cases}$$

The potential energy of the system is

$$V=M_1gc+M_2g(c+c_1\alpha+c_2\beta+c_3\gamma),$$

or, retaining only small quantities of the second order,

$$V/g=b_3{}^2(\tfrac{3}{16}R_1M_1-\tfrac{5}{16}R_2M_2)-\tfrac{5}{8}M_2R_2b_3\beta_3+\tfrac{3}{16}\beta_3{}^2M_2R_2+c_1{}^2(\tfrac{3}{16}R_1M_1-\tfrac{5}{16}R_2M_2)$$
$$-\tfrac{5}{8}M_2R_2c_1\gamma_1+\tfrac{3}{16}M_2R_2\gamma_1{}^2.$$

If now we express the coordinates l, m, n of any particle of the upper or lower hemisphere in terms of its coordinates relative to the axes Z_2xyz and $Z_1\xi\eta\zeta$ respectively, and form the sum $\frac{1}{2}\Sigma m(\dot{l}^2+\dot{m}^2+\dot{n}^2)$ for each hemisphere, neglecting terms above the second order of small quantities, and remembering that the principal moments of inertia of a hemisphere of mass M and radius R at its centre of gravity are $\frac{2}{5}MR^2$, $\frac{83}{320}MR^2$, $\frac{83}{320}MR^2$, we find for the kinetic energy of the system the value T, where

$$2T=\tfrac{2}{5}\dot{\alpha}_2{}^2(M_1R_1{}^2+M_2R_2{}^2)+\tfrac{4}{5}\dot{\alpha}_2\dot{a}_2M_2R_2{}^2+\tfrac{2}{5}\dot{a}_2{}^2M_2R_2{}^2$$
$$+\dot{b}_3{}^2\{\tfrac{13}{20}R_1{}^2M_1+M_2(\tfrac{13}{20}R_2{}^2+\tfrac{5}{4}R_1R_2+R_1{}^2)\}+2\dot{b}_3\dot{\beta}_3M_2R_2(\tfrac{13}{20}R_2+\tfrac{5}{8}R_1)+\tfrac{13}{20}\dot{\beta}_3{}^2M_2R_2{}^2$$
$$+\dot{c}_1{}^2\{\tfrac{13}{20}R_1{}^2M_1+M_2(\tfrac{13}{20}R_2{}^2+\tfrac{5}{4}R_1R_2+R_1{}^2)\}+2\dot{c}_1\dot{\gamma}_1M_2R_2(\tfrac{5}{8}R_1+\tfrac{13}{20}R_2)+\tfrac{13}{20}\dot{\gamma}_1{}^2M_2R_2{}^2.$$

The equations of motion evidently separate into three distinct sets, consisting of

(i) Equations for the coordinates α_2 and a_2: these coordinates give rise to no terms in V, and do not correspond to vibrations in the stricter sense; in fact, the equilibrium is not disturbed if either of the hemispheres is turned through any angle about its axis of revolution. We can therefore neglect these equations.

(ii) Equations involving the coordinates b_3 and β_3.

(iii) Equations for the coordinates c_1 and γ_1; these are exactly the same as the equations for b_3 and β_3, so we need consider only the latter.

The equations for b_3 and β_3 are, *in extenso*,

$$\{\tfrac{13}{20}R_1^2M_1+M_2(R_1^2+\tfrac{5}{4}R_1R_2+\tfrac{13}{20}R_2^2)\}\ddot{b}_3+M_2R_2(\tfrac{5}{8}R_1+\tfrac{13}{20}R_2)\ddot{\beta}_3$$
$$+g(\tfrac{3}{8}R_1M_1-\tfrac{5}{8}R_2M_2)b_3-\tfrac{5}{8}gM_2R_2\beta_3=0,$$
$$(\tfrac{5}{8}R_1+\tfrac{13}{20}R_2)\ddot{b}_3+\tfrac{13}{20}R_2\ddot{\beta}_3-\tfrac{5}{8}gb_3+\tfrac{3}{8}g\beta_3=0.$$

The corresponding determinantal equation for λ, where $2\pi/\sqrt{\lambda}$ is a period, is

$$\begin{vmatrix}\{\tfrac{13}{20}R_1^2M_1+M_2(R_1^2+\tfrac{4}{5}R_1R_2+\tfrac{13}{20}R_2^2)\}\lambda-g(\tfrac{3}{8}R_1M_1-\tfrac{5}{8}R_2M_2) & M_2R_2(\tfrac{5}{8}R_1+\tfrac{13}{20}R_2)\lambda+\tfrac{5}{8}gM_2R_2 \\ (\tfrac{5}{8}R_1+\tfrac{13}{20}R_2)\lambda+\tfrac{5}{8}g & \tfrac{13}{20}R_2\lambda-\tfrac{3}{8}g\end{vmatrix}=0.$$

This is a quadratic equation in λ: it is easily found that its roots are positive if

$$9R_1M_1>40R_2M_2,$$

and *this is the condition for stability of the equilibrium.*

The vibrations of non-holonomic systems about a state of steady motion are most conveniently discussed by use of the equations of motion given in § 88. The method will be illustrated by the following example.

Example. A solid of revolution has an equatorial plane of symmetry, and is rolling with angular velocity n round its axis in steady motion on a perfectly rough horizontal plane, the equatorial plane of the solid being vertical. This motion being slightly disturbed, to find the period of a vibration.

Let G be the centre of gravity of the solid, and let (C, A) be its moments of inertia about the axis and about a line through G perpendicular to the axis. Take as moving axes of reference $Gxyz$, where Gz is the axis of the solid, Gy is perpendicular to the plane through Gz and the point of contact (so Gy is horizontal), and Gx is normal to the plane Gyz. Let F, F', R be the components of the force acting on the solid at the point of contact, F being in the plane Gxz, F' being parallel to Gy, and R being normal to the plane. Let $(\theta_1, \theta_2, \theta_3)$ and $(\omega_1, \omega_2, \omega_3)$ denote as usual the components of angular velocity of the axes and of the body respectively, and let (u, v, w) be the components of the velocity of G, parallel to the moving axes. Further, let ρ be the radius of curvature of the meridian of the solid at the equator, a the radius of its equatorial circle, θ the angle made by Gz with the vertical, and ϕ the angle between Gy and its undisturbed direction. Then we have

$$\theta_1=\omega_1=-\dot{\phi}\sin\theta,\quad \theta_2=\omega_2=\dot{\theta},\quad \theta_3=\dot{\phi}\cos\theta,$$

and the kinetic energy is

$$T=\tfrac{1}{2}M(u^2+v^2+w^2)+\tfrac{1}{2}A(\omega_1^2+\omega_2^2)+\tfrac{1}{2}C\omega_3^2.$$

The equations of § 18 therefore give, if P is the point of contact, PK the perpendicular from this point on the axis, and GN the perpendicular from G on the horizontal plane,

$$\begin{cases} M(\dot{u}-v\theta_3+w\theta_2) &= F\cos\theta-(R-Mg)\sin\theta, \\ M(\dot{v}-w\theta_1+u\theta_3) &= F', \\ M(\dot{w}-u\theta_2+v\theta_1) &= (R-Mg)\cos\theta+F\sin\theta, \\ A\dot{\omega}_1-A\omega_2\theta_3+C\omega_3\theta_2 &= -F'.GK, \\ A\dot{\omega}_2-C\omega_3\theta_1+A\omega_1\theta_3 &= -F.GN-R.NP, \\ C\dot{\omega}_3 &= F'.PK. \end{cases}$$

In these equations, GK and NP are measured positively parallel to the positive direction of the axis of z and the horizontal projection of this direction respectively.

The conditions of no sliding at P are

$$\begin{cases} u\cos\theta + w\sin\theta - GN\,.\,\omega_2 = 0, \\ v + PK\,.\,\omega_3 - GK\,.\,\omega_1 = 0, \end{cases}$$

and the condition of contact of the body and plane is

$$w\cos\theta - u\sin\theta = \frac{d}{dt}(-GK\cos\theta + PK\sin\theta).$$

These equations determine the motion in the general case, when the disturbance from steady motion is not supposed to be small. When this latter assumption is made, we have

$$\theta = \frac{\pi}{2} + \chi, \quad \omega_3 = n + \varpi, \quad v = -an + \eta,$$

where χ, ϖ, η are small; and F, F', u, w, ω_1, ω_2, θ_1, θ_2, θ_3 are small, while R is nearly equal to Mg. Moreover we have $NP = (\rho - a)\chi$. The equations therefore become

$$\begin{cases} M(\dot{u} + an\theta_3) = -R + Mg, \\ M\dot{\eta} = F', \\ M(\dot{w} - an\theta_1) = F, \\ A\dot{\omega}_1 + Cn\theta_2 = 0, \\ A\dot{\omega}_2 - Cn\theta_1 = -Fa - Mg(\rho - a)\chi, \\ C\dot{\varpi} = F'a, \\ w - a\omega_2 = 0, \\ \eta + a\varpi = 0, \end{cases}$$

where $\qquad \omega_1 = \theta_1 = -\dot{\phi}, \quad \omega_2 = \theta_2 = \dot{\chi}, \quad \theta_3 = 0.$

Eliminating F, F', R, and replacing θ_1, θ_2, θ_3, ω_1, ω_2 by their values, the equations become

$$\begin{cases} A\ddot{\phi} - Cn\dot{\chi} = 0, \\ A\ddot{\chi} + (C + Ma^2)n\dot{\phi} + Mg(\rho - a)\chi + Ma\dot{w} = 0, \\ C\dot{\varpi} = Ma\dot{\eta}, \\ w = a\dot{\chi}, \\ \eta = -a\varpi. \end{cases}$$

From the third and fifth of these equations we see that $\dot{\varpi}$ and $\dot{\eta}$ are zero, and therefore ϖ and η are constants. The other three equations give, on eliminating w,

$$\begin{cases} A\ddot{\phi} - Cn\dot{\chi} = 0, \\ (Ma^2 + A)\ddot{\chi} + (C + Ma^2)n\dot{\phi} + Mg(\rho - a)\chi = 0, \end{cases}$$

and therefore the equation for the determination of χ is

$$A(A + Ma^2)\ddot{\chi} + \{MgA(\rho - a) + Cn^2(C + Ma^2)\}\chi = 0;$$

this equation shews that the period of a vibration is

$$2\pi\left\{\frac{A(A + Ma^2)}{MgA(\rho - a) + Cn^2(C + Ma^2)}\right\}^{\frac{1}{2}}.$$

91. *Dissipative systems; frictional forces.*

We now proceed to the consideration of systems for which the principle of conservation of dynamical energy is not valid, the energy of the system being continually changed into some other form (e.g. heat) which is not recognised in dynamics. We shall first consider *frictional systems.*

If two rigid bodies which are not perfectly smooth are in contact, the reaction between them at the point of contact may be resolved into a component along the common normal to their surfaces at the point, which is called the *normal pressure*, and a component in the common tangent-plane, which is called the *frictional force.* The frictional force is determined by the following law*, which has been established experimentally: *The bodies will not slide on each other, provided the frictional force required for the prevention of sliding does not exceed μ times the normal pressure, where μ is a constant called the "limiting coefficient of friction," which depends only on the material of which the surfaces in contact are composed. If on the other hand the frictional force required to prevent sliding is greater than μ times the normal pressure, there will be sliding at the point of contact, and the frictional force called into play will be μ times the normal pressure.*

Painlevé has pointed out that the four hypotheses—(1) that the above laws of friction hold, (2) that there exist rigid bodies, (3) that the normal pressure between bodies cannot be negative, (4) that all accelerations and tensions are finite—taken together lead in some cases to contradictions of the fundamental laws of dynamics. For a discussion on this subject, cf. *Comptes Rendus*, CXL. (1905), pp. 635, 702, 847: *ibid.* CXLI. (1905), pp. 310, 401, 546; *Zeitschrift für M. u. P.* LVIII. (1909), p. 186.

The following examples illustrate the motion of systems involving frictional forces.

Example 1. *Motion of a particle on a rough fixed plane curve.*

Consider the motion of a particle which is constrained to move in a rough fixed tube of small bore, in the form of a plane curve, under forces which depend solely on its position in the tube. Let $f(s)$ and $g(s)$ denote the components of force per unit mass acting on the particle in direction of the tangent and normal to the tube, where s is the distance of the particle from some fixed point of the tube, measured along the arc in the direction in which the particle is moving; and let R be the normal reaction per unit mass, and μ the coefficient of friction.

Since the components of acceleration of the particle along the tangent and normal are $v\,dv/ds$ and v^2/ρ, where v is the velocity of the particle and ρ the radius of curvature of the tube, we have

$$v\frac{dv}{ds}=f(s)-\mu R, \qquad \frac{v^2}{\rho}=g(s)+R.$$

Eliminating R, we have

$$\frac{dv^2}{ds}+\frac{2\mu}{\rho}v^2=2f(s)+2\mu g(s).$$

Integrating, we have

$$v^2=ce^{-2\mu\phi}+2e^{-2\mu\phi}\int^s e^{2\mu\phi}\{f(s)+\mu g(s)\}\,ds,$$

where $\phi=\int ds/\rho$, and c is a constant depending on the initial circumstances of the motion.

* The discovery that the friction is proportional to the normal pressure was made by G. Amontons, *Paris Mém.*, année 1699, p. 206.

The right-hand side of this equation is a known function of s, say $=F(s)$. Then we have

$$\left(\frac{ds}{dt}\right)^2 = F(s),$$

so the relation between s and t is

$$t - t_0 = \int^s \{F(s)\}^{-\frac{1}{2}} ds.$$

This equation represents the solution of the problem.

Example 2. A circular hoop of mass M stands on rough ground, and a particle of mass m is attached to the end of the horizontal diameter. To find whether the hoop will roll or slide.

Let us investigate the rolling motion, assumed possible, and so determine whether the friction required to produce this motion is, or is not, greater than the maximum friction actually available, i.e. μ times the corresponding normal pressure. Let θ be the angle turned through by the hoop from the commencement of the motion, and let x and y be the coordinates of the centre of gravity of the system, referred to horizontal and vertical (downward) axes through its own initial position, so that

$$x = a\theta - \frac{ma}{M+m}(1-\cos\theta), \qquad y = \frac{ma}{M+m}\sin\theta,$$

where a is the radius of the hoop.

The kinetic and potential energies are

$$\begin{cases} T = Ma^2\dot{\theta}^2 + ma^2\dot{\theta}^2(1-\sin\theta), \\ V = -mga\sin\theta. \end{cases}$$

The Lagrangian equation of motion is therefore

$$\frac{d}{dt}[2a^2\dot{\theta}\{M+m(1-\sin\theta)\}] + ma^2\dot{\theta}^2\cos\theta = mga\cos\theta.$$

For the initial motion, this equation gives

$$2a\ddot{\theta}(M+m) = mg,$$

so initially we have

$$\ddot{x} = a\ddot{\theta} = \frac{mg}{2(M+m)}, \qquad \ddot{y} = \frac{m}{M+m}a\ddot{\theta} = \frac{m^2g}{2(M+m)^2}.$$

But if F be the frictional force and R the normal pressure, we have

$$F = (M+m)\ddot{x}, \qquad R = (M+m)(-\ddot{y}+g),$$

so initially we have

$$\frac{F}{R} = \frac{\ddot{x}}{-\ddot{y}+g} = \frac{m(M+m)}{2M^2+4Mm+m^2}.$$

The hoop will therefore roll or slide according as the coefficient of friction is not less, or less than

$$\frac{m(M+m)}{2M^2+4Mm+m^2}.$$

Example 3. A particle moves under gravity on a rough cycloid whose plane is vertical and whose base is horizontal: if ϕ be the inclination of the tangent at any point to the horizontal, so that the equation of the cycloid can be written

$$s = 4a\sin\phi,$$

and if $\tan\epsilon$ be the coefficient of friction, shew that the motion is given by the equation

$$ce^{\phi\tan\epsilon}\sin(\phi+\epsilon) = \cos\left(\sqrt{\frac{g}{a}}\frac{t}{2\cos\epsilon}\right),$$

where c is a constant.

92. *Resisting forces depending on the velocity.*

A different type of dissipative system is illustrated by the motion of a projectile in the air, as the resistance of the air depends on the velocity of the projectile. No general rule can be formulated for the solution of problems involving forces of this kind: a case of practical interest, however, namely the motion of a projectile under the influence of gravity and of a resistance varying as some power of the projectile's velocity, can be integrated in the following manner.

For low velocities (below 100 ft./sec.) the resistance of the air to a projectile is nearly proportional to the square of the velocity. For high velocities (say 2000 ft./sec.) the resistance is approximately a linear function of the velocity.

At time t let v be the velocity of the projectile, kv^n the resistance per unit mass, θ the inclination of the path to the horizontal, and ρ the radius of curvature of the path. The components of acceleration of the projectile along the tangent and normal to its path are $v\,dv/ds$ and v^2/ρ; and hence the equations of motion are

$$\begin{cases} v\,dv/ds = -g\sin\theta - kv^n, \\ v^2/\rho = g\cos\theta. \end{cases}$$

Dividing the first equation by the second, we obtain

$$\frac{1}{v^{n+1}}\frac{dv}{d\theta} - \frac{\tan\theta}{v^n} = \frac{k}{g\cos\theta},$$

or

$$\frac{d}{d\theta}\left(\frac{1}{v^n}\right) + \frac{1}{v^n}\frac{d}{d\theta}(n\log\sec\theta) = -\frac{nk}{g}\sec\theta.$$

Integrating, we have

$$(1/v^n)\sec^n\theta + \text{Constant} = -(nk/g)\int \sec^{n+1}\theta\,d\theta.$$

This equation gives v in terms of θ. To obtain t, the equation $v^2 = \rho g\cos\theta$ gives

$$gt = -\int v\sec\theta\,d\theta,$$

and as v is a known function of θ, this equation gives t as a function of θ. The rectangular coordinates (x, y) of the particle can now be found from the equations

$$x = \int v\cos\theta\,dt, \qquad y = \int v\sin\theta\,dt.$$

The solution of the problem is thus reduced to quadratures.

Resisting forces proportional respectively to v, v^2, and $av + bv^2$ were considered by Newton, *Principia*, Book II. §§ 1, 2, 3. The case of a resistance proportional to any power of the velocity was then examined by John Bernoulli* in 1711.

* *Opera*, I. p. 502.

D'Alembert* shewed that if gu denotes the ratio of the resistance to the mass of the projectile, the integration can be effected in the four cases

$$u = a + bv^n,$$
$$u = a + b\log v,$$
$$u = av^n + R + bv^{-n},$$
$$u = a(\log v)^n + R\log v + b,$$

where a, b, n are arbitrary constants and R is another constant depending on them.

Siacci† obtained many more integrable cases, of which the following may be mentioned:

$$\log \int v\,du = \tfrac{1}{2}c\int \frac{du}{1 + a(u-1)^c} - \tfrac{1}{2}c\,\frac{du}{1 + b(u+1)^c} + C,$$

where a, b, c, C are arbitrary constants: this equation defines v in terms of u, the number of terms involved being finite when c is rational.

Poisson pointed out in 1806‡ that the theory of *singular solutions* of differential equations has applications in Dynamics, notably in the case of a particle under a resisting force. If a particle is moving in a straight line under a resisting force varying as the square root of the velocity, the equation of motion is

$$dv/dt = -av^{\frac{1}{2}}.$$

The initial velocity being c^2, the motion is represented by the *general integral*

$$v = (c - \tfrac{1}{2}at)^2$$

so long as $t < 2c/a$, after which it is represented by the *singular solution* $v = 0$.

Example 1. A heavy particle falls vertically from rest at the origin in a medium whose resistance varies directly as the velocity. Shew that the distance traversed in time t is

$$\frac{gt}{\mu} - \frac{g}{\mu^2} + \frac{g}{\mu^2}e^{-\mu t},$$

where μv is the resistance per unit mass.

Example 2. A heavy particle falls vertically from rest at the origin in a medium whose resistance varies as the square of the velocity: shew that the distance traversed in time t is

$$\frac{1}{\mu}\log\cosh(\sqrt{g\mu}\,t),$$

where μv^2 denotes the resistance per unit mass.

93. *Rayleigh's dissipation-function.*

When a system is subject to external resisting forces which are directly proportional to the velocities of their points of application, it is possible to express the equations of motion of the system in general coordinates in terms of the kinetic and potential energies and of a single new function.

For let the energy lost to the system by the action of the resisting force which is applied to a particle m of the system, whose coordinates are (x, y, z), in an arbitrary displacement $(\delta x, \delta y, \delta z)$ be

$$k_x\dot{x}\delta x + k_y\dot{y}\delta y + k_z\dot{z}\delta z,$$

* *Traité de l'équilibre et du mouvement des fluides*, Paris, 1744.

† *Comptes Rendus*, CXXXII. (1901), p. 1175.

‡ *Journal de l'École Polyt.* VI. (Cahier 13), p. 60.

where k_x, k_y, k_z are functions of x, y, z only. The equations of motion of the typical particle m will therefore be

$$\begin{cases} m\ddot{x} = -k_x\dot{x} + X, \\ m\ddot{y} = -k_y\dot{y} + Y, \\ m\ddot{z} = -k_z\dot{z} + Z, \end{cases}$$

where X, Y, Z are the components of the total force (external and molecular) on the particle, except the force of resistance.

Now let a function F be defined by the equations

$$F = \tfrac{1}{2}\Sigma\,(k_x\dot{x}^2 + k_y\dot{y}^2 + k_z\dot{z}^2),$$

where the summation is extended over all the particles of the system; so that F, which is called the *dissipation-function*, represents half the rate at which energy is being lost to the system by the action of the resisting forces; and let $(q_1, q_2, \ldots, q_n)$ be coordinates specifying the configuration of the system.

Multiplying the equations of motion of the particle m by $\partial x/\partial q_r$, $\partial y/\partial q_r$, $\partial z/\partial q_r$, respectively, and summing for all the particles of the system, we have

$$\Sigma m\left(\ddot{x}\frac{\partial x}{\partial q_r} + \ddot{y}\frac{\partial y}{\partial q_r} + \ddot{z}\frac{\partial z}{\partial q_r}\right) = -\Sigma\left(k_x\dot{x}\frac{\partial x}{\partial q_r} + k_y\dot{y}\frac{\partial y}{\partial q_r} + k_z\dot{z}\frac{\partial z}{\partial q_r}\right) + \Sigma\left(X\frac{\partial x}{\partial q_r} + Y\frac{\partial y}{\partial q_r} + Z\frac{\partial z}{\partial q_r}\right).$$

As in § 26, we have

$$\Sigma m\left(\ddot{x}\frac{\partial x}{\partial q_r} + \ddot{y}\frac{\partial y}{\partial q_r} + \ddot{z}\frac{\partial z}{\partial q_r}\right) = \frac{d}{dt}\left(\frac{\partial T}{\partial \dot{q}_r}\right) - \frac{\partial T}{\partial q_r},$$

where T is the kinetic energy; and

$$\Sigma\left(X\frac{\partial x}{\partial q_r} + Y\frac{\partial y}{\partial q_r} + Z\frac{\partial z}{\partial q_r}\right) = Q_r,$$

where $Q_1\delta q_1 + Q_2\delta q_2 + \ldots + Q_n\delta q_n$ denotes the work done by the external forces (excluding the resistances) in an arbitrary infinitesimal displacement; while we have

$$-\Sigma\left(k_x\dot{x}\frac{\partial x}{\partial q_r} + k_y\dot{y}\frac{\partial y}{\partial q_r} + k_z\dot{z}\frac{\partial z}{\partial q_r}\right) = -\Sigma\left(k_x\dot{x}\frac{\partial \dot{x}}{\partial \dot{q}_r} + k_y\dot{y}\frac{\partial \dot{y}}{\partial \dot{q}_r} + k_z\dot{z}\frac{\partial \dot{z}}{\partial \dot{q}_r}\right) = -\frac{\partial F}{\partial \dot{q}_r}.$$

It follows that *the equations of motion of the system in terms of the coordinates* $(q_1, q_2, \ldots, q_n)$ *can be written in the form*

$$\frac{d}{dt}\left(\frac{\partial T}{\partial \dot{q}_r}\right) - \frac{\partial T}{\partial q_r} + \frac{\partial F}{\partial \dot{q}_r} = Q_r \qquad (r = 1, 2, \ldots, n).$$

Example. If the resisting forces depend on the relative (as opposed to the absolute) velocity of their points of application, so that the forces acting on two particles (x_1, y_1, z_1) and (x_2, y_2, z_2) have the components

$$-k_x(\dot{x}_1-\dot{x}_2), \quad -k_y(\dot{y}_1-\dot{y}_2), \quad -k_z(\dot{z}_1-\dot{z}_2)$$

and

$$-k_x(\dot{x}_2-\dot{x}_1), \quad -k_y(\dot{y}_2-\dot{y}_1), \quad -k_z(\dot{z}_2-\dot{z}_1)$$

respectively, shew that the equations in general coordinates can be formed with the expression

$$\tfrac{1}{2}\Sigma\{k_x(\dot{x}_1-\dot{x}_2)^2+k_y(\dot{y}_1-\dot{y}_2)^2+k_z(\dot{z}_1-\dot{z}_2)^2\}$$

as a dissipation-function.

94. *Vibrations of dissipative systems.*

If a dynamical system is specified by its kinetic energy function, potential energy function, and dissipation function, methods similar to those of Chapter VII can be applied in order to determine the nature of the small vibrations of the system about an equilibrium-configuration.

For simplicity we shall consider a system with two degrees of freedom. As in § 76, we find that for the vibrational problem the kinetic energy and dissipation function can be taken as homogeneous quadratic functions of the velocities, and the potential energy as a homogeneous quadratic function of the coordinates, the coefficients in these functions being constants. Taking as coordinates those variables which would be normal coordinates if there were no dissipation function, we can write these three functions in the form

$$\begin{cases} T=\tfrac{1}{2}(\dot{q}_1^2+\dot{q}_2^2), \\ F=\tfrac{1}{2}(a\dot{q}_1^2+2h\dot{q}_1\dot{q}_2+b\dot{q}_2^2), \\ V=\tfrac{1}{2}(\lambda_1 q_1^2+\lambda_2 q_2^2), \end{cases}$$

where λ_1 and λ_2 will be supposed positive, so that the equilibrium would be stable if there were no dissipative forces.

The equations of motion are

$$\frac{d}{dt}\left(\frac{\partial T}{\partial \dot{q}_r}\right)-\frac{\partial T}{\partial q_r}+\frac{\partial F}{\partial \dot{q}_r}+\frac{\partial V}{\partial q_r}=0 \qquad (r=1, 2),$$

or

$$\ddot{q}_1+a\dot{q}_1+h\dot{q}_2+\lambda_1 q_1=0,$$

$$\ddot{q}_2+h\dot{q}_1+b\dot{q}_2+\lambda_2 q_2=0.$$

If we attempt to find a particular solution of these equations in the form

$$q_1=Ae^{pt}, \quad q_2=Be^{pt},$$

on substituting these values in the differential equations we have

$$A(p^2+ap+\lambda_1)+Bhp=0,$$

$$Ahp+B(p^2+bp+\lambda_2)=0,$$

from which it follows that p must be a root of the equation

$$(p^2+ap+\lambda_1)(p^2+bp+\lambda_2)-h^2p^2=0.$$

We shall suppose the dissipative forces to be comparatively small, so that squares of the quantities a, h, b can be neglected; on this supposition, the roots of the last equation are readily found to be

$$p_1 = i\sqrt{\lambda_1} - \tfrac{1}{2}a, \quad p_2 = i\sqrt{\lambda_2} - \tfrac{1}{2}b.$$

Corresponding to the root p_1 we have, from the second of the equations connecting A and B,

$$\frac{B}{A} = \frac{ih\sqrt{\lambda_1}}{\lambda_1 - \lambda_2}.$$

A particular solution of the differential equations is therefore given by

$$\begin{cases} q_1 = (\lambda_1 - \lambda_2)\, e^{-\frac{1}{2}at}(\cos\sqrt{\lambda_1}t + i\sin\sqrt{\lambda_1}t), \\ q_2 = h\sqrt{\lambda_1} e^{-\frac{1}{2}at}(i\cos\sqrt{\lambda_1}t - \sin\sqrt{\lambda_1}t), \end{cases}$$

and a second particular solution is obtained by changing i to $-i$ in these expressions. It follows that two independent real particular solutions of the differential equations are

$$\begin{cases} q_1 = (\lambda_1 - \lambda_2)\, e^{-\frac{1}{2}at}\cos\sqrt{\lambda_1}t \\ q_2 = -h\sqrt{\lambda_1} e^{-\frac{1}{2}at}\sin\sqrt{\lambda_1}t \end{cases} \quad \text{and} \quad \begin{cases} q_1 = (\lambda_1 - \lambda_2)\, e^{-\frac{1}{2}at}\sin\sqrt{\lambda_1}t, \\ q_2 = h\sqrt{\lambda_1} e^{-\frac{1}{2}at}\cos\sqrt{\lambda_1}t, \end{cases}$$

and therefore the most general real solution involving $e^{p_1 t}$ is

$$\begin{cases} q_1 = (\lambda_1 - \lambda_2)\, Ae^{-\frac{1}{2}at}\sin(\sqrt{\lambda_1}t + \epsilon), \\ q_2 = h\sqrt{\lambda_1} Ae^{-\frac{1}{2}at}\sin\left(\sqrt{\lambda_1}t + \frac{\pi}{2} + \epsilon\right), \end{cases}$$

where A and ϵ are real arbitrary constants. This represents one of the normal modes of vibration of the system. Adding to this the corresponding solution in $e^{p_2 t}$, we have finally the general solution of the vibrational problem, namely

$$\begin{cases} q_1 = (\lambda_1 - \lambda_2)\, Ae^{-\frac{1}{2}at}\sin(\sqrt{\lambda_1}t + \epsilon) + h\sqrt{\lambda_2} Be^{-\frac{1}{2}bt}\sin\left(\sqrt{\lambda_2}t + \frac{\pi}{2} + \gamma\right), \\ q_2 = h\sqrt{\lambda_1} Ae^{-\frac{1}{2}at}\sin\left(\sqrt{\lambda_1}t + \frac{\pi}{2} + \epsilon\right) + (\lambda_2 - \lambda_1)\, Be^{-\frac{1}{2}bt}\sin(\sqrt{\lambda_2}t + \gamma), \end{cases}$$

where A, B, ϵ, γ are four constants which must be determined from the initial circumstances of the motion.

Now we suppose the dissipative forces such that energy is being continually lost to the system, so that F is a positive definite form, and therefore a and b are positive. The last equations therefore shew that the vibration gradually dies away, on account of the presence of the factors $e^{-\frac{1}{2}at}$ and $e^{-\frac{1}{2}bt}$: the periods of the normal vibrations are (neglecting squares of a, h, b) the same as if the dissipative forces were absent; and in a normal vibration, the amplitude of oscillation of one of the coordinates is small compared with the amplitude of oscillation of the other coordinate, while the phases of the vibration in the two coordinates at any instant differ by a quarter-period.

A similar analysis leads to corresponding results for systems with more than two degrees of freedom; supposing that the dissipative forces are small and that the dissipation function and potential energy are positive definite forms, we find that the periods of the normal vibrations are (neglecting squares of the coefficients in the dissipation function) unaltered by the presence of the dissipative forces, but that the vibration gradually dies away: and if $(q_1, q_2, \ldots, q_n)$ are the normal coordinates of the system when the dissipative forces are absent, there is a normal vibration of the system when the dissipative forces are present, in which the amplitude of the vibrations in $q_2, q_3, \ldots, q_n$ is small compared with the amplitude of the vibration in q_1, and the phase of the vibrations in $q_2, q_3, \ldots, q_n$ differs by a quarter-period from the phase of the vibration in q_1.

Example. Discuss the vibrations of a system which is acted on by periodic external forces which have the same period as one of the normal modes of free vibration of the system; shewing the importance of dissipative forces (even where small) in this case.

95. *Impact.*

Another mode in which energy may be lost* to a dynamical system is by the *collision* of bodies which belong to the system; a collision generally results in a decrease of dynamical energy.

The analytical discussion of collisions is based on the following experimental law†. *When two bodies collide, the values of the relative velocity of the surfaces in contact (estimated normally to the surfaces) at instants immediately after and immediately before the impact bear a definite ratio to each other: this ratio depends only on the material of which the bodies are composed.*

This ratio will in general be denoted by $-e$. When e is zero, the bodies are said to be *inelastic*.

The general problem of impact reduces therefore to a problem in impulsive motion in which the unknown impulsive force at the point of contact of the bodies is to be determined by the condition that the change in relative normal velocity of the bodies satisfies the above law.

96. *Loss of kinetic energy in impact.*

We shall now find the loss of kinetic energy when two perfectly smooth bodies impinge on each other.

Let m typify the mass of a particle of either body, and let (u_0, v_0, w_0) and (u, v, w) denote its components of velocity before and after the impact, and

* I.e. lost to the system considered as a dynamical system: the energy is not annihilated, but appears in some other manifestation, e.g. heat.

† The laws of impact were discovered in 1668 by John Wallis (*Phil. Trans.* No. 43, p. 864) and Christopher Wren (*ibid.* p. 867).

let (U, V, W) be the components of the total impulsive force (external and molecular) on this particle. The equations of impulsive motion (§ 35) give

$$m(u-u_0)=U, \quad m(v-v_0)=V, \quad m(w-w_0)=W.$$

Multiplying these equations by $(u+eu_0)$, $(v+ev_0)$, $(w+ew_0)$ respectively, adding, and summing for all the particles of both bodies, we have

$$\Sigma m\{(u-u_0)(u+eu_0)+(v-v_0)(v+ev_0)+(w-w_0)(w+ew_0)\}$$
$$=\Sigma\{U(u+eu_0)+V(v+ev_0)+W(w+ew_0)\}.$$

Now so far as molecular impulses are concerned, we have

$$\Sigma(Uu+Vv+Ww)=0, \quad \text{and} \quad \Sigma(Uu_0+Vv_0+Ww_0)=0,$$

since the impulsive forces which correspond to each other in virtue of the law of Action and Reaction will give contributions to these sums which mutually destroy each other.

Also, since the part of $(u+eu_0)$ due to the normal component of velocity has the same value for each of the particles in contact at the point where the impact takes place (in virtue of the law of impact) it follows that the impulsive force between the bodies does not contribute to the sum $\Sigma U(u+eu_0)$, and similarly does not contribute to the sums $\Sigma V(v+ev_0)$ and $\Sigma W(w+ew_0)$.

We have therefore

$$\Sigma\{U(u+eu_0)+V(v+ev_0)+W(w+ew_0)\}=0,$$

and consequently

$$\Sigma m\{(u-u_0)(u+eu_0)+(v-v_0)(v+ev_0)+(w-w_0)(w+ew_0)\}=0,$$

or

$$\Sigma m(u^2+v^2+w^2)-\Sigma m(u_0{}^2+v_0{}^2+w_0{}^2)$$
$$=-\frac{1-e}{1+e}\Sigma m\{(u-u_0)^2+(v-v_0)^2+(w-w_0)^2\}.$$

This equation can be expressed by the statement that ***the kinetic energy lost in the impact is*** $(1-e)/(1+e)$ ***times the kinetic energy of that motion which would have to be compounded with the motion at the instant before the impact in order to produce the motion at the instant after the impact.***

97. *Examples of impact.*

The impulsive change of motion consequent on the collision of two free rigid bodies in space can be most simply determined by the following considerations.

The motion of each body before or after impact is specified by six quantities (e.g. the three components of velocity of its centre of gravity and the three components of angular velocity of the body about axes through its centre of gravity). The total number of equations required to determine the impulsive change of motion is therefore twelve. Of these, six are immediately furnished by the condition that the angular momentum of each body about any axis through the point of contact is unchanged (since the impulsive forces act at this point); another equation is obtained from the condition that the momentum of the system in the direction normal to the surfaces in contact is unchanged (since the normal impulsive forces on the two bodies at the point of contact are equal and opposite), and another by the experimental law of impact. If the bodies are perfectly smooth, the remaining four equations can be derived from the condition that the linear momentum of each body in any direction tangential to the surfaces in contact is unchanged (since there is no tangential impulse if the bodies are smooth): if on the other hand the bodies are perfectly or imperfectly rough, the condition that the linear momentum of the system in any direction tangential to the surfaces in contact is unchanged gives two equations; if the bodies are perfectly rough, the condition that the relative velocity of the bodies in any tangential direction after the impact is zero gives the other two: while if the bodies are imperfectly rough, the coefficient of friction between the surfaces in contact being μ, the remaining two equations are given by the conditions that

(α) the relative velocity in any tangential direction is zero after the impact, provided the tangential component of the impulse required for this does not exceed μ times the normal component of the impulse;

(β) if the last condition is not satisfied, there is a tangential impulse equal to μ times the normal impulse between the bodies.

In all cases, therefore, the required twelve equations can be found.

If the motion takes place in a plane, or if one of the bodies is fixed, this procedure is still valid after making some obvious modifications.

The following examples illustrate these principles:

Example 1. *An inelastic sphere of mass m falls with velocity V on a perfectly rough inelastic inclined plane of mass M and angle a, which rests on a smooth horizontal plane. Shew that the vertical velocity of the centre of the sphere immediately after the impact is*

$$\frac{5(M+m)\,V\sin^2 a}{7M+2m+5m\sin^2 a}.\qquad\text{(Coll. Exam.)}$$

Let U be the velocity of the plane after impact, u the velocity of the sphere parallel to and relative to the plane, ω the angular velocity of the sphere, and a its radius.

The equation of horizontal momentum gives

$$m(u\cos a-U)=MU.$$

The kinematical condition at the point of contact is $a\omega=u$.

The condition that the angular momentum of the sphere about the point of contact shall be the same before and after impact is

$$mVa\sin a=\tfrac{2}{5}ma^2\omega+ma(u-U\cos a).$$

These three equations give, on eliminating ω and U,

$$u \sin \alpha = \frac{5(M+m) V \sin^2 \alpha}{7M + 2m + 5m \sin^2 \alpha},$$

which is the result stated.

Example 2. A sphere of radius a rotating with angular velocity Ω about an axis inclined at an angle α to the vertical and moving, in the vertical plane containing that axis, with velocity V in a direction making an angle α with the horizon, strikes a perfectly rough horizontal plane. If the plane be tangentially inelastic, find the angle which the vertical plane containing the new direction of motion makes with the old.

Take rectangular axes $Oxyz$, where O is the point of contact, Oz is vertical, and yOz is the initial plane of motion; and let ω_1 and ω_2 be the components of angular velocity about Ox and Oy respectively after the impact, and M the mass of the sphere.

Equating the initial and final angular momenta about Ox, we have

$$MaV \cos \alpha = \tfrac{7}{5} Ma^2 \omega_1.$$

Equating the initial and final angular momenta about Oy, we have

$$\tfrac{2}{5} Ma^2 \Omega \sin \alpha = \tfrac{7}{5} Ma^2 \omega_2.$$

The tangent of the inclination of the new plane of motion to the plane yOz is (on account of the perfect roughness of the plane) ω_2/ω_1, and this is therefore equal to

$$\frac{\tfrac{2}{5} Ma^2 \Omega \sin \alpha}{MaV \cos \alpha}$$

or

$$\tfrac{2}{5} a (\Omega / V) \tan \alpha.$$

Example 3. A perfectly rough circular disc of mass M and radius c impinges upon a rod of mass m and length 2a capable of turning freely about a pivot at its centre. If the point of impact is distant b from the centre of the rod, and the direction of motion of the centre of the disc makes angles α, β with the rod before and after collision, shew that

$$2(3Mb^2 + ma^2) \tan \beta = 3(ema^2 - 3Mb^2) \tan \alpha. \qquad \text{(Coll. Exam.)}$$

Let V denote the initial velocity of the disc, and let v denote its final velocity and Ω its final angular velocity.

Since there is no sliding at the point of contact, we have

$$v \cos \beta + c\Omega = 0.$$

Denoting by ω the final angular velocity of the rod, and by I the normal impulse between the rod and disc, the equation of the motion of the rod is

$$Ib = \tfrac{1}{3} ma^2 \omega.$$

The equation of impulsive motion of the disc in the direction normal to the rod is

$$M(v \sin \beta + V \sin \alpha) = I,$$

and the law of impact gives the relation

$$v \sin \beta + b\omega = eV \sin \alpha.$$

Equating the initial and final angular momenta of the disc about the point of contact, we have

$$V \cos \alpha = v \cos \beta - \tfrac{1}{2} c\Omega.$$

Eliminating v, Ω, I, ω from these equations, we have

$$2 \tan \beta (3Mb^2 + ma^2) = 3 \tan \alpha (mea^2 - 3Mb^2),$$

which is the result stated.

Example 4. *A circular hoop, in motion without rotation in its own plane, impinges on a rough fixed straight-edged obstacle in the plane. The velocity of the centre of the hoop before impact is V, in a direction making an angle a with the edge, and the coefficient of friction is* μ. *To find the impulsive change of motion.*

Let u and v denote the components of velocity of the centre of the hoop after the impact, parallel and perpendicular to the edge, and let ω be the angular velocity, M the mass and a the radius of the hoop.

Equating the angular momenta about the point of contact before and after the impact, we have

$$-Ma^2\omega + Mau = MVa\cos\alpha.$$

The law of impact gives the equation

$$v = eV\sin\alpha.$$

Since the plane is rough, $u + a\omega$ is zero after the impact, provided the frictional impulse required for this does not exceed μ times the normal impulse: but if this condition is not satisfied, the frictional impulse is μ times the normal impulse.

Let F be the frictional and R the normal impulse: then we have

$$M(u - V\cos\alpha) = -F, \quad M(v + V\sin\alpha) = R, \quad Ma^2\omega = -aF.$$

We have therefore $\qquad R = M(1+e)\,V\sin\alpha,$

and if $u + a\omega$ is zero, we shall have

$$F = \tfrac{1}{2}MV\cos\alpha.$$

The quantity $u + a\omega$ will therefore be zero after the impact, provided

$$\mu \geqslant \cot\alpha/2\,(1+e);$$

and if μ does not satisfy this inequality, we shall have

$$F = \mu M(1+e)\,V\sin\alpha.$$

Thus finally, if $\mu \geqslant \cot\alpha/2\,(1+e)$, the motion is determined by the equations

$$u = V\cos\alpha + a\omega, \quad v = eV\sin\alpha, \quad u + a\omega = 0,$$

while if $\mu < \cot\alpha/2\,(1+e)$, the motion is determined by the equations

$$u = V\cos\alpha + a\omega, \quad v = eV\sin\alpha, \quad a\omega = -\mu(1+e)\,V\sin\alpha.$$

MISCELLANEOUS EXAMPLES.

1. A perfectly rough sphere of radius a is made to rotate about a vertical diameter, which is fixed, with a constant angular velocity n. A uniform sphere of radius b is placed on it at a point distant $a\alpha$ from the highest point: investigate the motion and determine in any position the angular velocity of the sphere. Shew that the sphere will leave the rotating sphere when the point of contact is at an angular distance θ from the vertex, where

$$\cos\theta = \frac{10}{17}\cos\alpha + \frac{4}{119}\,\frac{a^2n^2\sin^2\alpha}{(a+b)\,g}.$$

(Camb. Math. Tripos, Part I, 1889.)

2. A rough sphere of radius a rolls under gravity on the surface of a cone of revolution which is compelled to turn about its vertical axis with uniform angular velocity n, its vertex being uppermost; if α be the semi-vertical angle of the cone, $r\sin\alpha$ be the distance of the centre of the sphere from the axis of the cone, ψ be the angle turned

through, relatively to the cone, by the vertical plane containing the centre of the sphere, and ω_3 be the rate of rotation of the sphere about the common normal, prove that

$$7\dot{r}^2+\frac{2+5\sin^2\alpha}{49}\left(\frac{A}{r}+nr+B\right)^2-10gr\cos\alpha=C,$$

$$a(\omega_3-n\sin\alpha)=\frac{\cos\alpha}{7}\left(\frac{A}{r}+nr\right)+\frac{(2+5\sin^2\alpha)B}{14\cos\alpha},$$

$$(7\dot{\psi}-6n)r^2=A,$$

where A, B, C are determinate constants. (Camb. Math. Tripos, Part I, 1897.)

3 A homogeneous solid of revolution of mass M with a plane circular base of radius c rolls without slipping with its edge in contact with a rough horizontal plane. Shew that θ, ω, Ω are determined by the equations

$$Mac\frac{d}{d\theta}(\Omega\cos^2\theta)-Mc^2\Omega\cos^2\theta=(C+Mc^2)\cos\theta\frac{d\omega}{d\theta},$$

$$\{A(C+Mc^2)-M^2a^2c^2\}\frac{d}{d\theta}(\Omega\cos^2\theta)+C(C+Mc^2)\,\omega\cos\theta-Mac\,C\Omega\cos^2\theta=0,$$

$$(A+Mc^2)\,\dot{\theta}^2+A\Omega^2\cos^2\theta-2Mac\,\omega\Omega\cos\theta+(C+Mc^2)\,\omega^2+2Mg\,(a\sin\theta+c\cos\theta)=\text{Constant},$$

where θ is the inclination of the axis of the body to the horizon, Ω the angular velocity of the vertical plane containing its axis, ω the angular velocity of the body about its axis, A the moment of inertia of the body about a diameter of its base, C the moment of inertia of the body about its axis and a the distance of the centre of gravity from the base. (Camb. Math. Tripos, Part I, 1898.)

4. A wheel with $4n$ spokes arranged symmetrically rolls with its axis horizontal on a perfectly rough horizontal plane. If the wheel and spokes be made of a fine heavy wire, prove that the condition for stability is

$$V^2>\frac{3}{4}\,\frac{2n+\pi}{4n+3\pi}\,ga,$$

where a is the radius of the wheel and V its velocity. (Coll. Exam.)

5. A body rolls under gravity on a fixed horizontal plane. If this plane be taken as plane of yz, shew that

$$\Sigma m\,\{(y-y_A)\,\dot{z}-(z-z_A)\,\dot{y}\}=\text{Constant},$$

where (x, y, z) are the coordinates of a particle m and (x_A, y_A, z_A) of the point of contact, and the summation is extended over all the particles of the body. (Neumann.)

6. One portion of a horizontal plane is perfectly smooth and the other portion is perfectly rough. A uniform heavy ellipsoid of semi-axes (a, b, c) has its b-axis vertical and moves with velocity v in the direction of its a-axis along the smooth portion of the plane towards the rough. Shew that, if

$$v^2<2g\,\frac{b^2+k^2}{b^2}\,(a-b),$$

the ellipsoid will return to the smooth portion, k being the radius of gyration about the c-axis, and that the motion will then consist of an oscillation about a steady state of motion.

In the special case $a=2b$, shew that after the return of the ellipsoid to the smooth portion, the b-axis can never make an angle with the vertical which is greater than $\arctan\sqrt{\frac{5}{7}}$. (Coll. Exam.)

7. A shell in the form of a prolate spheroid whose centre of gravity is at its centre contains a symmetrical gyrostat, which rotates with angular velocity ω about its axis and whose centre and axis coincide with those of the spheroid. Shew that in the steady motion of the spheroid on a perfectly rough horizontal plane, when its centre describes a circle of radius c with angular velocity Ω, the inclination α of the axis to the vertical is given by

$$\{Mbc\,(a\cot\alpha+b)-Ab\cos\alpha+C\,(a\sin\alpha+c)\}\,\Omega^2+C'b\omega\Omega-Mgb\,(a-b\cot\alpha)=0,$$

where M is the mass of the shell and gyrostat, A the moment of inertia of the shell and gyrostat together about a line through their centre perpendicular to their axis, C, C' those of the shell and gyrostat respectively about the axis, a the distance measured parallel to the axis of the point of contact of the shell and plane from the centre and b its distance from the axis. (Camb. Math. Tripos, Part I, 1899

8. A uniform perfectly rough sphere of radius a starting from rest rolls down under gravity between two non-intersecting straight rods at right angles to each other whose shortest distance apart is $2c$ and which are equally inclined at an angle α to the vertical. If ρ_0, ρ_0' are the original distances of the points of contact from the points where the shortest distance intersects the rods and ρ, ρ' their distances at a subsequent time when the velocity is V, shew that

$$V^2=16c^2\,\frac{\{\rho^2\rho'^2-a^2\,(\rho^2+\rho'^2)\}}{16c^4-(\rho^2-\rho'^2)^2}\,\frac{\dot\rho\dot\rho'}{\rho\rho'},$$

and that

$$V^2\left\{\frac{28a^2-20c^2-5\rho^2-5\rho'^2}{4a^2-4c^2-\rho^2-\rho'^2}\right\}=10g\left\{(\rho-\rho_0+\rho'-\rho_0')\cos\alpha+\frac{1}{4c}(\rho^2-\rho_0^{\,2}-\rho'^2+\rho_0'^2)\sqrt{-\cos 2\alpha}\right\}.$$

(Camb. Math. Tripos, Part I, 1889.)

9. A particle moves under gravity on a rough helix whose axis is vertical. If a be the radius and γ the angle of the helix, shew that the velocity v and arc described s can be expressed in terms of a parameter θ by the equations

$$\left\{\begin{aligned}-\frac{2}{a}\cos\gamma\,.\,s&=\int\frac{(1+\theta^2)\,d\theta}{\theta\,\{\mu\cos\gamma+\theta\,(\mu\cos\gamma+2\sin\gamma)\}},\\ v^2&=\frac{ga}{2\cos\gamma}\left(\theta-\frac{1}{\theta}\right).\end{aligned}\right.$$

10. A particle is projected horizontally with velocity u so as to slide on a rough inclined plane. Investigate the motion.

Prove that if

$$2\geqslant 2\mu\cot\alpha>1,$$

the particle approaches asymptotically a line of greatest slope at distance

$$\frac{u^2}{g}\,.\,\frac{2\mu\cos\alpha}{4\mu^2\cos^2\alpha-\sin^2\alpha},$$

where μ is the coefficient of friction, and α is the inclination of the plane.

(Coll. Exam.)

11. A rough cycloidal tube has its axis vertical and vertex uppermost. If a be the radius of the generating circle and a particle be projected from the vertex with velocity $\sqrt{4ag}\sin\alpha$, shew that it will reach the cusp with velocity equal to

$$[4ag\cos^2\alpha\,\{1-2\sin\alpha\,e^{-(\frac{1}{2}\pi-\alpha)\tan\alpha}\}]^{\frac{1}{2}},$$

where α is the angle of friction. (Coll. Exam.)

12. A heavy rod of length $2a$ is moving in a vertical plane so that one end is in contact with a rough vertical wall and the other end moves along the ground supposed to be equally rough; and the coefficient of friction for each of the rough surfaces is $\tan \epsilon$. Shew that the inclination of the rod to the vertical at any time is given by

$$\ddot{\theta}(k^2+a^2\cos 2\epsilon)-a^2\dot{\theta}^2\sin 2\epsilon=ag\sin(\theta-2\epsilon).$$ (Coll. Exam.)

13. A thin spherical shell rests upon a horizontal plane and contains a particle of finite mass which is initially at its lowest point. The coefficient of friction between the particle and the shell is given, that between the shell and the plane being practically infinite. Motion in two dimensions is set up by applying to the shell an impulse which gives it an angular velocity Ω. Obtain an equation for the angle through which the shell has rolled when the particle begins to slip. (Coll. Exam.)

14. A circular disc of radius a is placed in a vertical plane touching a uniform rough (μ) board which can turn freely about a horizontal axis in the upper surface of the board through its centre of gravity, the point of contact of the disc being at a distance b from this axis. A string, parallel to the surface of the board, is attached to the point of the disc furthest from the board and to an arm perpendicular to the board at the axis, and rigidly connected to the board. The centre of gravity of the board and arm lies in the axis. The system starts from rest in that position in which the centre of the disc lies in the horizontal plane through the axis. Shew that slipping will take place between the disc and the board, when the board makes an angle θ with the vertical given by

$$\tan\theta=\frac{A+a^2+6\mu ab+3b^2}{2\mu A+7\mu a^2+\frac{5}{2}ab},$$

where A is the moment of inertia of the board about the axis divided by the mass of the disc. (Coll. Exam.)

15. A hoop is projected with velocity V down a plane of inclination α, the coefficient of friction being μ ($>\tan\alpha$). It has initially such a backward spin Ω that after a time t_1 it starts moving uphill and continues to do so for a time t_2, after which it once more descends. Shew that, if the motion take place in a vertical plane at right angles to the given inclined plane, then

$$(t_1+t_2)g\sin\alpha=a\Omega-V.$$ (Coll. Exam.)

16. A ring of radius a is fixed on a smooth horizontal table; a second ring is placed on the table inside the first and in contact with it, and is projected with velocity V, but without rotation, in a direction parallel to the tangent at the point of contact. Find the time that elapses before slipping ceases between the rings if the coefficient of friction between them is μ, and prove that the point of contact will in this time describe an arc of length $(a\log 2)/\mu$.

Discuss the motion that will ensue if at the moment slipping ceases the fixed ring be released and left free to move, and prove that during the time that the inner ring rolls half round the outer one the centre of the latter will be displaced a distance

$$\frac{m}{M+m}(a-b)(\pi^2+4)^{\frac{1}{2}},$$

where m, M are the masses of the inner and outer rings and b is the radius of the inner ring. (Camb. Math. Tripos, Part I, 1900.)

17. In the vertical motion of a heavy particle descending in a medium whose resistance varies as the square of the velocity, shew that the quantity

$$e^{-k\alpha}+e^{k\beta},$$

where kv^2 is the resistance, and α and β are the distances described in two successive equal intervals τ of time, depends only on τ and is independent of the initial velocity. (Coll. Exam.)

18. Prove that a heavy particle, let fall from rest in a medium in which the resistance varies as the square of the velocity, will acquire a velocity $U \tanh(gt/U)$, and describe a space $U^2 \log\cosh(gt/U)/g$ in a time t, where U denotes the terminal velocity in the medium.

Shew also that, for the complete trajectory of a projectile in such a medium, the angle θ between the asymptotes is given by

$$U^2/V^2 = \text{arcsinh} \cot\theta + \cot\theta \operatorname{cosec}\theta,$$

where V is the velocity when the projectile moves horizontally. (Coll. Exam.)

19. Shew that the horizontal and vertical coordinates (x, y) of a particle moving under gravity in a medium of which the resistance is R satisfy the equation

$$\frac{d^3y}{dx^3} + \frac{2gR}{v^4 \cos^3\phi} = 0,$$

v being the velocity and ϕ the inclination of the tangent to the horizontal. (Coll. Exam.)

20. A particle is moving, under gravity, in a medium in which the resistance varies as the velocity. Shew that the equation of the trajectory referred to the vertical asymptote and a line parallel to the direction of motion when the velocity was infinite, can be written in the form

$$y = b \log(x/a). \qquad \text{(Coll Exam.)}$$

21. Prove that in the motion of a projectile through a resisting medium which causes a retardation kv^3, where k is very small and the particle is projected horizontally with velocity V, the approximate equation of the path is (neglecting k^2)

$$y = \frac{gx^2}{2V^2} + \frac{kgx^3}{3V}\left(1 + \frac{g^2x^2}{10V^4}\right),$$

the axis of x being in the direction of projection and the axis of y vertically downwards. (Coll. Exam.)

22. A particle moves in a straight line under no forces in a medium whose resistance is $(v^2 - v^3 \log s)/s$, where v is the velocity and s the distance from a given point in the line. Shew that the connexion between s and t is given by an equation of the form

$$t = a + \tfrac{1}{2}cs^2 + s\log s,$$

where a and c are constants.

23. A particle is moving in a resisting medium under a central attraction; shew that, if R be the retardation due to the resistance of the medium, and v the velocity, the rate of description of areas by the radius vector to the fixed centre of force varies as

$$e^{-\int \frac{R}{v} dt}. \qquad \text{(Coll. Exam.)}$$

24. Prove that in a resisting medium, a particle can describe a parabola under the action of a force to the focus which varies as the distance, provided the resistance at a point, where the velocity is v, be $k\{v(v - v_0)\}^{\frac{3}{2}}$; where v_0 is the velocity at the vertex. Determine k. (Coll. Exam.)

25. A particle moves in a resisting medium under a force P tending to a fixed centre. If R be the resistance, shew that

$$\frac{d}{ds}\left\{Pp^3 \frac{dr}{dp}\right\} = -2Rp^2,$$

r being the radius vector and p the perpendicular on the tangent.

If $u=1/r$, $P=\mu u^2$, and $R=kv^2$, and we neglect k^2 and higher powers, shew that the differential equation to the path is

$$p\frac{d^2p}{du^2}-3\left(\frac{dp}{du}\right)^2=\frac{2\mu k}{h^2}\frac{u^2}{(1-p^2u^2)^{\frac{1}{2}}},$$

h being a certain constant. (Coll. Exam.)

26. A particle is moving under a central force $\phi(r)$ repelling it from the origin, in a resisting medium which imposes a retarding force equal to k times the velocity. Prove that the orbit is given by the equations

$$r^2\dot{\theta}=he^{-kt},\quad \ddot{r}+k\dot{r}-h^2r^{-3}e^{-2kt}=\phi(r),$$

where h is a constant quantity. (Coll. Exam.)

27. A particle is moving in a circle under a force of attraction to an interior point varying as the distance; the resistance of the medium is equal to its density multiplied by the square of the velocity. Shew that the density at any point is proportional to the tangent of the angle between the lines joining it to the centre of force and the centre of the circle. (Coll. Exam.)

28. A rod of length a is rotating about one extremity, which is fixed, under the action of no forces except the resistance of the atmosphere. Supposing the retarding effect of the resistance on a small element of length dx to be $Adx\,.\,(\text{velocity})^2$, shew that the angular velocity at the time t is given by

$$\frac{1}{\omega}-\frac{1}{\Omega}=\frac{Aa^4}{4Mk^2}t,$$

where Mk^2 is the moment of inertia about the fixed extremity, and Ω is a constant. (Coll. Exam.)

29. A smooth oval disc of mass M, turning on a smooth horizontal table with angular velocity ω but without any translational velocity, strikes a smooth horizontal rod of mass m at its middle point. Prove that the angular velocity is diminished in the ratio

$$(M+m)k^2-mex^2 : (M+m)k^2+mx^2,$$

where e is the coefficient of elasticity, x the distance of the centre of gravity from the normal at the point of impact and k the radius of gyration about a vertical axis through the centre of gravity. (Coll. Exam.)

30. Two rods, each of length a and mass m, are jointed together at their upper ends and the system falls symmetrically, with its plane vertical, on to a smooth inelastic plane. Just before impact the joint has a velocity V and each rod has an angular velocity Ω, tending to increase its inclination α to the horizon. Shew that the impulse between each rod and the plane is

$$m(k^2+c^2\sin^2\alpha)(V+a\Omega\cos\alpha)/\{k^2+c^2+a(a-2c)\cos^2\alpha\},$$

where c is the distance of the centre of gravity of each rod from the joint and mk^2 is the moment of inertia of each rod about its centre of gravity. (Coll. Exam.)

31. Three equal uniform rods AB, BC, CD, each of length $2a$, and hinged at B and C, are in one straight line and moving with a given velocity in a horizontal plane at right angles to their lengths. The ends A and D meet simultaneously two fixed inelastic obstacles, reducing A and D to rest. Determine when they will form an equilateral triangle, and shew that $\frac{1}{5}$ of the original momentum is destroyed by the impacts. (Coll. Exam.)

32. A smooth uniform cube is free to turn about a horizontal axis passing through the centres of two opposite faces and is at rest with two faces horizontal; an equal and similar cube is dropped with velocity u and without rotation so as to strike the former along a line parallel to the fixed axis and at a distance c from the vertical plane containing it, prove that the angular velocity imparted to the lower cube is

$$\frac{(1+e)\,cu}{c^2+k^2+a^2(1-\sin 2\alpha)},$$

where α is the inclination to the horizon of the lower face of the falling cube, $2a$ is the length of an edge, k the radius of gyration and e the coefficient of restitution.

Find also the motion of the upper cube immediately after the impact.

(Coll. Exam.)

33. A perfectly elastic circular disc of mass M and radius c impinges without rotation upon a rod of mass m and length $2a$ which is free to turn about a pivot at its centre, the point of impact being at a distance b from the pivot. Prove that if the component of the velocity of the centre of the disc normal to the rod be halved by the impact, $Mb^2=ma^2$, the friction being sufficient to prevent sliding. (Coll. Exam.)

34. A perfectly rough sphere of radius a is projected horizontally with a velocity V from a point at a height h above a horizontal plane. The sphere has also initially an angular velocity Ω about its horizontal diameter perpendicular to the plane of its motion. Shew that before it ceases to bound on the plane it passes over a horizontal distance

$$\frac{2\sqrt{2}}{7}\cdot\sqrt{\frac{h}{g}}\cdot\frac{e}{1-e}(5V+2a\Omega),$$

where e is the coefficient of elasticity, and the distance is reckoned from the first point of contact.

Compare the final with the initial kinetic energy. (Coll. Exam.)

35. A homogeneous elastic sphere (coefficient of elasticity e) is projected against a perfectly rough vertical wall so that its centre moves in a vertical plane at right angles to the wall. If the initial components of the velocity of its centre are u and v, and its initial angular velocity (Ω) is about an axis perpendicular to the vertical plane, find the subsequent motion after impinging on the wall, and shew that if its centre returns to its original position the coordinates of the point of impact referred to this point are

$$\frac{2eu}{g}\cdot\frac{(7e+5)\,v+2a\Omega}{7+10e+7e^2}+a,$$

and
$$\frac{2e}{g}\,\frac{\{(7e+5)\,v+2a\Omega\}\,\{v\,(7+5e)-2ae\Omega\}}{(7+10e+7e^2)^2},$$

where a is the radius of the sphere. (Coll. Exam.)

CHAPTER IX

THE PRINCIPLES OF LEAST ACTION AND LEAST CURVATURE

98. *The trajectories of a dynamical system.*

The chief object of investigation in Dynamics is the gradual change in time of the coordinates $(q_1, q_2, \ldots, q_n)$ which specify the configuration of a dynamical system. When the system has three (or less than three) degrees of freedom, there is often a gain in clearness when we avail ourselves of a geometrical representation of the problem: if a point be taken whose rectangular coordinates referred to fixed axes are the coordinates (q_1, q_2, q_3) of the given dynamical system, the path of this point in space can be regarded as illustrating the successive states of the system. In the same way when $n > 3$ we can still regard the motion of the system as represented by the path of a point whose coordinates are $(q_1, q_2, \ldots, q_n)$ in space of n dimensions; this path is called the *trajectory* of the system, and its introduction makes it natural to use geometrical terms such as "intersection," "adjacent," etc., when speaking of the relations of different states or types of motion in the system.

99. *Hamilton's principle, for conservative holonomic systems.*

Consider any conservative holonomic dynamical system whose configuration at any instant is specified by n independent coordinates $(q_1, q_2, \ldots, q_n)$, and let L be the kinetic potential which characterises its motion. Let a given arc AB in space of n dimensions represent part of a trajectory of the system, and let CD be part of an adjacent arc which is not necessarily a trajectory: it would however of course be possible to make CD a trajectory by subjecting the system to additional constraints. Let t be the time at which the representative point $(q_1, q_2, \ldots, q_n)$ occupies any position P on AB: we shall suppose each point on CD correlated to some value of the time, so that there will be a point Q on CD (or on the arc of which CD is a portion) which corresponds to the same value t as P does. As the arc CD is described, the correlated value of t will be supposed to vary continuously in the same sense. A moving point which describes the arc CD will therefore pass through positions corresponding to a continuous sequence of values of $q_1, q_2, \ldots, q_n, t$, and consequently to each point on CD there will correspond a set of values of $\dot{q}_1, \dot{q}_2, \ldots, \dot{q}_n$.

We shall denote by δ the variation by which we pass from a point of AB to that point of CD which is correlated to the same value of the time, and shall denote by t_0, t_1, $t_0+\Delta t_0$, $t_1+\Delta t_1$ the values of t which correspond to the terminal points A, B, C, D respectively, and by L_R the value of the function L at any point R of either arc.

If now we form the difference of the values of the integral

$$\int L(\dot{q}_1, \dot{q}_2, \ldots, \dot{q}_n, q_1, q_2, \ldots, q_n, t)\, dt,$$

taken along the arcs AB and CD respectively, we have

$$\begin{aligned}\int_{CD} Ldt - \int_{AB} Ldt &= L_B\Delta t_1 - L_A\Delta t_0 + \int_{t_0}^{t_1} \delta L dt \\ &= L_B\Delta t_1 - L_A\Delta t_0 + \int_{t_0}^{t_1} \sum_{r=1}^{n} \left(\frac{\partial L}{\partial \dot{q}_r}\delta\dot{q}_r + \frac{\partial L}{\partial q_r}\delta q_r\right) dt \\ &= L_B\Delta t_1 - L_A\Delta t_0 + \int_{t_0}^{t_1} \sum_{r=1}^{n} \left\{\frac{\partial L}{\partial \dot{q}_r}\delta\dot{q}_r + \frac{d}{dt}\left(\frac{\partial L}{\partial \dot{q}_r}\right)\delta q_r\right\} dt,\end{aligned}$$

by Lagrange's equations,

$$\begin{aligned}&= L_B\Delta t_1 - L_A\Delta t_0 + \int_{t_0}^{t_1} \frac{d}{dt}\left(\sum_{r=1}^{n} \frac{\partial L}{\partial \dot{q}_r}\delta q_r\right) dt \\ &= L_B\Delta t_1 - L_A\Delta t_0 + \left(\sum_{r=1}^{n} \frac{\partial L}{\partial \dot{q}_r}\delta q_r\right)_B - \left(\sum_{r=1}^{n} \frac{\partial L}{\partial \dot{q}_r}\delta q_r\right)_A.\end{aligned}$$

But if $(\Delta q_r)_B$ denote the increment of q_r in passing from B to D, we have

$$(\Delta q_r)_B = (\delta q_r)_B + (\dot{q}_r)_B\Delta t_1,$$

and similarly if $(\Delta q_r)_A$ denote the increment of q_r in passing from A to C, we have

$$(\Delta q_r)_A = (\delta q_r)_A + (\dot{q}_r)_A\Delta t_0,$$

and consequently

$$\int_{CD} Ldt - \int_{AB} Ldt = \left[\sum_{r=1}^{n} \frac{\partial L}{\partial \dot{q}_r}\Delta q_r + \left(L - \sum_{r=1}^{n} \frac{\partial L}{\partial \dot{q}_r}\dot{q}_r\right)\Delta t\right]_A^B.$$

Suppose now that C coincides with A, and D coincides with B, and that the times correlated to C and D are t_0 and t_1 respectively, so that Δq_1, Δq_2, ..., Δq_n, Δt, are zero at A and B: then the last equation becomes

$$\int_{CD} Ldt - \int_{AB} Ldt = 0,$$

which shews that *the integral* $\int Ldt$ *has a stationary value for any part of an actual trajectory AB, as compared with neighbouring paths CD which have the same terminal points as the actual trajectory and for which the time has the same terminal values.* This result is called *Hamilton's principle**.

* Hamilton, *Phil. Trans.* 1834, p. 307; *ibid.* 1835, p. 95.

If the kinetic potential L does not contain the time explicitly, we can evidently replace the condition that the time is to have the same terminal values by the condition that the total time of description is to be the same for AB as for CD, since $\sum_{r=1}^{n} \dot{q}_r \frac{\partial L}{\partial \dot{q}_r} - L$, which represents the total energy of the system, is in this case constant.

Helmholtz, *J. für Math.* c. p. 151, remarked that the conditions for a stationary value of

$$\int \left\{ L(\theta_1, \theta_2, \ldots, \theta_n, q_1, \ldots, q_n) + \sum_r \frac{\partial L}{\partial \theta_r} (\dot{q}_r - \theta_r) \right\} dt$$

(where the q's and θ's are regarded as independent variables) are

$$\theta_r = \dot{q}_r, \qquad 0 = \frac{\partial L}{\partial q_r} - \frac{d}{dt}\left(\frac{\partial L}{\partial \theta_r}\right), \qquad (r = 1, 2, \ldots, n),$$

so that we again obtain Lagrange's equations.

100. *The principle of Least Action for conservative holonomic systems.*

Suppose now that the dynamical system considered is such that the kinetic potential does not involve the time explicitly, so that the integral of energy

$$\sum_{r=1}^{n} \dot{q}_r \frac{\partial L}{\partial \dot{q}_r} - L = h$$

exists. Taking as before AB to be part of a trajectory and CD to be part of any adjacent arc, to the successive points of which values of the time are so correlated as to satisfy an equation of the form

$$\sum_{r=1}^{n} \dot{q}_r \frac{\partial L}{\partial \dot{q}_r} - L = h + \Delta h,$$

where Δh is a small constant, we have

$$\int_{CD} \left(\sum_{r=1}^{n} \dot{q}_r \frac{\partial L}{\partial \dot{q}_r} \right) dt - \int_{AB} \left(\sum_{r=1}^{n} \dot{q}_r \frac{\partial L}{\partial \dot{q}_r} \right) dt$$

$$= \int_{CD} (h + \Delta h)\, dt - \int_{AB} h dt + \int_{CD} L dt - \int_{AB} L dt$$

$$= (h + \Delta h)(t_1 + \Delta t_1 - t_0 - \Delta t_0) - h(t_1 - t_0) + \left[\sum_{r=1}^{n} \frac{\partial L}{\partial \dot{q}_r} \Delta q_r - h \Delta t \right]_A^B$$

$$= \left[\sum_{r=1}^{n} \frac{\partial L}{\partial \dot{q}_r} \Delta q_r + t \Delta h \right]_A^B .$$

If therefore we suppose that C coincides with A and D coincides with B, and that Δh is zero, we shall have

$$\int_{CD} \left(\sum_{r=1}^{n} \dot{q}_r \frac{\partial L}{\partial \dot{q}_r} \right) dt = \int_{AB} \left(\sum_{r=1}^{n} \dot{q}_r \frac{\partial L}{\partial \dot{q}_r} \right) dt,$$

which shews that *the integral* $\int\left(\sum_{r=1}^{n}\dot{q}_r\frac{\partial L}{\partial\dot{q}_r}\right)dt$ *has a stationary value for any part of an actual trajectory, as compared with neighbouring paths between the same termini for which the time is correlated to the coordinates in such a way as to satisfy the same equation of energy.* This is called the *principle of Least Action*, the integral

$$\int\left(\sum_{r=1}^{n}\dot{q}_r\frac{\partial L}{\partial\dot{q}_r}\right)dt$$

being called the *Action*.

In natural problems, for which L is the difference of a kinetic energy T, homogeneous of the second degree in the velocities, and a potential energy V, independent of the velocities, we have (§ 41)

$$\sum_{r=1}^{n}\dot{q}_r\frac{\partial L}{\partial\dot{q}_r}=2T,$$

and the stationary integral can therefore in this case be written $\int T\,dt$.

The Principle of Least Action originated in Maupertuis' attempt (*Mém. de l'Acad.*, 1744, p. 417) to obtain for the corpuscular theory of light a theorem analogous to Fermat's "Principle of Least Time." Maupertuis' principle was established by Euler (Addit. II. p. 309 of his *Methodus inveniendi lineas curvas*, 1744) for the case of a single particle under a central force, and by Lagrange (*Miscell. Taurin.* II. (1760–1), *Oeuvres*, I. p. 365) for much more general problems

Example 1. Shew that the principle of Least Action can be extended to systems for which the integral of energy does not exist, in the following form. Let the expression $\sum_{r=1}^{n}\dot{q}_r\frac{\partial L}{\partial\dot{q}_r}-L$ be denoted by h; then the integral

$$\int\left(\sum_{r=1}^{n}\dot{q}_r\frac{\partial L}{\partial\dot{q}_r}+t\frac{dh}{dt}\right)dt$$

has a stationary value for any part of an actual trajectory, as compared with other paths between the same terminal points for which h has the same terminal values.

Example 2. If a dynamical system which possesses an integral of energy is reduced to a system of lower order as in § 42, shew that the principle of Least Action for the original system is identical with Hamilton's principle for the reduced system.

101. *Extension of Hamilton's principle to non-conservative dynamical systems.*

We shall now extend Hamilton's principle to holonomic dynamical systems in which the forces are no longer supposed to be conservative.

Let T denote the kinetic energy of such a system, and let $\sum_{r=1}^{n}Q_r\delta q_r$ denote the work done on the system by the external forces in an arbitrary displacement $(\delta q_1, \delta q_2, \ldots, \delta q_n)$; the equations of motion of the system are therefore

$$\frac{d}{dt}\left(\frac{\partial T}{\partial\dot{q}_r}\right)-\frac{\partial T}{\partial q_r}=Q_r \qquad (r=1, 2, \ldots, n).$$

Let α denote a part of a trajectory of the system, and let β be an adjacent arc having the same terminals, the times correlated to the path β at the terminals being the same as the values t_0 and t_1 of the time at the terminals in the trajectory α; then if δ denotes the variation by which we pass from a position on α to the contemporaneous position on β, we have

$$\begin{aligned}\int_{t_0}^{t_1}\left(\delta T + \sum_{r=1}^{n} Q_r\delta q_r\right)dt &= \int_{t_0}^{t_1}\sum_{r=1}^{n}\left(\frac{\partial T}{\partial \dot{q}_r}\delta\dot{q}_r + \frac{\partial T}{\partial q_r}\delta q_r + Q_r\delta q_r\right)dt\\ &= \int_{t_0}^{t_1}\sum_{r=1}^{n}\left\{\frac{\partial T}{\partial \dot{q}_r}\delta\dot{q}_r + \frac{d}{dt}\left(\frac{\partial T}{\partial \dot{q}_r}\right)\delta q_r\right\}dt\\ &= \int_{t_0}^{t_1}\frac{d}{dt}\left(\sum_{r=1}^{n}\frac{\partial T}{\partial \dot{q}_r}\delta q_r\right)dt\\ &= \left[\sum_{r=1}^{n}\frac{\partial T}{\partial \dot{q}_r}\delta q_r\right]_{t_0}^{t_1}\\ &= 0.\end{aligned}$$

This result

$$\int_{t_0}^{t_1}\left(\delta T + \sum_{r=1}^{n} Q_r\delta q_r\right)dt = 0$$

is (like the theorem of § 99, which is really a particular case of it) known as *Hamilton's principle.*

102. *Extension of Hamilton's principle and the principle of Least Action to non-holonomic systems*.*

We shall now shew that Hamilton's principle, when suitably formulated, is true even for dynamical systems which are not holonomic.

Consider a non-holonomic conservative system, in which the variations of the n coordinates $(q_1, q_2, \ldots, q_n)$ are connected by m non-integrable kinematical equations

$$A_{1k}dq_1 + A_{2k}dq_2 + \ldots + A_{nk}dq_n + T_k dt = 0 \qquad (k = 1, 2, \ldots, m)$$

where $A_{11}, A_{12}, \ldots, A_{nm}, T_1, \ldots, T_m$, are given functions of $q_1, q_2, \ldots, q_n$: so that if L denotes the kinetic potential, the motion is determined (§ 87) by the n equations

$$\frac{d}{dt}\left(\frac{\partial L}{\partial \dot{q}_r}\right) - \frac{\partial L}{\partial q_r} = \lambda_1 A_{r1} + \lambda_2 A_{r2} + \ldots + \lambda_m A_{rm} \qquad (r = 1, 2, \ldots, n),$$

together with the above kinematical equations; the unknown quantities being

$$q_1, q_2, \ldots, q_n, \lambda_1, \lambda_2, \ldots, \lambda_m.$$

Let AB be part of a trajectory of the system, and let CD be a path derived from AB by displacements consistent with the instantaneous kine-

* Cf. Hölder, *Gött. Nach.* 1896, p. 122, Voss, *Gött. Nach.* 1900, p. 322, Hamel, *Math. Ann.* cxi. (1935), p. 94.

matical equations, i.e. the above kinematical equations with the terms $T_k dt$ omitted; this path CD will not in general be itself a path whose continuous description would satisfy the kinematical conditions, so CD is really a kinematically impossible path.

It may naturally be asked why we do not take CD to be a kinematically possible path: the answer to which is, that in that case the displacements from AB to CD would not be displacements consistent with the kinematical equations: for in non-holonomic systems, if two adjacent possible configurations are given, the displacement from one to the other is not in general a possible displacement; there are infinitely more possible adjacent positions than there are possible displacements from the given position.

Proceeding as in the proof of Hamilton's principle given in § 99, δ denoting as usual a displacement from a point of AB to the contemporaneous point on CD, we have

$$\int_{CD} L dt - \int_{AB} L dt = L_B \Delta t_1 - L_A \Delta t_0 + \int_{t_0}^{t_1} \sum_{r=1}^{n} \left(\frac{\partial L}{\partial \dot{q}_r} \delta \dot{q}_r + \frac{\partial L}{\partial q_r} \delta q_r \right) dt$$

$$= L_B \Delta t_1 - L_A \Delta t_0 + \int_{t_0}^{t_1} \sum_{r=1}^{n} \left\{ \frac{\partial L}{\partial \dot{q}_r} \delta \dot{q}_r + \frac{d}{dt}\left(\frac{\partial L}{\partial \dot{q}_r}\right) \delta q_r - (\lambda_1 A_{r1} + \ldots + \lambda_m A_{rm}) \delta q_r \right\} dt.$$

Since the displacements obey the relations

$$A_{1k} \delta q_1 + A_{2k} \delta q_2 + \ldots + A_{nk} \delta q_n = 0,$$

it follows that the terms of the type $\lambda_s A_{rs} \delta q_r$ in the integral annul each other, so we have

$$\int_{CD} L dt - \int_{AB} L dt = L_B \Delta t_1 - L_A \Delta t_0 + \int_{t_0}^{t_1} \sum_{r=1}^{n} \left\{ \frac{\partial L}{\partial \dot{q}_r} \delta \dot{q}_r + \frac{d}{dt}\left(\frac{\partial L}{\partial \dot{q}_r}\right) \delta q_r \right\} dt.$$

From this point the proof proceeds as in § 99. We thus obtain the result that *Hamilton's principle applies to every dynamical system, whether holonomic or not. In every case the varied path considered is to be derived from the actual orbit by displacements which do not violate the kinematical equations representing the constraints; but it is only for holonomic systems that the varied motion is a possible motion; so that if we compare the actual motion with adjacent motions which obey the kinematical equations of constraint, Hamilton's principle is true only for holonomic systems.*

The same remarks obviously apply to the principle of Least Action, and to Hamilton's principle as applied to non-conservative systems.

103. *Are the stationary integrals actual minima? Kinetic foci.*

So far we have only shewn that the integrals which occur in Hamilton's principle and the principle of Least Action are *stationary* for the trajectories as compared with adjacent paths. The question now arises, whether they are actually *maxima* or *minima*.

We shall select for consideration the principle of Least Action, and for convenience of exposition shall suppose the number of degrees of freedom in the dynamical system to be two, the motion being defined by a kinetic energy

$$T = \tfrac{1}{2} a_{11}(q_1, q_2)\, \dot{q}_1^2 + a_{12}(q_1, q_2)\, \dot{q}_1 \dot{q}_2 + \tfrac{1}{2} a_{22}(q_1, q_2)\, \dot{q}_2^2,$$

and a potential energy

$$V = \psi(q_1, q_2).$$

The discussion can be extended without difficulty to Hamilton's principle, and to systems with any number of degrees of freedom. The principle of Least Action, as applied to the above system, is (§ 100) that the integral

$$\int (a_{11}\dot{q}_1^2 + 2a_{12}\dot{q}_1\dot{q}_2 + a_{22}\dot{q}_2^2)\, dt$$

has a stationary value for an actual trajectory as compared with other paths between the same termini for which dt is connected with the differentials of the coordinates by the same equation of energy

$$T + V = h.$$

This latter equation gives

$$a_{11}\dot{q}_1^2 + 2a_{12}\dot{q}_1\dot{q}_2 + a_{22}\dot{q}_2^2 = 2(h - \psi),$$

or

$$dt = \{2(h - \psi)\}^{-\frac{1}{2}} (a_{11}\, dq_1^2 + 2a_{12}\, dq_1 dq_2 + a_{22}\, dq_2^2)^{\frac{1}{2}},$$

so the stationary integral can be taken to be

$$I = \int (h - \psi)^{\frac{1}{2}} (a_{11} + 2a_{12} q_2' + a_{22} q_2'^2)^{\frac{1}{2}}\, dq_1,$$

where q_2' stands for dq_2/dq_1; this integral is to be taken between terminals, at each of which the values of q_1 and q_2 are given.

Writing this equation

$$I = \int f(q_1, q_2, q_2')\, dq_1,$$

we shall discuss the discrimination of its maxima and minima (which was first effected by Jacobi) by a method suggested by Culverwell*.

Consider any number of paths adjacent to the actual trajectory. These paths will be supposed to have the same terminals, and to be continuous, but their directions may have abrupt changes at any finite number of points. For such a path let $(q_1, q_2 + \delta q_2)$ be a point corresponding to a point (q_1, q_2) on the actual trajectory; we shall frequently write $\alpha\phi$ for δq_2, where α is a small constant the order of which determines the order of magnitude of the quantities we are dealing with, and ϕ is zero at the terminal points.

* *Proc. Lond. Math. Soc.* XXIII. (1892), p. 241.

Let the expansion of the function

$$f(q_1, q_2 + \alpha\phi, q_2' + \alpha\phi')$$

in ascending powers of α be

$$f(q_1, q_2, q_2') + \alpha(U_0\phi + U_1\phi') + \tfrac{1}{2}\alpha^2(U_{00}\phi^2 + 2U_{01}\phi\phi' + U_{11}\phi'^2) + \ldots;$$

let δI denote the terms involving α in the first degree in

$$\int f(q_1, q_2 + \alpha\phi, q_2' + \alpha\phi')\, dq_1,$$

and let $\delta^2 I$ denote the terms in α^2.

When the range of integration is small, and its terminals are fixed, the value of ϕ' at any point is large compared with the value of ϕ. For since ϕ is zero at the terminals, we have

$$\phi = \int_P^{q_1} \phi'\, dq_1,$$

where P and R denote the terminals. If therefore β be the numerically greatest value of ϕ' between P and R, it follows that ϕ can never exceed $(q_{1(R)} - q_{1(P)})\,\beta$, and consequently by taking the range sufficiently small the ratio of ϕ to ϕ' can be diminished indefinitely.

Thus if the range is very small, the most important term in $\delta^2 I$ is $\frac{1}{2}\int U_{11}\phi'^2 dq_1$; and as the sign of this is always the same as that of U_{11} (the sign of dq_1 is taken to be positive), we see that for small ranges, I is a maximum or minimum according as U_{11} is negative or positive. Now

$$U_{11} = \frac{\partial^2 f}{\partial q_2'^2} = (h - \psi)^{\frac{1}{2}}(a_{11} + 2a_{12}q_2' + a_{22}q_2'^2)^{-\frac{3}{2}}(a_{11}a_{22} - a_{12}^2),$$

and this is positive, since the kinetic energy is a positive definite form and therefore $a_{11}a_{22} - a_{12}^2$ is positive. We thus have the result that *for small ranges the Action is a minimum for the actual trajectory.*

Now consider any point A on an actual trajectory, and let another actual trajectory be drawn through A making a very small angle with the first. If this intersects the first trajectory again, say at a point B, then the limiting position of the point B when the angle between the trajectories diminishes indefinitely is called the *kinetic focus* of A on the first trajectory, or the point *conjugate* to A.

We shall now shew that for finite ranges the Action is a minimum, provided the final point is not beyond the kinetic focus of the initial point.

For let P and Q be the terminals; we have seen that if Q is very near to P, the quantity $\delta^2 I$ is always positive and of order α^2 compared with the value of I for the limits P and Q. It is therefore evident that as we remove

Q further from P, the quantity $\delta^2 I$ cannot become capable of a negative value until after Q has passed through the point for which $\delta^2 I$ can vanish for a suitably chosen value of $\alpha\phi$.

Suppose then that PBQ is an arc of an actual trajectory, Q being the first point for which it is possible to draw a varied curve PHQ for which $\delta^2 I$ is zero; we shall shew that the varied curve PHQ must itself be a trajectory. For if it is not a trajectory between two of its own points A and C (supposed near each other), let a trajectory ADC be drawn between these points. Then the integral taken along ADC is less than that taken along AHC, so the integral taken along $PADCQ$ is less than that along PHQ, which by hypothesis is equal to that along PBQ. Hence $\delta^2 I$ along $PADCQ$ is negative, and therefore Q cannot be the first point for which, as we proceed from P, the variation ceases to be positive; which is contrary to what has been proved. It follows that $PAHCQ$ is a trajectory, and Q is the kinetic focus of P. Hence *the Action is a true minimum, provided that in passing along the trajectory the final point is reached before the kinetic focus of the initial point.*

Lastly we shall consider the case in which the kinetic focus of the initial point is reached before we arrive at the final point. Suppose, with the notation just used, that the initial and final points are P and R; and let two points E and F be taken, the former on the curve PHQ and the latter on the arc QR; these points being taken so close together that the trajectory EGF joining them gives a true minimum. Since the integral taken along EGF is less than that along EQF, it follows that the integral taken along $PEGFR$ is less than that along $PEQR$; but the latter is equal to that along $PBQR$, since both integrals are equal from P to Q; and therefore the integral along $PBQR$ is not a minimum; but it is not a maximum, since the integral taken along any small part of it is a minimum. Hence *when the kinetic focus of the initial point is reached before we arrive at the final point, the Action is neither a maximum nor a minimum.*

A simple example illustrative of the results obtained in this article is furnished by the motion of a particle under no forces on a smooth sphere. The trajectories are great-circles on the sphere, and the Action taken along any path (whether a trajectory or not) is proportional to the length of the path. The kinetic focus of any point A is the diametrically opposite point A' on the sphere, since any two great-circles through A intersect again at A'. The theorems of this article amount therefore in this case to the statement that an arc of a great-circle joining any two points A and B on the sphere is the shortest distance from A to B when (and only when) the point A' diametrically opposite to A does not lie on the arc, i.e. when the arc in question is less than half a great-circle.

104. *Representation of the motion of dynamical systems by means of geodesics.*

The principle of Least Action leads to an interesting transformation of the motion of natural dynamical systems with two degrees of freedom.

Let the kinetic energy of such a system be

$$\tfrac{1}{2}\{a_{11}(q_1, q_2)\dot{q}_1^2 + 2a_{12}(q_1, q_2)\dot{q}_1\dot{q}_2 + a_{22}(q_1, q_2)\dot{q}_2^2\},$$

and let its potential energy be $\psi(q_1, q_2)$. By § 100, the orbits corresponding to that family of solutions for which the total energy is h are given by the condition that

$$\int (a_{11}\dot{q}_1^2 + 2a_{12}\dot{q}_1\dot{q}_2 + a_{22}\dot{q}_2^2)\,dt$$

is stationary for any part of an actual orbit, as compared with any other arc between the same terminals for which dt is connected with the differentials of the coordinates by the relation

$$\tfrac{1}{2}(a_{11}\dot{q}_1^2 + 2a_{12}\dot{q}_1\dot{q}_2 + a_{22}\dot{q}_2^2) + \psi(q_1, q_2) = h.$$

The integral

$$\int (h-\psi)^{\frac{1}{2}}(a_{11}dq_1^2 + 2a_{12}dq_1dq_2 + a_{22}dq_2^2)^{\frac{1}{2}}$$

is therefore stationary. But this integral expresses the principle of Least Action for the motion of a particle under no forces on any surface whose linear element is given by the equation

$$ds^2 = (h-\psi)(a_{11}dq_1^2 + 2a_{12}dq_1dq_2 + a_{22}dq_2^2),$$

and is therefore the defining condition of the geodesics on this surface. Consequently *the equations of the orbits in the given dynamical system are the same as the equations of the geodesics on this surface.*

Example 1. Shew that the parabolic orbits of a free heavy projectile correspond to the geodesics on a certain surface of revolution.

Example 2. Shew that the orbits described under a central attractive force $\phi'(r)$ in a plane correspond to geodesics on a surface of revolution, the equation of whose meridian-curve is $z=f(\rho)$, where

$$f'(\rho) = \{(\rho\,dr/r\,d\rho)^2 - 1\}^{\frac{1}{2}}$$

and where r and ρ are connected by the relation $\rho^2 = r^2\{-\phi(r)+h\}$.

105. *The least-curvature principle of Gauss and Hertz.*

We shall now discuss a principle which, like Hamilton's principle, can be used to define the orbits of a dynamical system, but which does not involve the sign of integration.

In any dynamical system (whether holonomic or non-holonomic) let (x_r, y_r, z_r) be the coordinates of a typical particle m_r at time t, and (X_r, Y_r, Z_r) the components of the external force which acts on the particle. Consider the function

$$\Sigma m_r\left\{\left(\ddot{x}_r - \frac{X_r}{m_r}\right)^2 + \left(\ddot{y}_r - \frac{Y_r}{m_r}\right)^2 + \left(\ddot{z}_r - \frac{Z_r}{m_r}\right)^2\right\},$$

where the summation is extended over all the particles of the system, and where $(\ddot{x}_r, \ddot{y}_r, \ddot{z}_r)$ refer to any kinematically possible path for which the coordinates and velocities at the instant considered are the same as in some actual trajectory. This function substantially represents what was called by Gauss the *constraint* and by Hertz (who however considered primarily the case in which the external forces are zero) the *curvature** of the kinematically possible path considered. In what follows Hertz's terminology will be used.

We shall shew that *of all paths consistent with the constraints (which are supposed to do no work), the actual trajectory is that which has the least curvature*†.

In the simple case of a single particle moving on a smooth surface under no external forces, this result clearly reduces to the statement that the curvature in space (in the ordinary sense of the term) of the orbit is the least which is consistent with the condition that the particle is to remain on the surface.

To establish this result, let the equations which express the constraints (using x_r to typify any one of the three coordinates of any particle) be

$$\sum_r x_{kr}\, dx_r = 0 \qquad (k = 1, 2, \ldots, m),$$

where the coefficients x_{kr} are given functions of the coordinates. Differentiating these relations, we have

$$\sum_r x_{kr}\ddot{x}_r + \sum_r \sum_s \frac{\partial x_{kr}}{\partial x_s} \dot{x}_r \dot{x}_s = 0 \qquad (k = 1, 2, \ldots, m).$$

Let $\ddot{x}_r$ be a typical component of acceleration in the path considered (which is supposed to be kinematically possible, but is not necessarily the actual trajectory), and let $\ddot{x}_{r0}$ be the corresponding component of acceleration in the actual trajectory. Subtracting the preceding equation, considered as relating to the actual trajectory, from the same equation, considered as relating to the kinematically possible path, we have (since the velocities are the same in the two paths)

$$\sum_r x_{kr}(\ddot{x}_r - \ddot{x}_{r0}) = 0 \qquad (k = 1, 2, \ldots, m).$$

This equation shews that a small displacement of the system, in which the displacement δx_r of the coordinate x_r is proportional to $(\ddot{x}_r - \ddot{x}_{r0})$, is consistent with the equations of constraint, i.e. is a possible displacement.

The components of the forces exercised by the constraints are typified by

* Strictly speaking, the square root of this function, and not the function itself, was called the curvature by Hertz.

† Gauss, *Crelle's Journal*, IV. (1829), p. 232; *Werke*, v. p. 23. Gauss measured the *constraint* by "the sum of the masses of the particles, each multiplied by the square of its deviation from unconstrained motion." The above analytical expression for it was first given by H. Scheffler, *Zeitschrift für Math.* III. (1858), p. 197. Hertz's theory is given in his *Mechanik*. Cf. Gugino, *Lincei Rend.* IX. (1929), p. 1090.

$(m_r\ddot{x}_{r0} - X_r)$: and in any possible displacement the forces of constraint do no work. We have therefore

$$\sum_r (m_r\ddot{x}_{r0} - X_r)(\ddot{x}_r - \ddot{x}_{r0}) = 0,$$

an equation which may be written in the form

$$\sum_r m_r\left(\ddot{x}_r - \frac{X_r}{m_r}\right)^2 = \sum_r m_r\left(\ddot{x}_{r0} - \frac{X_r}{m_r}\right)^2 + \sum_r m_r(\ddot{x}_r - \ddot{x}_{r0})^2,$$

or (reverting to the use of y's and z's)

$$\sum_r m_r\left\{\left(\ddot{x}_r - \frac{X_r}{m_r}\right)^2 + \left(\ddot{y}_r - \frac{Y_r}{m_r}\right)^2 + \left(\ddot{z}_r - \frac{Z_r}{m_r}\right)^2\right\}$$
$$= \sum_r m_r\left\{\left(\ddot{x}_{r0} - \frac{X_r}{m_r}\right)^2 + \left(\ddot{y}_{r0} - \frac{Y_r}{m_r}\right)^2 + \left(\ddot{z}_{r0} - \frac{Z_r}{m_r}\right)^2\right\}$$
$$+ \sum_r m_r\{(\ddot{x}_r - \ddot{x}_{r0})^2 + (\ddot{y}_r - \ddot{y}_{r0})^2 + (\ddot{z}_r - \ddot{z}_{r0})^2\}.$$

Since the terms in the last summation on the right-hand side are all positive, it follows that

$$\sum_r m_r\left\{\left(\ddot{x}_r - \frac{X_r}{m_r}\right)^2 + \left(\ddot{y}_r - \frac{Y_r}{m_r}\right)^2 + \left(\ddot{z}_r - \frac{Z_r}{m_r}\right)^2\right\}$$
$$> \sum_r m_r\left\{\left(\ddot{x}_{r0} - \frac{X_r}{m_r}\right)^2 + \left(\ddot{y}_{r0} - \frac{Y_r}{m_r}\right)^2 + \left(\ddot{z}_{r0} - \frac{Z_r}{m_r}\right)^2\right\},$$

which establishes the result stated.

106. *Expression of the curvature of a path in terms of generalised coordinates.*

Lipschitz has shewn* that the curvature of a kinematically possible path in a holonomic dynamical system with n degrees of freedom can be expressed in terms of the derivates of the n independent coordinates which define the position of the system.

Let $(q_1, q_2, \ldots, q_n)$ be the coordinates; let $(\ddot{q}_1, \ddot{q}_2, \ldots, \ddot{q}_n)$ be the accelerations of these coordinates in any kinematically possible path, and let $(\ddot{q}_{10}, \ddot{q}_{20}, \ldots, \ddot{q}_{n0})$ be the accelerations in the actual trajectory which corresponds to the same values of $(q_1, q_2, \ldots, q_n, \dot{q}_1, \dot{q}_2, \ldots, \dot{q}_n)$. Using x_r to typify any one of the three rectangular coordinates of any particle m_r, and X_r to typify the corresponding component of force, the Gauss-Hertz curvature of the path is $\sum_r m_r(\ddot{x}_r - X_r/m_r)^2$; and it has been shewn in the last article that this can be written in the form

$$\sum_r m_r(\ddot{x}_{r0} - X_r/m_r)^2 + \sum_r m_r(\ddot{x}_r - \ddot{x}_{r0})^2.$$

* *Journal für Math.* LXXXII. p. 323. Cf. also Wassmuth, *Wien. Sitz.* CIV. (1895); and for further work connected with the principle of Least Curvature see Leitinger, *Wien. Sitz.* CXVI. (1908), p. 1321 and Schenkl, *Wien. Sitz.* CXXII. (1913), p. 721.

The first of these summations is the same for all the paths considered, since it depends only on the actual trajectory: we can therefore omit it without causing the whole expression to lose its minimum-property, and we can call the remaining summation $\sum_r m_r(\ddot{x}_r - \ddot{x}_{r0})^2$ the *curvature* of the path.

Let the kinetic energy be

$$T = \tfrac{1}{2}\sum_k \sum_l a_{kl}\dot{q}_k\dot{q}_l,$$

where the quantities a_{kl} are given functions of $(q_1, q_2, \ldots, q_n)$; let D denote the determinant formed of the quantities a_{kl}, and let A_{kl} denote the minor of a_{kl} in this determinant.

From the equation

$$\sum_r m_r\dot{x}_r^2 = \sum_k \sum_l a_{kl}\dot{q}_k\dot{q}_l$$

we have

$$a_{kl} = \sum_r m_r \frac{\partial x_r}{\partial q_k}\frac{\partial x_r}{\partial q_l}.$$

Now

$$\ddot{x}_r = \sum_k \frac{\partial x_r}{\partial q_k}\ddot{q}_k + \sum_k \sum_l \frac{\partial^2 x_r}{\partial q_k \partial q_l}\dot{q}_k\dot{q}_l,$$

and consequently, since the coordinates and velocities are the same for all the paths considered, we have

$$\ddot{x}_r - \ddot{x}_{r0} = \sum_k \frac{\partial x_r}{\partial q_k}(\ddot{q}_k - \ddot{q}_{k0}).$$

But if we write

$$S_k = \frac{d}{dt}\left(\frac{\partial T}{\partial \dot{q}_k}\right) - \frac{\partial T}{\partial q_k} - \sum_r \frac{\partial x_r}{\partial q_k}X_r \qquad (k = 1, 2, \ldots, n),$$

since this expression is zero for the actual trajectory, we have

$S_k =$ the difference of the values of $\frac{d}{dt}\left(\frac{\partial T}{\partial \dot{q}_k}\right)$ for the path considered and the actual trajectory,

or

$$S_k = \sum_l a_{kl}(\ddot{q}_l - \ddot{q}_{l0}) \qquad (k = 1, 2, \ldots, n),$$

whence we have

$$\ddot{q}_k - \ddot{q}_{k0} = \frac{1}{D}\sum_l A_{kl}S_l \qquad (k = 1, 2, \ldots, n);$$

and consequently

$$\ddot{x}_r - \ddot{x}_{r0} = \frac{1}{D}\sum_k \sum_l \frac{\partial x_r}{\partial q_k}A_{kl}S_l.$$

The curvature, $\sum_r m_r(\ddot{x}_r - \ddot{x}_{r0})^2$, is therefore

$$\frac{1}{D^2}\sum_r \sum_k \sum_l \sum_i \sum_j m_r \frac{\partial x_r}{\partial q_k}\frac{\partial x_r}{\partial q_i}A_{kl}A_{ij}S_lS_j,$$

or

$$\frac{1}{D^2}\sum_k \sum_l \sum_i \sum_j a_{ki}A_{kl}A_{ij}S_lS_j.$$

But by a well-known property of determinants, we have

$$\sum_i \sum_k a_{ki} A_{kl} A_{ij} = D A_{lj},$$

and therefore finally *the curvature can be expressed in terms of the coordinates* $(q_1, q_2, \ldots, q_n)$ *and their derivates in the form*

$$\frac{1}{D} \sum_l \sum_j A_{lj} S_j S_l.$$

107. *Appell's equations.*

The Gauss-Hertz law of Least Curvature is the basis of a form in which Appell has proposed* to write the general differential equations of dynamics. This form, as will be seen, is equally applicable to holonomic and non-holonomic systems.

Consider any dynamical system; let

$$A_{1k} dq_1 + A_{2k} dq_2 + \ldots + A_{nk} dq_n + T_k dt = 0 \quad (k = 1, 2, \ldots, m)$$

be the non-integrable equations connecting the variations of the generalised coordinates $q_1, q_2, \ldots, q_n$; in holonomic systems these equations will of course be non-existent.

Let S denote the function $\frac{1}{2} \sum_k m_k (\ddot{x}_k^2 + \ddot{y}_k^2 + \ddot{z}_k^2)$, where m_k typifies the mass of a particle of the system, whose rectangular coordinates at time t are (x_k, y_k, z_k). By means of the equations which define the position of the particles at any time in terms of the coordinates $(q_1, q_2, \ldots, q_n)$, it is possible to express S in terms of $(q_1, q_2, \ldots, q_n)$ and the first and second derivates of these variables with respect to the time. Moreover, by use of the equations of constraint we can express m of the velocities $(\dot{q}_1, \dot{q}_2, \ldots, \dot{q}_n)$ in terms of the others: let the coordinates corresponding to these latter be denoted by $(p_1, p_2, \ldots, p_{n-m})$. By differentiating these relations we can express $\ddot{q}_1, \ddot{q}_2, \ldots, \ddot{q}_n$ in terms of the quantities $\ddot{p}_1, \ddot{p}_2, \ldots, \ddot{p}_{n-m}, \dot{p}_1, \dot{p}_2, \ldots, \dot{p}_{n-m}, q_1, q_2, \ldots, q_n$, and hence S can be expressed in terms of this last set of variables.

Now any small displacement which is consistent with the constraints can be defined by the changes $(\delta p_1, \delta p_2, \ldots, \delta p_{n-m})$ in the quantities $(p_1, p_2, \ldots, p_{n-m})$; let $\sum_{r=1}^{n-m} P_r \delta p_r$ denote the work done by the external forces in such a displacement. As in § 26, we have

$$\sum_k m_k \left(\ddot{x}_k \frac{\partial x_k}{\partial p_r} + \ddot{y}_k \frac{\partial y_k}{\partial p_r} + \ddot{z}_k \frac{\partial z_k}{\partial p_r} \right) = P_r.$$

* *Journal für Math.* CXXI. (1900), p. 310. Cf. Seeger, *J. Wash. Acad. Sci.* XX. (1930), p. 481.

Let the equation which expresses the change in x_k in terms of the changes in $(p_1, p_2, \ldots, p_{n-m})$ be

$$\delta x_k = \sum_{r=1}^{n-m} \pi_r \delta p_r,$$

where $(\pi_1, \pi_2, \ldots, \pi_{n-m})$ are known functions of the coordinates: the equations of this type are of course non-integrable. From this we have $\partial x_k/\partial p_r = \pi_r$, and so the equation which expresses $\dot{x}_k$ in terms of

$$(\dot{p}_1, \dot{p}_2, \ldots, \dot{p}_{n-m})$$

will be of the form

$$\dot{x}_k = \sum_{r=1}^{n-m} \pi_r \dot{p}_r + \alpha,$$

where α denotes some function of the coordinates. Differentiating this equation, we have

$$\ddot{x}_k = \sum_{r=1}^{n-m} \pi_r \ddot{p}_r + \sum_{r=1}^{n-m} \frac{d\pi_r}{dt} \dot{p}_r + \frac{d\alpha}{dt},$$

whence $\dfrac{\partial \ddot{x}_k}{\partial \ddot{p}_r} = \pi_r = \dfrac{\partial x_k}{\partial p_r}$.

It follows that

$$\begin{aligned} P_r &= \sum_k m_k \left(\ddot{x}_k \frac{\partial x_k}{\partial p_r} + \ddot{y}_k \frac{\partial y_k}{\partial p_r} + \ddot{z}_k \frac{\partial z_k}{\partial p_r} \right) \\ &= \sum_k m_k \left(\ddot{x}_k \frac{\partial \ddot{x}_k}{\partial \ddot{p}_r} + \ddot{y}_k \frac{\partial \ddot{y}_k}{\partial \ddot{p}_r} + \ddot{z}_k \frac{\partial \ddot{z}_k}{\partial \ddot{p}_r} \right) \\ &= \partial S/\partial \ddot{p}_r, \end{aligned}$$

and therefore *the equations of a dynamical system, whether holonomic or not, can be expressed in the form*

$$\frac{\partial S}{\partial \ddot{p}_r} = P_r \qquad (r = 1, 2, \ldots, n-m),$$

where S denotes the function $\frac{1}{2}\sum m_k (\ddot{x}_k^2 + \ddot{y}_k^2 + \ddot{z}_k^2)$, *and* $(p_1, p_2, \ldots, p_{n-m})$ *are coordinates equal in number to the degrees of freedom of the system**.

It is evident that the result is valid even if the quantities $p_1, \ldots, p_{n-m}$ are not true coordinates, but are quasi-coordinates.

Example. Obtain from Appell's equations the equations

$$\begin{cases} A\dot{\omega}_1 - (B - C)\,\omega_2\omega_3 = L, \\ B\dot{\omega}_2 - (C - A)\,\omega_3\omega_1 = M, \\ C\dot{\omega}_3 - (A - B)\,\omega_1\omega_2 = N, \end{cases}$$

for the motion of a rigid body one of whose points is fixed; where $(\omega_1, \omega_2, \omega_3)$ are the components of angular velocity of the body resolved along its own principal axes of inertia at the fixed point, (A, B, C) are the principal moments of inertia, and (L, M, N) are the moments of the external forces about the principal axes.

* On the connexion of these equations with the Principle of Least Action, cf. H. Brell, *Wien. Sitz.* CXXII. (1913), p. 933.

108. *Bertrand's theorem.*

A theorem in impulsive motion, which belongs to the same group of results as the least-curvature principle of Gauss and Hertz, is due to Sturm * and may be stated thus: *If a given set of impulses is applied to different points of a system (whether holonomic or non-holonomic) in motion, the kinetic energy of the resulting motion is greater than the kinetic energy of the motion which the system would acquire under the action of the same impulses and constraints and of any additional constraints due to the reactions of perfectly smooth or perfectly rough fixed surfaces, or rigid connexions between particles of the system.*

For let m be the mass of a typical particle of the system, and let (u, v, w) (u', v', w'), (u_1, v_1, w_1) denote the components of velocity of this particle before the application of the impulses, after the application of the impulses, and in the comparison motion, respectively.

Let (X, Y, Z) denote the components of the external impulse acting on the particle: (X', Y', Z') the components of the impulse due to the constraints of the system: and $(X' + X_1, Y' + Y_1, Z' + Z_1)$ the components of the impulse due to the constraints in the comparison motion.

The equations of impulsive motion are

$$m(u' - u) = X + X', \quad m(v' - v) = Y + Y', \quad m(w' - w) = Z + Z',$$
$$m(u_1 - u) = X + X' + X_1, \quad m(v_1 - v) = Y + Y' + Y_1, \quad m(w_1 - w) = Z + Z' + Z_1.$$

Subtracting, we have

$$m(u_1 - u') = X_1, \qquad m(v_1 - v') = Y_1, \qquad m(w_1 - w') = Z_1.$$

Multiply these last equations by u_1, v_1, w_1, respectively, add, and sum for all the particles of the system; we thus have

$$\Sigma m \{(u_1 - u') u_1 + (v_1 - v') v_1 + (w_1 - w') w_1\} = \Sigma (X_1 u_1 + Y_1 v_1 + Z_1 w_1).$$

Now from the nature of the constraints, it follows that finite forces acting on all the particles of the system and proportional to the impulsive forces (X_1, Y_1, Z_1), would on the whole do no work in a displacement whose components are proportional to the quantities (u_1, v_1, w_1); and therefore we have

$$\Sigma (X_1 u_1 + Y_1 v_1 + Z_1 w_1) = 0,$$

or
$$\Sigma m \{(u_1 - u') u_1 + (v_1 - v') v_1 + (w_1 - w') w_1\} = 0;$$

this equation can be written in the form

$$\Sigma m (u'^2 + v'^2 + w'^2) - \Sigma m (u_1^2 + v_1^2 + w_1^2) = \Sigma m \{(u' - u_1)^2 + (v' - v_1)^2 + (w' - w_1)^2\},$$

* Sturm, *Comptes Rendus*, XIII. (1841), p. 1046. It is always known as *Bertrand's theorem.*

which shews that

$$\tfrac{1}{2}\Sigma m\,(u'^2 + v'^2 + w'^2) > \tfrac{1}{2}\Sigma m\,(u_1{}^2 + v_1{}^2 + w_1{}^2),$$

and so establishes Bertrand's theorem.

The theorem may readily be extended to the case when the forces are not impulsive but continuous: in this case the increase of kinetic energy per unit of time is diminished by the introduction of fresh constraints that do not affect the potential energy.

The following result, due to Lord Kelvin and generally known as *Thomson's theorem* *, can easily be established by a proof of the same character as the above: *If any number of points of a dynamical system are suddenly set in motion with prescribed velocities, the kinetic energy of the resulting motion is less than that of any other kinematically possible motion which the system can take with the prescribed velocities, the excess being the energy of the motion which must be compounded with either to produce the other.*

Lord Rayleigh has remarked† that the theorems of Thomson and Bertrand may both be comprehended in the statement that the introduction of fresh constraints increases the inertia, or moment of inertia, of a system.

Example. A framework of $(n-1)$ equal rhombuses, each with one diagonal in the same continuous straight line, and two open ends, each of which is half of a rhombus, is formed by $2n$ equal rods which are freely jointed in pairs at the corners of all the rhombuses. Impulses P perpendicular to and towards the line of the diagonals are applied to the two free extremities of one open end; shew that the initial velocity, parallel to the diagonal, of the extremities of the other open end is

$$\frac{3P}{m}\frac{\sin a \cos a}{\cos^2 a + n^2 \sin^2 a},$$

where m is the mass of each rod, and $2a$ is the angle between each pair of rods at the points of crossing. (Camb. Math. Tripos, Part I, 1896.)

Miscellaneous Examples.

1. If the problem of determining the motion of a particle on a surface whose linear element is given by the equation

$$ds^2 = Edu^2 + 2F du\, dv + G dv^2,$$

under the action of forces such that the potential energy is $V(u, v)$, can be solved, shew that the problem of determining the motion of a particle on a surface whose linear element is given by

$$ds^2 = V(u, v)\,(Edu^2 + 2Fdu\, dv + Gdv^2),$$

under forces derivable from a potential energy $1/V(u, v)$, can also be solved.

(Darboux.)

2. If in two dynamical systems in which the kinetic energies are respectively $\Sigma a_{ik}\dot{q}_i\dot{q}_k$ and $\Sigma b_{ik}\dot{q}_i\dot{q}_k$, and the potential energies are respectively U and V, the trajectories

* Thomson and Tait's *Natural Philosophy*, § 317.

† *Theory of Sound*, Vol. I. p. 100.

are the same curves, though described with different velocities, so that the relations between the coordinates $(q_1, q_2, \ldots, q_n)$ are the same in the two problems, shew that

$$V = \frac{aU+\beta}{\gamma U+\delta},$$

where a, β, γ, δ are constants, and that

$$\Sigma b_{ik}\, dq_i\, dq_k = (\gamma U+\delta)\, \Sigma a_{ik}\, dq_i\, dq_k. \qquad \text{(Painlevé.)}$$

3. If all the trajectories of a particle in a plane, described under forces such that the potential energy of the particle is $V(x, y)$, with a value h of the constant of energy, are subjected to a transformation

$$x = \phi(X, Y), \qquad y = \psi(X, Y),$$

where ϕ and ψ are conjugate functions of (X, Y), shew that the new curves so obtained are the trajectories of a particle acted on by forces derivable from the potential energy

$$[V\{\phi(X, Y), \psi(X, Y)\} - h]\left\{\left(\frac{\partial\phi}{\partial X}\right)^2 + \left(\frac{\partial\phi}{\partial Y}\right)^2\right\},$$

with a zero value of the constant of energy. (Goursat.)

4. If T and V denote respectively the kinetic and potential energies of a dynamical system, shew that

$$2\frac{d^2V}{dt^2} + \Sigma m(\ddot{x}^2 + \ddot{y}^2 + \ddot{z}^2)$$

differs from

$$\Sigma \frac{1}{m}\left\{\left(m\ddot{x} + \frac{\partial V}{\partial x}\right)^2 + \left(m\ddot{y} + \frac{\partial V}{\partial y}\right)^2 + \left(m\ddot{z} + \frac{\partial V}{\partial z}\right)^2\right\}$$

by a quantity which does not involve the accelerations; and hence that

$$\frac{d^2T}{dt^2} - \tfrac{1}{2}\Sigma m(\ddot{x}^2 + \ddot{y}^2 + \ddot{z}^2)$$

is a maximum when the accelerations have the values corresponding to the actual motion, as compared with all motions which are consistent with the constraints and satisfy the same integral of energy, and which have the same values of the coordinates and velocities at the instant considered; provided the constraints do no work. (Förster.)

CHAPTER X

HAMILTONIAN SYSTEMS AND THEIR INTEGRAL-INVARIANTS

109. *Hamilton's form of the equations of motion.*

We shall now obtain for the differential equations of motion of a conservative holonomic dynamical system a form which constitutes the basis of most of the advanced theory of Dynamics.

Let $(q_1, q_2, \ldots, q_n)$ be the coordinates and $L(q_1, q_2, \ldots, q_n, \dot{q}_1, \dot{q}_2, \ldots, \dot{q}_n, t)$ the kinetic potential of the system, so that the equations of motion in the Lagrangian form are

$$\frac{d}{dt}\left(\frac{\partial L}{\partial \dot{q}_r}\right) - \frac{\partial L}{\partial q_r} = 0 \qquad (r = 1, 2, \ldots, n).$$

Write

$$\frac{\partial L}{\partial \dot{q}_r} = p_r \qquad (r = 1, 2, \ldots, n),$$

so that

$$\dot{p}_r = \frac{\partial L}{\partial q_r} \qquad (r = 1, 2, \ldots, n).$$

From the former of these sets of equations we can regard either of the sets of quantities $(\dot{q}_1, \dot{q}_2, \ldots, \dot{q}_n)$ or $(p_1, p_2, \ldots, p_n)$ as functions of the other set.

If δ denote the increment in any function of the variables

$$(q_1, q_2, \ldots, q_n, p_1, p_2, \ldots, p_n) \quad \text{or} \quad (q_1, q_2, \ldots, q_n, \dot{q}_1, \dot{q}_2, \ldots, \dot{q}_n)$$

due to small changes in the arguments, we have

$$\delta L = \sum_{r=1}^{n} \left(\frac{\partial L}{\partial q_r}\,\delta q_r + \frac{\partial L}{\partial \dot{q}_r}\,\delta \dot{q}_r\right)$$

$$= \sum_{r=1}^{n} (\dot{p}_r \delta q_r + p_r \delta \dot{q}_r)$$

$$= \delta \sum_{r=1}^{n} p_r \dot{q}_r + \sum_{r=1}^{n} (\dot{p}_r \delta q_r - \dot{q}_r \delta p_r),$$

or

$$\delta\left\{\sum_{r=1}^{n} p_r \dot{q}_r - L\right\} = \sum_{r=1}^{n} (\dot{q}_r \delta p_r - \dot{p}_r \delta q_r).$$

Thus if the quantity $\sum_{r=1}^{n} p_r\dot{q}_r - L$, when expressed in terms of

$$(q_1, q_2, \ldots, q_n, p_1, p_2, \ldots, p_n, t),$$

be denoted by H, we have

$$\delta H = \sum_{r=1}^{n} (\dot{q}_r \delta p_r - \dot{p}_r \delta q_r) \quad \ldots\ldots\ldots\ldots\ldots\ldots\ldots(1),$$

or

$$\frac{dq_r}{dt} = \frac{\partial H}{\partial p_r}, \quad \frac{dp_r}{dt} = -\frac{\partial H}{\partial q_r} \quad (r = 1, 2, \ldots, n) \quad \ldots(2).$$

The motion of the dynamical system may be regarded as defined by these equations, which are said to be in the *Hamiltonian* or *canonical* form; the dependent variables are $(q_1, q_2, \ldots, q_n, p_1, p_2, \ldots, p_n)$, and the system consists of $2n$ equations, each of the first order; whereas the Lagrangian system consists of n equations, each of the second order.

The Hamiltonian form was introduced by Hamilton in 1834*. In part he had been anticipated by the great French mathematicians: for Poisson in 1809† had taken the step of introducing a function

$$\sum_{r=1}^{n} p_r\dot{q}_r - T$$

and expressing it in terms of $(q_1, q_2, \ldots, q_n, p_1, \ldots, p_n)$, and had actually derived half of Hamilton's equations: while Lagrange in 1810‡ had obtained a particular set of equations (for the variation of elements) in the Hamiltonian form, the disturbing function taking the place of the function H. Moreover the theory of non-linear partial differential equations of the first order had led to systems of ordinary differential equations possessing this form: for, as was shewn by Pfaff§ in 1814–15 and by Cauchy|| in 1819 (completing the earlier work of Lagrange and Monge), the equations of the characteristics of a partial differential equation

$$f(x_1, x_2, \ldots, x_n, p_1, p_2, \ldots, p_n) = 0,$$

where

$$p_s = \frac{\partial z}{\partial x_s},$$

are

$$\frac{dx_1}{\partial f/\partial p_1} = \frac{dx_2}{\partial f/\partial p_2} = \ldots = \frac{dx_n}{\partial f/\partial p_n} = \frac{dp_1}{-\partial f/\partial x_1} = \ldots = \frac{dp_n}{-\partial f/\partial x_n}.$$

Hamilton's investigation was extended to the cases when the kinetic potential contains the time, etc. by Ostrogradsky ¶ in 1848–50 and by Donkin** in 1854.

The equation (1) above is often called the *Hamiltonian form of the equation of virtual work.* It may be written in the more symmetrical form

$$\delta \left(\sum_{r=1}^{n} p_r dq_r - H dt \right) = d \left(\sum_{r=1}^{n} p_r \delta q_r - H \delta t \right),$$

* *Brit. Ass. Rep.* 1834, p. 513; *Phil. Trans.* 1835, p. 95.

† *Journal de l'École polyt.* VIII. (Cahier XV), (1809), p. 266.

‡ *Mém. de l'Inst.* 1809, p. 343.

§ *Berlin Abhand.* 1814–15, p. 76.

|| *Bull. soc. philomath.* 1819, p. 10.

¶ *Mélanges de l'Acad. de St.-Pét.* Oct. 1848; *Mém. de l'Acad. de St.-Pét.* VI. (1850), p. 385.

** *Phil. Trans.* 1854, p. 71.

which directly suggests the importance of the differential form

$$\sum_{r=1}^{n} p_r dq_r - Hdt$$

in connexion with the differential equations of dynamics: cf. § 137 below.

When the kinetic potential L does not involve t explicitly, the Hamiltonian function H will evidently likewise not involve t explicitly, and the system will possess (§ 41) an integral of energy, namely

$$\sum_{r=1}^{n} \dot{q}_r \frac{\partial L}{\partial \dot{q}_r} - L = h,$$

where h is a constant. This equation can be written

$$H(q_1, q_2, \ldots, q_n, p_1, p_2, \ldots, p_n) = h,$$

and *this is the integral of energy, which is possessed by the dynamical system when the function H does not involve the time explicitly.* For natural problems, it follows at once from § 41 that H is the sum of the kinetic and potential energies of the system.

Example. Shew that the equations of motion of the simple pendulum are

$$\frac{dq}{dt} = \frac{\partial H}{\partial p}, \quad \frac{dp}{dt} = -\frac{\partial H}{\partial q},$$

where

$$H = \tfrac{1}{2}p^2 l^{-2} - gl \cos q,$$

and where q denotes the angle made by the pendulum with the vertical at time t, l is the length of the pendulum, and the mass of the bob is taken as unity.

110. *Equations arising from the Calculus of Variations.*

From the preceding chapter it appears that the whole science of Dynamics can be based on the stationary character of certain integrals, namely those which occur in Hamilton's principle and the principle of Least Action: similarly the differential equations of most physical problems can be regarded as arising in problems of the Calculus of Variations.

Thus, the problem of finding the state of thermal equilibrium in an isotropic conducting body, when the points of its surface are kept at given temperatures, can be formulated as follows: to find, among all functions V having given values at the surface, that one which makes the value of the integral

$$\iiint \left\{ \left(\frac{\partial V}{\partial x}\right)^2 + \left(\frac{\partial V}{\partial y}\right)^2 + \left(\frac{\partial V}{\partial z}\right)^2 \right\} dx\, dy\, dz,$$

integrated throughout the surface, a minimum.

We shall now shew that *all the differential equations which arise from problems in the Calculus of Variations, with one independent variable, can be expressed in the Hamiltonian form*.*

* Cf. Ostrogradsky, *Mém. de l'Acad. de St.-Pét.* VI. (1850), p. 385.

Suppose, for clearness, that there are two dependent variables; the proof is equally applicable to any number of variables.

Let $L(t, y, \dot{y}, \ddot{y}, \ldots, \overset{(m)}{y}, z, \dot{z}, \ddot{z}, \ldots, \overset{(n)}{z})$ be a function of the independent variable t, the dependent variables y, z, and their derivates up to orders m, n, respectively.

The conditions that the integral

$$\int L(t, y, \dot{y}, \ldots, \overset{(m)}{y}, z, \dot{z}, \ldots, \overset{(n)}{z})\, dt$$

may be stationary, can, by the ordinary procedure of the Calculus of Variations, be written in the form

$$\begin{cases} 0 = \dfrac{\partial L}{\partial y} - \dfrac{d}{dt}\left(\dfrac{\partial L}{\partial \dot{y}}\right) + \ldots + (-1)^m \dfrac{d^m}{dt^m}\left(\dfrac{\partial L}{\partial \overset{(m)}{y}}\right), \\ 0 = \dfrac{\partial L}{\partial z} - \dfrac{d}{dt}\left(\dfrac{\partial L}{\partial \dot{z}}\right) + \ldots + (-1)^n \dfrac{d^n}{dt^n}\left(\dfrac{\partial L}{\partial \overset{(n)}{z}}\right). \end{cases}$$

Now write

$$\begin{cases} p_1 = \dfrac{\partial L}{\partial \dot{y}} - \dfrac{d}{dt}\left(\dfrac{\partial L}{\partial \ddot{y}}\right) + \ldots\ldots\ldots\ldots + (-1)^{m-1} \dfrac{d^{m-1}}{dt^{m-1}}\left(\dfrac{\partial L}{\partial \overset{(m)}{y}}\right); \\ p_2 = \dfrac{\partial L}{\partial \ddot{y}} - \dfrac{d}{dt}\left(\dfrac{\partial L}{\partial \dddot{y}}\right) + \ldots + (-1)^{m-2} \dfrac{d^{m-2}}{dt^{m-2}}\left(\dfrac{\partial L}{\partial \overset{(m)}{y}}\right), \\ \ldots\ldots\ldots\ldots\ldots\ldots\ldots\ldots\ldots\ldots\ldots\ldots\ldots\ldots \\ p_m = \dfrac{\partial L}{\partial \overset{(m)}{y}}, \\ p_{m+1} = \dfrac{\partial L}{\partial \dot{z}} - \dfrac{d}{dt}\left(\dfrac{\partial L}{\partial \ddot{z}}\right) + \ldots\ldots\ldots\ldots + (-1)^{n-1} \dfrac{d^{n-1}}{dt^{n-1}}\left(\dfrac{\partial L}{\partial \overset{(n)}{z}}\right), \\ p_{m+2} = \dfrac{\partial L}{\partial \ddot{z}} - \ldots\ldots\ldots\ldots + (-1)^{n-2} \dfrac{d^{n-2}}{dt^{n-2}}\left(\dfrac{\partial L}{\partial \overset{(n)}{z}}\right), \\ \ldots\ldots\ldots\ldots\ldots\ldots\ldots\ldots\ldots\ldots\ldots\ldots\ldots\ldots \\ p_{m+n} = \dfrac{\partial L}{\partial \overset{(n)}{z}}, \end{cases}$$

and write

$$q_1 = y, \quad q_2 = \dot{y}, \quad \ldots, \quad q_m = \overset{(m-1)}{y}, \quad q_{m+1} = z, \quad q_{m+2} = \dot{z}, \quad \ldots, \quad q_{m+n} = \overset{(n-1)}{z}.$$

Then if

$$H = -L + p_1 q_2 + p_2 q_3 + \ldots + p_{m-1} q_m + p_m \overset{(m)}{y} + p_{m+1} q_{m+2} + \ldots$$
$$+ p_{m+n-1} q_{m+n} + p_{m+n} \overset{(n)}{z},$$

(where H is supposed expressed as a function of $(t, q_1, \ldots, q_{m+n}, p_1, \ldots, p_{m+n})$, the quantities $\overset{(m)}{y}$ and $\overset{(n)}{z}$ being eliminated by use of the equations $p_m = \partial L/\partial \overset{(m)}{y}$, $p_{m+n} = \partial L/\partial \overset{(n)}{z}$), and if δ denote an increment due to small changes in the arguments $q_1, q_2, \ldots, q_{m+n}, p_1, p_2, \ldots, p_{m+n}$, we have

$$\delta H = -\sum_{r=0}^{m-1} \frac{\partial L}{\partial \overset{(r)}{y}} \delta q_{r+1} - \frac{\partial L}{\partial \overset{(m)}{y}} \delta \overset{(m)}{y} - \sum_{r=0}^{n-1} \frac{\partial L}{\partial \overset{(r)}{z}} \delta q_{m+r+1} - \frac{\partial L}{\partial \overset{(n)}{z}} \delta \overset{(n)}{z}$$

$$+ \sum_{r=1}^{m-1} p_r \delta q_{r+1} + p_m \delta \overset{(m)}{y} + \sum_{r=1}^{m-1} q_{r+1} \delta p_r + \overset{(m)}{y} \delta p_m$$

$$+ \sum_{r=m+1}^{m+n-1} p_r \delta q_{r+1} + p_{m+n} \delta \overset{(n)}{z} + \sum_{r=m+1}^{m+n-1} q_{r+1} \delta p_r + \overset{(n)}{z} \delta p_{m+n}.$$

Using the relations

$$\frac{\partial L}{\partial y} = \dot{p}_1, \quad \frac{\partial L}{\partial \dot{y}} = \dot{p}_2 + p_1, \quad \frac{\partial L}{\partial \ddot{y}} = \dot{p}_3 + p_2, \quad \ldots, \quad \frac{\partial L}{\partial \overset{(m)}{y}} = p_m, \text{ etc.,}$$

this becomes
$$\delta H = -\sum_{r=1}^{m+n} \dot{p}_r \delta q_r + \sum_{r=1}^{m+n} \dot{q}_r \delta p_r.$$

Thus, if H is expressed in terms of the variables

$$(t, p_1, p_2, \ldots, p_{m+n}, q_1, q_2, \ldots, q_{m+n}),$$

we have
$$\frac{dq_r}{dt} = \frac{\partial H}{\partial p_r} \qquad \frac{dp_r}{dt} = -\frac{\partial H}{\partial q_r} \qquad (r = 1, 2, \ldots, m+n),$$

and *the differential equations of the problem are thus expressed in the Hamiltonian form.*

The systems of differential equations which arise in the problems of the Calculus of Variations are often called *isoperimetrical* systems.

111. *Integral-invariants.*

The nature of Hamiltonian systems of differential equations is fundamentally connected with the properties of certain expressions to which Poincaré* has given the name *integral-invariants.*

Consider any system of ordinary differential equations

$$\frac{dx_1}{dt} = X_1, \qquad \frac{dx_2}{dt} = X_2, \qquad \ldots, \qquad \frac{dx_n}{dt} = X_n,$$

where $X_1, X_2, \ldots, X_n$ are given functions of $x_1, x_2, \ldots, x_n; t$. We may regard these equations as defining the motion of a point whose coordinates are $(x_1, x_2, \ldots, x_n)$ in space of n dimensions.

* *Acta Math.* XIII. (1890).

If now we consider a group of such points, which occupy a p-dimensional region ζ_0 at the beginning of the motion, they will at any subsequent time t occupy another p-dimensional region ζ. A p-tuple integral taken over ζ is called an *integral-invariant,* if it has the same value at all times t; the number p is called the *order* of the integral-invariant.

Thus, in the motion of an incompressible fluid, the integral which represents the volume of the fluid, when the integration is extended over all the elements of fluid which were contained initially in any given region, is an integral-invariant; since the total volume occupied by these elements does not vary with the time.

Example 1. Consider the dynamical problem of determining the motion of a particle in a plane under no forces: let (x, y) be the coordinates of the particle, and (u, v) its components of velocity. The equations of motion may be written

$$\dot{x}=u, \quad \dot{y}=v, \quad \dot{u}=0, \quad \dot{v}=0.$$

The quantity

$$I=\int(\delta x - t\delta u),$$

where the integration is taken, in the four-dimensional space in which (x, y, u, v) are coordinates, along the curvilinear arc which is the locus at time t of points which were initially on some given curvilinear arc in the space, is an integral-invariant. For the solution of the dynamical problem is given by the equations

$$u=a, \quad v=b, \quad x=at+c, \quad y=bt+d,$$

where a, b, c, d are constants: and therefore we have

$$I=\int(t\delta a+\delta c-t\delta a)$$
$$=\int\delta c,$$

and this is independent of t.

Example 2. In the plane motion of a particle whose coordinates are (x, y) and whose velocity-components are (u, v), under the influence of a centre of force at the origin whose attraction is directly proportional to the distance, shew that

$$\int(u\delta x - x\delta u)$$

is an integral-invariant.

112. *The variational equations.*

The integral-invariants of a given system of differential equations furnish integrals of another system of differential equations which can be derived from these.

For let the given system of equations be

$$\frac{dx_r}{dt}=X_r(x_1, x_2, \ldots, x_n, t) \qquad (r=1, 2, \ldots, n).$$

Let $(x_1, x_2, \ldots, x_n)$ and $(x_1+\delta x_1, x_2+\delta x_2, \ldots, x_n+\delta x_n)$ be the values of the dependent variables at time t in two neighbouring solutions of this set of equations; where $(\delta x_1, \delta x_2, \ldots, \delta x_n)$ are infinitesimal quantities. Then we have

$$\frac{d}{dt}(x_r+\delta x_r)=X_r(x_1+\delta x_1, x_2+\delta x_2, \ldots, x_n+\delta x_n, t) \qquad (r=1, 2, \ldots, n),$$

and consequently,

$$\frac{d}{dt}\,\delta x_r = \frac{\partial X_r}{\partial x_1}\,\delta x_1 + \frac{\partial X_r}{\partial x_2}\,\delta x_2 + \ldots + \frac{\partial X_r}{\partial x_n}\,\delta x_n \qquad (r = 1, 2, \ldots, n).$$

These last n equations, together with the original n equations, may be regarded as a set of $2n$ equations in which $(x_1, x_2, \ldots, x_n, \delta x_1, \delta x_2, \ldots, \delta x_n)$ are the dependent variables.

Now if

$$\int \sum_r F_r(x_1, x_2, \ldots, x_n)\,\delta x_r$$

denotes an integral-invariant of the original system, the quantity

$$\frac{d}{dt}\left\{\sum_r F_r(x_1, x_2, \ldots, x_n)\,\delta x_r\right\}$$

must, since the path of integration is quite arbitrary, be zero in virtue of precisely this extended system of differential equations; and therefore

$$\sum_r F_r(x_1, x_2, \ldots, x_n)\,\delta x_r = \text{constant}$$

must be an integral of these equations: so that *to an integral-invariant of order one of the original system of equations there corresponds an integral of the extended system of equations, and vice versa.*

If a *particular* solution $(x_1, x_2, \ldots, x_n)$ of the original equations is known, we can substitute the corresponding values $(x_1, x_2, \ldots, x_n)$ in the extended differential equations, and so obtain n linear differential equations to determine $(\delta x_1, \delta x_2, \ldots, \delta x_n)$, i.e. to determine the solutions of the original equations which are adjacent to the known particular solution. These n equations are called the *variational* equations.

113. *Integral-invariants of order one.*

Let us now find the conditions to be satisfied in order that

$$\int (M_1\delta x_1 + M_2\delta x_2 + \ldots + M_n\delta x_n),$$

where $(M_1, M_2, \ldots, M_n)$ are functions of $(x_1, x_2, \ldots, x_n, t)$, may be an integral-invariant of order one of the system of differential equations

$$dx_r/dt = X_r(x_1, x_2, \ldots, x_n, t) \qquad (r = 1, 2, \ldots, n).$$

We must have

$$\frac{d}{dt}(M_1\delta x_1 + M_2\delta x_2 + \ldots + M_n\delta x_n) = 0,$$

where the derivates of $(\delta x_1, \delta x_2, \ldots, \delta x_n)$ are to be determined by the extended system of differential equations introduced in the last article; and therefore

$$\sum_{r=1}^{n}\left(\frac{dM_r}{dt}\,\delta x_r + M_r\frac{d\,.\,\delta x_r}{dt}\right)=0,$$

or
$$\sum_{r=1}^{n}\left(\frac{\partial M_r}{\partial t}\,\delta x_r+\sum_{k=1}^{n}\frac{\partial M_r}{\partial x_k}X_k\,\delta x_r+M_r\sum_{k=1}^{n}\frac{\partial X_r}{\partial x_k}\,\delta x_k\right)=0.$$

Since $(\delta x_1, \delta x_2, \ldots, \delta x_n)$ are independent, the coefficient of each quantity δx_r in this equation must be zero: and consequently *the conditions for integral-invariancy are*

$$\frac{\partial M_r}{\partial t}+\sum_{k=1}^{n}\frac{\partial M_r}{\partial x_k}X_k+\sum_{k=1}^{n}M_k\frac{\partial X_k}{\partial x_r}=0 \qquad (r=1, 2, \ldots, n).$$

Corollary 1. If an integral of the differential equations, say

$$F(x_1, x_2, \ldots, x_n, t)=\text{constant},$$

is known, we can at once determine an integral-invariant.

For we have

$$\begin{aligned}\frac{\partial}{\partial t}\left(\frac{\partial F}{\partial x_r}\right)+\sum_{k=1}^{n}\frac{\partial}{\partial x_k}\left(\frac{\partial F}{\partial x_r}\right)X_k+\sum_{k=1}^{n}\frac{\partial F}{\partial x_k}\frac{\partial X_k}{\partial x_r}&=\frac{\partial}{\partial x_r}\left(\frac{\partial F}{\partial t}+\sum_{k=1}^{n}\frac{\partial F}{\partial x_k}X_k\right)\\&=\frac{\partial}{\partial x_r}\left(\frac{dF}{dt}\right)\\&=0,\end{aligned}$$

and therefore *the expression*

$$\int\left(\sum_{r=1}^{n}\frac{\partial F}{\partial x_r}\,\delta x_r\right)$$

is an integral-invariant.

Corollary 2. The converse of Corollary 1 is also true, namely that *if* $\int\left(\sum_{r=1}^{n}\frac{\partial U}{\partial x_r}\,\delta x_r\right)$ *is an integral-invariant of the differential equations, where U is a given function of the variables, then an integral of the system can be found.*

For we have

$$\begin{aligned}0&=\frac{\partial}{\partial t}\left(\frac{\partial U}{\partial x_r}\right)+\sum_{k=1}^{n}X_k\frac{\partial}{\partial x_k}\left(\frac{\partial U}{\partial x_r}\right)+\sum_{k=1}^{n}\frac{\partial U}{\partial x_k}\frac{\partial X_k}{\partial x_r}\\&=\frac{\partial}{\partial x_r}\left(\frac{\partial U}{\partial t}+\sum_{k=1}^{n}\frac{\partial U}{\partial x_k}X_k\right),\end{aligned}$$

and consequently the expression

$$\frac{\partial U}{\partial t}+\sum_{k=1}^{n}\frac{\partial U}{\partial x_k}X_k,$$

which is a given function of $(x_1, x_2, \ldots, x_n, t)$, is independent of $(x_1, x_2, \ldots, x_n)$; let its value be $\phi(t)$: this is a known quantity.

Then we have $$dU/dt = \phi(t),$$

or $$U - \int \phi(t)\, dt = \text{constant};$$

and this is an integral of the system.

114. *Relative integral-invariants.*

Hitherto we have only considered those integral-invariants which have the invariantive property when the domain of the initial values, over which the integration is taken, is quite arbitrary; these are sometimes called *absolute* integral-invariants. We shall now consider integrals which have the invariantive property only when the domain over which the integration is taken is a *closed* manifold (using the language of n-dimensional geometry); these are called *relative* integral-invariants.

The theory of relative integral-invariants can be reduced to that of absolute integral-invariants in the following way.

Let $$\int (M_1 \delta x_1 + M_2 \delta x_2 + \ldots + M_n \delta x_n)$$

be a relative integral-invariant of the equations

$$dx_r/dt = X_r \qquad (r = 1, 2, \ldots, n),$$

where $(M_1, M_2, \ldots, M_n, X_1, X_2, \ldots, X_n)$ are functions of $(x_1, x_2, \ldots, x_n, t)$; so that this expression is invariable with respect to t when the integration is taken, in the space in which $(x_1, x_2, \ldots, x_n)$ are coordinates, round the closed curve which is the locus at time t of points which were initially situated on some definite closed curve in the space.

By Stokes' theorem, this integral is equivalent to the integral

$$\iint \sum_{i,j} \left(\frac{\partial M_i}{\partial x_j} - \frac{\partial M_j}{\partial x_i} \right) \delta x_i \delta x_j,$$

where the integration is now taken over a diaphragm bounded by the curve; this diaphragm can be taken to be the locus at time t of points which were originally situated on a definite diaphragm bounded by the initial position of the closed curve: and since the diaphragm is not a closed surface, this integral is an *absolute* integral-invariant of order two of the equations.

Similarly, by a generalisation of Stokes' theorem, any relative integral invariant of order p is equivalent to an absolute integral-invariant of order $(p + 1)$.

115. *A relative integral-invariant which is possessed by all Hamiltonian systems.*

Consider now the case in which the system of differential equations is a Hamiltonian system, so that it can be written

$$\frac{dq_r}{dt} = \frac{\partial H}{\partial p_r}, \qquad \frac{dp_r}{dt} = -\frac{\partial H}{\partial q_r} \qquad (r = 1, 2, \ldots, n),$$

where H is a given function of $(q_1, q_2, \ldots, q_n, p_1, p_2, \ldots, p_n, t)$.

For this system let

$$\Omega = \int L\,dt$$

denote Hamilton's integral, so that L is the kinetic potential; let

$$(\alpha_1, \alpha_2, \ldots, \alpha_n, \beta_1, \beta_2, \ldots, \beta_n)$$

be the initial values of the variables

$$(q_1, q_2, \ldots, q_n, p_1, p_2, \ldots, p_n)$$

respectively, and let δ denote the variation from a point of one orbit to the contemporaneous point of an adjacent orbit. By § 99, we have

$$\delta\Omega = \sum_{r=1}^{n} p_r \delta q_r - \sum_{r=1}^{n} \beta_r \delta\alpha_r.$$

Let C_0 denote any closed curve in the space of $2p$ dimensions in which $(q_1, q_2, \ldots, q_n, p_1, p_2, \ldots, p_n)$ are coordinates, and let C denote the closed curve which is the locus at time t of the points which are initially on C_0. Integrating the last equation round the set of trajectories which pass from C_0 to C, we have

$$\int_C \sum_{r=1}^{n} p_r \delta q_r = \int_{C_0} \sum_{r=1}^{n} \beta_r \delta\alpha_r,$$

and this equation shews that *the quantity* $\int \sum_{r=1}^{n} p_r \delta q_r$ *is a relative integral-invariant of any Hamiltonian system of differential equations.*

116. *On systems which possess the relative integral-invariant* $\int \Sigma p\,\delta q$.

We shall next study the converse problem suggested by the result of the last article, namely that of determining all the systems of differential equations which possess the relative integral-invariant $\int \sum_{r=1}^{n} p_r \delta q_r$, where $(q_1, q_2, \ldots, q_n)$ are half the dependent variables, and $(p_1, p_2, \ldots, p_n)$ are the other half.

Consider then a system of ordinary differential equations of order $2n$ in which the variables can be separated into two sets, $(q_1, q_2, \ldots, q_n)$ and $(p_1, p_2, \ldots, p_n)$, such that

$$\int (p_1 \delta q_1 + p_2 \delta q_2 + \ldots + p_n \delta q_n)$$

is a relative integral-invariant of the equations, and consequently by Stokes' theorem

$$\iint (\delta p_1 \delta q_1 + \delta p_2 \delta q_2 + \ldots + \delta p_n \delta q_n)$$

is an absolute integral-invariant.

Let the system of differential equations be

$$\frac{dq_r}{dt} = Q_r, \qquad \frac{dp_r}{dt} = P_r \qquad (r = 1, 2, \ldots, n),$$

where $(Q_1, Q_2, \ldots, Q_n, P_1, P_2, \ldots, P_n)$ are given functions of

$$(q_1, q_2, \ldots, q_n, p_1, p_2, \ldots, p_n, t).$$

As the domain of integration of the absolute integral-invariant is of two dimensions, we can suppose that each point in it is specified by two quantities λ and μ, which do not vary with the time but are characteristic of the trajectory on which the point in question lies. The absolute integral-invariant can therefore be written in the form

$$\iint \left(\sum_{i=1}^{n} \frac{\partial (q_i, p_i)}{\partial (\lambda, \mu)} \right) d\lambda \, d\mu,$$

and as λ and μ do not vary with the time, we must have

$$\frac{d}{dt} \sum_{i=1}^{n} \frac{\partial (q_i, p_i)}{\partial (\lambda, \mu)} = 0,$$

or

$$\sum_{i=1}^{n} \left\{ \frac{\partial (Q_i, p_i)}{\partial (\lambda, \mu)} + \frac{\partial (q_i, P_i)}{\partial (\lambda, \mu)} \right\} = 0,$$

or

$$\sum_{i=1}^{n} \sum_{k=1}^{n} \left\{ \frac{\partial Q_i}{\partial q_k} \frac{\partial (q_k, p_i)}{\partial (\lambda, \mu)} + \frac{\partial Q_i}{\partial p_k} \frac{\partial (p_k, p_i)}{\partial (\lambda, \mu)} + \frac{\partial P_i}{\partial q_k} \frac{\partial (q_i, q_k)}{\partial (\lambda, \mu)} + \frac{\partial P_i}{\partial p_k} \frac{\partial (q_i, p_k)}{\partial (\lambda; \mu)} \right\} = 0.$$

Owing to the complete arbitrariness of the domain of integration and the choice of λ and μ, the coefficients of $\frac{\partial q_k}{\partial \lambda} \frac{\partial p_i}{\partial \mu}$, $\frac{\partial q_i}{\partial \lambda} \frac{\partial q_k}{\partial \mu}$, and $\frac{\partial p_k}{\partial \lambda} \frac{\partial p_i}{\partial \mu}$ in this equation must vanish separately. We thus obtain

$$\left.\begin{aligned} \frac{\partial Q_i}{\partial q_k} + \frac{\partial P_k}{\partial p_i} &= 0 \\ \frac{\partial P_i}{\partial q_k} - \frac{\partial P_k}{\partial q_i} &= 0 \\ \frac{\partial Q_i}{\partial p_k} - \frac{\partial Q_k}{\partial p_i} &= 0 \end{aligned}\right\} \qquad (i, k = 1, 2, \ldots, n).$$

These equations shew that a function $H(q_1, q_2, \ldots, q_n, p_1, p_2, \ldots, p_n, t)$ exists such that

$$Q_r = \partial H/\partial p_r, \qquad P_r = -\partial H/\partial q_r \qquad (r = 1, 2, \ldots, n);$$

and thus we have the result that *if a system of equations*

$$\frac{dq_r}{dt} = Q_r, \qquad \frac{dp_r}{dt} = P_r \qquad (r = 1, 2, \ldots, n)$$

possesses the relative integral-invariant

$$\int (p_1 \delta q_1 + p_2 \delta q_2 + \ldots + p_n \delta q_n),$$

then the equations have the Hamiltonian form

$$\frac{dq_r}{dt} = \frac{\partial H}{\partial p_r}, \quad \frac{dp_r}{dt} = -\frac{\partial H}{\partial q_r} \qquad (r = 1, 2, \ldots, n);$$

this is the converse of the theorem of the last article.

Corollary. If

$$\int (p_1 \delta q_1 + p_2 \delta q_2 + \ldots + p_n \delta q_n)$$

is a relative integral-invariant of a system of equations

$$dq_r/dt = Q_r, \qquad dp_r/dt = P_r \qquad (r = 1, 2, \ldots, k),$$

where k is greater than n, it follows in the same way that the equations for $(q_1, q_2, \ldots, q_n, p_1, p_2, \ldots, p_n)$ form a Hamiltonian system

$$\frac{dq_r}{dt} = \frac{\partial H}{\partial p_r}, \quad \frac{dp_r}{dt} = -\frac{\partial H}{\partial q_r} \qquad (r = 1, 2, \ldots, n),$$

where H is a function of $(q_1, q_2, \ldots, q_n, p_1, p_2, \ldots, p_n, t)$ only, not involving $(q_{n+1}, q_{n+2}, \ldots, q_k, p_{n+1}, \ldots, p_k)$.

117. *The expression of integral-invariants in terms of integrals.*

If the solution of a system of differential equations

$$\frac{dx_r}{dt} = X_r(x_1, x_2, \ldots, x_n, t) \qquad (r = 1, 2, \ldots, n)$$

is known, the absolute and relative integral-invariants of the system may easily be constructed.

Thus, let

$$y_1(x_1, x_2, \ldots, x_n, t) = c_1, \; y_2(x_1, x_2, \ldots, x_n, t) = c_2, \; \ldots, \; y_n(x_1, x_2, \ldots, x_n, t) = c_n,$$

where $c_1, c_2, \ldots, c_n$ are constants, be n integrals of the system; the absolute integral-invariants of order one are evidently given by the formula

$$\int (N_1 \delta y_1 + N_2 \delta y_2 + \ldots + N_n \delta y_n),$$

where $(N_1, N_2, \ldots, N_n)$ are any functions of $(y_1, y_2, \ldots, y_n)$ which do not involve t: and the relative integral-invariants of order one are given by the formula

$$\int (N_1 \delta y_1 + N_2 \delta y_2 + \ldots + N_n \delta y_n + \delta F),$$

where F is any function of $(x_1, x_2, \ldots, x_n, t)$, since the term $\int \delta F$ vanishes when the domain of integration is closed.

It follows from this that *any system of differential equations possesses an infinite number of absolute and relative integral-invariants of the first order.*

118. *The theorem of Lie and Koenigs.*

The preceding results enable us to establish a theorem due to Lie* and Koenigs† on the reduction of any system of ordinary differential equations to the Hamiltonian form.

Let
$$\frac{dx_r}{dt} = X_r \qquad (r = 1, 2, \ldots, k)$$

be the given system of equations, and let

$$\int (\xi_1 \delta x_1 + \xi_2 \delta x_2 + \ldots + \xi_k \delta x_k)$$

be any relative or absolute integral-invariant of order one of this system, where $\xi_1, \xi_2, \ldots, \xi_k$ are given functions of the variables: we have seen in the last article that an infinite number of such integral-invariants exist.

Now let the differential form

$$\xi_1 \delta x_1 + \xi_2 \delta x_2 + \ldots + \xi_k \delta x_k$$

be reduced to the canonical form

$$p_1 \delta q_1 + p_2 \delta q_2 + \ldots + p_n \delta q_n - \delta \Omega,$$

where
$$(p_1, p_2, \ldots, p_n, q_1, q_2, \ldots, q_n, \Omega)$$

are independent functions of $(x_1, x_2, \ldots, x_k)$, in number not greater than k, and where Ω may be zero‡. Let $(u_1, u_2, \ldots, u_{k-2n})$ be a set of other functions of $(x_1, x_2, \ldots, x_k)$, such that $(u_1, u_2, \ldots, u_{k-2n}, q_1, q_2, \ldots, q_n, p_1, p_2, \ldots, p_n)$ are a set of k independent functions of $(x_1, x_2, \ldots, x_k)$; and suppose that the

* *Archiv for Math. og Natur.* II. (1877), p. 10.

† *Comptes Rendus*, CXXI. (1895), p. 875.

‡ The proof of the possibility of this reduction (which however requires in general the solution of a number of ordinary differential equations) will be found in any treatise on Pfaff's problem.

system of differential equations, when expressed in terms of these k functions as independent variables, becomes

$$dq_r/dt = Q_r, \qquad dp_r/dt = P_r \qquad (r = 1, 2, \ldots, n),$$
$$du_s/dt = U_s \qquad (s = 1, 2, \ldots, k-2n),$$

where $(Q_1, Q_2, \ldots, Q_n, P_1, P_2, \ldots, P_n, U_1, U_2, \ldots, U_{k-2n})$ are functions of the new variables.

The expression

$$\int (p_1 \delta q_1 + p_2 \delta q_2 + \ldots + p_n \delta q_n)$$

is an integral-invariant (relative or absolute) of this system, since integral-invariancy is a property unaffected by such transformations as have been performed: and consequently it follows (§ 116) that the first $2n$ equations have the form

$$\frac{dq_r}{dt} = \frac{\partial H}{\partial p_r}, \qquad \frac{dp_r}{dt} = -\frac{\partial H}{\partial q_r} \qquad (r = 1, 2, \ldots, n),$$

where H is a function of $(q_1, q_2, \ldots, q_n, p_1, p_2, \ldots, p_n, t)$ only. *The given system of differential equations is thus reduced to a Hamiltonian system of order $2n$, together with the $(k-2n)$ additional equations*

$$\frac{du_s}{dt} = U_s \qquad (s = 1, 2, \ldots, k-2n).$$

119. *The Last Multiplier.*

Before proceeding to discuss integral-invariants of higher order than those hitherto considered, we shall introduce the conception, introduced by Jacobi* in 1844, of the *Last Multiplier* of a system of equations.

Let
$$\frac{dx_1}{X_1} = \frac{dx_2}{X_2} = \ldots = \frac{dx_n}{X_n} = \frac{dx}{X},$$

where $(X_1, X_2, \ldots, X_n, X)$ are given functions of the variables $(x_1, x_2, \ldots, x_n, x)$, be a given system of equations: and suppose that $(n-1)$ integrals of this system are known, say

$$f_r(x_1, x_2, \ldots, x_n, x) = a_r \qquad (r = 1, 2, \ldots, n-1).$$

From these equations let $(x_1, x_2, \ldots, x_{n-1})$ be expressed as functions of x_n and x: then there remains only the solution of the equation of the first order

$$\frac{dx_n}{X_n'} = \frac{dx}{X'}$$

to be effected; in which accents are used to denote that $(x_1, x_2, \ldots, x_{n-1})$ have been replaced in X_n and X by the values thus obtained.

* *Crelle's Journal*, XXVII. p. 199, XXIX. pp. 213, 333.

We shall shew that *the integral of this equation is*

$$\int \frac{M'}{\Delta'} (X' dx_n - X_n' dx) = \text{constant},$$

where M denotes any solution of the partial differential equation

$$\frac{\partial}{\partial x_1} (MX_1) + \frac{\partial}{\partial x_2} (MX_2) + \dots + \frac{\partial}{\partial x_n} (MX_n) + \frac{\partial}{\partial x} (MX) = 0,$$

and Δ denotes the Jacobian

$$\frac{\partial (f_1, f_2, \dots, f_{n-1})}{\partial (x_1, x_2, \dots, x_{n-1})}.$$

The function M is called the *Last Multiplier* of the system of differential equations.

For the proof of this theorem, we shall require the following lemma:

If a system of differential equations

$$dx_r/dt = X_r \qquad (r = 1, 2, \dots, n)$$

is transformed by change of variables into another system

$$dy_r/dt = Y_r \qquad (r = 1, 2, \dots, n),$$

then
$$\sum_{r=1}^{n} \frac{\partial X_r}{\partial x_r} = \frac{1}{D} \sum_{r=1}^{n} \frac{\partial (DY_r)}{\partial y_r},$$

where D denotes the Jacobian

$$\frac{\partial (x_1, x_2, \dots, x_n)}{\partial (y_1, y_2, \dots, y_n)}.$$

To prove this, we have

$$\sum_{r=1}^{n} \frac{\partial X_r}{\partial x_r} = \sum_{r=1}^{n} \frac{\partial}{\partial x_r} \left(\sum_{k=1}^{n} Y_k \frac{\partial x_r}{\partial y_k} \right)$$

$$= \sum_{r=1}^{n} \sum_{s=1}^{n} \frac{\partial y_s}{\partial x_r} \frac{\partial}{\partial y_s} \left(\sum_{k=1}^{n} Y_k \frac{\partial x_r}{\partial y_k} \right)$$

$$= \sum_{r=1}^{n} \sum_{s=1}^{n} \sum_{k=1}^{n} \frac{\partial y_s}{\partial x_r} \left(Y_k \frac{\partial^2 x_r}{\partial y_s \partial y_k} + \frac{\partial Y_k}{\partial y_s} \frac{\partial x_r}{\partial y_k} \right).$$

In this expression the coefficient of $\partial Y_k/\partial y_s$ is $\sum_{r=1}^{n} \frac{\partial y_s}{\partial x_r} \frac{\partial x_r}{\partial y_k}$, which is zero or unity according as s is different from, or equal to, k. Also $\partial y_s/\partial x_r = A_{rs}/D$, where A_{rs} denotes the minor of $\partial x_r/\partial y_s$ in the determinant D: so the coefficient of Y_k in the above expression, which is

$$\sum_{r=1}^{n} \sum_{s=1}^{n} \frac{\partial y_s}{\partial x_r} \frac{\partial^2 x_r}{\partial y_s \partial y_k},$$

may be written

$$\frac{1}{D}\sum_{r=1}^{n}\sum_{s=1}^{n} A_{rs}\frac{\partial^2 x_r}{\partial y_s \partial y_k}, \text{ or } \frac{1}{D}\sum_{r=1}^{n}\frac{\partial(x_1, x_2, \ldots, x_{r-1}, \partial x_r/\partial y_k, x_{r+1}, \ldots, x_n)}{\partial(y_1, y_2, \ldots, y_n)},$$

or

$$\frac{1}{D}\frac{\partial D}{\partial y_k}.$$

We have therefore

$$\sum_{r=1}^{n}\frac{\partial X_r}{\partial x_r} = \sum_{k=1}^{n}\frac{\partial Y_k}{\partial y_k} + \sum_{k=1}^{n} Y_k \cdot \frac{1}{D}\frac{\partial D}{\partial y_k}$$

$$= \frac{1}{D}\sum_{k=1}^{n}\frac{\partial(DY_k)}{\partial y_k},$$

which establishes the lemma.

Now in the original problem write

$$\frac{dx_1}{X_1} = \frac{dx_2}{X_2} = \ldots = \frac{dx_n}{X_n} = \frac{dx}{X} = dt,$$

and consider the change of variables from

$$(x_1, x_2, \ldots, x_n, x) \text{ to } (a_1, a_2, \ldots, a_{n-1}, x_n, x):$$

by the lemma, we have

$$\frac{\partial X_1}{\partial x_1} + \frac{\partial X_2}{\partial x_2} + \ldots + \frac{\partial X_n}{\partial x_n} + \frac{\partial X}{\partial x} = \Delta\left\{\frac{\partial}{\partial x_n}\left(\frac{X_n'}{\Delta'}\right) + \frac{\partial}{\partial x}\left(\frac{X'}{\Delta'}\right)\right\},$$

so the quantity M, which is a solution of the equation

$$\frac{1}{M}\frac{dM}{dt} + \frac{\partial X_1}{\partial x_1} + \frac{\partial X_2}{\partial x_2} + \ldots + \frac{\partial X_n}{\partial x_n} + \frac{\partial X}{\partial x} = 0,$$

satisfies the equation

$$\frac{1}{\Delta M}\frac{dM}{dt} + \frac{\partial}{\partial x_n}\left(\frac{X_n'}{\Delta'}\right) + \frac{\partial}{\partial x}\left(\frac{X'}{\Delta'}\right) = 0,$$

or

$$\frac{\partial}{\partial x_n}\left(\frac{X_n' M'}{\Delta'}\right) + \frac{\partial}{\partial x}\left(\frac{X' M'}{\Delta'}\right) = 0,$$

which shews that the expression

$$\frac{M'}{\Delta'}(X' dx_n - X_n' dx)$$

is the perfect differential of some function of x_n and x; this establishes the theorem of the Last Multiplier.

Boltzmann and Larmor's hydrodynamical representation of the Last Multiplier.

The theorem of the Last Multiplier may also be made apparent by physical considerations. For simplicity we shall take the number of variables to be three, so that the differential equations may be written

$$\frac{dx}{u} = \frac{dy}{v} = \frac{dz}{w},$$

where (u, v, w) are given functions of (x, y, z); and the last multiplier M satisfies the equation

$$\frac{\partial}{\partial x}(Mu)+\frac{\partial}{\partial y}(Mv)+\frac{\partial}{\partial z}(Mw)=0.$$

This equation shews that in the hydrodynamical problem of the steady motion of a fluid in which (u, v, w) are the velocity-components at the point (x, y, z), the equation of continuity is satisfied when M is taken as the density of the fluid at the point (x, y, z).

Now let
$$\phi(x, y, z)=C$$
be an integral of the differential equations; then the flow will take place between the surfaces represented by this equation; thus we can consider separately the flow in the two-dimensional sheet between consecutive surfaces C and $C+\delta C$. The flow through the gap between any two given points P and Q on C must be the same whatever be the arc joining P and Q across which it is estimated: and since the flow across arcs PR and RQ together is the same as that across PQ, we see that the flow across an arc joining P and Q must be expressible in the form $f(Q)-f(P)$. So if ds denotes an element of this arc, and τ the (variable) thickness of the sheet, so that $\tau=\{(\partial\phi/\partial x)^2+(\partial\phi/\partial y)^2+(\partial\phi/\partial z)^2\}^{-\frac{1}{2}}\,.\,\delta C$, and if ξ denotes the velocity-component perpendicular to ds, we have

$$\int_P^Q M\xi\tau\, ds=f(Q)-f(P),$$

so that $M\xi\tau\, ds$ is the perfect differential of a function of position. But it is easily seen that this expression can be written in the form $M\delta C(v\,dx-u\,dy)/(\partial\phi/\partial z)$; and consequently

$$\frac{M(v\,dx-u\,dy)}{\partial\phi/\partial z}$$

is a perfect differential; this is the theorem of the last multiplier for the case considered.

We readily find for ξds the value

$$(\phi_x{}^2+\phi_y{}^2+\phi_z{}^2)^{-\frac{1}{2}}\,.\begin{vmatrix} dx & dy & dz \\ u & v & w \\ \phi_x & \phi_y & \phi_z \end{vmatrix};$$

so the theorem really states that $M(\phi_x{}^2+\phi_y{}^2+\phi_z{}^2)^{-1}$ *is an integrating factor of the equation*

$$\begin{vmatrix} dx & dy & dz \\ u & v & w \\ \phi_x & \phi_y & \phi_z \end{vmatrix}=0.$$

This, as was remarked by Appell (*Comptes Rendus*, CLV. (1912), p. 878), is a symmetrical form of the theorem of the Last Multiplier.

120. *Derivation of an integral from two multipliers.*

Suppose now that two distinct solutions M and N of the partial differential equation of the last multiplier have been obtained, so that

$$\left(X_1\frac{\partial}{\partial x_1}+X_2\frac{\partial}{\partial x_2}+\ldots+X_n\frac{\partial}{\partial x_n}+X\frac{\partial}{\partial x}\right)\log M+\frac{\partial X_1}{\partial x_1}+\frac{\partial X_2}{\partial x_2}+\ldots+\frac{\partial X_n}{\partial x_n}+\frac{\partial X}{\partial x}=0,$$

and

$$\left(X_1\frac{\partial}{\partial x_1}+X_2\frac{\partial}{\partial x_2}+\ldots+X_n\frac{\partial}{\partial x_n}+X\frac{\partial}{\partial x}\right)\log N+\frac{\partial X_1}{\partial x_1}+\frac{\partial X_2}{\partial x_2}+\ldots+\frac{\partial X_n}{\partial x_n}+\frac{\partial X}{\partial x}=0.$$

Subtracting these equations, we have

$$\left(X_1 \frac{\partial}{\partial x_1} + X_2 \frac{\partial}{\partial x_2} + \ldots + X_n \frac{\partial}{\partial x_n} + X \frac{\partial}{\partial x}\right) \log \frac{M}{N} = 0\,;$$

but this is the condition that the equation

$$\log (M/N) = \text{constant}$$

shall be an integral of the system

$$\frac{dx_1}{X_1} = \frac{dx_2}{X_2} = \ldots = \frac{dx_n}{X_n} = \frac{dx}{X},$$

and we have therefore the theorem that *the quotient of two last multipliers of a system of differential equations is an integral of the system.*

The reader who is acquainted with the theory of infinitesimal transformations will be able to prove without difficulty that if the equation

$$X_1 \frac{\partial f}{\partial x_1} + X_2 \frac{\partial f}{\partial x_2} + \ldots + X_n \frac{\partial f}{\partial x_n} + X \frac{\partial f}{\partial x} = 0$$

admits the infinitesimal transformations

$$\xi_{i1} \frac{\partial f}{\partial x_1} + \xi_{i2} \frac{\partial f}{\partial x_2} + \ldots + \xi_{in} \frac{\partial f}{\partial x_n} + \xi_i \frac{\partial f}{\partial x} \qquad (i = 1, 2, \ldots, n),$$

then the reciprocal of the determinant

$$\begin{vmatrix} X_1 & X_2 & \ldots\ldots & X_n & X \\ \xi_{11} & \xi_{12} & \ldots\ldots & \xi_{1n} & \xi_1 \\ \ldots & \ldots & \ldots & \ldots & \ldots \\ \ldots & \ldots & \ldots & \ldots & \ldots \\ \xi_{n1} & \xi_{n2} & \ldots\ldots & \xi_{nn} & \xi_n \end{vmatrix}$$

is a last multiplier.

121. *Application of the last multiplier to Hamiltonian systems: use of a single known integral.*

If the system of differential equations considered is a Hamiltonian system, we have evidently $\sum_r \partial X_r / \partial x_r = 0$, and consequently $M = 1$ is a solution of the partial differential equation which determines the last multiplier; so *the last multiplier of a Hamiltonian system of equations is unity.*

From this result we can deduce a theorem which enables us to integrate completely any conservative holonomic dynamical system with two degrees of freedom when one integral is known in addition to the integral of energy.

Let the system be

$$\frac{dq_1}{\dfrac{\partial H}{\partial p_1}} = \frac{dq_2}{\dfrac{\partial H}{\partial p_2}} = \frac{dp_1}{-\dfrac{\partial H}{\partial q_1}} = \frac{dp_2}{-\dfrac{\partial H}{\partial q_2}} = dt,$$

and in addition to the integral of energy $H(q_1, q_2, p_1, p_2) = h$, let an integral $V(q_1, q_2, p_1, p_2) = c$ be known. From the theorem of the last multiplier it follows that

$$\int \frac{1}{\dfrac{\partial (V, H)}{\partial (p_1, p_2)}} \left\{ \frac{\partial H}{\partial p_2} dq_1 - \frac{\partial H}{\partial p_1} dq_2 \right\} = \text{constant}$$

is another integral; where, in the integrand, p_1 and p_2 are supposed to be replaced by their values in terms of q_1 and q_2 obtained from the known integrals H and V.

But if we suppose that the result of solving the equations $H=h$ and $V=c$ for p_1 and p_2 is represented by the equations

$$\begin{cases} p_1 = f_1(q_1, q_2, h, c), \\ p_2 = f_2(q_1, q_2, h, c), \end{cases}$$

then we have identically

$$\begin{cases} \dfrac{\partial H}{\partial p_1}\dfrac{\partial f_1}{\partial c} + \dfrac{\partial H}{\partial p_2}\dfrac{\partial f_2}{\partial c} = 0, \\ \dfrac{\partial V}{\partial p_1}\dfrac{\partial f_1}{\partial c} + \dfrac{\partial V}{\partial p_2}\dfrac{\partial f_2}{\partial c} = 1, \end{cases}$$

and therefore

$$\frac{\partial f_1}{\partial c} = \frac{\partial H/\partial p_2}{\dfrac{\partial(V, H)}{\partial(p_1, p_2)}}, \qquad \frac{\partial f_2}{\partial c} = \frac{-\partial H/\partial p_1}{\dfrac{\partial(V, H)}{\partial(p_1, p_2)}},$$

so *the theorem of the last multiplier can be expressed by the statement that*

$$\int\left(\frac{\partial f_1}{\partial c}dq_1 + \frac{\partial f_2}{\partial c}dq_2\right)$$

is an integral.

This result leads directly to the theorem already mentioned, which may be thus stated*: *If in the dynamical system defined by the equations*

$$\frac{dq_r}{dt} = \frac{\partial H}{\partial p_r}, \qquad \frac{dp_r}{dt} = -\frac{\partial H}{\partial q_r} \qquad (r = 1, 2),$$

the integral of energy is $H(q_1, q_2, p_1, p_2) = h$, *and if* $V(q_1, q_2, p_1, p_2) = c$ *denotes any other integral not involving the time, then the expression* $p_1dq_1 + p_2dq_2$, *where* p_1 *and* p_2 *have the values found from these integrals, is the exact differential of a function* $\theta(q_1, q_2, h, c)$; *and the remaining integrals of the system are*

$$\frac{\partial\theta}{\partial c} = \text{constant}, \quad \textit{and} \quad \frac{\partial\theta}{\partial h} = t + \text{constant}.$$

This amounts to saying that if any singly-infinite family of orbits is selected (e.g. the orbits which issue from a point $q_1 = \alpha_1$, $q_2 = \alpha_2$) which have

* This theorem is really an application of the well-known method for the solution of a partial differential equation of the first order, the equations of the dynamical system being the equations of the characteristics of the partial differential equation. As a dynamical theorem, it was published for a simple case (motion of a single particle) by Jacobi in 1836 (*Comptes Rendus*, III. p. 59), and for the general case given here by Poisson in 1837 (*J. de M.* II. p. 317) and Liouville in 1840 (*J. de M.* V. p. 351).

the same energy, so that to any point (q_1, q_2) there correspond definite values of p_1 and p_2 (namely the values of p_1 and p_2 corresponding to the orbit which passes through the point q_1, q_2 and belongs to the family), then the value of the integral $\int p_1 dq_1 + p_2 dq_2$ taken along any arc joining two definite points (q_{10}, q_{20}) and (q_{11}, q_{21}) is independent of the arc chosen.

To complete the proof, we have on differentiating the equations $H = h$ and $V = c$,

$$\begin{cases} \dfrac{\partial H}{\partial q_1} + \dfrac{\partial H}{\partial p_1}\dfrac{\partial f_1}{\partial q_1} + \dfrac{\partial H}{\partial p_2}\dfrac{\partial f_2}{\partial q_1} = 0, \\[2ex] \dfrac{\partial V}{\partial q_1} + \dfrac{\partial V}{\partial p_1}\dfrac{\partial f_1}{\partial q_1} + \dfrac{\partial V}{\partial p_2}\dfrac{\partial f_2}{\partial q_1} = 0, \end{cases}$$

and consequently

$$\frac{\partial f_2}{\partial q_1} = \frac{\dfrac{\partial (V, H)}{\partial (q_1, p_1)}}{\dfrac{\partial (V, H)}{\partial (p_1, p_2)}}, \quad \text{and similarly} \quad \frac{\partial f_1}{\partial q_2} = \frac{\dfrac{\partial (V, H)}{\partial (p_2, q_2)}}{\dfrac{\partial (V, H)}{\partial (p_1, p_2)}}.$$

But since $V = c$ is an integral, we have

$$\frac{\partial V}{\partial q_1}\dot{q}_1 + \frac{\partial V}{\partial p_1}\dot{p}_1 + \frac{\partial V}{\partial q_2}\dot{q}_2 + \frac{\partial V}{\partial p_2}\dot{p}_2 = 0,$$

or

$$\frac{\partial (V, H)}{\partial (q_1, p_1)} + \frac{\partial (V, H)}{\partial (q_2, p_2)} = 0,$$

and therefore

$$\frac{\partial f_2}{\partial q_1} - \frac{\partial f_1}{\partial q_2} = 0.$$

This equation shews that $f_1 dq_1 + f_2 dq_2$ is the perfect differential of some function $\theta(q_1, q_2, h, c)$: and the result derived above from the theory of the last multiplier shews that $\partial\theta/\partial c = \text{constant}$ is an integral.

Moreover, we have

$$dt = \frac{dq_1}{\partial H/\partial p_1} = \frac{dq_2}{\partial H/\partial p_2},$$

and therefore

$$dt = \frac{\dfrac{\partial V}{\partial p_2} dq_1 - \dfrac{\partial V}{\partial p_1} dq_2}{\dfrac{\partial (V, H)}{\partial (p_2, p_1)}}.$$

But obtaining $\partial f_1/\partial h$ and $\partial f_2/\partial h$ in the same way as $\partial f_1/\partial c$ and $\partial f_2/\partial c$ were found, we have

$$\frac{\partial f_1}{\partial h} = \frac{\partial V/\partial p_2}{\dfrac{\partial (V, H)}{\partial (p_2, p_1)}} \quad \text{and} \quad \frac{\partial f_2}{\partial h} = \frac{\partial V/\partial p_1}{\dfrac{\partial (V, H)}{\partial (p_1, p_2)}}.$$

Consequently $$dt = \frac{\partial f_1}{\partial h} dq_1 + \frac{\partial f_2}{\partial h} dq_2,$$

or $$t = \frac{\partial \theta}{\partial h} + \text{constant},$$

which completes the proof of the theorem.

Example. In the problem of two centres of gravitation (§ 53), if (r, r') denote the radii vectores to the centres of force, and (θ, θ') the angles formed by r, r' with the line joining the centres of force, obtain the integral

$$r^2 r'^2 \dot{\theta}\dot{\theta}' - 2c(\mu \cos\theta + \mu' \cos\theta') = \text{constant},$$

and hence complete the solution by the above theorem.

122. *Integral-invariants whose order is equal to the order of the system.*

The theory of the last multiplier of a system of differential equations is connected with that of the integral-invariants whose order is equal to the order of the system.

Let $$\frac{dx_r}{dt} = X_r \qquad (r = 1, 2, \ldots, k),$$

where $(X_1, X_2, \ldots, X_k)$ are given functions of $(x_1, x_2, \ldots, x_k, t)$, be a system of ordinary differential equations; and let us find the condition which must be satisfied in order that

$$\iiint \ldots \int M \,\delta x_1 \,\delta x_2 \ldots \delta x_k$$

may be an integral-invariant, where M is a function of the variables.

Let $(c_1, c_2, \ldots, c_k)$ be any set of constants of integration of these equations, so that, by solving the equations, $(x_1, x_2, \ldots, x_k)$ can be expressed in terms of $(c_1, c_2, \ldots, c_k, t)$. Then we have

$$\iiint \ldots \int M \,\delta x_1 \,\delta x_2 \ldots \delta x_k = \iiint \ldots \int M \frac{\partial (x_1, x_2, \ldots, x_k)}{\partial (c_1, c_2, \ldots, c_k)} \delta c_1 \,\delta c_2 \ldots \delta c_k,$$

and therefore the condition of integral-invariancy is

$$\frac{d}{dt}\left\{ M \frac{\partial (x_1, x_2, \ldots, x_k)}{\partial (c_1, c_2, \ldots, c_k)} \right\} = 0,$$

or $$\frac{dM}{dt} \frac{\partial (x_1, x_2, \ldots, x_k)}{\partial (c_1, c_2, \ldots, c_k)} + M \sum_{r=1}^{k} \frac{\partial (x_1, x_2, \ldots, x_{r-1}, X_r, x_{r+1}, \ldots, x_k)}{\partial (c_1, c_2, \ldots, c_k)} = 0,$$

or $$\frac{dM}{dt} \frac{\partial (x_1, x_2, \ldots, x_k)}{\partial (c_1, c_2, \ldots, c_k)} + M \sum_{r=1}^{k} \frac{\partial X_r}{\partial x_r} \frac{\partial (x_1, x_2, \ldots, x_k)}{\partial (c_1, c_2, \ldots, c_k)} = 0,$$

or $$\frac{dM}{dt} + M \sum_{r=1}^{k} \frac{\partial X_r}{\partial x_r} = 0,$$

which shews that *M must be a last multiplier of the system of equations.*

This result gives immediately the theorem that *for a dynamical system whose motion is determined by the equations*

$$\frac{dq_r}{dt} = \frac{\partial H}{\partial p_r}, \quad \frac{dp_r}{dt} = -\frac{\partial H}{\partial q_r} \qquad (r = 1, 2, \ldots, n),$$

where H is any function of $(q_1, q_2, \ldots, q_n, p_1, p_2, \ldots, p_n, t)$, *the expression*

$$\iiint \ldots \int \delta q_1 \delta q_2 \ldots \delta q_n \delta p_1 \delta p_2 \ldots \delta p_n$$

is an integral-invariant; since in this case unity is a last multiplier. This theorem is of importance in the applications of dynamics to thermodynamics.

Example. For a system with two degrees of freedom, let the energy-integral when solved for p_1 take the form

$$H'(q_1, q_2, p_2, h) + p_1 = 0.$$

Shew that, for trajectories which correspond to the same value of the constant of energy, the quantity

$$\frac{\partial H'}{\partial h} \delta q_1 \delta q_2 \delta p_2$$

is independent of t and also of the choice of coordinates: and hence shew that the trajectories of the problem can be represented as the stream-lines in the steady motion of a fluid whose density is $\partial H'/\partial h$.

123. *Reduction of differential equations to the Lagrangian form.*

Another question to which the theory of the last multiplier can be applied is the following: To find under what conditions a given system of ordinary differential equations of the second order

$$\ddot{q}_k = f_k(\dot{q}_1, \dot{q}_2, \ldots, \dot{q}_n, q_1, q_2, \ldots, q_n) \qquad (k = 1, 2, \ldots, n)$$

is equivalent to a Lagrangian system

$$\frac{d}{dt}\left(\frac{\partial L}{\partial \dot{q}_r}\right) - \frac{\partial L}{\partial q_r} = 0 \qquad (r = 1, 2, \ldots, n),$$

where L is a function of $(\dot{q}_1, \dot{q}_2, \ldots, \dot{q}_n, q_1, q_2, \ldots, q_n, t)$.

If these two systems are equivalent, the equations

$$\sum_{k=1}^{n} \left(\frac{\partial^2 L}{\partial \dot{q}_r \partial \dot{q}_k} \ddot{q}_k + \frac{\partial^2 L}{\partial \dot{q}_r \partial q_k} \dot{q}_k \right) + \frac{\partial^2 L}{\partial \dot{q}_r \partial t} - \frac{\partial L}{\partial q_r} = 0 \qquad (r = 1, 2, \ldots, n)$$

must evidently reduce to identities when the quantities $\ddot{q}_k$ are replaced by the expressions f_k; and therefore *the required condition is that a function L shall exist satisfying the simultaneous partial differential equations*

$$\sum_{k=1}^{n} \left(\frac{\partial^2 L}{\partial \dot{q}_r \partial \dot{q}_k} f_k + \frac{\partial^2 L}{\partial \dot{q}_r \partial q_k} \dot{q}_k \right) + \frac{\partial^2 L}{\partial \dot{q}_r \partial t} - \frac{\partial L}{\partial q_r} = 0 \qquad (r = 1, 2, \ldots, n),$$

where $(\dot{q}_1, \dot{q}_2, \ldots, \dot{q}_n, q_1, q_2, \ldots, q_n, t)$ *are regarded as the independent variables.*

When $n=1$, the question can be solved in terms of the last multiplier. For the equation satisfied by L is then

$$\frac{\partial^2 L}{\partial \dot{q}^2} f + \frac{\partial^2 L}{\partial \dot{q} \partial q} \dot{q} + \frac{\partial^2 L}{\partial \dot{q} \partial t} - \frac{\partial L}{\partial q} = 0,$$

from which we have

$$-\frac{\partial}{\partial \dot{q}} \left(\frac{\partial^2 L}{\partial \dot{q}^2} f \right) = \frac{\partial}{\partial \dot{q}} \left(\frac{\partial^2 L}{\partial \dot{q} \partial q} \dot{q} + \frac{\partial^2 L}{\partial \dot{q} \partial t} - \frac{\partial L}{\partial q} \right)$$

$$= \frac{\partial^3 L}{\partial \dot{q}^2 \partial q} \dot{q} + \frac{\partial^3 L}{\partial \dot{q}^2 \partial t},$$

and therefore if we write $\partial^2 L / \partial \dot{q}^2 = M$, the function M satisfies the equation

$$\frac{\partial}{\partial \dot{q}} (Mf) + \frac{\partial}{\partial q} (M\dot{q}) + \frac{\partial M}{\partial t} = 0\,;$$

but this is the equation defining the last multiplier M of the system of equations

$$dt = \frac{dq}{\dot{q}} = \frac{d\dot{q}}{f(\dot{q}, q, t)},$$

and therefore *when* $n=1$, *the determination of the function L reduces to the determination of the last multiplier of the system.*

124. *Case in which the kinetic energy is quadratic in the velocities.*

When $n>1$, the most important case is that in which each of the functions f_r consists of a part F_r which is homogeneous and of the second degree in $(\dot{q}_1, \dot{q}_2, \ldots, \dot{q}_n)$ and a part G_r which does not involve $(\dot{q}_1, \dot{q}_2, \ldots, \dot{q}_n)$, and it is required to determine whether the equations

$$\ddot{q}_r = F_r + G_r \qquad (r=1, 2, \ldots, n)$$

are equivalent to a system

$$\frac{d}{dt}\left(\frac{\partial T}{\partial \dot{q}_r}\right) - \frac{\partial T}{\partial q_r} = Q_r \qquad (r=1, 2, \ldots, n),$$

where T is homogeneous and of the second degree in $(\dot{q}_1, \dot{q}_2, \ldots, \dot{q}_n)$ and also involves the variables $(q_1, q_2, \ldots, q_n)$, and $(Q_1, Q_2, \ldots, Q_n)$ are functions of $(q_1, q_2, \ldots, q_n)$ only.

The value of T is clearly not dependent on $(G_1, G_2, \ldots, G_n)$, and therefore we can consider the problem in which $(G_1, G_2, \ldots, G_n)$ are zero, i.e. the problem of finding a function T such that the equations

$$\ddot{q}_r = F_r \qquad (r=1, 2, \ldots, n)$$

are equivalent to the system

$$\frac{d}{dt}\left(\frac{\partial T}{\partial \dot{q}_r}\right) - \frac{\partial T}{\partial q_r} = 0 \qquad (r=1, 2, \ldots, n).$$

The condition for this is the existence of a function T satisfying the partial differential equations

$$\sum_{k=1}^{n} \frac{\partial^2 T}{\partial \dot{q}_r \partial \dot{q}_k} F_k + \sum_{k=1}^{n} \frac{\partial^2 T}{\partial \dot{q}_r \partial q_k} \dot{q}_k - \frac{\partial T}{\partial q_r} = 0 \qquad (r=1, 2, \ldots, n).$$

Since F_k is homogeneous, we have $\sum_{s=1}^{n} \dot{q}_s \partial F_k / \partial \dot{q}_s = 2F_k$, and therefore

$$\sum_{k=1}^{n} \frac{\partial^2 T}{\partial \dot{q}_r \partial \dot{q}_k} F_k = \tfrac{1}{2} \sum_{s=1}^{n} \sum_{k=1}^{n} \dot{q}_s \frac{\partial F_k}{\partial \dot{q}_s} \frac{\partial^2 T}{\partial \dot{q}_r \partial \dot{q}_k} \qquad (r=1, 2, \ldots, n)$$

$$= \sum_{s=1}^{n} \dot{q}_s \frac{\partial}{\partial \dot{q}_r} \left(\tfrac{1}{2} \sum_{k=1}^{n} \frac{\partial F_k}{\partial \dot{q}_s} \frac{\partial T}{\partial \dot{q}_k} \right) - \sum_{s=1}^{n} \dot{q}_s \cdot \tfrac{1}{2} \sum_{k=1}^{n} \frac{\partial^2 F_k}{\partial \dot{q}_s \partial \dot{q}_r} \frac{\partial T}{\partial \dot{q}_k}.$$

But since $\partial F/\partial \dot{q}_r$ is homogeneous, we have

$$\frac{\partial F_k}{\partial \dot{q}_r} = \sum_{s=1}^{n} \dot{q}_s \frac{\partial^2 F_k}{\partial \dot{q}_r \partial \dot{q}_s},$$

and therefore

$$\sum_{k=1}^{n} \frac{\partial^2 T}{\partial \dot{q}_r \partial \dot{q}_k} F_k = \sum_{s=1}^{n} \dot{q}_s \frac{\partial}{\partial \dot{q}_r}\left(\tfrac{1}{2}\sum_{k=1}^{n} \frac{\partial F_k}{\partial \dot{q}_s}\frac{\partial T}{\partial \dot{q}_k}\right) - \tfrac{1}{2}\sum_{s=1}^{n} \frac{\partial F_k}{\partial \dot{q}_r}\frac{\partial T}{\partial \dot{q}_k}.$$

The equations to be satisfied by T may consequently be written

$$\sum_{s=1}^{n} \dot{q}_s \frac{\partial}{\partial \dot{q}_r}\left(\tfrac{1}{2}\sum_{k=1}^{n} \frac{\partial F_k}{\partial \dot{q}_s}\frac{\partial T}{\partial \dot{q}_k}\right) - \tfrac{1}{2}\sum_{k=1}^{n} \frac{\partial F_k}{\partial \dot{q}_r}\frac{\partial T}{\partial \dot{q}_k} + \sum_{s=1}^{n} \frac{\partial^2 T}{\partial \dot{q}_r \partial q_s}\dot{q}_s - \frac{\partial T}{\partial q_r} = 0 \qquad (r=1, 2, \ldots, n),$$

or

$$\sum_{s=1}^{n} \dot{q}_s \frac{\partial}{\partial \dot{q}_r}\left(\tfrac{1}{2}\sum_{k=1}^{n} \frac{\partial F_k}{\partial \dot{q}_s}\frac{\partial T}{\partial \dot{q}_k} + \frac{\partial T}{\partial q_s}\right) - \left(\tfrac{1}{2}\sum_{k=1}^{n} \frac{\partial F_k}{\partial \dot{q}_r}\frac{\partial T}{\partial \dot{q}_k} + \frac{\partial T}{\partial q_r}\right) = 0 \qquad (r=1, 2, \ldots, n),$$

and evidently these may be replaced by the equations

$$\tfrac{1}{2}\sum_{k=1}^{n} \frac{\partial F_k}{\partial \dot{q}_r}\frac{\partial T}{\partial \dot{q}_k} + \frac{\partial T}{\partial q_r} = 0 \qquad (r=1, 2, \ldots, n).$$

Thus, writing f_r for $(F_r + G_r)$, we have the theorem that *if the system of equations*

$$\ddot{q}_r = f_r \qquad (r=1, 2, \ldots, n),$$

where f_r consists of a part which is homogeneous of degree two in the velocities and a part which does not involve the velocities, is reducible to the form

$$\frac{d}{dt}\left(\frac{\partial T}{\partial \dot{q}_r}\right) - \frac{\partial T}{\partial q_r} = Q_r \qquad (r=1, 2, \ldots, n),$$

then T must be an integral of the system

$$\tfrac{1}{2}\sum_{k=1}^{n} \frac{\partial f_k}{\partial \dot{q}_r}\frac{\partial T}{\partial \dot{q}_k} + \frac{\partial T}{\partial q_r} = 0 \qquad (r=1, 2, \ldots, n).$$

MISCELLANEOUS EXAMPLES.

1. In the problem of two centres of gravitation, the distance between the centres of force is $2c$, and the semi-major axes of the two conics which pass through the moving particle and have their foci at the centre of force are (q_1, q_2). Writing

$$p_1 = \frac{q_1^2 - q_2^2}{q_1^2 - c^2}\frac{dq_1}{dt}, \quad p_2 = \frac{q_1^2 - q_2^2}{c^2 - q_2^2}\frac{dq_2}{dt},$$

shew that the equations of motion are

$$\frac{dq_r}{dt} = \frac{\partial H}{\partial p_r}, \quad \frac{dp_r}{dt} = -\frac{\partial H}{\partial q_r} \qquad (r=1, 2),$$

where

$$H = \tfrac{1}{2}\frac{q_1^2 - c^2}{q_1^2 - q_2^2}p_1^2 + \tfrac{1}{2}\frac{c^2 - q_2^2}{q_1^2 - q_2^2}p_2^2 - \frac{\mu_1}{q_1 - q_2} - \frac{\mu_2}{q_1 + q_2},$$

and μ_1 and μ_2 are constants.

2. Shew that

$$\iiiint \Sigma\, \delta q_i \delta p_i \delta q_j \delta p_j,$$

where the summation is extended over the $\tfrac{1}{2}n(n-1)$ combinations of the indices i and j, is an integral-invariant of any Hamiltonian system in which $(q_1, q_2, \ldots, q_n, p_1, p_2, \ldots, p_n)$ are the variables. (Poincaré.)

3. In the problem defined by the equations

$$\frac{dq_r}{dt}=\frac{\partial H}{\partial p_r},\quad \frac{dp_r}{dt}=-\frac{\partial H}{\partial q_r}\qquad (r=1, 2),$$

where

$$H=q_1p_1-q_2p_2-aq_1^2+bq_2^2,$$

shew that

$$\frac{p_2-bq_2}{q_1}=\text{constant}$$

is an integral; and hence by the theorem of § 121 obtain the two remaining integrals

$$\begin{cases} q_1q_2 = \text{constant}, \\ \log q_1 = t+\text{constant}. \end{cases}$$

4. If M is a last multiplier of a system of differential equations

$$\frac{dx_1}{X_1}=\frac{dx_2}{X_2}=\ldots=\frac{dx_n}{X_n}=\frac{dx}{X},$$

of which the equation

$$f(x_1, x_2, \ldots, x_n, x)=\text{Constant}$$

is a known integral, and if an accent annexed to a function of $x_1, x_2, \ldots, x_n, x$ is used to indicate that x_n has been replaced in the function by its values found from this integral, shew that $M'/(\partial f/\partial x_n)'$ is a last multiplier of the reduced system

$$\frac{dx_1}{X_1'}=\frac{dx_2}{X_2'}=\ldots=\frac{dx_{n-1}}{X'_{n-1}}=\frac{dx}{X'}.\qquad \text{(Jacobi.)}$$

5. If $\theta_1=$ Constant, $\theta_2=$ Constant, ..., $\theta_n=$ Constant are a set of integrals of the equations

$$\frac{dx}{X}=\frac{dx_1}{X_1}=\frac{dx_2}{X_2}=\ldots=\frac{dx_n}{X_n},$$

shew that

$$\frac{1}{X}\frac{\partial(\theta_1, \theta_2, \ldots, \theta_n)}{\partial(x_1, x_2, \ldots, x_n)}$$

is a last multiplier.

6. Let $(u_1, u_2, \ldots, u_n)$ be n dependent variables, and let $I_1, I_2, \ldots, I_n$ be a set of linear differential expressions defined by the equations

$$I_r=\sum_{k=1}^{n}\{p_{rk}(t)\,u_k+q_{rk}(t)\,\dot{u}_k+r_{rk}(t)\,\ddot{u}_k\}\qquad (r=1, 2, \ldots, n).$$

If $(v_1, v_2, \ldots, v_n)$ are functions of t such that

$$v_1I_1+v_2I_2+\ldots+v_nI_n$$

is an exact differential, shew that the functions $(v_1, v_2, \ldots, v_n)$ satisfy a set of n linear differential equations, which will be called the system *adjoint* to the system of linear differential equations

$$I_r=0\qquad (r=1, 2, \ldots, n).$$

If F_r denotes the expression

$$\frac{d}{dt}\left(\frac{\partial L}{\partial \dot{q}_r}\right)-\frac{\partial L}{\partial q_r}\qquad (r=1, 2, \ldots, n),$$

where L is any given function of $(\dot{q}_1, \dot{q}_2, \ldots, \dot{q}_n, q_1, q_2, \ldots, q_n, t)$, shew that the system of linear differential equations

$$\sum_k\left(\frac{\partial F_r}{\partial q_k}u_k+\frac{\partial F_r}{\partial \dot{q}_k}\dot{u}_k+\frac{\partial F_r}{\partial \ddot{q}_k}\ddot{u}_k\right)=0\qquad (r=1, 2, \ldots, n)$$

is adjoint to itself.

Shew that the converse of this latter theorem is also true. (Hirsch.)

CHAPTER XI

THE TRANSFORMATION-THEORY OF DYNAMICS

125. *Hamilton's Characteristic Function and Contact-Transformations.*

We have seen* that the integration of a dynamical system which is soluble by quadratures can generally be effected by transforming it into another dynamical system with fewer degrees of freedom. We shall in the present chapter investigate the general theory which underlies this procedure, and, indeed, underlies the solution of all dynamical systems.

The origin of the method is to be found in a celebrated memoir on optics, which was presented to the Royal Irish Academy by Hamilton in 1824†: the principles there introduced were afterwards transferred by their discoverer to the field of dynamics.

In order to follow Hamilton's thought, we must refer to the connexion between dynamics and optics—a connexion which is perhaps less obvious in our day than in his, when the corpuscular theory of light was still widely held. If a ray of light traverses an optically heterogeneous but isotropic medium, the refractive index at any point (x, y, z) being μ, the path of a ray may be determined by Fermat's Principle‡, namely that the integral

$$\int \mu (x, y, z)\, ds$$

has a stationary value when the integration is taken along the actual ray joining two given terminal points, as compared with neighbouring paths joining them. If on the other hand we consider the motion of a free particle of unit mass in a conservative field of force where its potential energy is $\phi (x, y, z)$, and its constant of energy is h, the path of the particle may be determined by the Principle of Least Action (§ 100), which in this case asserts that the integral

$$\int \{h - \phi (x, y, z)\}^{\frac{1}{2}} ds$$

has a stationary value for the actual trajectory as compared with neighbouring paths joining the same terminal points. Comparing these two statements, we

* Cf. Chapter III, §§ 38–42.

† *Trans. R. Irish Acad.* xv. (1828), p. 69; xvi. (1830), pp. 4, 93; xvii. (1837), p. 1.

‡ Cf. my *History of the Theories of Aether and Electricity*, pp. 9–10, 102–3.

see that *the trajectories of the particle in the dynamical problem are the same as the paths of the rays in the optical problem*, provided a suitable correspondence

$$\mu = (h - \phi)^{\frac{1}{2}}$$

is set up between the potential-energy function in the one case and the refractive index in the other.

In the corpuscular theory of light, this was regarded as furnishing the *explanation* of the optical phenomena, the ray of light being conceived as a procession of rapidly-moving corpuscles. But the statement in itself is true whatever hypothesis regarding light be adopted: and therefore it supplies a means of connecting dynamics with the *undulatory* hypothesis. This idea is the starting-point of Hamilton's theory.

When the undulatory hypothesis is adopted, we have the choice of two different methods of discussing the propagation of light mathematically: the first is to consider *rays*, the second is to consider *wave-fronts*. The latter method, which was introduced by Huygens in 1690, may be thus explained.

Consider a wave-front, or locus of disturbance in an optical medium, as it exists at a definite instant t, having the form of a surface σ. Each element of this wave-front may be regarded as the source of a secondary wave, propagated outwards from it; so that at a subsequent instant t', the disturbance originating in any point (x, y, z) of the original wave-front will extend over a surface. To obtain the equation of this surface, we observe that the time taken by light to travel through the medium from an arbitrary point (x, y, z) to another arbitrary point (x', y', z') depends only on the six quantities (x, y, z, x', y', z'): let it be denoted by $V(x, y, z, x', y', z')$. This function $V(x, y, z, x', y', z')$ was called by Hamilton the *characteristic function* for the medium in question. A disturbance which originates at a point (x, y, z) of the original wave-front at the instant t will therefore at the instant t' extend over the surface whose equation in the coordinates (x', y', z') is

$$V(x, y, z, x', y', z') = t' - t \quad \text{.......................(1).}$$

Now according to the principle of wave-propagation laid down by Huygens, the wave-front which represents the whole disturbance at the instant t' is the envelope of the secondary waves which arise from the various elements of the original wave-front. Call this new wave-front Σ; and denote the direction-cosines of the normal to the wave-front σ at (x, y, z) by (l, m, n), and the direction-cosines of the normal to the wave-front Σ at the corresponding point* (x', y', z') by (l', m', n'): these are the

* The point (x', y', z') is said to correspond to (x, y, z) if the secondary wave propagated from (x, y, z) touches the envelope Σ at (x', y', z').

direction-cosines of the rays at (x', y', z') and (x, y, z) respectively, since in an isotropic medium the ray is normal to the wave-front*. Then since Σ is the envelope of the surfaces V corresponding to points on σ, the equation

$$\frac{\partial V}{\partial x} dx + \frac{\partial V}{\partial y} dy + \frac{\partial V}{\partial z} dz = 0$$

must be satisfied by all those values of the ratios $dx : dy : dz$ which correspond to directions in the tangent-plane to σ, i.e. which satisfy the relation

$$l\,dx + m\,dy + n\,dz = 0.$$

Hence we have

$$\frac{1}{l}\frac{\partial V}{\partial x} = \frac{1}{m}\frac{\partial V}{\partial y} = \frac{1}{n}\frac{\partial V}{\partial z} \quad \text{.........................(2).}$$

Moreover, since (l', m', n') are the direction-cosines of the normal to the surface V at the point (x', y', z'), we have

$$\frac{1}{l'}\frac{\partial V}{\partial x'} = \frac{1}{m'}\frac{\partial V}{\partial y'} = \frac{1}{n'}\frac{\partial V}{\partial z'} \quad \text{.........................(3).}$$

Now a ray of light which passes through the point (x, y, z) in the direction (l, m, n) at time t passes through the point (x', y', z') in the direction (l', m', n') at time t': and equations (1), (2), (3), together with the equation

$$l'^2 + m'^2 + n'^2 = 1 \quad \text{...............................(4),}$$

are six equations, from which we can determine the six quantities (x', y', z', l', m', n') in terms of (x, y, z, l, m, n). Thus *by these equations the behaviour of rays of light in the medium is completely specified in terms of the single function* $V(x, y, z, x', y', z')$. It will be observed that they are not differential equations, but that they give directly, in the integrated form, the changes in any system of rays after a finite interval of propagation through the medium. It is evident therefore that all problems in optics depend on the determination of Hamilton's characteristic function $V(x, y, z, x', y', z')$ for the optical medium or system of media through which the rays pass.

From the point of view of Pure Mathematics, we regard the change from the set of variables (x, y, z, l, m, n) to the set of variables (x', y', z', l', m', n'), or (to express it geometrically) from the surfaces σ to the surfaces Σ, as a *transformation*. The function V is thus to be regarded as determining a transformation of space which changes any surface σ into a new surface Σ. It is evident that if two surfaces σ and σ' touch at a point, the corresponding transformed surfaces Σ and Σ' also touch at the corresponding point: on this account the transformation has been called by S. Lie a *contact-transformation*. Thus *any function* $V(x, y, z, x', y', z')$ *defines a contact-transformation, which*

* For simplicity we are supposing that the medium, though optically heterogeneous, is isotropic. Hamilton considered also the more general case of a crystalline medium.

transforms any wave-front σ into the wave-front Σ which is derived from σ by propagation through the medium in the interval of time $t'-t$.

A simple example of a contact-transformation is the well-known geometrical transformation known as *reciprocation.* In order to find the reciprocal of any given surface σ, with respect to a given surface, we correlate to every point (x, y, z) on σ a plane, namely the polar plane of (x, y, z) with respect to the quadric. When the point (x, y, z) takes all possible positions on the surface σ, the plane envelopes a surface Σ, which is the reciprocal of σ. The transformation from σ to Σ is evidently a contact-transformation. In this case Hamilton's function V is linear with respect to (x, y, z) and also with respect to (x', y', z').

Proceeding now with Hamilton's problem, equations (2) and (3) may be written

$$\frac{\partial V}{\partial x} = \kappa l, \quad \frac{\partial V}{\partial y} = \kappa m, \quad \frac{\partial V}{\partial z} = \kappa n,$$

$$\frac{\partial V}{\partial x'} = \lambda l', \quad \frac{\partial V}{\partial y'} = \lambda m', \quad \frac{\partial V}{\partial z'} = \lambda n',$$

where κ and λ are quantities not as yet determined. They can however be readily found. For the equations may be written

$$dV = \kappa (l dx + m dy + n dz) + \lambda (l' dx' + m' dy' + n' dz') \quad \text{.........(5)}.$$

Now by proceeding a small distance ds' along the ray at (x', y', z'), we increase V by the time which light takes to travel along ds'. But if the units are so chosen that the velocity of light in free aether is unity, then the velocity of light in the medium at (x', y', z') is $1/\mu'$, where μ' denotes the refractive index at this point. Thus the time taken by light to describe ds' is $\mu' ds'$, or $\mu' (l'^2 + m'^2 + n'^2)\, ds'$, or $\mu' (l' dx' + m' dy' + n' dz')$. Comparing this with equation (5), we see that $\lambda = \mu'$. Similarly $\kappa = -\mu$, where μ denotes the refractive index at (x, y, z). Thus Hamilton's general formula becomes

$$dV = \mu' (l' dx' + m' dy' + n' dz') - \mu (l dx + m dy + n dz).$$

If we write

$$\mu l = \xi, \quad \mu m = \eta, \quad \mu n = \zeta, \quad \mu' l' = \xi', \quad \mu' m' = \eta', \quad \mu' n' = \zeta',$$

this takes the form

$$dV = \xi' dx' + \eta' dy' + \zeta' dz' - \xi dx - \eta dy - \zeta dz \quad \text{...........(6)}.$$

The quantities (ξ, η, ζ), (ξ', η', ζ') were called by Hamilton the *components of normal slowness of propagation* of the wave at (x, y, z) and (x', y', z') respectively.

Consider now the particular case in which the interval of time $(t'-t)$ between the two positions σ and Σ of the same wave-front is very small:

denote it by Δt. In this case the contact-transformation is said to be *infinitesimal*. Write

$$\left.\begin{array}{lll} x' = x + \alpha\Delta t, & y' = y + \beta\Delta t, & z' = z + \gamma\Delta t, \\ \xi' = \xi + u\Delta t, & \eta' = \eta + v\Delta t, & \zeta' = \zeta + w\Delta t, \\ & V = W\Delta t & \end{array}\right\} \quad \ldots\ldots\ldots\ldots(7).$$

Then equation (6) becomes

$$\begin{aligned} dW\,.\,\Delta t &= (\xi + u\Delta t)(dx + d\alpha\,.\,\Delta t) + (\eta + v\Delta t)(dy + d\beta\,.\,\Delta t) \\ &\qquad + (\zeta + w\Delta t)(dz + d\gamma\,.\,\Delta t) - \xi dx - \eta dy - \zeta dz \\ &= u\Delta t dx + v\Delta t dy + w\Delta t dz + \xi\Delta t d\alpha + \eta\Delta t d\beta + \zeta\Delta t d\gamma, \end{aligned}$$

or $\quad dW = udx + vdy + wdz + \xi d\alpha + \eta d\beta + \zeta d\gamma,$

or $\quad d(\xi\alpha + \eta\beta + \zeta\gamma - W) = \alpha d\xi + \beta d\eta + \gamma d\zeta - udx - vdy - wdz.$

Thus if we denote the function $\xi\alpha + \eta\beta + \zeta\gamma - W$ by H, and suppose H expressed as a function of $x, y, z, \xi, \eta, \zeta$, we have

$$dH = \alpha d\xi + \beta d\eta + \gamma d\zeta - udx - vdy - wdz \quad \ldots\ldots\ldots\ldots(8).$$

Now evidently, from (7), in the limit u becomes $\frac{d\xi}{dt}$, α becomes $\frac{dx}{dt}$, etc. Thus we have

$$dH = \frac{dx}{dt}d\xi + \frac{dy}{dt}d\eta + \frac{dz}{dt}d\zeta - \frac{d\xi}{dt}dx - \frac{d\eta}{dt}dy - \frac{d\zeta}{dt}dz,$$

so the rates of increase of the six variables $(x, y, z, \xi, \eta, \zeta)$ are given by the equations

$$\left.\begin{array}{lll} \dfrac{dx}{dt} = \dfrac{\partial H}{\partial \xi}, & \dfrac{dy}{dt} = \dfrac{\partial H}{\partial \eta}, & \dfrac{dz}{dt} = \dfrac{\partial H}{\partial \zeta} \\ \dfrac{d\xi}{dt} = -\dfrac{\partial H}{\partial x}, & \dfrac{d\eta}{dt} = -\dfrac{\partial H}{\partial y}, & \dfrac{d\zeta}{dt} = -\dfrac{\partial H}{\partial z} \end{array}\right\} \quad \ldots\ldots\ldots\ldots(9),$$

and *this is a Hamiltonian system of equations, such as occurs in dynamics. Our investigation shews that it may be regarded as representing an infinitesimal contact-transformation, that is to say, the motion of a wave-front from one position to a position indefinitely near it.* The integrals of this Hamiltonian system are the equations (1), (2), (3), (4) above: they represent a finite contact-transformation, that is to say, the motion of a wave-front from one position to the position which it acquires after a finite interval of time. Thus we see how *by using the ideas of the undulatory theory of light, Hamilton was able to obtain an integrated form for the differential equations of dynamics, depending on a single unknown function.*

126. *Contact-transformations in space of any number of dimensions.*

The rest of the present chapter will be concerned with the application of Hamilton's ideas, described in the preceding article, to the general case of a dynamical system with any number of degrees of freedom, and the connexion of the results with certain theorems due to Lagrange, Poisson, Pfaff, and Jacobi.

We shall first define a contact-transformation in n-dimensional space, using for this purpose a generalisation of equation (6) of the last article. Let $(q_1, q_2, \ldots, q_n, p_1, p_2, \ldots, p_n)$ be a set of $2n$ variables, and let

$$(Q_1, Q_2, \ldots, Q_n, P_1, P_2, \ldots, P_n)$$

be $2n$ other variables which are defined in terms of them by $2n$ equations. If the equations connecting the two sets of variables are such that the differential form

$$P_1 dQ_1 + P_2 dQ_2 + \ldots + P_n dQ_n - p_1 dq_1 - p_2 dq_2 - \ldots - p_n dq_n$$

is, when expressed in terms of $(q_1, q_2, \ldots, q_n, p_1, p_2, \ldots, p_n)$ and their differentials, the perfect differential of a function of $(q_1, q_2, \ldots, q_n, p_1, p_2, \ldots, p_n)$, then the change from the set of variables $(q_1, q_2, \ldots, q_n, p_1, p_2, \ldots, p_n)$ to the other set $(Q_1, Q_2, \ldots, Q_n, P_1, P_2, \ldots, P_n)$ is called a *contact-transformation.*

It may be observed that this is different in form from the definition which is most convenient when contact-transformations are studied with a view to their applications in geometry and in the theory of partial differential equations: the latter definition may be stated thus: a contact-transformation is a transformation from a set of $(2n+1)$ variables $(q_1, q_2, \ldots, q_n, p_1, p_2, \ldots, p_n, z)$ to another set $(Q_1, Q_2, \ldots, Q_n, P_1, P_2, \ldots, P_n, Z)$, for which the equation

$$dZ - P_1 dQ_1 - P_2 dQ_2 - \ldots - P_n dQ_n = \rho\,(dz - p_1 dq_1 - p_2 dq_2 - \ldots - p_n dq_n)$$

is satisfied, where ρ denotes some function of $(q_1, q_2, \ldots, q_n, p_1, p_2, \ldots, p_n, z)$.

If the n variables $(Q_1, Q_2, \ldots, Q_n)$ are functions of $(q_1, q_2, \ldots, q_n)$ only, the contact-transformation from the variables $(q_1, q_2, \ldots, q_n, p_1, \ldots, p_n)$ to the variables $(Q_1, Q_2, \ldots, Q_n, P_1, \ldots, P_n)$ is called an *extended point-transformation*, the equations which connect $(q_1, q_2, \ldots, q_n)$ with $(Q_1, Q_2, \ldots, Q_n)$ being in this case said to define a *point-transformation.*

From the definition it is clear that the result of performing two contact-transformations in succession is to obtain a change of variables which is itself a contact-transformation. It is also evident that if the transformation from $(q_1, q_2, \ldots, q_n, p_1, \ldots, p_n)$ to $(Q_1, Q_2, \ldots, Q_n, P_1, \ldots, P_n)$ is a contact-transformation, then the transformation from $(Q_1, Q_2, \ldots, Q_n, P_1, P_2, \ldots, P_n)$ to $(q_1, q_2, \ldots, q_n, p_1, p_2, \ldots, p_n)$ is also a contact-transformation; this is generally expressed by saying that *the inverse of a contact-transformation is a contact-transformation.* This, together with the foregoing, shews that *contact-transformations possess the group-property.*

Example 1. Shew that the transformation defined by the equations

$$\begin{cases} Q = (2q)^{\frac{1}{2}} e^k \cos p, \\ P = (2q)^{\frac{1}{2}} e^{-k} \sin p, \end{cases}$$

is a contact-transformation.

In this case we have

$$P\,dQ - p\,dq = (2q)^{\frac{1}{2}} \sin p\,\{(2q)^{-\frac{1}{2}} \cos p\,dq - (2q)^{\frac{1}{2}} \sin p\,dp\} - p\,dq$$
$$= d\,(q \sin p \cos p - qp),$$

which is a perfect differential.

Example 2. Shew that the transformation

$$\begin{cases} Q = \log\left(\dfrac{1}{q} \sin p\right), \\ P = q \cot p, \end{cases}$$

is a contact-transformation.

Example 3. Shew that the transformation

$$\begin{cases} Q = \log(1 + q^{\frac{1}{2}} \cos p), \\ P = 2\,(1 + q^{\frac{1}{2}} \cos p)\, q^{\frac{1}{2}} \sin p, \end{cases}$$

is a contact-transformation.

We shall now obtain the explicit analytical expression of a contact-transformation.

Let the transformation from variables $(q_1, q_2, \ldots, q_n, p_1, \ldots, p_n)$ to variables $(Q_1, Q_2, \ldots, Q_n, P_1, \ldots, P_n)$ be a contact-transformation, so that

$$\sum_{r=1}^{n} (P_r dQ_r - p_r dq_r) = dW,$$

where dW is a complete differential.

From the equations which define $(Q_1, Q_2, \ldots, Q_n, P_1, \ldots, P_n)$ in terms of $(q_1, q_2, \ldots, q_n, p_1, \ldots, p_n)$ it may be possible to eliminate $(P_1, P_2, \ldots, P_n, p_1, \ldots, p_n)$ completely, so as to obtain one or more relations between the variables

$$(Q_1, Q_2, \ldots, Q_n, q_1, \ldots, q_n);$$

let the number of such relations be k, and let them be denoted by

$$\Omega_r(q_1, q_2, \ldots, q_n, Q_1, \ldots, Q_n) = 0 \quad (r = 1, 2, \ldots, k) \ldots (\text{A}).$$

The meaning of these relations may be illustrated by reverting to the geometrical theory of contact-transformations in ordinary three-dimensional space, when there are three cases to consider:

(*α*) There may be only a single relation between the new and old coordinates, say

$$\Omega(x, y, z, x', y', z') = 0.$$

When (x, y, z) are given, this equation, regarded as the locus of a point (x', y', z'), represents a surface; so that each point (x, y, z) is transformed into a *surface*, which we may call an Ω-surface: and any arbitrary surface σ is transformed into a surface Σ which is the envelope of the Ω-surfaces corresponding to the individual points of σ. This is the general case, and is the only one we considered in § 125.

(*β*) There may be two relations of this kind, say

$$\Omega_1(x, y, z, x', y', z') = 0, \qquad \Omega_2(x, y, z, x', y', z') = 0.$$

If (x, y, z) are given, these two equations in (x', y', z') represent a curve: so each point (x, y, z) is transformed into a *curve*, which we may call a K-curve: and any arbitrary surface σ is transformed into a surface Σ which is the envelope of the K-curves corresponding to the individual points of σ.

(γ) There may be three relations of this kind, say

$$\Omega_1(x, y, z, x', y', z')=0, \qquad \Omega_2(x, y, z, x', y', z')=0, \qquad \Omega_3(x, y, z, x', y', z')=0,$$

in which case each point (x, y, z) is transformed into a *point* (x', y', z'), and any arbitrary surface σ is transformed into a surface Σ which is the locus of the points corresponding to the individual points of σ.

Since the variations $(dq_1, dq_2, \ldots, dq_n, dQ_1, \ldots, dQ_n)$ in the equation

$$\sum_{r=1}^{n} (P_r dQ_r - p_r dq_r) = dW$$

are conditioned only by the relations

$$\frac{\partial \Omega_r}{\partial q_1} dq_1 + \frac{\partial \Omega_r}{\partial q_2} dq_2 + \ldots + \frac{\partial \Omega_r}{\partial q_n} dq_n + \frac{\partial \Omega_r}{\partial Q_1} dQ_1 + \ldots + \frac{\partial \Omega_r}{\partial Q_n} dQ_n = 0 \qquad (r = 1, 2, \ldots, k),$$

we must have

$$\left.\begin{aligned} P_r &= \frac{\partial W}{\partial Q_r} + \lambda_1 \frac{\partial \Omega_1}{\partial Q_r} + \ldots + \lambda_k \frac{\partial \Omega_k}{\partial Q_r} \\ p_r &= -\frac{\partial W}{\partial q_r} - \lambda_1 \frac{\partial \Omega_1}{\partial q_r} - \ldots - \lambda_k \frac{\partial \Omega_k}{\partial q_r} \end{aligned}\right\} \quad (r = 1, 2, \ldots, n) \ldots\ldots (\mathrm{B}),$$

where $(\lambda_1, \lambda_2, \ldots, \lambda_k)$ are undetermined multipliers and where W is a function of $(q_1, q_2, \ldots, q_n, Q_1, Q_2, \ldots, Q_n)$. The equations (A) and (B) are $(2n + k)$ equations to determine the $(2n + k)$ quantities

$$(Q_1, \ldots, Q_n, P_1, \ldots, P_n, \lambda_1, \ldots, \lambda_k)$$

in terms of $(q_1, \ldots, q_n, p_1, \ldots, p_n)$. *These equations may therefore be regarded as explicitly formulating the contact-transformation, in terms of the functions* $(W, \Omega_1, \Omega_2, \ldots, \Omega_k)$ *which characterise the transformation.*

Conversely, if $(W, \Omega_1, \Omega_2, \ldots, \Omega_k)$ are any $(k + 1)$ functions of the variables $(q_1, q_2, \ldots, q_n, Q_1, \ldots, Q_n)$, where $k \leqslant n$, and if

$$(Q_1, Q_2, \ldots, Q_n, P_1, \ldots, P_n, \lambda_1, \ldots, \lambda_k)$$

are defined in terms of $(q_1, q_2, \ldots, q_n, p_1, \ldots, p_n)$ by the equations*

$$\left\{\begin{aligned} &\Omega_r(q_1, q_2, \ldots, q_n, Q_1, Q_2, \ldots, Q_n) = 0 && (r = 1, 2, \ldots, k), \\ &P_r = \frac{\partial W}{\partial Q_r} + \lambda_1 \frac{\partial \Omega_1}{\partial Q_r} + \ldots + \lambda_k \frac{\partial \Omega_k}{\partial Q_r} && (r = 1, 2, \ldots, n), \\ &p_r = -\frac{\partial W}{\partial q_r} - \lambda_1 \frac{\partial \Omega_1}{\partial q_r} - \ldots - \lambda_k \frac{\partial \Omega_k}{\partial q_r} && (r = 1, 2, \ldots, n), \end{aligned}\right.$$

* These equations were first given in Jacobi's *Vorlesungen über Dynamik* (1866), p. 470, where their utility in the transformation of partial differential equations of the first order (to which dynamical problems can be reduced) was indicated. Their place in the theory of contact-transformations was pointed out by Lie.

then *the transformation from* $(q_1, q_2, \ldots, q_n, p_1, \ldots, p_n)$ *to* $(Q_1, Q_2, \ldots, Q_n, P_1, \ldots, P_n)$ *is a contact-transformation*; for the expression

$$\sum_{r=1}^{n} (P_r dQ_r - p_r dq_r)$$

becomes, in virtue of these equations, dW, and so is a perfect differential.

Example. If $Q=(2q)^{\frac{1}{2}} k^{-\frac{1}{2}} \cos p, \quad P=(2q)^{\frac{1}{2}} k^{\frac{1}{2}} \sin p,$

shew that
$$P=\frac{\partial W}{\partial Q}, \quad p=-\frac{\partial W}{\partial q},$$

where
$$W=\tfrac{1}{2} Q (2qk - k^2Q^2)^{\frac{1}{2}} - q \arccos \{k^{\frac{1}{2}} Q/(2q)^{\frac{1}{2}}\},$$

so that the transformation from (q, p) to (Q, P) is a contact-transformation.

127. *The bilinear covariant of a general differential form.*

Now let $(x_1, x_2, \ldots, x_n)$ be any set of n variables, and consider a differential form

$$X_1 dx_1 + X_2 dx_2 + \ldots + X_n dx_n,$$

where $(X_1, X_2, \ldots, X_n)$ denote any functions of $(x_1, x_2, \ldots, x_n)$; a form of this kind is called a *Pfaff's expression** in the variables $(x_1, x_2, \ldots, x_n)$. Let this expression be denoted by θ_d, and write

$$\theta_\delta = X_1 \delta x_1 + X_2 \delta x_2 + \ldots + X_n \delta x_n,$$

where δ is the symbol of an independent set of increments. Then we have

$$\begin{aligned} \delta\theta_d - d\theta_\delta &= \delta (X_1 dx_1 + X_2 dx_2 + \ldots + X_n dx_n) - d (X_1 \delta x_1 + X_2 \delta x_2 + \ldots + X_n \delta x_n) \\ &= \delta X_1 dx_1 + \ldots + \delta X_n dx_n + X_1 \delta dx_1 + \ldots + X_n \delta dx_n \\ &\qquad - dX_1 \delta x_1 - \ldots - dX_n \delta x_n - X_1 d\delta x_1 - \ldots - X_n d\delta x_n. \end{aligned}$$

Using the relations $\delta dx_r = d\delta x_r$, which exist since the variations d and δ are independent, and replacing dX_r, δX_r by

$$\frac{\partial X_r}{\partial x_1} dx_1 + \ldots + \frac{\partial X_r}{\partial x_n} dx_n, \quad \frac{\partial X_r}{\partial x_1} \delta x_1 + \ldots + \frac{\partial X_r}{\partial x_n} \delta x_n \text{ respectively,}$$

we have
$$\delta\theta_d - d\theta_\delta = \sum_{i=1}^{n} \sum_{j=1}^{n} a_{ij} dx_i \delta x_j,$$

where a_{ij} denotes the quantity $\partial X_i/\partial x_j - \partial X_j/\partial x_i$.

Let $(y_1, y_2, \ldots, y_n)$ be a new set of variables derived from $(x_1, x_2, \ldots, x_n)$ by some transformation; let the differential form when expressed in terms of these variables be

$$Y_1 dy_1 + Y_2 dy_2 + \ldots + Y_n dy_n.$$

* Pfaff's celebrated memoir on these expressions was presented to the Berlin Academy in 1815: *Abhandl. Akad. der Wiss.* 1814-15, p. 76.

and let the quantity $\partial Y_i/\partial y_j - \partial Y_j/\partial y_i$ be denoted by b_{ij}. Then since the expression $\delta\theta_d - d\theta_\delta$ has obviously the same value whatever be the variables in terms of which it is expressed, we have

$$\sum_{i=1}^{n}\sum_{j=1}^{n} a_{ij}dx_i\delta x_j = \sum_{i=1}^{n}\sum_{j=1}^{n} b_{ij}dy_i\delta y_j.$$

The expression $\Sigma a_{ij}dx_i\delta x_j$ is, on account of this equation, called the *bilinear covariant* of the form $\sum_{r=1}^{n} X_r dx_r$.

128. *The conditions for a contact-transformation expressed by means of the bilinear covariant.*

Let $(Q_1, Q_2, \ldots, Q_n, P_1, \ldots, P_n)$ be variables connected with $(q_1, q_2, \ldots, q_n, p_1, \ldots, p_n)$ by a contact transformation, so that $\sum_{r=1}^{n} P_r dQ_r$ differs from $\sum_{r=1}^{n} p_r dq_r$ by an exact differential.

It is clear from the last article that the bilinear covariant of a differential form is not affected by the addition of an exact differential to the form, since it depends only on the quantities $\partial X_i/\partial x_j - \partial X_j/\partial x_i$, which are all zero when the form is an exact differential: and we have shewn that the bilinear covariant of a form is transformed by any transformation into the bilinear covariant of the transformed form. It follows that the bilinear covariants of the forms $\sum_{r=1}^{n} P_r dQ_r$ and $\sum_{r=1}^{n} p_r dq_r$ are equal, i.e. that

$$\sum_{r=1}^{n} (\delta P_r dQ_r - dP_r \delta Q_r) = \sum_{r=1}^{n} (\delta p_r dq_r - dq_r \delta p_r);$$

so that *if the transformation from*

$$(q_1, q_2, \ldots, q_n, p_1, \ldots, p_n) \text{ to } (Q_1, Q_2, \ldots, Q_n, P_1, \ldots, P_n)$$

is a contact-transformation, the expression

$$\sum_{r=1}^{n} (\delta p_r dq_r - dq_r \delta p_r)$$

is invariant under the transformation.

Example. For the transformation defined by the equations

$$Q = (2q)^{\frac{1}{2}} k^{-\frac{1}{2}} \cos p, \qquad P = (2q)^{\frac{1}{2}} k^{\frac{1}{2}} \sin p,$$

we have

$$\begin{cases} dP = (2q)^{-\frac{1}{2}} k^{\frac{1}{2}} \sin p\, dq + (2q)^{\frac{1}{2}} k^{\frac{1}{2}} \cos p\, dp, \\ \delta Q = (2q)^{-\frac{1}{2}} k^{-\frac{1}{2}} \cos p\, \delta q - (2q)^{\frac{1}{2}} k^{-\frac{1}{2}} \sin p\, \delta p, \\ \delta P = (2q)^{-\frac{1}{2}} k^{\frac{1}{2}} \sin p\, \delta q + (2q)^{\frac{1}{2}} k^{\frac{1}{2}} \cos p\, \delta p, \\ dQ = (2q)^{-\frac{1}{2}} k^{-\frac{1}{2}} \cos p\, dp - (2q)^{\frac{1}{2}} k^{-\frac{1}{2}} \sin p\, dp. \end{cases}$$

By multiplication we have

$$dP\delta Q - \delta P dQ = -\sin^2 p\,(dq\,\delta p - \delta q\,dp) + \cos^2 p\,(dp\,\delta q - \delta p\,dq)$$
$$= dp\,\delta q - \delta p\,dq,$$

and consequently the transformation is a contact-transformation.

129. *The conditions for a contact-transformation in terms of Lagrange's bracket-expressions.*

We shall now give another form to the conditions that a transformation from variables $(q_1, q_2, \ldots, q_n, p_1, \ldots, p_n)$ to variables $(Q_1, Q_2, \ldots, Q_n, P_1, \ldots, P_n)$ may be a contact-transformation.

If $(q_1, q_2, \ldots, q_n, p_1, \ldots, p_n)$ are any functions of two variables (u, v) (and possibly of any number of other variables), the expression

$$\sum_{r=1}^{n}\left(\frac{\partial q_r}{\partial u}\frac{\partial p_r}{\partial v} - \frac{\partial p_r}{\partial u}\frac{\partial q_r}{\partial v}\right)$$

is called a *Lagrange's bracket-expression**. and is usually denoted by the symbol $[u, v]$.

If now $(q_1, q_2, \ldots, q_n, p_1, \ldots, p_n)$ are any functions of $2n$ variables $(Q_1, Q_2, \ldots, Q_n, P_1, \ldots, P_n)$, then in the expression

$$\sum_{r=1}^{n}(dp_r\delta q_r - \delta p_r dq_r)$$

we can replace dp_r by

$$\frac{\partial p_r}{\partial Q_1}dQ_1 + \frac{\partial p_r}{\partial Q_2}dQ_2 + \ldots + \frac{\partial p_r}{\partial Q_n}dQ_n + \frac{\partial p_r}{\partial P_1}dP_1 + \ldots + \frac{\partial p_r}{\partial P_n}dP_n,$$

and similarly for the other quantities; we thus obtain, on collecting terms,

$$\sum_{r=1}^{n}(dp_r\delta q_r - \delta p_r dq_r) = \sum_{k,\,l}[u_k, u_l](du_l\delta u_k - \delta u_l du_k),$$

where the summation on the right-hand side is taken over all pairs of variables (u_k, u_l) in the set $(Q_1, Q_2, \ldots, Q_n, P_1, \ldots, P_n)$.

But if the transformation from the variables $(q_1, q_2, \ldots, q_n, p_1, \ldots, p_n)$ to the variables $(Q_1, Q_2, \ldots, Q_n, P_1, \ldots, P_n)$ is a contact-transformation, we have

$$\sum_{r=1}^{n}(dp_r\delta q_r - \delta p_r dq_r) = \sum_{r=1}^{n}(dP_r\delta Q_r - \delta P_r dQ_r),$$

and this holds for all types of variation δ and d of the quantities; comparing with the above equation, we have therefore

$$\begin{cases} [P_i, P_k] = 0, \quad [Q_i, Q_k] = 0 & (i, k = 1, 2, \ldots, n), \\ \qquad\qquad [Q_i, P_k] = 0 & (i, k = 1, 2, \ldots, n;\ i \gtrless k), \\ \qquad\qquad [Q_i, P_i] = 1 & (i = 1, 2, \ldots, n). \end{cases}$$

* Lagrange, *Mém. de l'Institut de France*, année 1808: reprinted *Oeuvres*, VI. p. 713.

These may be regarded as partial differential equations which must be satisfied by $(q_1, q_2, \ldots, q_n, p_1, \ldots, p_n)$, *considered as functions of*

$$(Q_1, Q_2, \ldots, Q_n, P_1, \ldots, P_n)$$

in order that the transformation from one set of variables to the other may be a contact-transformation. These equations represent in an explicit form the conditions implied in the invariance of the expression

$$\sum_{r=1}^{n} (dp_r \delta q_r - \delta p_r dq_r).$$

130. *Poisson's bracket-expressions.*

We shall next introduce another class of bracket-expressions which are intimately connected with those of Lagrange.

If u and v are any two functions of a set of variables $(q_1, q_2, \ldots, q_n, p_1, \ldots, p_n)$, the expression

$$\sum_{r=1}^{n} \left(\frac{\partial u}{\partial q_r} \frac{\partial v}{\partial p_r} - \frac{\partial u}{\partial p_r} \frac{\partial v}{\partial q_r} \right)$$

is called the *Poisson's bracket-expression** of the functions u and v, and is denoted by the symbol (u, v).

Suppose now that $(u_1, u_2, \ldots, u_{2n})$ are $2n$ independent functions of the variables $(q_1, q_2, \ldots, q_n, p_1, \ldots, p_n)$, so that conversely $(q_1, q_2, \ldots, q_n, p_1, \ldots, p_n)$ are functions of $(u_1, u_2, \ldots, u_{2n})$. There will evidently be some connexion between the Poisson-brackets (u_r, u_s) and the Lagrange-brackets $[u_r, u_s]$: this connexion we shall now investigate.

We have

$$\sum_{t=1}^{2n} (u_t, u_r) [u_t, u_s] = \sum_{t=1}^{2n} \sum_{i=1}^{n} \sum_{j=1}^{n} \left(\frac{\partial u_t}{\partial q_i} \frac{\partial u_r}{\partial p_i} - \frac{\partial u_t}{\partial p_i} \frac{\partial u_r}{\partial q_i} \right) \left(\frac{\partial q_j}{\partial u_t} \frac{\partial p_j}{\partial u_s} - \frac{\partial p_j}{\partial u_t} \frac{\partial q_j}{\partial u_s} \right)$$

Now multiply out the right-hand side, remembering that

$$\sum_{t=1}^{2n} \frac{\partial u_t}{\partial q_i} \frac{\partial q_j}{\partial u_t} \quad \text{and} \quad \sum_{t=1}^{2n} \frac{\partial u_t}{\partial p_i} \frac{\partial p_j}{\partial u_t}$$

are each zero if $i \lesseqgtr j$ and unity if $i = j$; and that

$$\sum_{t=1}^{2n} \frac{\partial u_t}{\partial q_i} \frac{\partial p_j}{\partial u_t} \quad \text{and} \quad \sum_{t=1}^{2n} \frac{\partial u_t}{\partial p_i} \frac{\partial q_j}{\partial u_t}$$

are each zero; the equation becomes

$$\sum_{t=1}^{2n} (u_t, u_r) [u_t, u_s] = \sum_{i=1}^{n} \left(\frac{\partial u_r}{\partial p_i} \frac{\partial p_i}{\partial u_s} + \frac{\partial u_r}{\partial q_i} \frac{\partial q_i}{\partial u_s} \right),$$

and consequently

$$\sum_{t=1}^{2n} (u_t, u_r) [u_t, u_s] = 0 \text{ when } r \gtrless s,$$

while

$$\sum_{t=1}^{2n} (u_t, u_r) [u_t, u_r] = 1.$$

* Poisson, *Journal de l'École polytech.* VIII. (Cahier 15), (1809), p. 266.

But these are the conditions which must be satisfied in order that the two determinants

$$\begin{vmatrix} [u_1, u_1] & [u_1, u_2] & \ldots & [u_1, u_{2n}] \\ [u_2, u_1] & [u_2, u_2] & \ldots & [u_2, u_{2n}] \\ \ldots & \ldots & \ldots & \ldots \\ \ldots & \ldots & \ldots & \ldots \\ [u_{2n}, u_1] & \ldots & \ldots & [u_{2n}, u_{2n}] \end{vmatrix} \quad \text{and} \quad \begin{vmatrix} (u_1, u_1) & (u_2, u_1) & \ldots & (u_{2n}, u_1) \\ (u_1, u_2) & (u_2, u_2) & \ldots & (u_{2n}, u_2) \\ \ldots & \ldots & \ldots & \ldots \\ \ldots & \ldots & \ldots & \ldots \\ (u_1, u_{2n}) & \ldots & \ldots & (u_{2n}, u_{2n}) \end{vmatrix}$$

may be *reciprocal*, i.e. that any element in the one should be equal to the minor of the corresponding element in the other, divided by this latter determinant; the product of the two determinants being unity; and thus *the connexion between the Lagrange-brackets and the Poisson-brackets is expressed by the fact that the determinants formed from them are reciprocal.*

Example 1. If f, ϕ, ψ are any three functions of $(q_1, q_2, \ldots, q_n, p_1, \ldots, p_n)$, shew that

$$((f, \phi), \psi)+((\phi, \psi), f)+((\psi, f), \phi)=0.$$

Example 2. If F, Φ are functions of $(f_1, f_2, \ldots, f_k)$, which in turn are functions of $(q_1, q_2, \ldots, q_n, p_1, \ldots, p_n)$, shew that

$$(F, \Phi)= \sum_{r,s} \left(\frac{\partial F}{\partial f_r}\frac{\partial \Phi}{\partial f_s} - \frac{\partial F}{\partial f_s}\frac{\partial \Phi}{\partial f_r}\right)(f_r, f_s),$$

where the summation is taken over all combinations f_r, f_s.

131. *The conditions for a contact-transformation expressed by means of Poisson's bracket-expressions.*

Now let $(Q_1, Q_2, \ldots, Q_n, P_1, \ldots, P_n)$ denote $2n$ functions of $2n$ variables $(q_1, q_2, \ldots, q_n, p_1, \ldots, p_n)$; we shall shew that *the conditions which must be satisfied in order that the transformation from one set of variables to the other may be a contact-transformation may be written in the form*

$$\begin{cases} (P_i, P_j)=0, \quad (Q_i, Q_j)=0 & (i, j=1, 2, \ldots, n), \\ (Q_i, P_j)=0 & (i, j=1, 2, \ldots, n;\ i \lesseqgtr j), \\ (Q_i, P_i)=1 & (i=1, 2, \ldots, n). \end{cases}$$

For we have seen in § 129 that the conditions for a contact-transformation are expressed by the equations

$$\begin{cases} [P_i, P_j]=0, \quad [Q_i, Q_j]=0 & (i, j=1, 2, \ldots, n), \\ [Q_i, P_j]=0 & (i, j=1, 2, \ldots, n;\ i \lesseqgtr j), \\ [Q_i, P_i]=1 & (i=1, 2, \ldots, n). \end{cases}$$

Hence the relations

$$\sum_{t=1}^{2n} (u_t, u_r)[u_t, u_s]=0 \qquad (r \gtrless s)$$

of the last article become

$$(Q_i, Q_j) = 0 \quad (P_i, P_j) = 0 \qquad (i, j = 1, 2, \ldots, n),$$
$$(P_j, Q_i) = 0 \qquad (i, j = 1, 2, \ldots, n;\ i \gtrless j),$$

while the relations

$$\sum_{t=1}^{2n} (u_t, u_r)[u_t, u_r] = 1$$

give

$$(Q_i, P_i) = 1 \qquad (i = 1, 2, \ldots, n);$$

the theorem is thus established.

Example 1. If $(Q_1, Q_2, \ldots, Q_n, P_1, \ldots, P_n)$ are connected with $(q_1, q_2, \ldots, q_n, p_1, \ldots, p_n)$ by a contact-transformation, shew that

$$\sum_{r=1}^{n} \left(\frac{\partial\phi}{\partial Q_r}\frac{\partial\psi}{\partial P_r} - \frac{\partial\phi}{\partial P_r}\frac{\partial\psi}{\partial Q_r}\right) = \sum_{r=1}^{n} \left(\frac{\partial\phi}{\partial q_r}\frac{\partial\psi}{\partial p_r} - \frac{\partial\phi}{\partial p_r}\frac{\partial\psi}{\partial q_r}\right),$$

so that the Poisson-brackets of any two functions ϕ and ψ with respect to the two sets of variables are equal.

Example 2. If $(Q_1, Q_2, \ldots, Q_n)$ are given functions of $(q_1, q_2, \ldots, q_n, p_1, \ldots, p_n)$, and satisfy the partial differential equations

$$(Q_r, Q_s) = 0 \qquad (r, s = 1, 2, \ldots, n),$$

shew that n other functions $(P_1, P_2, \ldots, P_n)$ can be found such that the transformation from $(q_1, q_2, \ldots, q_n, p_1, \ldots, p_n)$ to $(Q_1, Q_2, \ldots, Q_n, P_1, \ldots, P_n)$ is a contact-transformation. (Lie.)

132. *The sub-groups of Mathieu transformations and extended point-transformations.*

If within a group of transformations there exists a set of transformations such that the result of performing in succession two transformations of the set is always equivalent to a transformation which also belongs to the set, this set of transformations is said to form a *sub-group* of the group.

A sub-group of the general group of contact-transformations is evidently constituted by those transformations for which the equation

$$\sum_{r=1}^{n} P_r dQ_r = \sum_{r=1}^{n} p_r dq_r$$

is satisfied. These transformations have been studied by Mathieu*.

They are essentially the same as the transformations called "homogeneous contact-transformations in $(q_1, q_2, \ldots, q_n, p_1, \ldots, p_n)$" by Lie.

In this case, we see from § 126 that $(Q_1, Q_2, \ldots, Q_n, P_1, \ldots, P_n)$ are to be obtained by eliminating $(\lambda_1, \lambda_2, \ldots, \lambda_k)$ from the $(2n + k)$ equations

$$\begin{cases} \Omega_r(q_1, q_2, \ldots, q_n, Q_1, \ldots, Q_n) = 0 & (r = 1, 2, \ldots, k), \\ P_r = \lambda_1 \dfrac{\partial\Omega_1}{\partial Q_r} + \lambda_2 \dfrac{\partial\Omega_2}{\partial Q_r} + \ldots + \lambda_k \dfrac{\partial\Omega_k}{\partial Q_r} & (r = 1, 2, \ldots, n), \\ p_r = -\lambda_1 \dfrac{\partial\Omega_1}{\partial q_r} - \lambda_2 \dfrac{\partial\Omega_2}{\partial q_r} - \ldots - \lambda_k \dfrac{\partial\Omega_k}{\partial q_r} & (r = 1, 2, \ldots, n). \end{cases}$$

* *Journal de Math.* xix. (1874), p. 265.

From the form of these equations it is evident that if $(p_1, p_2, \ldots, p_n)$ are each multiplied by any quantity μ, the effect is to multiply $(P_1, P_2, \ldots, P_n)$ each by μ; and therefore $(P_1, P_2, \ldots, P_n)$ must be homogeneous of the first degree (though not necessarily integral) in $(p_1, p_2, \ldots, p_n)$.

A sub-group within the group of Mathieu transformations is constituted by those transformations for which $(P_1, P_2, \ldots, P_n)$ are not only homogeneous of the first degree in $(p_1, p_2, \ldots, p_n)$ but also integral, i.e. linear, in them; so that we have equations of the form

$$P_r = \sum_{k=1}^{n} p_k f_{rk}(q_1, q_2, \ldots, q_n) \qquad (r = 1, 2, \ldots, n).$$

Substituting in the equation

$$\sum_{r=1}^{n} P_r dQ_r - \sum_{r=1}^{n} p_r dq_r = 0,$$

and equating to zero the coefficient of p_k, we have

$$\sum_{r=1}^{n} f_{rk}(q_1, q_2, \ldots, q_n)\, dQ_r = dq_k \qquad (k = 1, 2, \ldots, n),$$

so $(q_1, q_2, \ldots, q_n)$ are functions of $(Q_1, Q_2, \ldots, Q_n)$ only, and

$$f_{rk} = \partial q_k / \partial Q_r \qquad (r, k = 1, 2, \ldots, n).$$

It follows that *transformations of this kind are obtained by assigning n arbitrary relations connecting the variables* $(q_1, q_2, \ldots, q_n)$ *with the variables* $(Q_1, Q_2, \ldots, Q_n)$, *and then determining* $(P_1, P_2, \ldots, P_n)$ *from the equations*

$$P_r = \sum_{k=1}^{n} p_k \frac{\partial q_k}{\partial Q_r} \qquad (r = 1, 2, \ldots, n).$$

These transformations are extended point-transformations (§ 126).

Example. If
$$\sum_{r=1}^{n} P_r dQ_r = \sum_{r=1}^{n} p_r dq_r,$$
shew that
$$\sum_{k=1}^{n} p_k \frac{\partial Q_r}{\partial p_k} = 0, \qquad \sum_{k=1}^{n} p_k \frac{\partial P_r}{\partial p_k} = P_r.$$

133. *Infinitesimal contact-transformations.*

We shall now consider transformations in which the new variables $(Q_1, Q_2, \ldots, Q_n, P_1, \ldots, P_n)$ differ from the original variables $(q_1, q_2, \ldots q_n, p_1, \ldots, p_n)$ by quantities which are infinitesimal. Let these differences be denoted by $(\Delta q_1, \Delta q_2, \ldots, \Delta q_n, \Delta p_1, \ldots, \Delta p_n)$, where

$$\left.\begin{aligned} \Delta q_r &= \phi_r(q_1, q_2, \ldots, q_n, p_1, \ldots, p_n)\, \Delta t \\ \Delta p_r &= \psi_r(q_1, q_2, \ldots, q_n, p_1, \ldots, p_n)\, \Delta t \end{aligned}\right\} \qquad (r = 1, 2, \ldots, n),$$

and Δt is an arbitrary infinitesimal constant; so that

$$\left.\begin{aligned} Q_r &= q_r + \Delta q_r = q_r + \phi_r \Delta t \\ P_r &= p_r + \Delta p_r = p_r + \psi_r \Delta t \end{aligned}\right\} \qquad (r = 1, 2, \ldots, n),$$

and the transformation is specified by the functions

$$(\phi_1, \phi_2, \ldots, \phi_n, \psi_1, \psi_2, \ldots, \psi_n).$$

Now suppose that the transformation is a contact-transformation. Then we have

$$\sum_{r=1}^{n} (P_r dQ_r - p_r dq_r) = dW,$$

where W is some function of $(q_1, q_2, \ldots, q_n, p_1, \ldots, p_n)$; or

$$\sum_{r=1}^{n} \{(p_r + \psi_r \Delta t)(dq_r + d\phi_r \cdot \Delta t) - p_r dq_r\} = dW,$$

or

$$\Delta t \sum_{r=1}^{n} (\psi_r dq_r + p_r d\phi_r) = dW.$$

It is evident that the function W must contain Δt as a factor: writing $W = U\Delta t$, where U is some function of $(q_1, q_2, \ldots, q_n, p_1, \ldots, p_n)$, the equation becomes

$$\sum_{r=1}^{n} (\psi_r dq_r + p_r d\phi_r) = dU.$$

Hence we have

$$\sum_{r=1}^{n} (\psi_r dq_r - \phi_r dp_r) = d\left(U - \sum_{r=1}^{n} p_r \phi_r\right)$$

$$= -dK(q_1, q_2, \ldots, q_n, p_1, \ldots, p_n) \text{ say,}$$

and therefore

$$\phi_r = \frac{\partial K}{\partial p_r}, \qquad \psi_r = -\frac{\partial K}{\partial q_r} \qquad (r = 1, 2, \ldots, n).$$

Thus *the most general infinitesimal contact-transformation is defined by the equations*

$$Q_r = q_r + \frac{\partial K}{\partial p_r}\Delta t, \qquad P_r = p_r - \frac{\partial K}{\partial q_r}\Delta t \qquad (r = 1, 2, \ldots, n),$$

where K is an arbitrary function of $(q_1, q_2, \ldots, q_n, p_1, \ldots, p_n)$, and Δt is an arbitrary infinitesimal quantity independent of $(q_1, q_2, \ldots, q_n, p_1, \ldots, p_n)$.

The increment in any function $f(q_1, q_2, \ldots, q_n, p_1, \ldots, p_n)$ when its arguments $(q_1, q_2, \ldots, q_n, p_1, \ldots, p_n)$ are subjected to this transformation is

$$\sum_{r=1}^{n} \left(\frac{\partial f}{\partial q_r}\frac{\partial K}{\partial p_r} - \frac{\partial f}{\partial p_r}\frac{\partial K}{\partial q_r}\right)\Delta t,$$

or

$$(f, K)\,\Delta t;$$

on this account the Poisson-bracket (f, K) is said to be the *symbol* of the most general infinitesimal transformation of the infinite group which consists of all contact-transformations of the $2n$ variables $(q_1, q_2, \ldots, q_n, p_1, \ldots, p_n)$.

134. *The resulting new view of dynamics.*

The theorem established in the last article enables us to extend to all conservative holonomic dynamical systems, whatever be the number of degrees of freedom, the conception which was formulated at the end of § 125 for certain simple systems. For the motion is expressed (§ 109) by equations of the type

$$\frac{dq_r}{dt} = \frac{\partial H}{\partial p_r}, \qquad \frac{dp_r}{dt} = -\frac{\partial H}{\partial q_r} \qquad (r = 1, 2, \ldots, n),$$

and from the last article it follows that we can interpret these equations as implying that the transformation from the values of the variables at time t to their values at time $t + dt$ is an infinitesimal contact-transformation. *The whole course of a dynamical system can thus be regarded as the gradual self-unfolding of a contact-transformation.* This result is really a generalisation of the statement that *the paths of the rays in a pencil of light can be specified by the gradual propagation of a wave-front.* Taken in conjunction with the group-property of contact-transformations, it is the foundation of the transformation-theory of dynamical systems.

From this it is evident that if $(q_1, q_2, \ldots, q_n, p_1, \ldots, p_n)$ are the variables in a dynamical system, and $(\alpha_1, \alpha_2, \ldots, \alpha_n, \beta_1, \ldots, \beta_n)$ are their respective values at some selected epoch $t = t_0$, the equations which express $(q_1, q_2, \ldots, q_n, p_1, \ldots, p_n)$ in terms of $(\alpha_1, \alpha_2, \ldots, \alpha_n, \beta_1, \ldots, \beta_n, t)$ (and which constitute the solution of the differential equations of motion) express a contact-transformation from $(\alpha_1, \alpha_2, \ldots, \alpha_n, \beta_1, \ldots, \beta_n)$ to $(q_1, q_2, \ldots, q_n, p_1, \ldots, p_n)$; in this t is regarded merely as a parameter occurring in the equations which define the transformation.

135. *Helmholtz's reciprocal theorem.*

Since the values of the variables $(q_1, q_2, \ldots, q_n, p_1, \ldots, p_n)$ of a dynamical system at time t are derivable by a contact-transformation from their values $(\alpha_1, \alpha_2, \ldots, \alpha_n, \beta_1, \ldots, \beta_n)$ at time t_0, we have (§ 128)

$$\sum_{i=1}^{n} (\Delta p_i \delta q_i - \delta p_i \Delta q_i) = \sum_{i=1}^{n} (\Delta \beta_i \delta \alpha_i - \delta \beta_i \Delta \alpha_i),$$

where the symbols Δ and δ refer to increments arrived at by passages from a given orbit to two different adjacent orbits respectively.

Now suppose that δ refers to the increments obtained in passing to that orbit which is defined by the values

$$(\alpha_1, \alpha_2, \ldots, \alpha_n, \beta_1, \beta_2, \ldots, \beta_{r-1}, \beta_r + \delta\beta_r, \beta_{r+1}, \ldots, \beta_n)$$

at time t_0; and let Δ refer to the increment obtained in passing to that orbit which is defined by the values

$$(q_1, q_2, \ldots, q_n, p_1, \ldots, p_{s-1}, p_s + \Delta p_s, p_{s+1}, \ldots, p_n)$$

at time t_1; then the above equation becomes

$$\Delta p_s \delta q_s = -\delta\beta_r \Delta\alpha_r,$$

so the increment in q_s due to an increment in β_r (when $\alpha_1, \alpha_2, \ldots, \alpha_n$, $\beta_1, \ldots, \beta_{r-1}, \beta_{r+1}, \ldots, \beta_n$ are not varied) is equal to the increment (with sign reversed) in α_r corresponding to an increment in p_s (when $q_1, q_2, \ldots, q_n, p_1, \ldots, p_{s-1}, p_{s+1}, \ldots, p_n$ are not varied) equal to the previous increment in β_r.

This result can for many systems be physically interpreted, as was observed by Helmholtz*; for a small impulse applied to a system can be conveniently measured by the resulting change in one of the momenta $(p_1, \ldots, p_n)$, and the change in α_r due to a change in p_s can be realised in the *reversed motion*, i.e. the motion which starts from some given position with each of the velocities corresponding to that position changed in sign, so that the subsequent history of the system is the same as its previous history, but performed in reverse order. We can therefore state the theorem broadly thus: *the change produced in any interval by a small initial impulse of any type in the coordinate of any other (or of the same) type, in the direct motion, is equal to the change produced in the same interval of the reversed motion in the coordinate of the first type by an equal small initial impulse of the second type*†.

Example. In elliptic motion under a centre of force in the centre, if a small velocity δv in the direction of the normal be communicated to the particle as it is passing through either extremity of the major axis, shew that the tangential deviation produced after a quarter-period is $\mu^{-\frac{1}{2}}\delta v$, where μ is the constant of force. Shew also that a tangential velocity δv, communicated at the extremity of the minor axis, produces after a quarter-period an equal normal deviation $\mu^{-\frac{1}{2}}\delta v$. (Lamb.)

136. *Jacobi's theorem on the transformation of a given dynamical system into another dynamical system.*

It appears from § 116 that if a Hamiltonian system of differential equations

$$\frac{dq_r}{dt} = \frac{\partial H}{\partial p_r}, \qquad \frac{dp_r}{dt} = -\frac{\partial H}{\partial q_r} \qquad (r = 1, 2, \ldots, n)$$

is transformed by change of variables, the system of differential equations so obtained will still have the Hamiltonian form

$$\frac{dQ_r}{dt} = \frac{\partial K}{\partial P_r}, \qquad \frac{dP_r}{dt} = -\frac{\partial K}{\partial Q_r} \qquad (r = 1, 2, \ldots, n),$$

provided the new variables $(Q_1, Q_2, \ldots, Q_n, P_1, \ldots, P_n)$ are such that

$$\int P_1 \delta Q_1 + P_2 \delta Q_2 + \ldots + P_n \delta Q_n$$

is an integral-invariant (relative or absolute) of the system.

* *Journal für Math.* C. (1886).

† Cf. Lamb, *Proc. Lond. Math. Soc.* XIX. (1898), p. 144.

A transformation of this kind is, in general, *special to the problem considered,* i.e. it transforms the given Hamiltonian system into another Hamiltonian system, but it will not necessarily transform any other arbitrarily chosen Hamiltonian system into a Hamiltonian system. Among these transformations however are included transformations which have the property of conserving the Hamiltonian form of *any* dynamical system to which they may be applied: these may be obtained in the following way.

We have seen (§ 115) that

$$\int \sum_{r=1}^{n} p_r \delta q_r$$

is a relative integral-invariant of any Hamiltonian system. Let $(Q_1, Q_2, \ldots, Q_n, P_1, \ldots, P_n)$ be a set of $2n$ variables obtained from $(q_1, q_2, \ldots, q_n, p_1, \ldots, p_n)$ by a contact-transformation, so that

$$\sum_{r=1}^{n} P_r dQ_r - \sum_{r=1}^{n} p_r dq_r = dW,$$

where dW denotes an exact differential. The equations which define the transformation may involve the time, so that $(Q_1, Q_2, \ldots, Q_n, P_1, \ldots, P_n)$ are functions of $(q_1, q_2, \ldots, q_n, p_1, \ldots, p_n, t)$; but in the variation denoted by d in this equation the time is not supposed to be varied: if t is supposed to vary, the equation becomes

$$\sum_{r=1}^{n} P_r dQ_r - \sum_{r=1}^{n} p_r dq_r = dW + U dt,$$

where U denotes some function of the variables.

Now the variation denoted by δ in the integral-invariant is a variation from a point of one orbit to the contemporaneous point of an adjacent orbit; if therefore we regard the variables as functions of $(a_1, a_2, \ldots, a_{2n}, t)$, where $(a_1, a_2, \ldots, a_{2n})$ are the constants of integration which occur in the solution of the equations of motion, the variation δ is one in which $(a_1, a_2, \ldots, a_{2n})$ are varied but t is not varied: we have consequently, as a special case of the last equation,

$$\sum_{r=1}^{n} P_r \delta Q_r - \sum_{r=1}^{n} p_r \delta q_r = \delta W,$$

and therefore

$$\int \sum_{r=1}^{n} P_r \delta Q_r$$

is a relative integral-invariant; so the transformed system of differential equations, in which $(Q_1, Q_2, \ldots, Q_n, P_1, \ldots, P_n)$ are taken as dependent variables, will have the Hamiltonian form and can be written

$$\frac{dQ_r}{dt} = \frac{\partial K}{\partial P_r}, \quad \frac{dP_r}{dt} = -\frac{\partial K}{\partial Q_r} \qquad (r = 1, 2, \ldots, n),$$

where K is some function of $(Q_1, Q_2, \ldots, Q_n, P_1, \ldots, P_n, t)$.

Hence *a contact-transformation of the variables* $(q_1, q_2, \ldots, q_n, p_1, \ldots, p_n)$ *of any dynamical system conserves the Hamiltonian form of the equations of the system**. In the case of an ordinary "change of variables" in the dynamical system, in which $(Q_1, Q_2, \ldots, Q_n)$ are functions of $(q_1, q_2, \ldots, q_n)$ only, the contact-transformation is merely an extended point-transformation.

Example. Shew that the contact-transformation defined by the equations

$$q=(2Q)^{\frac{1}{2}} k^{-\frac{1}{2}} \cos P, \qquad p= 2Q)^{\frac{1}{2}} k^{\frac{1}{2}} \sin P,$$

changes the system

$$\frac{dq}{dt}=\frac{\partial H}{\partial p}, \quad \frac{dp}{dt}=-\frac{\partial H}{\partial q},$$

where

$$H=\tfrac{1}{2}(p^2+k^2q^2),$$

into the system

$$\frac{dQ}{dt}=\frac{\partial K}{\partial P}, \quad \frac{dP}{dt}=-\frac{\partial K}{\partial Q},$$

where

$$K=kQ.$$

137. *Representation of a dynamical problem by a differential form.*

The reason for the importance of contact-transformations in connexion with dynamical problems is more clearly seen by the introduction of a certain differential form which is invariantively related to the problem.

Let any differential form with $(2n+1)$ independent variables $(x_1, x_2, \ldots, x_{2n+1})$ be

$$X_1 dx_1 + X_2 dx_2 + \ldots + X_{2n+1} dx_{2n+1};$$

we have seen (§ 127) that its bilinear covariant

$$\sum_{i=1}^{2n+1} \sum_{j=1}^{2n+1} a_{ij}\, dx_i \delta x_j,$$

where a_{ij} denotes the quantity $(\partial X_i/\partial x_j - \partial X_j/\partial x_i)$, is invariantively related to the form. If we equate to zero the coefficients of $\delta x_1, \delta x_2, \ldots, \delta x_{2n+1}$, we obtain the system of $(2n+1)$ equations

$$\sum_{i=1}^{2n+1} a_{i1}\, dx_i = 0, \quad \sum_{i=1}^{2n+1} a_{i2}\, dx_i = 0, \quad \ldots, \quad \sum_{i=1}^{2n+1} a_{i,2n+1}\, dx_i = 0.$$

Since the determinant of the quantities a_{ij} is skew-symmetric and of odd order, it is zero, and these equations are therefore mutually compatible. They are known as the *first Pfaff's system of equations* corresponding to the differential form $\sum_{r=1}^{2n+1} X_r dx_r$, and from the mode of their formation are invariantively connected with it; that is to say, if any change of variables is made, the new variables $(y_1, y_2, \ldots, y_{2n+1})$ being given functions of $(x_1, x_2, \ldots, x_{2n+1})$, and if the differential form be changed by this transformation to

$$\sum_{r=1}^{2n+1} Y_r dy_r,$$

* This important theorem was first given by Jacobi, *Comptes Rendus*, v. (1837), p. 61.

and if $\sum_{i=1}^{2n+1} b_{i1} dy_i = 0, \quad \sum_{i=1}^{2n+1} b_{i2} dy_i = 0, \quad \ldots, \quad \sum_{i=1}^{2n+1} b_{i,2n+1} dy_i = 0$

be the first Pfaff's system derived from the differential form

$$\sum_{r=1}^{2n+1} Y_r dy_r,$$

then this system is equivalent to the system

$$\sum_{i=1}^{2n+1} a_{i1} dx_i = 0, \quad \sum_{i=1}^{2n+1} a_{i2} dx_i = 0, \quad \ldots, \quad \sum_{i=1}^{2n+1} a_{i,2n+1} dx_i = 0.$$

Consider now the special differential form

$$p_1 dq_1 + p_2 dq_2 + \ldots + p_n dq_n - H dt$$

in the $(2n+1)$ variables $(q_1, q_2, \ldots, q_n, p_1, \ldots, p_n, t)$, where H is any function of $(q_1, q_2, \ldots, q_n, p_1, \ldots, p_n, t)$. Forming the corresponding quantities a_{ij}, we find that the first Pfaff's system of differential equations of this differential form is

$$-dp_r - \frac{\partial H}{\partial q_r} dt = 0 \qquad (r = 1, 2, \ldots, n),$$

$$dq_r - \frac{\partial H}{\partial p_r} dt = 0 \qquad (r = 1, 2, \ldots, n),$$

$$dH - \frac{\partial H}{\partial t} dt = 0.$$

Of these the last equation is a consequence of the others: and therefore the system of equations can be written

$$\frac{dq_r}{dt} = \frac{\partial H}{\partial p_r}, \quad \frac{dp_r}{dt} = -\frac{\partial H}{\partial q_r} \qquad (r = 1, 2, \ldots, n);$$

but these are the equations of motion of a dynamical system in which the Hamiltonian function is H. It follows that *the dynamical system whose Hamiltonian function is H is invariantively connected with the differential form*

$$p_1 dq_1 + p_2 dq_2 + \ldots + p_n dq_n - H dt,$$

inasmuch as the equations of motion of the dynamical system, in terms of any variables $(x_1, x_2, \ldots, x_{2n}, \tau)$ whatever, are the first Pfaff's system of the differential form

$$X_1 dx_1 + X_2 dx_2 + \ldots + X_{2n} dx_{2n} + T d\tau$$

which is derived from the form

$$p_1 dq_1 + p_2 dq_2 + \ldots + p_n dq_n - H dt$$

by the transformation from the variables $(q_1, q_2, \ldots, q_n, p_1, \ldots, p_n, t)$ to the variables $(x_1, x_2, \ldots, x_{2n}, \tau)$.

138. *The Hamiltonian function of the transformed equations.*

The result of the last article furnishes another proof of the theorem that the equations of dynamics

$$\frac{dq_r}{dt} = \frac{\partial H}{\partial p_r}, \qquad \frac{dp_r}{dt} = -\frac{\partial H}{\partial q_r} \qquad (r = 1, 2, \ldots, n)$$

conserve the Hamiltonian form under all contact-transformations of $(q_1, q_2, \ldots, q_n, p_1, \ldots, p_n)$, and moreover it enables us to find the Hamiltonian function K of the system thus obtained,

$$\frac{dQ_r}{dt} = \frac{\partial K}{\partial P_r}, \qquad \frac{dP_r}{dt} = -\frac{\partial K}{\partial Q_r} \qquad (r = 1, 2, \ldots, n).$$

For let the contact-transformation be defined by the equations

$$\begin{cases} \Omega_r = 0 & (r = 1, 2, \ldots, k), \\ P_r = \dfrac{\partial W}{\partial Q_r} + \lambda_1 \dfrac{\partial \Omega_1}{\partial Q_r} + \lambda_2 \dfrac{\partial \Omega_2}{\partial Q_r} + \ldots + \lambda_k \dfrac{\partial \Omega_k}{\partial Q_r} & (r = 1, 2, \ldots, n), \\ p_r = -\dfrac{\partial W}{\partial q_r} - \lambda_1 \dfrac{\partial \Omega_1}{\partial q_r} - \lambda_2 \dfrac{\partial \Omega_2}{\partial q_r} - \ldots - \lambda_k \dfrac{\partial \Omega_k}{\partial q_r} & (r = 1, 2, \ldots, n), \end{cases}$$

where $(\Omega_1, \Omega_2, \ldots, \Omega_k, W)$ are any functions of the variables $(q_1, q_2, \ldots, q_n, Q_1, Q_2, \ldots, Q_n, t)$.

From these equations we have identically

$$\sum_{r=1}^{n} p_r dq_r = \sum_{r=1}^{n} P_r dQ_r - \sum_{r=1}^{n} \left(\frac{\partial W}{\partial q_r} dq_r + \frac{\partial W}{\partial Q_r} dQ_r \right) - \sum_{s=1}^{k} \lambda_s \sum_{r=1}^{n} \left(\frac{\partial \Omega_s}{\partial q_r} dq_r + \frac{\partial \Omega_s}{\partial Q_r} dQ_r \right),$$

and hence (the symbol d denoting a variation in which all the variables, including t, are changed)

$$\sum_{r=1}^{n} p_r dq_r = \sum_{r=1}^{n} P_r dQ_r + \frac{\partial W}{\partial t} dt - dW + \sum_{s=1}^{k} \lambda_s \frac{\partial \Omega_s}{\partial t} dt,$$

or

$$\sum_{r=1}^{n} p_r dq_r - H dt = \sum_{r=1}^{n} P_r dQ_r - \left(H - \frac{\partial W}{\partial t} - \sum_{s=1}^{k} \lambda_s \frac{\partial \Omega_s}{\partial t} \right) dt - dW.$$

The perfect differential dW on the right-hand side can be neglected, since it does not affect the first Pfaff's system of the differential form: and hence *the contact-transformation transforms the system of equations*

$$\frac{dq_r}{dt} = \frac{\partial H}{\partial p_r}, \qquad \frac{dp_r}{dt} = -\frac{\partial H}{\partial q_r} \qquad (r = 1, 2, \ldots, n)$$

to the system

$$\frac{dQ_r}{dt} = \frac{\partial K}{\partial P_r}, \qquad \frac{dP_r}{dt} = -\frac{\partial K}{\partial Q_r} \qquad (r = 1, 2, \ldots, n),$$

where

$$K = H - \frac{\partial W}{\partial t} - \sum_{s=1}^{k} \lambda_s \frac{\partial \Omega_s}{\partial t},$$

K being supposed expressed in terms of $(Q_1, Q_2, \ldots, Q_n, P_1, \ldots, P_n, t)$.

139. *Transformations in which the independent variable is changed.*

The result of § 137 also enables us to determine those transformations of the whole set of $(2n+1)$ variables $(q_1, q_2, \ldots, q_n, p_1, \ldots, p_n, t)$ to new variables $(Q_1, Q_2, \ldots, Q_n, P_1, \ldots, P_n, T)$ by which any Hamiltonian system

$$\frac{dq_r}{dt} = \frac{\partial H}{\partial p_r}, \quad \frac{dp_r}{dt} = -\frac{\partial H}{\partial q_r} \qquad (r = 1, 2, \ldots, n)$$

is transformed into a system of the Hamiltonian form

$$\frac{dQ_r}{dT} = \frac{\partial K}{\partial P_r}, \quad \frac{dP_r}{dT} = -\frac{\partial K}{\partial Q_r} \qquad (r = 1, 2, \ldots, n).$$

For this is the same thing as finding the transformations which change the differential form

$$p_1 dq_1 + p_2 dq_2 + \ldots + p_n dq_n + h dt,$$

where the variables $(q_1, q_2, \ldots, q_n, p_1, \ldots, p_n, t, h)$ are connected by the equation

$$H(q_1, q_2, \ldots, q_n, p_1, \ldots, p_n, t) + h = 0,$$

into the differential form

$$P_1 dQ_1 + P_2 dQ_2 + \ldots + P_n dQ_n + k dT + \text{a perfect differential},$$

where the variables $(Q_1, Q_2, \ldots, Q_n, P_1, P_2, \ldots, P_n, T, k)$ are connected by the relation

$$K(Q_1, Q_2, \ldots, Q_n, P_1, \ldots, P_n, T) + k = 0.$$

But any contact-transformations of the $(2n+2)$ variables $(q_1, q_2, \ldots, q_n, t, p_1, \ldots, p_n, h)$ to new variables $(Q_1, Q_2, \ldots, Q_n, T, P_1, P_2, \ldots, P_n, k)$ will satisfy this condition; when the transformation has been assigned, the function K is obtained by substituting in the equation

$$H(q_1, q_2, \ldots, q_n, p_1, \ldots, p_n, t) + h = 0$$

the values of $(q_1, q_2, \ldots, q_n, t, p_1, \ldots, p_n, h)$ as functions of $(Q_1, \ldots, Q_n, T, P_1, \ldots, P_n, k)$, and then solving this equation for k, so that it takes the form

$$K(Q_1, Q_2, \ldots, Q_n, P_1, \ldots, P_n, T) + k = 0;$$

the required transformations are thereby completely determined.

140. *New formulation of the integration-problem.*

We have seen (§ 137) that if any change of variables is made in the dynamical system

$$\frac{dq_r}{dt} = \frac{\partial H}{\partial p_r}, \quad \frac{dp_r}{dt} = -\frac{\partial H}{\partial q_r} \qquad (r = 1, 2, \ldots, n),$$

the new differential equations will be the first Pfaff's system of the form which is derived from

$$p_1 dq_1 + p_2 dq_2 + \ldots + p_n dq_n - H dt$$

by the transformation.

Supposing that a transformation is found, defined by a set of equations

$$\left.\begin{aligned} q_r &= \phi_r (Q_1, Q_2, \ldots, Q_n, P_1, \ldots, P_n, t) \\ p_r &= \psi_r (Q_1, Q_2, \ldots, Q_n, P_1, \ldots, P_n, t) \end{aligned}\right\} \quad (r = 1, 2, \ldots, n),$$

which is such that the above differential form, when expressed in terms of the new variables, becomes

$$P_1 dQ_1 + P_2 dQ_2 + \ldots + P_n dQ_n - dT,$$

where dT is the perfect differential of some function of the variables $(Q_1, Q_2, \ldots, Q_n, P_1, \ldots, P_n, t)$; the corresponding first Pfaff's system of equations is

$$dQ_r = 0, \quad dP_r = 0 \qquad (r = 1, 2, \ldots, n),$$

and the integrals of these equations are

$$Q_r = \text{Constant}, \quad P_r = \text{Constant} \qquad (r = 1, 2, \ldots, n);$$

so *the equations*

$$\left.\begin{aligned} q_r &= \phi_r (Q_1, Q_2, \ldots, Q_n, P_1, \ldots, P_n, t) \\ p_r &= \psi_r (Q_1, Q_2, \ldots, Q_n, P_1, \ldots, P_n, t) \end{aligned}\right\} \quad (r = 1, 2, \ldots, n)$$

constitute the solution of the dynamical system, when the quantities $(Q_1, Q_2, \ldots, Q_n, P_1, \ldots, P_n)$ *are regarded as* $2n$ *arbitrary constants of integration.*

The integration-problem is thus reduced to the determination of a transformation for which the last term of the differential form becomes a perfect differential.

MISCELLANEOUS EXAMPLES.

1. Shew that the transformation defined by the equations

$$Q_1 = q_1^2 + \lambda^2 p_1^2, \qquad Q_2 = \frac{1}{2\lambda^2}(q_1^2 + q_2^2 + \lambda^2 p_1^2 + \lambda^2 p_2^2),$$

$$2\lambda P_1 = \arctan\left(\frac{q_1}{\lambda p_1}\right) - \arctan\left(\frac{q_2}{\lambda p_2}\right), \quad P_2 = \lambda \arctan\left(\frac{q_2}{\lambda p_2}\right),$$

is a contact-transformation, and that it reduces the dynamical system whose Hamiltonian function is $\frac{1}{2}(p_1^2 + p_2^2 + \lambda^{-2} q_1^2 + \lambda^{-2} q_2^2)$ to the dynamical system whose Hamiltonian function is Q_2

2. If $(x_1, x_2, \ldots, x_{2n})$ denote any functions of $(q_1, q_2, \ldots, q_n, p_1, \ldots, p_n)$, and if

$$p_1 dq_1 + p_2 dq_2 + \ldots + p_n dq_n = X_1 dx_1 + X_2 dx_2 + \ldots + X_{2n} dx_{2n};$$

if moreover a_{mn} denotes $\partial X_m / \partial x_n - \partial X_n / \partial x_m$, D denotes the determinant formed of the quantities a_{mn}, A_{ik} denotes the minor of a_{ik} in D, divided by D, and u and v denote arbitrary functions of the variables, shew that

$$\sum_{r=1}^{n} \left(\frac{\partial u}{\partial q_r}\frac{\partial v}{\partial p_r} - \frac{\partial u}{\partial p_r}\frac{\partial v}{\partial q_r}\right) = \sum_{i=1}^{n} \sum_{k=1}^{n} A_{ik} \frac{\partial v}{\partial x_i}\frac{\partial u}{\partial x_k}. \qquad \text{(Clebsch.)}$$

3. Shew that for any Hamiltonian system the integral-invariants

$$\iiint \ldots \int \delta q_1 \delta q_2 \ldots \delta q_n \delta p_1 \ldots \delta p_n$$

and

$$\iiint \ldots \int \delta Q_1 \delta Q_2 \ldots \delta Q_n \delta P_1 \ldots \delta P_n,$$

extended over corresponding domains, are equal if $(q_1, q_2, \ldots, q_n, p_1, \ldots, p_n)$ and $(Q_1, Q_2, \ldots, Q_n, P_1, \ldots, P_n)$ are connected by a contact-transformation.

4. Prove that the contact-transformation defined by the equations

$$\begin{cases} q_1 = \lambda_1^{-\frac{1}{2}} (2Q_1)^{\frac{1}{2}} \cos P_1 + \lambda_2^{-\frac{1}{2}} (2Q_2)^{\frac{1}{2}} \cos P_2, \\ q_2 = -\lambda_1^{-\frac{1}{2}} (2Q_1)^{\frac{1}{2}} \cos P_1 + \lambda_2^{-\frac{1}{2}} (2Q_2)^{\frac{1}{2}} \cos P_2, \\ p_1 = \frac{1}{2} (2\lambda_1 Q_1)^{\frac{1}{2}} \sin P_1 + \frac{1}{2} (2\lambda_2 Q_2)^{\frac{1}{2}} \sin P_2, \\ p_2 = -\frac{1}{2} (2\lambda_1 Q_1)^{\frac{1}{2}} \sin P_1 + \frac{1}{2} (2\lambda_2 Q_2)^{\frac{1}{2}} \sin P_2, \end{cases}$$

changes the system

$$\frac{dq_r}{dt} = \frac{\partial H}{\partial p_r}, \quad \frac{dp_r}{dt} = -\frac{\partial H}{\partial q_r} \qquad (r = 1, 2),$$

where

$$H = p_1^2 + p_2^2 + \tfrac{1}{8}\lambda_1^2 (q_1 - q_2)^2 + \tfrac{1}{8}\lambda_2^2 (q_1 + q_2)^2,$$

into the system

$$\frac{dQ_r}{dt} = \frac{\partial K}{\partial P_r}, \quad \frac{dP_r}{dt} = -\frac{\partial K}{\partial Q_r} \qquad (r = 1, 2),$$

where

$$K = \lambda_1 Q_1 + \lambda_2 Q_2.$$

Integrate this system, and hence integrate the original system.

CHAPTER XII

PROPERTIES OF THE INTEGRALS OF DYNAMICAL SYSTEMS

141. *Reduction of the order of a Hamiltonian system by use of the integral of energy.*

We have shewn in § 42 how the Lagrangian equations of motion of a conservative holonomic system can be reduced in order by use of the integral of energy of the system. We shall require the corresponding theorem for the equations of motion in their Hamiltonian form; this may be obtained as follows.

Consider a dynamical system with n degrees of freedom for which the Hamiltonian function H does not involve the time explicitly, so that

$$H + h = 0,$$

where h is a constant, is the integral of energy of the system.

Let this equation be solved for the variable p_1, so that it can be written

$$K(p_2, p_3, \ldots, p_n, q_1, \ldots, q_n, h) + p_1 = 0.$$

The differential form associated with the system is

$$p_1 dq_1 + p_2 dq_2 + \ldots + p_n dq_n + h dt,$$

where the variables $(q_1, q_2, \ldots, q_n, p_1, p_2, \ldots, p_n, h, t)$ are connected by the last equation: the differential form can therefore be written

$$p_2 dq_2 + p_3 dq_3 + \ldots + p_n dq_n + h dt - K(p_2, p_3, \ldots, p_n, q_1, \ldots, q_n, h)\, dq_1,$$

where we can regard $(q_1, q_2, \ldots, q_n, p_2, \ldots, p_n, h, t)$ as the $(2n+1)$ variables.

But the differential equations corresponding to this form are (§ 137)

$$\frac{dq_r}{dq_1} = \frac{\partial K}{\partial p_r}, \quad \frac{dp_r}{dq_1} = -\frac{\partial K}{\partial q_r} \qquad (r = 2, 3, \ldots, n),$$

$$\frac{dt}{dq_1} = \frac{\partial K}{\partial h}, \quad \frac{dh}{dq_1} = 0.$$

The last pair of equations can be separated from the rest of the system, since the first $(2n-2)$ equations do not involve t, and h is a constant.

The original differential equations can therefore be replaced by the reduced system

$$\frac{dq_r}{dq_1}=\frac{\partial K}{\partial p_r}, \quad \frac{dp_r}{dq_1}=-\frac{\partial K}{\partial q_r} \qquad (r=2, 3, \ldots, n),$$

which has only $(n-1)$ *degrees of freedom.*

This result is equivalent to that obtained in § 42, as can be shewn by direct transformation.

Example. Consider the system

$$\frac{dq_r}{dt}=\frac{\partial H}{\partial p_r}, \quad \frac{dp_r}{dt}=-\frac{\partial H}{\partial q_r} \qquad (r=1, 2),$$

where

$$H=\tfrac{1}{2}p_2^2+\frac{1}{2q_2^2}p_1^2-\frac{\mu}{2q_2^2},$$

μ being a constant; these are easily seen to be the equations of motion of a particle which is attracted to a fixed point with a force varying as the inverse cube of the distance: q_2 and q_1 are respectively the radius vector and vectorial angle of the particle referred to the centre of force.

Writing $H=-h$, and applying the theorem given above, the equations reduce to the system

$$\frac{dq_2}{dq_1}=\frac{\partial K}{\partial p_2}, \quad \frac{dp_2}{dq_1}=-\frac{\partial K}{\partial q_2},$$

where

$$K=-(\mu-q_2^2p_2^2-2hq_2^2)^{\frac{1}{2}}.$$

Since K does not involve q_1, the equation $K=$Constant is an integral of this last system, and we can therefore perform the same process again: writing $K=-k$, we have

$$p_2=\left(\frac{\mu-k^2}{q_2^2}-2h\right)^{\frac{1}{2}}=-L \text{ say},$$

and the system reduces to the single equation

$$\frac{dq_1}{dq_2}=\frac{\partial L}{\partial k}=\frac{k}{q_2^2}\left(\frac{\mu-k^2}{q_2^2}-2h\right)^{-\frac{1}{2}},$$

the integral of which (supposing $\mu<k^2$) is

$$q_2=(k^2-\mu)^{\frac{1}{2}}(-2h)^{-\frac{1}{2}}\sec\left\{\left(1-\frac{\mu}{k^2}\right)^{\frac{1}{2}}(q_1+\epsilon)\right\},$$

where ϵ is an arbitrary constant. This is the equation, in polar coordinates, of the orbit described by the particle.

142. *Hamilton's partial differential equation.*

If follows from § 138 that if a contact-transformation defined by the equations

$$P_r=-\frac{\partial W}{\partial Q_r}, \quad p_r=\frac{\partial W}{\partial q_r} \qquad (r=1, 2, \ldots, n),$$

where W denotes a given function of $(q_1, q_2, \ldots, q_n, Q_1, Q_2, \ldots, Q_n, t)$, is performed on the variables of a dynamical system defined by the equations

$$\frac{dq_r}{dt}=\frac{\partial H}{\partial p_r}, \quad \frac{dp_r}{dt}=-\frac{\partial H}{\partial q_r} \qquad (r=1, 2, \ldots, n),$$

the resulting system is

$$\frac{dQ_r}{dt} = \frac{\partial K}{\partial P_r}, \quad \frac{dP_r}{dt} = -\frac{\partial K}{\partial Q_r} \qquad (r = 1, 2, \ldots, n),$$

where

$$K = H + \partial W / \partial t.$$

If the function K is zero, the system will be said to be transformed into the *equilibrium-problem*. Now the function K will be zero, provided W is a function such that

$$\frac{\partial}{\partial t} W(q_1, q_2, \ldots, q_n, Q_1, \ldots, Q_n, t) + H(q_1, q_2, \ldots, q_n; p_1, \ldots, p_n, t) = 0,$$

i.e. provided W, considered as a function of the variables $(q_1, q_2, \ldots, q_n, t)$, satisfies the partial differential equation

$$\frac{\partial W}{\partial t} + H\left(q_1, q_2, \ldots, q_n, \frac{\partial W}{\partial q_1}, \frac{\partial W}{\partial q_2}, \ldots, \frac{\partial W}{\partial q_n}, t\right) = 0.$$

This is called *Hamilton's partial differential equation* associated with the given dynamical system. It was published by Hamilton in 1834*, being the extension to dynamics of the partial differential equation which he had discovered ten years previously in connexion with optics.

Suppose that a "complete integral" of this equation, i.e. a solution containing n arbitrary constants in addition to the additive constant, is known. Let $(\alpha_1, \alpha_2, \ldots, \alpha_n)$ be these arbitrary constants, so that the solution can be written $W(q_1, q_2, \ldots, q_n, \alpha_1, \alpha_2, \ldots, \alpha_n, t)$; and perform on the original dynamical system the contact-transformation from the variables $(q_1, q_2, \ldots, q_n, p_1, \ldots, p_n)$ to variables $(\alpha_1, \alpha_2, \ldots, \alpha_n, \beta_1, \beta_2, \ldots, \beta_n)$, defined by the equations

$$p_r = \frac{\partial W}{\partial q_r}, \quad \beta_r = -\frac{\partial W}{\partial \alpha_r} \qquad (r = 1, 2, \ldots, n).$$

Since W satisfies Hamilton's equation, the Hamiltonian function of the new system is zero, and consequently the equations of the system are

$$\frac{d\alpha_r}{dt} = 0, \quad \frac{d\beta_r}{dt} = 0 \qquad (r = 1, 2, \ldots, n),$$

so that $(\alpha_1, \alpha_2, \ldots, \alpha_n, \beta_1, \ldots, \beta_n)$ are constant throughout the motion. It follows that *if W denotes a complete integral of Hamilton's partial differential equation, containing n arbitrary constants $(\alpha_1, \alpha_2, \ldots, \alpha_n)$, then the equations*

$$\beta_r = -\frac{\partial W}{\partial \alpha_r}, \quad p_r = \frac{\partial W}{\partial q_r} \qquad (r = 1, 2, \ldots, n)$$

constitute the solution of the dynamical problem, since they express the variables $(q_1, q_2, \ldots, q_n, p_1, \ldots, p_n)$ in terms of t and $2n$ arbitrary constants $(\alpha_1, \alpha_2, \ldots, \alpha_n,$

* *Phil. Trans.* 1834, p. 247; *ibid.* 1835, p. 95.

$\beta_1, \ldots, \beta_n)$*. In this way the solution of any dynamical system with n degrees of freedom is made to depend on the solution of a single partial differential equation of the first order in $(n+1)$ independent variables.

It should however be observed that the converse of this theorem—namely the theorem that the solution of a partial differential equation such as Hamilton's depends on the solution of a set of ordinary differential equations (the differential equations of the characteristics), which in this case are of the Hamiltonian form, had been discovered by Pfaff and Cauchy (completing the earlier work of Lagrange and Monge) before Hamilton and Jacobi approached the subject from the dynamical side.

On the use that can be made of an *incomplete* integral of Hamilton's partial differential equation (i.e. one containing less than n arbitrary constants besides the additive constant), cf. Lehmann-Filhes, *Astr. Nach.* CLXV. (1904), col. 209.

It may be noted that Hamilton's partial differential equation is not applicable as it stands to non-holonomic systems: for an extension to such systems, cf. Quanjel, *Palermo Rendiconti*, XXII. (1906), p. 263.

The integration of Hamilton's equation by separation of variables is discussed by F. A. Dall' Acqua, *Math. Ann.* LXVI. (1908), p. 398.

Example. Consider the system

$$\frac{dq}{dt} = \frac{\partial H}{\partial p}, \quad \frac{dp}{dt} = -\frac{\partial H}{\partial q},$$

where

$$H = \tfrac{1}{2}p^2 - \frac{\mu}{q},$$

and μ is a constant. The Hamilton's equation corresponding to this system is

$$0 = \frac{\partial W}{\partial t} + \tfrac{1}{2}\left(\frac{\partial W}{\partial q}\right)^2 - \frac{\mu}{q};$$

a complete integral of this equation may be found in the following way. Assume

$$W = f(t) + \phi(q),$$

where f and ϕ are functions of their respective arguments: then we have

$$0 = f'(t) + \tfrac{1}{2}\{\phi'(q)\}^2 - \mu/q.$$

This equation can be satisfied by writing

$$f'(t) = (\mu/q) - \tfrac{1}{2}\{\phi'(q)\}^2 = \mu/a,$$

where a is a constant; which gives

$$f(t) = \mu t/a, \quad \phi(q) = (2\mu a)^{\frac{1}{2}} \arcsin (q/a)^{\frac{1}{2}} + \{2\mu q(a-q)/a\}^{\frac{1}{2}},$$

$$W = \mu t/a + (2\mu a)^{\frac{1}{2}} \arcsin (q/a)^{\frac{1}{2}} + \{2\mu q(a-q)/a\}^{\frac{1}{2}}.$$

The solution of the original problem is therefore given by the equations $\beta = -\partial W/\partial a$, $p = \partial W/\partial q$, where a and β are the two constants of integration.

143. *Hamilton's integral as a solution of Hamilton's partial differential equation.*

There are an infinite number of complete integrals of Hamilton's partial differential equation; and each one of them furnishes a contact-transformation

* This theorem is due to Jacobi, *Crelle's J.* XXVII. (1837), p. 97 and *Liouville's J.* III. (1837), pp. 60, 161.

from the variables $(q_1, q_2, \ldots, q_n, p_1, \ldots, p_n)$ of the dynamical system to variables $(\alpha_1, \alpha_2, \ldots, \alpha_n, \beta_1, \ldots, \beta_n)$, (the transformation involving t), such that the equations of motion of the system when expressed in terms of $(\alpha_1, \alpha_2, \ldots, \alpha_n, \beta_1, \ldots, \beta_n)$ become the equations of the equilibrium-problem, i.e. the quantities $(\alpha_1, \alpha_2, \ldots, \alpha_n, \beta_1, \ldots, \beta_n)$ are constants.

Among this infinite number of transformations there is one of special interest; namely that in which the quantities $(\alpha_1, \alpha_2, \ldots, \alpha_n, \beta_1, \ldots, \beta_n)$ are the *initial values* of $(q_1, q_2, \ldots, q_n, p_1, \ldots, p_n)$ respectively, i.e. their values at a time t_0, which is taken as an epoch from which the motion is estimated. In this case we can find in an explicit form the corresponding complete integral of Hamilton's partial differential equation.

For consider Hamilton's integral (§ 99)

$$\int_{t_0}^{t} L dt,$$

where L denotes the kinetic potential of the system. Suppose that δ denotes a variation due to small changes $(\delta\alpha_1, \delta\alpha_2, \ldots, \delta\alpha_n, \delta\beta_1, \ldots, \delta\beta_n)$ in the initial conditions.

Then (§ 99) we have

$$\delta \int_{t_0}^{t} L dt = \sum_{r=1}^{n} (p_r \delta q_r - \beta_r \delta\alpha_r).$$

It follows that if the quantity $\int_{t_0}^{t} L dt$, when the integration is performed, be expressed in terms of $(q_1, q_2, \ldots, q_n, \alpha_1, \ldots, \alpha_n, t)$, (we suppose this possible, i.e. we assume that it is not possible to eliminate $(\beta_1, \beta_2, \ldots, \beta_n, p_1, \ldots, p_n)$ from the relations connecting $(\alpha_1, \ldots, \alpha_n, \beta_1, \ldots, \beta_n, q_1, \ldots, q_n, p_1, \ldots, p_n)$, so as to obtain relations between $(q_1, \ldots, q_n, \alpha_1, \ldots, \alpha_n)$) and if the function thus obtained (which Hamilton called the *Principal Function*) be denoted by $W(q_1, q_2, \ldots, q_n, \alpha_1, \ldots, \alpha_n, t)$, then we shall have

$$\frac{\partial W}{\partial q_r} = p_r, \quad \frac{\partial W}{\partial \alpha_r} = -\beta_r \qquad (r = 1, 2, \ldots, n),$$

and therefore* *the transformation from*

$$(q_1, q_2, \ldots, q_n, p_1, \ldots, p_n) \text{ to } (\alpha_1, \alpha_2, \ldots, \alpha_n, \beta_1, \ldots, \beta_n)$$

is a contact-transformation, and the integral of the kinetic potential is the determining function of the transformation.

* Hamilton, *Phil. Trans.* 1834, p. 307; *ibid.* 1835, p. 95. In his earliest dynamical investigations, Hamilton used a "characteristic function" strictly analogous to the characteristic function which he had employed with such success in optics: this function being the Action integral, expressed in terms of the final and initial coordinates. He found however that this function, when employed in dynamics, involved the constant of energy, and so substituted for it the "principal function" described above.

Also we have

$$\frac{dW}{dt}=\frac{\partial W}{\partial t}+\sum_{r=1}^{n}\frac{\partial W}{\partial q_r}\frac{dq_r}{dt},$$

or

$$L=\frac{\partial W}{\partial t}+\sum_{r=1}^{n}p_r\dot{q}_r,$$

or

$$0=\frac{\partial W}{\partial t}+H,$$

and therefore *the integral of the kinetic potential satisfies the equation*

$$\frac{\partial W}{\partial t}+H\left(q_1, q_2, \ldots, q_n, \frac{\partial W}{\partial q_1}, \ldots, \frac{\partial W}{\partial q_n}, t\right)=0,$$

which is Hamilton's partial differential equation.

Some interesting developments in connexion with Hamilton's Principal Function are given by Conway and McConnell, *Proc. R. Irish Ac.* XLI. (1932), p. 18.

Example. Let $(\alpha_1, \alpha_2, \ldots, \alpha_n, \beta_1, \ldots, \beta_n)$ be the initial values (at time t_0) of $(q_1, q_2, \ldots, q_n, p_1, \ldots, p_n)$ respectively, in the dynamical system represented by the equations

$$\frac{dq_r}{dt}=\frac{\partial H}{\partial p_r}, \quad \frac{dp_r}{dt}=-\frac{\partial H}{\partial q_r} \qquad (r=1, 2, \ldots, n).$$

Suppose that from the relations connecting $(\alpha_1, \alpha_2, \ldots, \alpha_n, \beta_1, \ldots, \beta_n)$ with $(q_1, q_2, \ldots, q_n, p_1, \ldots, p_n)$ it is possible to eliminate $(\beta_1, \beta_2, \ldots, \beta_n, p_1, \ldots, p_n)$ entirely, so that a number (say m) of distinct relations exist between $(q_1, q_2, \ldots, q_n, \alpha_1, \ldots, \alpha_n)$; let these be solved for $(\alpha_1, \alpha_2, \ldots, \alpha_m)$, so as to take the form

$$F_r=f_r(q_1, \ldots, q_n, \alpha_{m+1}, \ldots, \alpha_n, t)-\alpha_r=0 \qquad (r=1, 2, \ldots, m),$$

and let V denote Hamilton's integral

$$\int_{t_0}^{t} L dt$$

for the system, expressed in terms of $(q_1, q_2, \ldots, q_n, \alpha_{m+1}, \ldots, \alpha_n)$. Establish the equations

$$p_r=\frac{\partial V}{\partial q_r}+\sum_{k=1}^{m}\lambda_k\frac{\partial F_k}{\partial q_r},$$

$$\beta_r=-\frac{\partial V}{\partial \alpha_r}-\sum_{k=1}^{m}\lambda_k\frac{\partial F_k}{\partial \alpha_r},$$

where $(\lambda_1, \lambda_2, \ldots, \lambda_m)$ are arbitrary; and shew that the function

$$W=V+\sum_{k=1}^{m}\lambda_k f_k$$

is an integral of the partial differential equation

$$\frac{\partial W}{\partial t}+H\left(q_1, q_2, \ldots, q_n, \frac{\partial W}{\partial q_1}, \ldots, \frac{\partial W}{\partial q_n}, t\right)=0.$$

144. *The connexion of integrals with infinitesimal transformations admitted by the system.*

Let

$$\frac{dq_r}{dt}=\frac{\partial H}{\partial p_r} \qquad \frac{dp_r}{dt}=-\frac{\partial H}{\partial q_r} \qquad (r=1, 2, \ldots, n)$$

be the equations of any dynamical system, and let

$$\phi(q_1, q_2, \ldots, q_n, p_1, \ldots, p_n, t) = \text{Constant}$$

denote any integral of the system; we shall shew that the knowledge of this integral enables us to find a particular solution of the variational equations (§ 112).

For the variational equation for δq_r is

$$\frac{d}{dt}\delta q_r = \frac{\partial^2 H}{\partial q_1 \partial p_r}\delta q_1 + \ldots + \frac{\partial^2 H}{\partial q_n \partial p_r}\delta q_n + \frac{\partial^2 H}{\partial p_1 \partial p_r}\delta p_1 + \ldots + \frac{\partial^2 H}{\partial p_n \partial p_r}\delta p_n;$$

but we have

$$\frac{\partial^2 H}{\partial q_1 \partial p_r}\frac{\partial \phi}{\partial p_1} + \ldots + \frac{\partial^2 H}{\partial q_n \partial p_r}\frac{\partial \phi}{\partial p_n} - \frac{\partial^2 H}{\partial p_1 \partial p_r}\frac{\partial \phi}{\partial q_1} - \ldots - \frac{\partial^2 H}{\partial p_n \partial p_r}\frac{\partial \phi}{\partial q_n}$$

$$= \frac{\partial}{\partial p_r}\left(-\sum_{k=1}^{n}\frac{dp_k}{dt}\frac{\partial \phi}{\partial p_k} - \sum_{k=1}^{n}\frac{dq_k}{dt}\frac{\partial \phi}{\partial q_k}\right) - \sum_{k=1}^{n}\frac{\partial H}{\partial q_k}\frac{\partial^2 \phi}{\partial p_k \partial p_r} + \sum_{k=1}^{n}\frac{\partial H}{\partial p_k}\frac{\partial^2 \phi}{\partial q_k \partial p_r}$$

$$= \frac{\partial}{\partial p_r}\left(-\frac{d\phi}{dt} + \frac{\partial \phi}{\partial t}\right) + \frac{d}{dt}\left(\frac{\partial \phi}{\partial p_r}\right) - \frac{\partial}{\partial t}\left(\frac{\partial \phi}{\partial p_r}\right)$$

$$= \frac{d}{dt}\left(\frac{\partial \phi}{\partial p_r}\right),$$

and hence the variational equations for $(\delta q_1, \delta q_2, \ldots, \delta q_n)$ are satisfied by the values

$$\delta q_r = \epsilon\frac{\partial \phi}{\partial p_r}, \qquad \delta p_r = -\epsilon\frac{\partial \phi}{\partial q_r} \qquad (r = 1, 2, \ldots, n),$$

where ϵ is a small constant. Similarly the variational equations for

$$(\delta p_1, \delta p_2, \ldots, \delta p_n)$$

can be shewn to be satisfied by these values; and hence *the equations*

$$\delta q_r = \epsilon\frac{\partial \phi}{\partial p_r}, \qquad \delta p_r = -\epsilon\frac{\partial \phi}{\partial q_r} \qquad (r = 1, 2, \ldots, n),$$

where ϵ is a small constant and ϕ is an integral of the original equations, constitute a solution of the variational equations.

This result can evidently be stated in the form: The infinitesimal contact-transformation of the variables $(q_1, q_2, \ldots, q_n, p_1, \ldots, p_n)$, which is defined by the equations

$$\delta q_r = \epsilon\frac{\partial \phi}{\partial p_r}, \qquad \delta p_r = -\epsilon\frac{\partial \phi}{\partial q_r} \qquad (r = 1, 2, \ldots, n),$$

transforms any orbit into an adjacent orbit, and therefore transforms the whole family of orbits into itself. Adopting the language of the group-theory, we say that the dynamical system *admits* this infinitesimal contact-transformation. We have therefore the theorem that *integrals of a dynamical system, and contact-transformations which change the system into itself, are substantially the same thing; any integral*

$$\phi(q_1, q_2, \ldots, q_n, p_1, \ldots, p_n, t) = \text{Constant}$$

corresponds to an infinitesimal transformation whose symbol (§ 133) *is the Poisson-bracket* (ϕ, f).

It will be observed that the ignoration of coordinates arises from the particular case of this theorem in which the integral is $p_r = \text{Constant}$, where q_r is the ignorable coordinate; the corresponding transformation is that which changes q_r without changing any of the other variables.

145. *Poisson's theorem.*

The last result leads to a theorem discovered by Poisson* in 1809, by means of which it is possible to construct from two known integrals of a dynamical system a third expression which is constant along any trajectory of the system, and which therefore (when it proves to be independent of the integrals already known) furnishes a new integral of the system.

Let $$\phi(q_1, q_2, \ldots, q_n, p_1, \ldots, p_n, t) = \text{Constant}$$
and $$\psi(q_1, q_2, \ldots, q_n, p_1, \ldots, p_n, t) = \text{Constant}$$
denote the two integrals which are supposed known. Consider the infinitesimal contact-transformation whose symbol is the Poisson-bracket (f, ψ); since ψ is an integral, this (§ 144) transforms every orbit into an adjacent orbit.

The increment of the function ϕ under this transformation is $\epsilon(\phi, \psi)$, where ϵ is a small constant; but since ϕ is an integral, ϕ has constant values along the original orbit and along the adjacent orbit: the value of (ϕ, ψ) must therefore be constant throughout the motion. We thus have Poisson's theorem, that *if* ϕ *and* ψ *are two integrals of the system, the Poisson-bracket* (ϕ, ψ) *is constant throughout the motion.*

If (ϕ, ψ), which is a function of the variables $(q_1, q_2, \ldots, q_n, p_1, \ldots, p_n, t)$, does not reduce to merely zero or a constant, and if moreover it is not expressible in terms of ϕ, ψ and such other integrals as are already known, then *the equation*

$$(\phi, \psi) = \text{Constant}$$

constitutes a new integral of the system†.

The following example will shew how Poisson's theorem can be applied to obtain new integrals of a dynamical system when two integrals are already known.

* *Journal de l'École polyt.* VIII. (1809), p. 266.

† A discussion of this theorem is given by Bertrand in Note VII to the third edition of Lagrange's *Méc. Anal.* (1853): cf. *Oeuvres de Lagrange*, t. XI. p. 484.

On the extension of Poisson's theorem to non-holonomic systems, cf. Dautheville, *Bull. de la Soc. math. de France*, XXXVII. (1909), p. 120.

Consider the motion of a particle of unit mass, whose rectangular coordinates are (q_1, q_2, q_3) and whose components of velocity are (p_1, p_2, p_3), which is free to move in space under the influence of a centre of force at the origin. The integrals of angular momentum about two of the axes are

$$p_3 q_2 - q_3 p_2 = \text{Constant},$$

and
$$p_1 q_3 - q_1 p_3 = \text{Constant}.$$

Let these be taken as the two known integrals ϕ and ψ; the Poisson-bracket (ϕ, ψ), which is

$$\sum_{r=1}^{3} \left(\frac{\partial \phi}{\partial q_r}\frac{\partial \psi}{\partial p_r} - \frac{\partial \psi}{\partial q_r}\frac{\partial \phi}{\partial p_r}\right),$$

becomes in this case

$$p_2 q_1 - q_2 p_1;$$

and in fact, the equation

$$p_2 q_1 - q_2 p_1 = \text{Constant}$$

is another integral of the motion, being the integral of angular momentum about the third axis.

146. *The constancy of Lagrange's bracket-expressions.*

The theorem of Poisson has, as might be expected, an analogue in the theory of Lagrange's bracket-expressions.

Let
$$u_r = a_r \qquad (r = 1, 2, \ldots, 2n)$$

denote $2n$ integrals of a dynamical system with n degrees of freedom; constituting the complete solution of the problem: the quantities u_r being given functions of the variables $(q_1, q_2, \ldots, q_n, p_1, \ldots, p_n, t)$, and the quantities a_r being arbitrary constants. By means of these equations we can express $(q_1, q_2, \ldots, q_n, p_1, \ldots, p_n)$ as functions of $(a_1, a_2, \ldots, a_{2n}, t)$, and form the Lagrange's bracket-expressions $[a_r, a_s]$, where a_r and a_s are any two of the quantities $(a_1, a_2, \ldots, a_{2n})$.

Since the transformation from the variables $(q_1, q_2, \ldots, q_n, p_1, \ldots, p_n)$ at time t to their values at time $t + dt$ is a contact-transformation, we have (§128)

$$\frac{d}{dt}\sum_{r=1}^{n} (\Delta q_r \delta p_r - \delta q_r \Delta p_r) = 0,$$

where the symbols Δ and δ refer to independent displacements from one trajectory to an adjacent trajectory. If now we take the symbol Δ to refer to a variation in which a_i only is varied, the rest of the quantities

$$(a_1, a_2, \ldots, a_{2n})$$

remaining unchanged, and take δ to refer to a variation in which a_j only is varied, the last equation becomes

$$\frac{d}{dt}\sum_{r=1}^{n}\left(\frac{\partial q_r}{\partial a_i}\frac{\partial p_r}{\partial a_j} - \frac{\partial q_r}{\partial a_j}\frac{\partial p_r}{\partial a_i}\right) = 0,$$

or
$$\frac{d}{dt}[a_i, a_j] = 0,$$
which shews that *the Lagrange-bracket* $[a_i, a_j]$ *has a constant value during the motion along any trajectory*; this theorem was given by Lagrange in 1808.

Lagrange's result, unlike Poisson's, does not enable us to find any new integrals; for we have to know all the integrals before we can form the Lagrange's bracket-expressions.

147. *Involution-systems.*

Let $(u_1, u_2, \ldots, u_r)$ denote r functions of $2n$ independent variables
$$(q_1, q_2, \ldots, q_n, p_1, \ldots, p_n);$$
if it is possible to express all the Poisson-brackets (u_i, u_k) as functions of $(u_1, u_2, \ldots, u_r)$, the functions $(u_1, u_2, \ldots, u_r)$ are said to form a *function-group**. Any function of $(u_1, u_2, \ldots, u_r)$ belongs to this group.

If the quantities (u_i, u_k) are all zero, the functions $(u_1, u_2, \ldots, u_r)$ are said to be *in involution*, or to form an *involution-system.*

Now suppose that $(u_1, u_2, \ldots, u_r)$ are functions in involution: and let $v=0$ and $w=0$ be any two equations which are consequences of the equations
$$u_1 = 0, \quad u_2 = 0, \ldots, u_r = 0;$$
we shall shew that v *and* w *satisfy the relation* $(v, w) = 0$.

For since $(u_1, u_2, \ldots, u_r)$ are in involution, each of the equations
$$u_1 = 0, \quad u_2 = 0, \ldots, u_r = 0$$
admits each of the r infinitesimal transformations whose symbols are
$$(u_1, f), (u_2, f), \ldots, (u_r, f);$$
and consequently the equation $v = 0$, being a consequence of these equations, must also admit these transformations; that is to say, we have
$$(u_k, v) = 0 \qquad (k = 1, 2, \ldots, r),$$
and therefore each of the equations
$$u_1 = 0, \quad u_2 = 0, \ldots, u_r = 0$$
admits the infinitesimal transformation whose symbol is (v, f). Since the equation $w = 0$ is a consequence of these equations, it follows that the equation $w = 0$ must also admit this transformation, and therefore we have
$$(v, w) = 0,$$
which establishes the result.

* Lie, *Math. Ann.* VIII. (1875). p. 215.

Hence we see that *if* $(u_1, u_2, \ldots, u_r)$ *are in involution, and the equations*

$$v_1 = 0, \quad v_2 = 0, \ldots, v_r = 0$$

are consequences of the equations

$$u_1 = 0, \quad u_2 = 0, \ldots, u_r = 0,$$

then the functions $(v_1, v_2, \ldots, v_r)$ *are in involution.*

148. *Solution of a dynamical problem when half the integrals are known.*

The result which was established for systems with two degrees of freedom in § 121 can now be extended to systems with any number of degrees of freedom. The theorem may be thus stated*: *If n distinct integrals*

$$\phi_r(q_1, q_2, \ldots, q_n, p_1, \ldots, p_n, t) = a_r \qquad (r = 1, 2, \ldots, n),$$

where $(a_1, a_2, \ldots, a_n)$ *are arbitrary constants, are known for the dynamical system*

$$\frac{dq_r}{dt} = \frac{\partial H}{\partial p_r}, \quad \frac{dp_r}{dt} = -\frac{\partial H}{\partial q_r} \qquad (r = 1, 2, \ldots, n),$$

where H is any given function of $(q_1, q_2, \ldots, q_n, p_1, \ldots, p_n, t)$, *and if the functions* $(\phi_1, \phi_2, \ldots, \phi_n)$ *are in involution, then on solving these integrals for* $(p_1, p_2, \ldots, p_n)$ *so as to obtain them in the form*

$$p_r = f_r(q_1, q_2, \ldots, q_n, a_1, a_2, \ldots, a_n, t) \quad (r = 1, 2, \ldots, n)$$

and substituting $(f_1, f_2, \ldots, f_n)$ *respectively for* $(p_1, p_2, \ldots, p_n)$ *in the expression*

$$p_1 dq_1 + p_2 dq_2 + \ldots + p_n dq_n - H dt,$$

the latter expression becomes a perfect differential: denoting it by

$$dV(q_1, q_2, \ldots, q_n, a_1, a_2, \ldots, a_n, t),$$

the remaining integrals of the system are

$$\frac{\partial V}{\partial a_r} = b_r \qquad (r = 1, 2, \ldots, n),$$

where $(b_1, b_2, \ldots, b_n)$ *are arbitrary constants.*

For since the functions $\phi_1 - a_1, \phi_2 - a_2, \ldots, \phi_n - a_n$ are in involution, it follows by the last article that the functions $p_1 - f_1, p_2 - f_2, \ldots, p_n - f_n$ are in involution, and therefore

$$(p_r - f_r, p_s - f_s) = 0 \qquad (r, s = 1, 2, \ldots, n),$$

or

$$\frac{\partial f_s}{\partial q_r} - \frac{\partial f_r}{\partial q_s} = 0 \qquad (r, s = 1, 2, \ldots, n).$$

* This theorem is essentially the application to Hamilton's partial differential equation of the well-known method for finding a Complete Integral of a non-linear partial differential equation of the first order. As a dynamical theorem it is due to Liouville, *Journal de Math.* xx. (1855), p. 137.

Also
$$-\frac{\partial H}{\partial q_r} = \frac{dp_r}{dt} = \frac{df_r}{dt}$$
$$= \frac{\partial f_r}{\partial t} + \sum_{s=1}^{n} \frac{\partial f_r}{\partial q_s}\frac{dq_s}{dt}$$
$$= \frac{\partial f_r}{\partial t} + \sum_{s=1}^{n} \frac{\partial f_s}{\partial q_r}\frac{\partial H}{\partial p_s},$$

and consequently
$$\frac{\partial f_r}{\partial t} = -\frac{\partial H}{\partial q_r} - \sum_{s=1}^{n} \frac{\partial H}{\partial p_s}\frac{\partial f_s}{\partial q_r}$$
$$= -\frac{\partial H_1}{\partial q_r},$$

where H_1 stands for the function H when expressed in terms of the arguments
$$(q_1, q_2, \ldots, q_n, a_1, \ldots, a_n, t).$$

The equations
$$\frac{\partial f_s}{\partial q_r} = \frac{\partial f_r}{\partial q_s}, \qquad \frac{\partial f_r}{\partial t} = -\frac{\partial H_1}{\partial q_r},$$

shew that
$$f_1 dq_1 + f_2 dq_2 + \ldots + f_n dq_n - H_1 dt$$
is the perfect differential of some function $V(q_1, q_2, \ldots, q_n, a_1, \ldots, a_n, t)$; which establishes the first part of the theorem.

If now the symbol d denote the total differential of the function V with respect to all its arguments, we have therefore
$$dV = f_1 dq_1 + f_2 dq_2 + \ldots + f_n dq_n - H_1 dt + \sum_r \frac{\partial V}{\partial a_r} da_r.$$

In this equation replace the quantities a_r by their values ϕ_r: we thus obtain an identity in $(q_1, q_2, \ldots, q_n, p_1, p_2, \ldots, p_n, t)$, namely
$$dV - \sum_r \frac{\partial V}{\partial a_r} d\phi_r = p_1 dq_1 + p_2 dq_2 + \ldots + p_n dq_n - Hdt,$$

where on the left-hand side of the equation we suppose that in dV and $\frac{\partial V}{\partial a_r}$ the quantities $(a_1, a_2, \ldots, a_n)$ are replaced by their values $(\phi_1, \phi_2, \ldots, \phi_n)$. This equation shews that the differential form
$$p_1 dq_1 + p_2 dq_2 + \ldots + p_n dq_n - Hdt,$$
when expressed in terms of the variables $(q_1, q_2, \ldots, q_n, \phi_1, \phi_2, \ldots, \phi_n, t)$, takes the form
$$-\sum_{r=1}^{n} \frac{\partial V}{\partial a_r} d\phi_r + dV,$$

and hence the differential equations of the original dynamical problem are equivalent to the first Pfaff's system of this differential form, namely
$$d(\partial V/\partial a_r) = 0, \qquad d\phi_r = 0 \qquad (r = 1, 2, \ldots, n).$$

The expressions $\partial V/\partial a_r$ are therefore constant throughout the motion, i.e. the equations

$$\partial V/\partial a_r = b_r \qquad (r = 1, 2, \ldots, n),$$

where $(b_1, b_2, \ldots, b_n)$ are new arbitrary constants, are integrals of the system; this completes the proof of the theorem.

Example. In the motion of a body under no forces with one point fixed, let (θ, ϕ, ψ) denote the three Eulerian angles which specify the position of the body relative to any fixed axes $OXYZ$ at the fixed point, (A, B, C) the principal moments of inertia of the body at the fixed point, a the constant of energy, a_1 the angular momentum about the fixed axis OZ, and a_2 the angular momentum about the normal to the invariable plane: and let $(\theta_1, \phi_1, \psi_1)$ denote $\partial T/\partial \dot\theta$, $\partial T/\partial \dot\phi$, $\partial T/\partial \dot\psi$ respectively. Obtain the equations

$$\begin{cases} \theta = \arctan\{(a_2{}^2 - a_1{}^2 - \theta_1{}^2)^{\frac{1}{2}}/a_1\} - \arctan\{(a_2{}^2 - \psi_1{}^2 - \theta_1{}^2)^{\frac{1}{2}}/\psi_1\}, \\ \phi_1 = -a_1, \\ \dfrac{\pi}{2} - \psi = \arctan\{\theta_1 (a_2{}^2 - \psi_1{}^2 - \theta_1{}^2)^{-\frac{1}{2}}\} + \arctan\left\{-\dfrac{A}{B}\dfrac{(2Ba - a_2{}^2)\,C + (C-B)\,\psi_1{}^2}{(2Aa - a_2{}^2)\,C + (C-A)\,\psi_1{}^2}\right\}^{\frac{1}{2}}. \end{cases}$$

Hence shew that

$$\theta d\theta_1 + \psi d\psi_1 + a_1 d\phi$$

is the perfect differential of a function V, and that the remaining integrals of the system are

$$\frac{\partial V}{\partial a} = b - t, \quad \frac{\partial V}{\partial a_1} = b_1, \quad \frac{\partial V}{\partial a_2} = b_2,$$

where b, b_1, b_2 are arbitrary constants. (Siacci.)

149. *Levi-Civita's theorem.*

Levi-Civita* has established a connexion between the integrals of a dynamical system and certain families of particular solutions of the equations of motion.

Consider first a system in which some of the coordinates are ignorable. Let $(q_1, q_2, \ldots, q_m)$ be the ignorable and $(q_{m+1}, \ldots, q_n)$ the non-ignorable coordinates; and let L denote the kinetic potential.

The integrals corresponding to the ignorable coordinates are

$$\partial L/\partial \dot q_r = \text{Constant} \qquad (r = 1, 2, \ldots, m),$$

and corresponding to these integrals there exists a class of particular solutions of the system, namely those *steady motions* (§ 83) in which $(\dot q_1, \dot q_2, \ldots, \dot q_m)$ have constant values which can be chosen arbitrarily, while $(q_{m+1}, q_{m+2}, \ldots, q_n)$ have constant values which are determined by the equations

$$\partial L/\partial q_r = 0 \qquad (r = m+1, m+2, \ldots, n);$$

there are ∞^{2m} of these particular solutions, since the m constant values of

* *Rend. dell' Acc. dei Lincei*, x. (1901), p. 3. Cf. Burgatti, *ibid.* xi. (1902), p. 309.

$(\dot{q}_1, \dot{q}_2, \ldots, \dot{q}_m)$ and the m initial values of $(q_1, q_2, \ldots, q_m)$ can be arbitrarily assigned. The theorem of Levi-Civita, to the consideration of which we shall now proceed, may be regarded as an extension of this result.

Let $$\frac{dq_r}{dt} = \frac{\partial H}{\partial p_r}, \quad \frac{dp_r}{dt} = -\frac{\partial H}{\partial q_r} \qquad (r = 1, 2, \ldots, n)$$

be the equations of motion of a dynamical system, the function H being supposed not to involve the time explicitly.

Let $$F_r(q_1, q_2, \ldots, q_n, p_1, \ldots, p_n) = 0 \qquad (r = 1, 2, \ldots, m) \ldots\text{(A)}$$

be a system of m relations, which when solved for $(p_1, p_2, \ldots, p_m)$ take the form

$$p_r = f_r(q_1, q_2, \ldots, q_n, p_{m+1}, \ldots, p_n) \qquad (r = 1, 2, \ldots, m) \ldots(\text{A}_1),$$

and which are *invariant relations* with respect to the Hamiltonian system, i.e. which are such that if we differentiate the relations (A_1) with respect to t, we obtain relations which are satisfied identically in virtue of the Hamiltonian equations and of the equations (A_1) themselves. These invariant relations include, as a particular case, integrals of the system: in this case, they will involve arbitrary constants.

Since the relations (A_1) are invariant relations, we have

$$-\frac{\partial H}{\partial q_r} = \frac{df_r}{dt} = -\sum_{j=m+1}^{n} \frac{\partial f_r}{\partial p_j}\frac{\partial H}{\partial q_j} + \sum_{j=1}^{n} \frac{\partial f_r}{\partial q_j}\frac{\partial H}{\partial p_j} \qquad (r = 1, 2, \ldots, m),$$

and writing

$$\{V, W\} = \sum_{j=m+1}^{n} \left(\frac{\partial V}{\partial p_j}\frac{\partial W}{\partial q_j} - \frac{\partial V}{\partial q_j}\frac{\partial W}{\partial p_j}\right),$$

this becomes

$$\frac{\partial H}{\partial q_r} + \{H, f_r\} + \sum_{s=1}^{m} \frac{\partial H}{\partial p_s}\frac{\partial f_r}{\partial q_s} = 0 \qquad (r = 1, 2, \ldots, m) \ldots(1);$$

this equation becomes an identity when for each of the quantities $(p_1, p_2, \ldots, p_m)$ we substitute the corresponding function f_r.

Moreover, we shall suppose that the relations (A) or (A_1) are in involution among themselves. This condition is expressed by the equations

$$\frac{\partial f_r}{\partial q_s} - \frac{\partial f_s}{\partial q_r} + \{f_r, f_s\} = 0 \qquad (r, s = 1, 2, \ldots, m) \ldots(2).$$

Let K denote the function obtained from H on replacing $(p_1, p_2, \ldots, p_m)$ by their values $(f_1, f_2, \ldots, f_m)$, so that

$$\left.\begin{aligned} \frac{\partial H}{\partial p_r} &= \frac{\partial K}{\partial p_r} - \sum_{s=1}^{m} \frac{\partial H}{\partial p_s}\frac{\partial f_s}{\partial p_r} \\ \frac{\partial H}{\partial q_r} &= \frac{\partial K}{\partial q_r} - \sum_{s=1}^{m} \frac{\partial H}{\partial p_s}\frac{\partial f_s}{\partial q_r} \end{aligned}\right\} \quad (r = m+1, m+2, \ldots, n) \ldots(3)$$

and $$\frac{\partial H}{\partial q_r} = \frac{\partial K}{\partial q_r} - \sum_{s=1}^{m} \frac{\partial H}{\partial p_s}\frac{\partial f_s}{\partial q_r} \qquad (r = 1, 2, \ldots, m) \ldots(4).$$

From (3) we have

$$\{H, f_r\} = \{K, f_r\} + \sum_{s=1}^{m} \frac{\partial H}{\partial p_s} \{f_r, f_s\} \qquad (r = 1, 2, \ldots, m),$$

and combining this with (4) we have

$$\frac{\partial H}{\partial q_r} + \{H, f_r\} = \frac{\partial K}{\partial q_r} + \{K, f_r\} + \sum_{s=1}^{m} \frac{\partial H}{\partial p_s}\left[-\frac{\partial f_s}{\partial q_r} + \{f_r, f_s\}\right].$$

Substituting in (1) this value of $\partial H/\partial q_r + \{H, f_r\}$, and using (2), we obtain the equations

$$\frac{\partial K}{\partial q_r} + \{K, f_r\} = 0 \qquad (r = 1, 2, \ldots, m) \ldots(5).$$

We shall now shew that the system of equations

$$\begin{cases} p_r = f_r(p_{m+1}, p_{m+2}, \ldots, p_n, q_1, \ldots, q_n) & (r = 1, 2, \ldots, m) \\ \dfrac{\partial K}{\partial p_r} = 0, \quad \dfrac{\partial K}{\partial q_r} = 0 & (r = m+1, m+2, \ldots, n) \ldots(\text{B}) \end{cases}$$

is invariant with respect to the Hamiltonian equations, i.e. that

$$\frac{d}{dt}\left(\frac{\partial K}{\partial p_r}\right) \text{ and } \frac{d}{dt}\left(\frac{\partial K}{\partial q_r}\right) \qquad (r = m+1, m+2, \ldots, n)$$

are zero in virtue of equations (A), (B), (1), (2), (3), (4), (5).

We have from the Hamiltonian equations

$$\left.\begin{aligned} \frac{d}{dt}\left(\frac{\partial K}{\partial p_r}\right) &= \left\{H, \frac{\partial K}{\partial p_r}\right\} + \sum_{s=1}^{m} \frac{\partial^2 K}{\partial p_r \partial q_s} \frac{\partial H}{\partial p_s} \\ \frac{d}{dt}\left(\frac{\partial K}{\partial q_r}\right) &= \left\{H, \frac{\partial K}{\partial q_r}\right\} + \sum_{s=1}^{m} \frac{\partial^2 K}{\partial q_r \partial q_s} \frac{\partial H}{\partial p_s} \end{aligned}\right\} \qquad (r = m+1, \ldots, n) \ldots(6),$$

and (5) gives on differentiation, using (B),

$$\left.\begin{aligned} \frac{\partial^2 K}{\partial p_r \partial q_s} + \left\{\frac{\partial K}{\partial p_r}, f_s\right\} &= 0 \\ \frac{\partial^2 K}{\partial q_r \partial q_s} + \left\{\frac{\partial K}{\partial q_r}, f_s\right\} &= 0 \end{aligned}\right\} \qquad (s = 1, 2, \ldots, m;\ r = m+1, m+2, \ldots, n) \ldots(7);$$

now taking account of (B), we have from (3)

$$\left.\begin{aligned} \frac{\partial H}{\partial p_r} &= -\sum_{s=1}^{m} \frac{\partial H}{\partial p_s} \frac{\partial f_s}{\partial p_r} \\ \frac{\partial H}{\partial q_r} &= -\sum_{s=1}^{m} \frac{\partial H}{\partial p_s} \frac{\partial f_s}{\partial q_r} \end{aligned}\right\} \qquad (r = m+1, m+2, \ldots, n)$$

and hence equations (6) become

$$\left.\begin{aligned} \frac{d}{dt}\left(\frac{\partial K}{\partial p_r}\right) &= \sum_{s=1}^{m} \frac{\partial H}{\partial p_s}\left[\frac{\partial^2 K}{\partial p_r \partial q_s} + \left\{\frac{\partial K}{\partial p_r}, f_s\right\}\right] \\ \frac{d}{dt}\left(\frac{\partial K}{\partial q_r}\right) &= \sum_{s=1}^{m} \frac{\partial H}{\partial p_s}\left[\frac{\partial^2 K}{\partial q_r \partial q_s} + \left\{\frac{\partial K}{\partial q_r}, f_s\right\}\right] \end{aligned}\right\} \qquad (r = m+1, m+2, \ldots, n),$$

or by (7),

$$\frac{d}{dt}\left(\frac{\partial K}{\partial p_r}\right)=0, \quad \frac{d}{dt}\left(\frac{\partial K}{\partial q_r}\right)=0 \quad (r=m+1,\, m+2,\, \ldots,\, n),$$

which proves that the system of equations (A) and (B) is invariant with respect to the Hamiltonian equations.

Now from the equations (A) and (B), let the variables

$$(p_1,\, p_2,\, \ldots,\, p_n,\, q_{m+1},\, \ldots,\, q_n)$$

be determined in terms of $(q_1, q_2, \ldots, q_m)$: from the invariant character of (A) and (B) it follows that on substituting these values in the Hamiltonian equations, we shall obtain m independent equations, namely those which express $(dq_1/dt, dq_2/dt, \ldots, dq_m/dt)$ in terms of $(q_1, q_2, \ldots, q_m)$, the others being identically satisfied: and the general solution of this system, which will contain m arbitrary constants, will give ∞^m particular solutions of the Hamiltonian equations. The solution of this system can, by making use of the integral of energy, be reduced to that of a system of order $(m-1)$: and thus we obtain Levi-Civita's theorem, which can be thus stated: *To any set of m invariant relations of a Hamiltonian system, which are in involution, there corresponds a family of* ∞^m *particular solutions of the Hamiltonian system, whose determination depends on the integration of a system of order* $(m-1)$.

If the invariant relations (A) are integrals of the system, they will contain another set of m arbitrary constants: and hence *to a set of m integrals of a Hamiltonian system, which are in involution, there corresponds in general a family of* ∞^{2m} *particular solutions of the system, which are obtained by integrating a system of order* $(m-1)$.

Example. For the dynamical system defined by the Hamiltonian function

$$H=q_1p_1-q_2p_2-aq_1^2+bq_2^2,$$

shew that the Levi-Civita particular solutions corresponding to the integral

$$(p_2-bq_2)/q_1=\text{Constant}$$

are given by the equations

$$q_1=0, \quad q_2=e^{-t+\epsilon}, \quad p_1=ae^{-t+\epsilon}, \quad p_2=be^{-t+\epsilon},$$

where ϵ is an arbitrary constant.

150 *Systems which possess integrals linear in the momenta.*

We shall now proceed to the consideration of systems which possess integrals of certain special kinds.

Suppose that a dynamical system, expressed by the equations

$$\frac{dq_r}{dt}=\frac{\partial H}{\partial p_r}, \quad \frac{dp_r}{dt}=-\frac{\partial H}{\partial q_r} \qquad (r=1,\, 2,\, \ldots,\, n),$$

has an integral which is linear and homogeneous in $(p_1, p_2, \ldots, p_n)$, say

$$f_1 p_1 + f_2 p_2 + \ldots + f_n p_n = \text{Constant},$$

where $(f_1, f_2, \ldots, f_n)$ are given functions of $(q_1, q_2, \ldots, q_n)$.

Consider the system of equations

$$\frac{dq_1}{f_1} = \frac{dq_2}{f_2} = \ldots = \frac{dq_n}{f_n},$$

which is of order $(n-1)$; suppose that the $(n-1)$ integrals which constitute its solution are

$$Q_r(q_1, q_2, \ldots, q_n) = \text{Constant} \qquad (r = 1, 2, \ldots, n-1):$$

and let Q_n be a function defined by the equation

$$Q_n = \int \frac{dq_1}{f_1},$$

where in the integrand the variables $(q_2, q_3, \ldots, q_n)$ are supposed replaced by their values in terms of $(q_1, Q_1, Q_2, \ldots, Q_{n-1})$ before the integration is carried out.

Then if the variables change in such a way that $(Q_1, Q_2, \ldots, Q_{n-1})$ remain constant and Q_n varies, it follows from the above equation that

$$\frac{dq_1}{f_1} = \frac{dq_2}{f_2} = \ldots = \frac{dq_n}{f_n} = dQ_n,$$

so that if $(Q_1, Q_2, \ldots, Q_n)$ are regarded as a set of new variables in terms of which $(q_1, q_2, \ldots, q_n)$ can be expressed, we shall have

$$\partial q_k / \partial Q_n = f_k \qquad (k = 1, 2, \ldots, n).$$

Suppose then that we consider the contact-transformation which is the extension of the point-transformation from the variables $(q_1, q_2, \ldots, q_n)$ to the variables $(Q_1, Q_2, \ldots, Q_n)$, so that the new variables $(P_1, P_2, \ldots, P_n)$ are defined (§ 132) by the equations

$$P_r = \sum_{k=1}^{n} p_k \frac{\partial q_k}{\partial Q_r} \qquad (r = 1, 2, \ldots, n).$$

By this transformation the differential equations of the dynamical system are changed into a new set of Hamiltonian equations

$$\frac{dQ_r}{dt} = \frac{\partial K}{\partial P_r}, \qquad \frac{dP_r}{dt} = -\frac{\partial K}{\partial Q_r} \qquad (r = 1, 2, \ldots, n),$$

and the known integral becomes

$$P_n = \text{Constant}.$$

Since $dP_n/dt = 0$, we have $\partial K / \partial Q_n = 0$, so the function K does not involve Q_n explicitly: and thus we obtain the result that *when a dynamical system possesses an integral which is linear and homogeneous in* $(p_1, p_2, \ldots, p_n)$, *there exists a point-transformation from the variables* $(q_1, q_2, \ldots, q_n)$ *to new variables* $(Q_1, Q_2, \ldots, Q_n)$, *which is such that the transformed Hamiltonian function does not involve* Q_n. The system as transformed possesses therefore an ignorable coordinate, and we have the theorem that *the only dynamical systems which possess integrals linear in the momenta are those which possess ignorable coordinates, or which can be transformed by an extended point-transformation into systems which possess ignorable coordinates.*

The converse of this theorem is evidently true.

This result might have been foreseen from the theorem (§ 144) that if

$$\phi(q_1, q_2, \ldots, q_n, p_1, \ldots, p_n, t) = \text{Constant}$$

is an integral of the system, then the differential equations of motion admit the infinitesimal transformation whose symbol is (ϕ, f). For when ϕ is linear and homogeneous in $(p_1, p_2, \ldots, p_n)$, this transformation is (§ 132) an extended point-transformation: if this point-transformation is transformed by change of variables so as to have the symbol $\partial f/\partial Q_n$, it is clear that the Hamiltonian function of the equations after transformation cannot involve Q_n explicitly.

Considering now in particular systems whose kinetic potential consists of a kinetic energy $T(\dot{q}_1, \dot{q}_2, \ldots, \dot{q}_n, q_1, \ldots, q_n)$ which is quadratic in the velocities $(\dot{q}_1, \dot{q}_2, \ldots, \dot{q}_n)$ and a potential energy $V(q_1, q_2, \ldots, q_n)$ which is independent of the velocities, we see that in order that an integral linear in the velocities may exist the system must possess an ignorable coordinate, or must be transformable by a point-transformation into a system which possesses an ignorable coordinate. But in either case the functions T and V evidently admit the same infinitesimal transformation, namely the transformation which, when the coordinates are so chosen that one of them is the ignorable coordinate, consists in increasing the ignorable coordinate by a small quantity and leaving the other coordinates and the velocities unaltered; and conversely, if T and V admit the same infinitesimal transformation, then there exists an integral linear in the velocities. This result is known as *Lévy's theorem*, having been published by Lévy* in 1878.

Example 1. Shew that if the differential equations of motion of a particle admit an integral linear in the components of momentum, the line of action of the force must belong to a linear complex.

(Cerruti, *Collect. math. in mem. D. Chelini*: cf. P. Grossi, *Palermo Rend.* XXIV. (1907), p. 25.)

Example 2. If the equations

$$\frac{d}{dt}\left(\frac{\partial T}{\partial \dot{q}_r}\right) - \frac{\partial T}{\partial q_r} = Q_r \qquad (r = 1, 2, \ldots, n),$$

* *Comptes Rendus*, LXXXVI.

where $T=\frac{1}{2}\sum_{i=1}^{n}\sum_{k=1}^{n} a_{ik}\dot{q}_i\dot{q}_k$, and where $(Q_1, Q_2, \ldots, Q_n, a_{11}, a_{12}, \ldots, a_{nn})$ are given functions of $(q_1, q_2, \ldots, q_n)$, possess an integral of the form

$$C_1\dot{q}_1+C_2\dot{q}_2+\ldots+C_n\dot{q}_n+C=\text{Constant},$$

where $(C_1, C_2, \ldots, C_n, C)$ are functions of $(q_1, q_2, \ldots, q_n)$, shew that it is possible to displace an invariable system in one direction from any one of its positions in the space S_n defined by the form

$$ds^2=\sum_{i=1}^{n}\sum_{k=1}^{n} a_{ik}dq_i dq_k.$$

Shew that for this a necessary and sufficient condition is that the ds^2 can be transformed in such a way that one of the variables becomes absent from the coefficients.

(Cerruti and Lévy.)

151. *Determination of the forces acting on a system for which an integral is known.*

Before proceeding to discuss systems which possess integrals quadratic in the velocities, we shall obtain a result due to Bertrand*, namely that in the motion of a dynamical system of given constitution, for which however the acting forces are unknown (it being supposed that the forces depend solely on the coordinates of their points of application, and not on the velocities of these points), we can discover the unknown forces provided one integral is known. Moreover, this integral cannot be chosen at random, but must satisfy certain conditions.

Let $(q_1, q_2, \ldots, q_n)$ be the n independent coordinates of the system, T the kinetic energy, and $(Q_1, Q_2, \ldots, Q_n)$ the unknown forces, which are supposed to depend only on $(q_1, q_2, \ldots, q_n)$; so the equations of motion are

$$\frac{d}{dt}\left(\frac{\partial T}{\partial \dot{q}_r}\right)-\frac{\partial T}{\partial q_r}=Q_r \qquad (r=1, 2, \ldots, n).$$

Let $\phi(\dot{q}_1, \dot{q}_2, \ldots, \dot{q}_n, q_1, \ldots, q_n, t)=\text{Constant}$

be an integral of the system; on differentiating it, we have

$$\sum_{r=1}^{n}\frac{\partial\phi}{\partial\dot{q}_r}\ddot{q}_r+\sum_{r=1}^{n}\frac{\partial\phi}{\partial q_r}\dot{q}_r+\frac{\partial\phi}{\partial t}=0.$$

Substituting in this equation for $(\ddot{q}_1, \ddot{q}_2, \ldots, \ddot{q}_n)$ their values as given by the equations of motion, we have a relation involving $(Q_1, Q_2, \ldots, Q_n)$ linearly. This relation, as it contains only the quantities $(q_1, \dot{q}_2, \ldots, \dot{q}_n, q_1, \ldots, q_n, t)$, to all of which we can assign arbitrary independent values, must be an identity: we can therefore differentiate it with respect to $(\dot{q}_1, \dot{q}_2, \ldots, \dot{q}_n)$, and so form n new equations which, likewise containing $(Q_1, Q_2, \ldots, Q_n)$ linearly, *will suffice in general to determine these unknown quantities.* The integral will

* *Journal de Math.* (i) XVII. (1852), p. 121.

relate to an actual system only when these values of $(Q_1, Q_2, \ldots, Q_n)$ satisfy the equation

$$\sum_{r=1}^{n} \frac{\partial \phi}{\partial \dot{q}_r} \ddot{q}_r + \sum_{r=1}^{n} \frac{\partial \phi}{\partial q_r} \dot{q}_r + \frac{\partial \phi}{\partial t} = 0;$$

the cases in which the equations for the determination of $(Q_1, Q_2, \ldots, Q_n)$ are not independent, so that $(Q_1, Q_2, \ldots, Q_n)$ are indeterminate, are those in which the integral is common to several distinct dynamical problems.

Example. If an integral of the equations of motion of a point in a plane is common to two different problems, shew that it is of the form

$$F(\phi', x, y, t) = \text{Constant},$$

where (x, y) are rectangular coordinates and ϕ' is the derivate with respect to t of a function $\phi(x, y)$ which, equated to a constant, represents the equations of a set of straight lines. (Bertrand.)

152. *Application to the case of a particle whose equations of motion possess an integral quadratic in the velocities.*

As an application of Bertrand's method, let it be required to find the nature of the potential energy function V in order that the equations of motion of a particle which is free to move in a plane under the action of conservative forces,

$$\ddot{x} = -\frac{\partial V}{\partial x}, \quad \ddot{y} = -\frac{\partial V}{\partial y},$$

may possess an integral (other than the integral of energy) of the form

$$P\dot{x}^2 + Q\dot{x}\dot{y} + R\dot{y}^2 + S\dot{y} + T\dot{x} + K = \text{Constant},$$

where P, Q, R, S, T, K are functions of x and y.

Differentiating the last equation, and substituting for $\ddot{x}$ and $\ddot{y}$ from the equations of motion, we have

$$\dot{x}^3 \frac{\partial P}{\partial x} + \dot{y}^3 \frac{\partial R}{\partial y} + \dot{x}^2\dot{y}\left(\frac{\partial P}{\partial y} + \frac{\partial Q}{\partial x}\right) + \dot{y}^2\dot{x}\left(\frac{\partial Q}{\partial y} + \frac{\partial R}{\partial x}\right) - 2P\dot{x}\frac{\partial V}{\partial x}$$

$$- Q\left(\dot{x}\frac{\partial V}{\partial y} + \dot{y}\frac{\partial V}{\partial x}\right) - 2R\dot{y}\frac{\partial V}{\partial y} + \frac{\partial S}{\partial y}\dot{y}^2 + \frac{\partial T}{\partial x}\dot{x}^2 + \dot{x}\dot{y}\left(\frac{\partial S}{\partial x} + \frac{\partial T}{\partial y}\right)$$

$$+ \frac{\partial K}{\partial x}\dot{x} + \frac{\partial K}{\partial y}\dot{y} - S\frac{\partial V}{\partial y} - T\frac{\partial V}{\partial x} = 0 \quad \ldots\ldots\ldots\ldots\ldots\ldots\ldots\ldots\text{(A)}.$$

Equating to zero the terms of the third degree in $\dot{x}$ and $\dot{y}$, we have

$$\frac{\partial P}{\partial x} = 0, \quad \frac{\partial R}{\partial y} = 0, \quad \frac{\partial P}{\partial y} + \frac{\partial Q}{\partial x} = 0, \quad \frac{\partial Q}{\partial y} + \frac{\partial R}{\partial x} = 0,$$

from which it is readily deduced that the terms of the second degree in the integral must have the form

$$(ay^2 + by + c)\dot{x}^2 + (ax^2 + b'x + c')\dot{y}^2 + (-2axy - b'y - bx + c_1)\dot{x}\dot{y},$$

where (a, b, c, b', c', c_1) are constants.

Equating to zero the terms of the second degree in $\dot{x}$ and $\dot{y}$ in equation (A), we have

$$\frac{\partial S}{\partial y}=0, \quad \frac{\partial T}{\partial x}=0, \quad \frac{\partial S}{\partial x}+\frac{\partial T}{\partial y}=0;$$

from these equations we deduce

$$S = mx + p, \quad T = -my + q,$$

where (m, p, q) are constants.

Equating to zero the terms independent of $\dot{x}$ and $\dot{y}$ in (A), we have

$$S\frac{\partial V}{\partial y}+T\frac{\partial V}{\partial x}=0,$$

or

$$\frac{\partial V}{\partial y}(mx+p)-\frac{\partial V}{\partial x}(my-q)=0.$$

This equation shews that if (m, p, q) are different from zero, the force is directed to a fixed centre of force, whose coordinates are $-p/m$ and q/m; we shall exclude this simple particular case, and hence it follows that the constants (m, p, q) must each be zero, so that the integral contains no terms of the first degree in $\dot{x}$, $\dot{y}$.

Equating to zero the terms linear in $\dot{x}$ and $\dot{y}$ in (A), we have

$$\begin{cases} -2P\dfrac{\partial V}{\partial x}-Q\dfrac{\partial V}{\partial y}+\dfrac{\partial K}{\partial x}=0, \\ -2R\dfrac{\partial V}{\partial y}-Q\dfrac{\partial V}{\partial x}+\dfrac{\partial K}{\partial y}=0. \end{cases}$$

Differentiating the former of these equations with respect to y, and the latter with respect to x, and equating the two values of $\dfrac{\partial^2 K}{\partial x \partial y}$ thus obtained, we have

$$2P\frac{\partial^2 V}{\partial x\partial y}+2\frac{\partial V}{\partial x}\frac{\partial P}{\partial y}+Q\frac{\partial^2 V}{\partial y^2}+\frac{\partial Q}{\partial y}\frac{\partial V}{\partial y}=2R\frac{\partial^2 V}{\partial x\partial y}+2\frac{\partial R}{\partial x}\frac{\partial V}{\partial y}+\frac{\partial Q}{\partial x}\frac{\partial V}{\partial x}+Q\frac{\partial^2 V}{\partial x^2},$$

and replacing P, Q, R by their values as found above, we have

$$\left(\frac{\partial^2 V}{\partial y^2}-\frac{\partial^2 V}{\partial x^2}\right)(-2axy-b'y-bx+c_1)+2\frac{\partial^2 V}{\partial x\partial y}(ay^2-ax^2+by-b'x+c-c')$$
$$+\frac{\partial V}{\partial x}(6ay+3b)+\frac{\partial V}{\partial y}(-6ax-3b')=0.$$

Darboux* has shewn that this partial differential equation for the function V can be integrated in the following way.

* *Archives Néerlandaises*, (ii) VI. p. 371 (1901).

Excluding the particular case in which the constant a is zero, we can always by change of axes reduce the given integral to the simpler form

$$\tfrac{1}{2}(x\dot{y} - y\dot{x})^2 + c\dot{x}^2 + c'\dot{y}^2 + K = \text{Constant},$$

which amounts to supposing that

$$a = \tfrac{1}{2}, \quad b = 0, \quad b' = 0, \quad c_1 = 0;$$

if moreover we replace $c - c'$ by $\frac{1}{2}c^2$, the partial differential equation for V becomes

$$xy\left(\frac{\partial^2 V}{\partial x^2} - \frac{\partial^2 V}{\partial y^2}\right) + (y^2 - x^2 + c^2)\frac{\partial^2 V}{\partial x \partial y} + 3y\frac{\partial V}{\partial x} - 3x\frac{\partial V}{\partial y} = 0.$$

To integrate this equation, we form the differential equation of the characteristics

$$xy\,(dy^2 - dx^2) + (x^2 - y^2 - c^2)\,dx\,dy = 0.$$

If in this equation we take x^2 and y^2 as new variables, it becomes a Clairaut's equation: we thus find that its integral is

$$(m+1)(mx^2 - y^2) - mc^2 = 0,$$

where m denotes the arbitrary constant. By a simple change of notation, we can write this integral in the form

$$\frac{x^2}{\alpha^2} + \frac{y^2}{\alpha^2 - c^2} = 1,$$

where the arbitrary constant is now α. This form puts in evidence the interesting fact that the characteristic curves of the partial differential equation are two families of confocal conics.

Taking then as new variables α and β the parameters of the confocal ellipses and hyperbolas, so that

$$x = \frac{\alpha\beta}{c}, \quad y = \frac{1}{c}\{(\alpha^2 - c^2)(c^2 - \beta^2)\}^{\frac{1}{2}},$$

it is known from the general theory that the partial differential equation will take the form

$$\frac{\partial^2 V}{\partial\alpha\,\partial\beta} + A\frac{\partial V}{\partial\alpha} + B\frac{\partial V}{\partial\beta} = 0,$$

where A and B are functions of α and β; in fact, on performing the change of variables we find

$$(\beta^2 - \alpha^2)\frac{\partial^2 V}{\partial\alpha\,\partial\beta} + 2\beta\frac{\partial V}{\partial\alpha} - 2\alpha\frac{\partial V}{\partial\beta} = 0,$$

which can be immediately integrated, giving

$$(\alpha^2 - \beta^2)\,V = f(\alpha) - \phi(\beta),$$

where f and ϕ are arbitrary functions of their arguments. It follows that *the only cases of the motion of a particle in a plane, under the action of conservative forces, which possess an integral quadratic in the velocities other than the integral of energy, are those for which the potential energy has the form*

$$V = \frac{f(\alpha) - \phi(\beta)}{\alpha^2 - \beta^2},$$

where α and β are the parameters of confocal ellipses and hyperbolas.

Since by differentiation we have

$$\dot{x}^2 + \dot{y}^2 = (\alpha^2 - \beta^2)\left(\frac{\dot{\alpha}^2}{\alpha^2 - c^2} + \frac{\dot{\beta}^2}{c^2 - \beta^2}\right),$$

the kinetic energy is

$$T = \tfrac{1}{2}(\alpha^2 - \beta^2)\left(\frac{\dot{\alpha}^2}{\alpha^2 - c^2} + \frac{\dot{\beta}^2}{c^2 - \beta^2}\right),$$

and an inspection of the forms of T and V shews that *these problems are of Liouville's class* (§ 43), *and are therefore integrable by quadratures.*

153. *General dynamical systems possessing integrals quadratic in the velocities.*

The complete determination of the explicit form of the most general dynamical system whose equations of motion possess an integral quadratic in the velocities (in addition to the integral of energy) has not yet been effected. It is obvious from § 43 that all dynamical systems which are of Liouville's type, or which are reducible to this type by a point-transformation, possess such integrals: and several more extended types have been determined*.

Example 1. Let

$$\phi_{kl}(q_k) \qquad (k, l = 1, 2, \ldots, n)$$

be n^2 functions depending solely on the arguments indicated, and let

$$\Phi = \sum_{k=1}^{n} \phi_{kl}\Phi_{kl} \qquad (l = 1, 2, \ldots, n)$$

denote the determinant formed by these functions. Shew that if the kinetic energy of a dynamical system is reducible to the form

$$\tfrac{1}{2}\sum_{k=1}^{n} \frac{\Phi}{\Phi_{k1}}\dot{q}_k^2,$$

and the potential energy is zero, there exists not only the integral of energy,

$$\sum_{k=1}^{n} \frac{\Phi}{\Phi_{k1}}\dot{q}_k^2 = a_1,$$

but also $(n-1)$ other integrals, homogeneous and of the second degree in the velocities, namely

$$\sum_{k=1}^{n} \frac{\Phi \cdot \Phi_{kl}}{\Phi^2_{k1}}\dot{q}_k^2 = a_l \qquad (l = 2, 3, \ldots, n),$$

where $(a_1, a_2, \ldots, a_n)$ are arbitrary constants: and that the problem is soluble by quadratures. (Stäckel.)

Example 2. Let the equations of motion of a dynamical system with two degrees of freedom be

$$\frac{d}{dt}\left(\frac{\partial T}{\partial \dot{q}_r}\right) - \frac{\partial T}{\partial q_r} = 0 \qquad (r = 1, 2),$$

where

$$T = \tfrac{1}{2}(a\dot{q}_1^2 + 2h\dot{q}_1\dot{q}_2 + b\dot{q}_2^2),$$

* Cf. G. di Pirro, *Annali di Mat.* XXIV. (1896), p. 315.

and (a, h, b) are any functions of the coordinates (q_1, q_2) : and let this system possess an integral

$$a'\dot{q}_1^2 + 2h'\dot{q}_1\dot{q}_2 + b'\dot{q}_2^2 = \text{Constant},$$

quadratic in the velocities and distinct from the equation of energy, where (a', h', b') are functions of the coordinates. If Δ and Δ' denote $(ab - h^2)$ and $(a'b' - h'^2)$ respectively, and if

$$T' = \tfrac{1}{2}\left(\frac{\Delta}{\Delta'}\right)^2 (a'q_1'^2 + 2h'q_1'q_2' + b'q_2'^2),$$

where q_r' stands for dq_r/dt', shew that the equations

$$\frac{d}{dt'}\left(\frac{\partial T'}{\partial q_r'}\right) - \frac{\partial T'}{\partial q_r} = 0 \qquad (r = 1, 2)$$

define the same relations between the coordinates (q_1, q_2) as the original equations of motion, and that one set of equations can be transformed to the other by the transformation

$$\Delta dt = \Delta' dt'.$$

Miscellaneous Examples.

1. A dynamical system is defined by its kinetic energy

$$\tfrac{1}{2}\Phi\left(\frac{\dot{q}_1^2}{\Phi_{11}} + \frac{\dot{q}_2^2}{\Phi_{21}} + \dots + \frac{\dot{q}_n^2}{\Phi_{n1}}\right),$$

(where Φ denotes the determinant

$$\begin{vmatrix} \phi_{11} & \phi_{12} & \cdots & \phi_{1n} \\ \phi_{21} & \phi_{22} & \cdots & \phi_{2n} \\ \cdots & \cdots & \cdots & \cdots \\ \cdots & \cdots & \cdots & \cdots \\ \phi_{n1} & \phi_{n2} & \cdots & \phi_{nn} \end{vmatrix}$$

in which the elements of the kth line are functions of q_k only, and Φ_{kl} denotes the minor of ϕ_{kl}), and by its potential energy

$$-\frac{\Psi}{\Phi},$$

where $$\Psi = \Phi_{11}\psi_1 + \Phi_{21}\psi_2 + \dots + \Phi_{n1}\psi_n,$$

and the quantity ψ_k denotes a function of q_k only. Shew that a complete integral of the Hamilton-Jacobi equation

$$\frac{\partial W}{\partial t} + \frac{1}{2\Phi}\left\{\Phi_{11}\left(\frac{\partial W}{\partial q_1}\right)^2 + \Phi_{21}\left(\frac{\partial W}{\partial q_2}\right)^2 + \dots + \Phi_{n1}\left(\frac{\partial W}{\partial q_n}\right)^2\right\} - \frac{\Psi}{\Phi} = 0$$

is $$W = -a_1 t + \sum_{i=1}^{n} \int \{a_1\phi_{i1} + a_2\phi_{i2} + \dots + a_n\phi_{in} + 2\psi_i\}^{\frac{1}{2}} dq_i,$$

where $(a_1, a_2, \dots, a_n)$ are arbitrary constants. (Goursat.)

2. If $$\phi(q_1, q_2, \dots, q_n, p_1, \dots, p_n, t) = \text{Constant}$$

is an integral of a dynamical system which possesses an integral of energy, shew that

$$\frac{\partial\phi}{\partial t} = \text{Constant}, \quad \frac{\partial^2\phi}{\partial t^2} = \text{Constant}, \text{ etc., are also integrals.}$$

3. A system of equations

$$\left.\begin{aligned}\frac{dq_r}{dt}&=A_r(q_1, q_2, \ldots, q_n, p_1, \ldots, p_n, t)\\ \frac{dp_r}{dt}&=B_r(q_1, q_2, \ldots, q_n, p_1, \ldots, p_n, t)\end{aligned}\right\} \qquad (r=1, 2, \ldots, n)$$

is such that if ϕ and ψ are any two integrals whatever, the Poisson-bracket (ϕ, ψ) is also an integral. Shew that the equations must have the Hamiltonian form

$$\frac{dq_r}{dt}=\frac{\partial H}{\partial p_r}, \quad \frac{dp_r}{dt}=-\frac{\partial H}{\partial q_r} \qquad (r=1, 2, \ldots, n).$$

(Korkine.)

4. If

$$\alpha_1=\text{Constant}, \quad \alpha_2=\text{Constant}, \quad \ldots, \quad \alpha_k=\text{Constant},$$
$$\beta_1=\text{Constant}, \quad \beta_2=\text{Constant}, \quad \ldots, \quad \beta_k=\text{Constant},$$

are any $2k$ integrals of a Hamiltonian system of differential equations, the variables being $(q_1, q_2, \ldots, q_n, p_1, \ldots, p_n)$, shew that

$$\sum_{\lambda_1, \ldots, \lambda_k} \Sigma \pm \frac{\partial \alpha_1}{\partial q_{\lambda_1}} \frac{\partial \alpha_2}{\partial q_{\lambda_2}} \cdots \frac{\partial \alpha_k}{\partial q_{\lambda_k}} \frac{\partial \beta_1}{\partial p_{\lambda_1}} \cdots \frac{\partial \beta_k}{\partial p_{\lambda_k}}=\text{Constant}$$

is also an integral. (Laurent.)

5. Let the expression

$$(H_1, H_2, \ldots, H_n)=\sum_{i=1}^{\nu} \frac{\partial (H_1, H_2, \ldots, H_n)}{\partial (x_{1i}, x_{2i}, \ldots, x_{ni})},$$

where $H_1, H_2, \ldots, H_n$ are functions of the $n\nu$ variables x_{ji} $(j=1, 2, \ldots, n;\ i=1, 2, \ldots, \nu)$ be called a *Poisson-bracket of the nth order*. If $G_1, G_2, \ldots, G_{h\nu}$ are $h\nu$ functions of $y_{11}, y_{12}, \ldots, y_{h\nu};\ x_{11}, x_{12}, \ldots, x_{k\nu};\ a_1, a_2, \ldots, a_{h\nu}$, where $(h+k=n)$, and if

$$P_i(G^n) \qquad \left(i=1, 2, \ldots, \binom{h\nu}{n}\right)$$

denotes all the Poisson-brackets formed from every n functions G, shew that

$$P_i(G^n)=0 \qquad \left(i=1, 2, \ldots, \binom{h\nu}{n}\right)$$

represents the necessary and sufficient conditions that the functions

$$y_{st}=F_{st}(x_{11}, x_{12}, \ldots, x_{k\nu};\ a_1, a_2, \ldots, a_{h\nu}) \qquad (s=1, 2, \ldots, h;\ t=1, 2, \ldots, \nu)$$

arising from the equations

$$G_i=0 \qquad (i=1, 2, \ldots, h\nu)$$

shall satisfy the simultaneous partial differential equations of the first order

$$P_i(y^h, F)=0, \qquad \left(i=1, 2, \ldots, \binom{h\nu}{n}\right)$$

where $P_i(y^h, F)$ denotes the expression which is obtained when we replace h of the functions F in $P_i(F^n)$ by as many y's. (Albeggiani.)

6. A particle of unit mass whose coordinates referred to fixed rectangular axes are (x, y) is free to move in a plane under forces derivable from a potential-energy function $f(x, y)$, the total energy being h. Shew that if the orthogonal trajectories of the curves

$$\frac{1}{h-f(x, y)}\left(\frac{\partial^2}{\partial x^2}+\frac{\partial^2}{\partial y^2}\right) \log \{h-f(x, y)\}=\text{Constant}$$

are orbits, the differential equations of motion of the particle possess an integral linear and homogeneous in the velocities $(\dot{x}, \dot{y})$.

7. The equations of motion of a free system of m particles are

$$\frac{d^2x_s}{dt^2}=X_s \qquad (s=1, 2, \ldots, 3m).$$

If an integral exists of the form

$$\sum_{s=1}^{3m} f_s\dot{x}_s - Ct = \text{Constant},$$

where $f_1, f_2, \ldots, f_{3m}$ are functions of $x_1, x_2, \ldots, x_{3m}$, and C is a constant, shew that this integral can be written

$$\sum_{s=1}^{3m} k_s\dot{x}_s + \sum_{r,s=1}^{3m} a_{rs}(x_s\dot{x}_r - x_r\dot{x}_s) - Ct = \text{Constant},$$

where the quantities k_s and a_{rs} are constants. (Pennacchietti.)

8. Two particles move on a surface under the action of different forces depending only on their respective positions: if their differential equations of motion have in common an integral independent of the time, shew that the surface is applicable on a surface of revolution. (Bertrand.)

CHAPTER XIII

THE REDUCTION OF THE PROBLEM OF THREE BODIES

154. *Introduction.*

The most celebrated of all dynamical problems is known as the *Problem of Three Bodies,* and may be enunciated as follows:

Three particles attract each other according to the Newtonian law, so that between each pair of particles there is an attractive force which is proportional to the product of the masses of the particles and the inverse square of their distance apart: they are free to move in space, and are initially supposed to be moving in any given manner; to determine their subsequent motion.

The practical importance of this problem arises from its applications to Celestial Mechanics: the bodies which constitute the solar system attract each other according to the Newtonian law, and (as they have approximately the form of spheres, whose dimensions are very small compared with the distances which separate them) it is usual to consider the problem of determining their motion in an ideal form, in which the bodies are replaced by particles of masses equal to the masses of the respective bodies and occupying the positions of their centres of gravity*.

The problem of three bodies cannot be solved in finite terms by means of any of the functions at present known to analysis. This difficulty has stimulated research to such an extent, that since the year 1750 over 800 memoirs, many of them bearing the names of the greatest mathematicians, have been published on the subject†. In the present chapter, we shall discuss the known integrals of the system and their application to the reduction of the problem to a dynamical problem with a lesser number of degrees of freedom.

* The motions of the bodies relative to their centres of gravity (in the consideration of which their sizes and shapes of course cannot be neglected) are discussed separately, e.g. in the Theory of Precession and Nutation. In some cases however (e.g. in the Theory of the Satellites of the Major Planets) the oblateness of one of the bodies exercises so great an effect, that the problem cannot be divided in this way.

† For the history of the Problem of Three Bodies, cf. A. Gautier, *Essai historique sur le problème des trois corps* (Paris, 1817): R. Grant, *History of Physical Astronomy from the earliest ages to the middle of the nineteenth century* (London, 1852): E. T. Whittaker, *Report on the progress of the solution of the Problem of Three Bodies* (Brit. Ass. Rep. 1899, p. 121): and E. O. Lovett, *Quart. Journ. Math.* XLII. (1911), p. 252, who discusses the memoirs of the period 1898–1908.

155. *The differential equations of the problem.*

Let P, Q, R denote the three particles, (m_1, m_2, m_3) their masses, and (r_{23}, r_{31}, r_{12}) their mutual distances. Take any fixed rectangular axes $Oxyz$, and let (q_1, q_2, q_3), (q_4, q_5, q_6), (q_7, q_8, q_9), be the coordinates of P, Q, R, respectively. The kinetic energy of the system is

$$T = \tfrac{1}{2}m_1(\dot{q}_1^2 + \dot{q}_2^2 + \dot{q}_3^2) + \tfrac{1}{2}m_2(\dot{q}_4^2 + \dot{q}_5^2 + \dot{q}_6^2) + \tfrac{1}{2}m_3(\dot{q}_7^2 + \dot{q}_8^2 + \dot{q}_9^2);$$

the force of attraction between m_1 and m_2 is $k^2 m_1 m_2 r_{12}^{-2}$, where k^2 is the constant of attraction: we shall suppose the units so chosen that k^2 is unity, so that this attraction becomes $m_1 m_2 r_{12}^{-2}$, and the corresponding term in the potential energy is $-m_1 m_2 r_{12}^{-1}$. The potential energy of the system is therefore

$$\begin{aligned} V &= -\frac{m_2 m_3}{r_{23}} - \frac{m_3 m_1}{r_{31}} - \frac{m_1 m_2}{r_{12}} \\ &= -m_2 m_3 \{(q_4 - q_7)^2 + (q_5 - q_8)^2 + (q_6 - q_9)^2\}^{-\frac{1}{2}} \\ &\quad - m_3 m_1 \{(q_7 - q_1)^2 + (q_8 - q_2)^2 + (q_9 - q_3)^2\}^{-\frac{1}{2}} \\ &\quad - m_1 m_2 \{(q_1 - q_4)^2 + (q_2 - q_5)^2 + (q_3 - q_6)^2\}^{-\frac{1}{2}}. \end{aligned}$$

The equations of motion of the system are

$$m_k \ddot{q}_r = -\partial V/\partial q_r \qquad (r = 1, 2, \ldots, 9),$$

where k denotes the integer part of $\frac{1}{3}(r + 2)$. This system consists of 9 differential equations, each of the 2nd order, and the system is therefore of order 18.

Writing
$$m_k \dot{q}_r = p_r \qquad (r = 1, 2, \ldots, 9),$$
and
$$H = \sum_{r=1}^{9} \frac{p_r^2}{2m_k} + V,$$

the equations take the Hamiltonian form

$$\frac{dq_r}{dt} = \frac{\partial H}{\partial p_r}, \quad \frac{dp_r}{dt} = -\frac{\partial H}{\partial q_r} \qquad (r = 1, 2, \ldots, 9),$$

and these are a set of 18 differential equations, each of the 1st order, for the determination of the variables $(q_1, q_2, \ldots, q_9, p_1, p_2, \ldots, p_9)$.

It was shewn by Lagrange* that this system can be reduced to a system which is only of the 6th order. That a reduction of this kind must be possible may be seen from the following considerations.

In the first place, since no forces act except the mutual attractions of the

* *Recueil des pieces qui ont remporté les prix de l'Acad. de Paris*, IX. (1772). Lagrange of course did not reduce the system to the Hamiltonian form. Cf. Bohlin, *Kongl. Sv. Vet.-Handl.* XLII. (1907), No. 9, for an improved Lagrangian reduction.

particles, the centre of gravity of the system moves in a straight line with uniform velocity. This fact is expressed by the 6 integrals

$$\begin{cases} p_1 + p_4 + p_7 = a_1, \\ p_2 + p_5 + p_8 = a_3, \\ p_3 + p_6 + p_9 = a_5, \end{cases}$$

$$\begin{cases} m_1 q_1 + m_2 q_4 + m_3 q_7 - (p_1 + p_4 + p_7)\, t = a_2, \\ m_1 q_2 + m_2 q_5 + m_3 q_8 - (p_2 + p_5 + p_8)\, t = a_4, \\ m_1 q_3 + m_2 q_6 + m_3 q_9 - (p_3 + p_6 + p_9)\, t = a_6, \end{cases}$$

where $a_1, a_2, \ldots, a_6$ are constants. It may be expected that the use of these integrals will enable us to depress the equations of motion from the 18th to the 12th order.

In the second place, the angular momentum of the three bodies round each of the coordinate axes is constant throughout the motion. This fact is analytically expressed by the equations

$$\begin{cases} q_1 p_2 - q_2 p_1 + q_4 p_5 - q_5 p_4 + q_7 p_8 - q_8 p_7 = a_7, \\ q_2 p_3 - q_3 p_2 + q_5 p_6 - q_6 p_5 + q_8 p_9 - q_9 p_8 = a_8, \\ q_3 p_1 - q_1 p_3 + q_6 p_4 - q_4 p_6 + q_9 p_7 - q_7 p_9 = a_9, \end{cases}$$

where a_7, a_8, a_9 are constants. By use of these three integrals we may expect to be able to depress further the equations of motion from the 12th to the 9th order. But when one of the coordinates which define the position of the system is taken to be the azimuth ϕ of one of the bodies with respect to some fixed axis (say the axis of z), and the other coordinates define the position of the system relative to the plane having this azimuth, the coordinate ϕ is an ignorable coordinate, and consequently the corresponding integral (which is one of the integrals of angular momentum above-mentioned) can be used to depress the order of the system by *two* units; the equations of motion can therefore, as a matter of fact, be reduced in this way to the 8th order. This fact (though contained implicity in Lagrange's memoir already cited) was first explicitly noticed by Jacobi* in 1843, and is generally referred to as the *elimination of the nodes*.

Lastly, it is possible again to depress the order of the equations by two units as in § 42, by using the integral of energy and eliminating the time. So finally *the equations of motion may be reduced to a system of the 6th order*.

* *Journ. für Math.* XXVI. p. 115. From the point of view of the theory of Partial Differential Equations, we may express the matter by saying that the integrals of angular momentum give rise to an involution-system, consisting of two functions which are in involution with each other and with H: and hence the Hamilton-Jacobi partial differential equation with 6 independent variables can be reduced to a partial differential equation with $6 - 2$ or 4 independent variables: this will be the Hamilton-Jacobi partial differential equation for the reduced system.

156. *Jacobi's equation.*

Jacobi*, in considering the motion of any number of free particles in space, which attract each other according to the Newtonian law, has introduced the function

$$\tfrac{1}{2}\sum_{i,j}\frac{m_i m_j}{M}r_{ij}^2,$$

where m_i and m_j are the masses of two typical particles of the system, r_{ij} is the distance between them at time t, M is the total mass of the particles, and the summation is extended over all pairs of particles in the system. This function, which has been used in researches concerning the stability of the system, will be called *Jacobi's function* and denoted by the symbol Φ.

We shall suppose the centre of gravity of the system to be at rest; let (x_i, y_i, z_i) be the coordinates of the particle m_i referred to fixed rectangular axes with the centre of gravity as origin. The kinetic energy of the system is

$$T=\tfrac{1}{2}\sum_i m_i(\dot{x}_i^2+\dot{y}_i^2+\dot{z}_i^2),$$

and consequently we have

$$2MT=(\sum_i m_i)\times\sum_i m_i(\dot{x}_i^2+\dot{y}_i^2+\dot{z}_i^2).$$

But
$$(\sum_i m_i)\times\sum_i m_i\dot{x}_i^2-(\sum_i m_i\dot{x}_i)^2=\sum_{i,j}m_i m_j(\dot{x}_i-\dot{x}_j)^2,$$

where the summation on the right-hand side is extended over every pair of particles in the system: and we have $\sum_i m_i\dot{x}_i=0$, in virtue of the properties of the centre of gravity.

Thus we have
$$T=\frac{1}{2M}\sum_{i,j}m_i m_j\{(\dot{x}_i-\dot{x}_j)^2+(\dot{y}_i-\dot{y}_j)^2+(\dot{z}_i-\dot{z}_j)^2\}$$
$$=\frac{1}{2M}\sum_{i,j}m_i m_j v_{ij}^2,$$

where v_{ij} denotes the velocity of the particle m_i relative to m

In the same way we can shew that

$$\tfrac{1}{2}\sum_i m_i(x_i^2+y_i^2+z_i^2)=\Phi.$$

If now V denotes the potential energy of the system, the arbitrary constant in V being determined by the condition that V is to be zero when the particles are at infinitely great distances from each other, we have

$$V=-\sum_{i,j}\frac{m_i m_j}{r_{ij}}.$$

The equations of motion of the particle m_i are

$$m_i\ddot{x}_i=-\frac{\partial V}{\partial x_i},\quad m_i\ddot{y}_i=-\frac{\partial V}{\partial y_i},\quad m_i\ddot{z}_i=-\frac{\partial V}{\partial z_i}.$$

Multiply these equations by x_i, y_i, z_i, respectively, add them, and sum for all the particles of the system: since V is homogeneous of degree -1 in the variables, we thus obtain

$$\sum_i m_i(x_i\ddot{x}_i+y_i\ddot{y}_i+z_i\ddot{z}_i)=V,$$

or
$$\frac{d^2}{dt^2}\tfrac{1}{2}\sum_i m_i(x_i^2+y_i^2+z_i^2)-2T=V,$$

or
$$\frac{d^2\Phi}{dt^2}=2T+V.$$

This is called *Jacobi's equation.*

* *Vorlesungen über Dyn.*, p. 22.

157. *Reduction to the 12th order, by use of the integrals of motion of the centre of gravity.*

We shall now proceed to carry out the reductions which have been described*. It will appear that it is possible to retain the Hamiltonian form of the equations throughout all the transformations.

Taking the equations of motion of the Problem of Three Bodies in the form obtained in § 155,

$$\frac{dq_r}{dt} = \frac{\partial H}{\partial p_r}, \quad \frac{dp_r}{dt} = -\frac{\partial H}{\partial q_r} \quad (r = 1, 2, \ldots, 9),$$

we have first to reduce this system from the 18th to the 12th order, by use of the integrals of motion of the centre of gravity. For this purpose we perform on the variables the contact-transformation defined by the equations

$$q_r = \frac{\partial W}{\partial p_r}, \quad p_r' = \frac{\partial W}{\partial q_r'} \quad (r = 1, 2, \ldots, 9),$$

where $$W = p_1 q_1' + p_2 q_2' + p_3 q_3' + p_4 q_4' + p_5 q_5' + p_6 q_6' + (p_1 + p_4 + p_7) q_7' + (p_2 + p_5 + p_8) q_8' + (p_3 + p_6 + p_9) q_9'.$$

Interpreting these equations, it is easily seen that (q_1', q_2', q_3') are the coordinates of m_1 relative to m_3, (q_4', q_5', q_6') are the coordinates of m_2 relative to m_3, (q_7', q_8', q_9') are the coordinates of m_3, (p_1', p_2', p_3') are the components of momentum of m_1, (p_4', p_5', p_6') are the components of momentum of m_2, and (p_7', p_8', p_9') are the components of momentum of the system.

The differential equations now become (§ 138)

$$\frac{dq_r'}{dt} = \frac{\partial H}{\partial p_r'}, \quad \frac{dp_r'}{dt} = -\frac{\partial H}{\partial q_r'} \quad (r = 1, 2, \ldots, 9),$$

where, on substitution of the new variables for the old, we have

$$\begin{aligned} H = &\left(\frac{1}{2m_1} + \frac{1}{2m_3}\right)(p_1'^2 + p_2'^2 + p_3'^2) + \left(\frac{1}{2m_2} + \frac{1}{2m_3}\right)(p_4'^2 + p_5'^2 + p_6'^2) \\ &+ \frac{1}{m_3}\{p_1'p_4' + p_2'p_5' + p_3'p_6' + \tfrac{1}{2}p_7'^2 + \tfrac{1}{2}p_8'^2 + \tfrac{1}{2}p_9'^2 - p_7'(p_1' + p_4') \\ &\qquad - p_8'(p_2' + p_5') - p_9'(p_3' + p_6')\} \\ &- m_2 m_3 \{q_4'^2 + q_5'^2 + q_6'^2\}^{-\frac{1}{2}} - m_3 m_1 \{q_1'^2 + q_2'^2 + q_3'^2\}^{-\frac{1}{2}} \\ &- m_1 m_2 \{(q_1' - q_4')^2 + (q_2' - q_5')^2 + (q_3' - q_6')^2\}^{-\frac{1}{2}}. \end{aligned}$$

* The contact-transformation used in § 157 is due to Poincaré, *C.R.* CXXIII. (1896); that used in § 158 is due to the author, and was originally published in the first edition of this work (1904). It appears worthy of note from the fact that it is an extended point-transformation, which shews that the reduction could be performed on the equations in their Lagrangian (as opposed to their Hamiltonian) form, by pure point-transformations. The second transformation in the alternative reduction (§ 160) is not an extended point-transformation. Another reduction of the Problem of Three Bodies can be constructed from the standpoint of Lie's Theory of Involution-systems and Distinguished Functions: cf. Lie, *Math. Ann.* VIII. p. 282. Cf. also Woronetz, *Bull. Univ. Kiej*, 1907, and Levi-Civita, *Atti del R. Ist. Veneto*, LXXIV. (1915), p. 907.

Since q_7', q_8', q_9' are altogether absent from H, they are ignorable coordinates: the corresponding integrals are

$$p_7' = \text{Constant}, \quad p_8' = \text{Constant}, \quad p_9' = \text{Constant}.$$

We can without loss of generality suppose these constants of integration to be zero, as this only means that the centre of gravity of the system is taken to be at rest: the reduced kinetic potential obtained by ignoration of coordinates will therefore be derived from the unreduced kinetic potential by replacing p_7', p_8', p_9' by zero, and the new Hamiltonian function will be derived from H in the same way. *The system of the 12th order, to which the equations of motion of the problem of three bodies have now been reduced, may therefore be written* (suppressing the accents to the letters)

$$\frac{dq_r}{dt} = \frac{\partial H}{\partial p_r}, \quad \frac{dp_r}{dt} = -\frac{\partial H}{\partial q_r} \qquad (r = 1, 2, \ldots, 6),$$

where

$$H = \left(\frac{1}{2m_1} + \frac{1}{2m_3}\right)(p_1^2 + p_2^2 + p_3^2) + \left(\frac{1}{2m_2} + \frac{1}{2m_3}\right)(p_4^2 + p_5^2 + p_6^2) + \frac{1}{m_3}(p_1p_4 + p_2p_5 + p_3p_6)$$
$$- m_2m_3\{q_4^2 + q_5^2 + q_6^2\}^{-\frac{1}{2}} - m_3m_1\{q_1^2 + q_2^2 + q_3^2\}^{-\frac{1}{2}} - m_1m_2\{(q_1 - q_4)^2 + (q_2 - q_5)^2 + (q_3 - q_6)^2\}^{-\frac{1}{2}}.$$

This system possesses an integral of energy,

$$H = \text{Constant},$$

and three integrals of angular momentum, namely

$$\begin{cases} q_2p_3 - q_3p_2 + q_5p_6 - q_6p_5 = A_1 \\ q_3p_1 - q_1p_3 + q_6p_4 - q_4p_6 = A_2 \\ q_1p_2 - q_2p_1 + q_4p_5 - q_5p_4 = A_3 \end{cases}$$

where A_1, A_2, A_3 are constants.

158. *Reduction to the 8th order, by use of the integrals of angular momentum and elimination of the nodes.*

The system of the 12th order obtained in the last article must now be reduced to the 8th order, by using the three integrals of angular momentum and by eliminating the nodes. This may be done in the following way.

Apply to the variables the contact-transformation defined by the equations

$$q_r = \frac{\partial W}{\partial p_r}, \quad p_r' = \frac{\partial W}{\partial q_r'} \qquad (r = 1, 2, \ldots, 6),$$

where

$$W = p_1(q_1' \cos q_5' - q_2' \cos q_6' \sin q_5') + p_2(q_1' \sin q_5' + q_2' \cos q_6' \cos q_5') + p_3q_2' \sin q_6'$$
$$+ p_4(q_3' \cos q_5' - q_4' \cos q_6' \sin q_5') + p_5(q_3' \sin q_5' + q_4' \cos q_6' \cos q_5') + p_6q_4' \sin q_6'.$$

It is readily seen that the new variables can be interpreted physically as follows:

In addition to the fixed axes $Oxyz$, take a new set of moving axes $Ox'y'z'$; Ox' is to be the intersection or *node* of the plane Oxy with the plane of the three bodies, Oy' is to be a line perpendicular to this in the plane of the three bodies, and Oz' is to be normal to the plane of the three bodies. Then (q_1', q_2') are the coordinates of m_1 relative to axes drawn through m_3 parallel to Ox', Oy'; (q_3', q_4') are the coordinates of m_2 relative to the same axes; q_5' is the angle between Ox' and Ox; q_6' is the angle between Oz' and Oz; p_1' and p_2' are the components of momentum of m_1 relative to the axes Ox', Oy'; p_3' and p_4' are the components of momentum of m_2 relative to the same axes; p_5' and p_6' are the angular momenta of the system relative to the axes Oz and Ox' respectively.

The equations of motion in terms of the new variables are (§ 138)

$$\frac{dq_r'}{dt} = \frac{\partial H}{\partial p_r'}, \quad \frac{dp_r'}{dt} = -\frac{\partial H}{\partial q_r'} \qquad (r = 1, 2, \ldots, 6),$$

where, on substitution in H of the new variables for the old, we have

$$\begin{aligned} H = {} & \left(\frac{1}{2m_1} + \frac{1}{2m_3}\right)\Big[p_1'^2 + p_2'^2 + \frac{1}{(q_2'q_3' - q_1'q_4')^2}\{(p_1'q_2' - p_2'q_1' + p_3'q_4' - p_4'q_3')\,q_4'\cot q_6' \\ & \qquad + p_5'q_4'\operatorname{cosec} q_6' + p_6'q_3'\}^2\Big] \\ & + \left(\frac{1}{2m_2} + \frac{1}{2m_3}\right)\Big[p_3'^2 + p_4'^2 + \frac{1}{(q_2'q_3' - q_1'q_4')^2}\{(p_1'q_2' - p_2'q_1' + p_3'q_4' - p_4'q_3')\,q_2'\cot q_6' \\ & \qquad + p_5'q_2'\operatorname{cosec} q_6' + p_6'q_1'\}^2\Big] \\ & + \frac{1}{m_3}\Big[p_1'p_3' + p_2'p_4' - \frac{1}{(q_2'q_3' - q_1'q_4')^2} \\ & \qquad \{(p_1'q_2' - p_2'q_1' + p_3'q_4' - p_4'q_3')\,q_4'\cot q_6' + p_5'q_4'\operatorname{cosec} q_6' + p_6'q_3'\} \\ & \qquad \{(p_1'q_2' - p_2'q_1' + p_3'q_4' - p_4'q_3')\,q_2'\cot q_6' + p_5'q_2'\operatorname{cosec} q_6' + p_6'q_1'\}\Big] \\ & - m_2m_3(q_3'^2 + q_4'^2)^{-\frac{1}{2}} - m_3m_1(q_1'^2 + q_2'^2)^{-\frac{1}{2}} - m_1m_2\{(q_1' - q_3')^2 + (q_2' - q_4')^2\}^{-\frac{1}{2}}. \end{aligned}$$

Now q_5' does not occur in H, and is therefore an ignorable coordinate; the corresponding integral is

$$p_5' = k, \qquad \text{where } k \text{ is a constant.}$$

The equation $dq_5'/dt = \partial H/\partial k$ can be integrated by a simple quadrature when the rest of the equations of motion have been integrated; the equations for q_5' and p_5' will therefore fall out of the system, which thus reduces to the system of the 10th order

$$\frac{dq_r'}{dt} = \frac{\partial H}{\partial p_r'}, \quad \frac{dp_r'}{dt} = -\frac{\partial H}{\partial q_r'} \qquad (r = 1, 2, 3, 4, 6),$$

where p_5' is to be replaced by the constant k wherever it occurs in H.

We have now made use of one of the three integrals of angular momentum (namely $p_5' = k$) and the elimination of the nodes: when the other two integrals of angular momentum are expressed in terms of the new variables, they become

$$\begin{cases} (p_2'q_1' - p_1'q_2' + p_4'q_3' - p_3'q_4') \sin q_5' \operatorname{cosec} q_6' - k \sin q_5' \cot q_6' + p_6' \cos q_5' = A_1, \\ -(p_2'q_1' - p_1'q_2' + p_4'q_3' - p_3'q_4') \cos q_5' \operatorname{cosec} q_6' + k \cos q_5' \cot q_6' + p_6' \sin q_5' = A_2. \end{cases}$$

The values of the constants A_1 and A_2 depend on the position of the fixed axes $Oxyz$; we shall choose the axis Oz to be the line of resultant angular momentum of the system, so that (cf. § 69) the constants A_1 and A_2 are zero: the special xy-plane thus introduced is called the *invariable plane* of the system. The two last equations then give

$$k \cos q_6' = p_2'q_1' - p_1'q_2' + p_4'q_3' - p_3'q_4',$$

$$p_6' = 0.$$

These equations determine q_6' and p_6' in terms of the other variables, and so can be regarded as replacing the equations

$$\frac{dq_6'}{dt} = \frac{\partial H}{\partial p_6'}, \quad \frac{dp_6'}{dt} = -\frac{\partial H}{\partial q_6'},$$

in the system. The system thus becomes

$$\frac{dq_r'}{dt} = \frac{\partial H}{\partial p_r'}, \quad \frac{dp_r'}{dt} = -\frac{\partial H}{\partial q_r'} \qquad (r = 1, 2, 3, 4),$$

where

$$H = \left(\frac{1}{2m_1} + \frac{1}{2m_3}\right)\left[p_1'^2 + p_2'^2 + \frac{q_4'^2}{(q_2'q_3' - q_1'q_4')^2} \{(p_1'q_2' - p_2'q_1' + p_3'q_4' - p_4'q_3') \cot q_6' + k \operatorname{cosec} q_6'\}^2\right]$$

$$+ \left(\frac{1}{2m_2} + \frac{1}{2m_3}\right)\left[p_3'^2 + p_4'^2 + \frac{q_2'^2}{(q_2'q_3' - q_1'q_4')^2} \{(p_1'q_2' - p_2'q_1' + p_3'q_4' - p_4'q_3') \cot q_6' + k \operatorname{cosec} q_6'\}^2\right]$$

$$+ \frac{1}{m_3}\left[p_1'p_3' + p_2'p_4' - \frac{q_2'q_4'}{(q_2'q_3' - q_1'q_4')^2} \{(p_1'q_2' - p_2'q_1' + p_3'q_4' - p_4'q_3') \cot q_6' + k \operatorname{cosec} q_6'\}^2\right]$$

$$- m_2 m_3 (q_3'^2 + q_4'^2)^{-\frac{1}{2}} - m_3 m_1 (q_1'^2 + q_2'^2)^{-\frac{1}{2}} - m_1 m_2 \{(q_1' - q_3')^2 + (q_2' - q_4')^2\}^{-\frac{1}{2}},$$

and where, *after the derivates of H have been formed*, q_6' is to be replaced by its value found from the equation

$$k \cos q_6' = p_2'q_1' - p_1'q_2' + p_4'q_3' - p_3'q_4'.$$

Now let H' be the function obtained when this value of q_6' is substituted in H; then if s denotes any one of the variables $q_1', q_2', q_3', q_4', p_1', p_2', p_3', p_4'$, we have

$$\frac{\partial H'}{\partial s} = \frac{\partial H}{\partial s} + \frac{\partial H}{\partial q_6'}\frac{\partial q_6'}{\partial s}.$$

But since $p_6' = 0$, we have $\partial H/\partial q_6' = \dot{p}_6' = 0$, and therefore

$$\frac{\partial H'}{\partial s} = \frac{\partial H}{\partial s};$$

in other words, we can make the substitution for q_6' in H *before* forming the derivates of H; and thus (suppressing the accents) *the equations of motion of the Problem of Three Bodies are reduced to the system of the 8th order*

$$\frac{dq_r}{dt} = \frac{\partial H}{\partial p_r}, \quad \frac{dp_r}{dt} = -\frac{\partial H}{\partial q_r} \qquad (r = 1, 2, 3, 4),$$

where

$$H = \left(\frac{1}{2m_1} + \frac{1}{2m_3}\right)(p_1^2 + p_2^2) + \left(\frac{1}{2m_2} + \frac{1}{2m_3}\right)(p_3^2 + p_4^2) + \frac{1}{m_3}(p_1 p_3 + p_2 p_4)$$
$$+ (q_2 q_3 - q_1 q_4)^{-2}\left\{\left(\frac{1}{2m_1} + \frac{1}{2m_3}\right) q_4^2 + \left(\frac{1}{2m_2} + \frac{1}{2m_3}\right) q_2^2 - \frac{q_2 q_4}{m_3}\right\}$$
$$\{k^2 - (p_2 q_1 - p_1 q_2 + p_4 q_3 - p_3 q_4)^2\}$$
$$- m_2 m_3 (q_3^2 + q_4^2)^{-\frac{1}{2}} - m_3 m_1 (q_1^2 + q_2^2)^{-\frac{1}{2}} - m_1 m_2 \{(q_1 - q_3)^2 + (q_2 - q_4)^2\}^{-\frac{1}{2}}.$$

Many of the quantities occurring in H have simple physical interpretations: thus $(q_2 q \;\; - q_1 q_4)$ is twice the area of the triangle formed by the bodies: and

$$\frac{2m_1 m_2 m_3}{m_1 + m_2 + m_3}\left\{\left(\frac{1}{2m_1} + \frac{1}{2m_3}\right) q_4^2 + \left(\frac{1}{2m_2} + \frac{1}{2m_3}\right) q_2^2 - \frac{1}{m_3} q_2 q_4\right\}$$

is the moment of inertia of the three bodies about the line in which the plane of the bodies meets the invariable plane through their centre of gravity.

It is also to be noted that this value of H differs from the value of H when k is zero by terms which do not involve the variables p_1, p_2, p_3, p_4: these terms in k can therefore be regarded as part of the potential energy, and we can say that the system differs from the corresponding system for which k is zero only by certain modifications in the potential energy. It may easily be shewn that when k is zero the motion takes place in a plane.

159. *Reduction to the 6th order.*

The equations of motion can now be reduced further from the 8th to the 6th order, by making use of the integral of energy

$$H = \text{Constant},$$

and eliminating the time. The theorem of § 141 shews that in performing this reduction the Hamiltonian form of the differential equations can be conserved. As the actual reduction is not required subsequently, it will not be given here in detail.

The Hamiltonian system of the 6th order thus obtained is, in the present state of our knowledge, the ultimate reduced form of the equations of motion of the general Problem of Three Bodies.

160. *Alternative reduction of the problem from the 18th to the 6th order.*

We shall now give another reduction* of the general problem of three bodies to a Hamiltonian system of the 6th order.

Let the original Hamiltonian system of equations of motion (§ 155) be transformed by the contact-transformation

$$q_r' = \frac{\partial W}{\partial p_r'}, \quad p_r = \frac{\partial W}{\partial q_r} \qquad (r = 1, 2, \ldots, 9),$$

where

$$\begin{aligned} W = p_1'(q_4 - q_1) + p_2'(q_5 - q_2) + p_3'(q_6 - q_3) \\ + p_4'\left(q_7 - \frac{m_1 q_1 + m_2 q_4}{m_1 + m_2}\right) + p_5'\left(q_8 - \frac{m_1 q_2 + m_2 q_5}{m_1 + m_2}\right) \\ + p_6'\left(q_9 - \frac{m_1 q_3 + m_2 q_6}{m_1 + m_2}\right) + p_7'(m_1 q_1 + m_2 q_4 + m_3 q_7) \\ + p_8'(m_1 q_2 + m_2 q_5 + m_3 q_8) + p_9'(m_1 q_3 + m_2 q_6 + m_3 q_9). \end{aligned}$$

The integrals of motion of the centre of gravity, when expressed in terms of the new variables, can be written

$$q_7' = q_8' = q_9' = p_7' = p_8' = p_9' = 0,$$

and consequently the transformed system is only of the 12th order: suppressing the accents in the new variables, it is

$$\frac{dq_r}{dt} = \frac{\partial H}{\partial p_r}, \quad \frac{dp_r}{dt} = -\frac{\partial H}{\partial q_r} \qquad (r = 1, 2, \ldots, 6),$$

where

$$\begin{aligned} H = \frac{1}{2\mu}(p_1^2 + p_2^2 + p_3^2) + \frac{1}{2\mu'}(p_4^2 + p_5^2 + p_6^2) - m_1 m_2 (q_1^2 + q_2^2 + q_3^2)^{-\frac{1}{2}} \\ - m_1 m_3 \left\{q_4^2 + q_5^2 + q_6^2 + \frac{2m_2}{m_1 + m_2}(q_1 q_4 + q_2 q_5 + q_3 q_6) \right. \\ \left. + \left(\frac{m_2}{m_1 + m_2}\right)^2 (q_1^2 + q_2^2 + q_3^2)\right\}^{-\frac{1}{2}} \\ - m_2 m_3 \left\{q_4^2 + q_5^2 + q_6^2 - \frac{2m_1}{m_1 + m_2}(q_1 q_4 + q_2 q_5 + q_3 q_6) \right. \\ \left. + \left(\frac{m_1}{m_1 + m_2}\right)^2 (q_1^2 + q_2^2 + q_3^2)\right\}^{-\frac{1}{2}}, \end{aligned}$$

and $\qquad \mu = \dfrac{m_1 m_2}{m_1 + m_2}, \qquad \mu' = \dfrac{m_3(m_1 + m_2)}{m_1 + m_2 + m_3}.$

The new variables may be interpreted physically in the following way: Let G be the centre of gravity of m_1 and m_2. Then (q_1, q_2, q_3) are the

* Due to Radau, *Annales de l'Éc. Norm. Sup.* v. (1868), p. 311.

projections of $m_1 m_2$ on the fixed axes, and (q_4, q_5, q_6) are the projections of Gm_3 on the axes. Further

$$\mu \frac{dq_r}{dt} = p_r \qquad (r = 1, 2, 3), \quad \text{and} \quad \mu' \frac{dq_r}{dt} = p_r \qquad (r = 4, 5, 6).$$

The new Hamiltonian system clearly represents the equations of motion of two particles, one of mass μ at a point whose coordinates are (q_1, q_2, q_3), and the other of mass μ' at a point whose coordinates are (q_4, q_5, q_6); these particles being supposed to move freely in space under the action of forces derivable from a potential energy represented by the terms in H which are independent of the p's. We have therefore replaced the Problem of Three Bodies by the problem of two bodies moving under this system of forces. This reduction, though substantially contained in Jacobi's* paper of 1843, was first explicitly stated by Bertrand† in 1852.

We shall suppose the axes so chosen that the plane of xy is the invariable plane for the motion of the particles μ and μ', i.e. so that the angular momentum of these particles about any line in the plane Oxy is zero.

Let the Hamiltonian system of the 12th order be transformed by the contact-transformation which is defined by the equations

$$q_r = \frac{\partial W}{\partial p_r}, \quad p_r' = \frac{\partial W}{\partial q_r'} \qquad (r = 1, 2, \ldots, 6),$$

where

$$W = (p_2 \sin q_5' + p_1 \cos q_5')\, q_1' \cos q_3' + q_1' \sin q_3' \{(p_2 \cos q_5' - p_1 \sin q_5')^2 + p_3^2\}^{\frac{1}{2}}$$
$$+ (p_5 \sin q_6' + p_4 \cos q_6')\, q_2' \cos q_4' + q_2' \sin q_4' \{(p_5 \cos q_6' - p_4 \sin q_6')^2 + p_6^2\}^{\frac{1}{2}}.$$

The new variables are easily seen to have the following physical interpretations: q_1' is the length of the radius vector from the origin to the particle μ, q_2' is the radius from the origin to μ', q_3' is the angle between q_1' and the intersection (or *node*) of the invariable plane with the plane through two consecutive positions of q_1' (which we shall call the *plane of instantaneous motion* of μ), q_4' is the angle between q_2' and the node of the invariable plane on the plane of instantaneous motion of μ', q_5' is the angle between Ox and the former of these nodes, q_6' is the angle between Ox and the latter of these nodes, p_1' is $\mu \dot{q}_1'$, p_2' is $\mu' \dot{q}_2'$, p_3' is the angular momentum of μ round the origin, p_4' is the angular momentum of μ' round the origin, p_5' is the angular momentum of μ round the normal at the origin to the invariable plane, and p_6' is the angular momentum of μ' round the same line.

The equations of motion in their new form are (§ 138)

$$\frac{dq_r'}{dt} = \frac{\partial H}{\partial p_r'}, \quad \frac{dp_r'}{dt} = -\frac{\partial H}{\partial q_r'} \qquad (r = 1, 2, \ldots, 6),$$

* *Journal für Math.* XXVI. p. 115. † *Journal de math.* XVII. p. 393.

where H is supposed expressed in terms of the new variables. Let this system be transformed by the contact-transformation

$$p_r'' = \frac{\partial W}{\partial q_r''}, \quad q_r' = \frac{\partial W}{\partial p_r'} \qquad (r = 1, 2, \ldots, 6),$$

where

$$W = q_5''(p_5' - p_6') + q_6''(p_5' + p_6') + q_1''p_1' + q_2''p_2' + q_3''p_3' + q_4''p_4'.$$

The equations of motion now become

$$\frac{dq_r''}{dt} = \frac{\partial H}{\partial p_r''}, \quad \frac{dp_r''}{dt} = -\frac{\partial H}{\partial q_r''} \qquad (r = 1, 2, \ldots, 6).$$

But H does not involve q_6'', as may be seen either by expressing H in terms of the new variables, or by observing that q_6'' depends on the arbitrarily chosen position of the axis Ox, while none of the other coordinates depend on this quantity. We have therefore

$$\dot{p}_6'' = -\partial H/\partial q_6'' = 0, \qquad \text{so } p_6'' = k,$$

where k is a constant; this is really one of the three integrals of angular momentum. Substituting k for p_6'' in H, the equation

$$\dot{q}_6'' = \partial H/\partial k$$

can be integrated by a quadrature when the rest of the equations have been solved: so the equations for p_6'' and q_6'' can be separated from the system, which reduces to the 10th order system

$$\frac{dq_r''}{dt} = \frac{\partial H}{\partial p_r''}, \quad \frac{dp_r''}{dt} = -\frac{\partial H}{\partial q_r''} \qquad (r = 1, 2, \ldots, 5).$$

We have still to use the two remaining integrals of angular momentum; these, when expressed in terms of the new variables, are readily found to be represented by

$$q_5'' = 90°, \qquad kp_5'' = p_3''^2 - p_4''^2;$$

no arbitrary constants of integration enter, owing to the fact that the plane of xy is the invariable plane.

The system may therefore be replaced by these two equations and the equations

$$\frac{dq_r''}{dt} = \frac{\partial H}{\partial p_r''}, \quad \frac{dp_r''}{dt} = -\frac{\partial H}{\partial q_r''} \qquad (r = 1, 2, 3, 4),$$

where, in this last set, q_5'' can be replaced by 90° before the derivates of H have been formed, and p_5'' is to be replaced by $(p_3''^2 - p_4''^2)/k$ after the derivates of H have been formed. Let H' denote the function derived from H by making this substitution for p_5'', and let s denote any one of the variables $q_1'', q_2'', q_3'', q_4'', p_1'', p_2'', p_3'', p_4''$; then we have

$$\frac{\partial H'}{\partial s} = \frac{\partial H}{\partial s} + \frac{\partial H}{\partial p_5''}\frac{\partial p_5''}{\partial s} = \frac{\partial H}{\partial s} + \dot{q}_5''\frac{\partial p_5''}{\partial s} = \frac{\partial H}{\partial s},$$

and it is therefore allowable to substitute for p_5'' in H *before* the derivates of H have been formed. The equations of motion are thus reduced to a system of the 8th order, which (suppressing the accents) may be written in the form

$$\frac{dq_r}{dt} = \frac{\partial H}{\partial p_r}, \quad \frac{dp_r}{dt} = -\frac{\partial H}{\partial q_r} \qquad (r = 1, 2, 3, 4),$$

where, effecting in H the transformations which have been indicated, we have

$$H = \frac{1}{2\mu}\left(p_1^2 + \frac{p_3^2}{q_1^2}\right) + \frac{1}{2\mu'}\left(p_2^2 + \frac{p_4^2}{q_2^2}\right) - m_1 m_2 q_1^{-1}$$

$$- m_1 m_3 \left\{q_2^2 - \frac{2m_2 q_1 q_2}{m_1 + m_2}\left(\cos q_3 \cos q_4 - \frac{k^2 - p_3^2 - p_4^2}{2p_3 p_4}\sin q_3 \sin q_4\right) + \frac{m_2^2}{(m_1 + m_2)^2} q_1^2\right\}^{-\frac{1}{2}}$$

$$- m_2 m_3 \left\{q_2^2 + \frac{2m_1 q_1 q_2}{m_1 + m_2}\left(\cos q_3 \cos q_4 - \frac{k^2 - p_3^2 - p_4^2}{2p_3 p_4}\sin q_3 \sin q_4\right) + \frac{m_1^2}{(m_1 + m_2)^2} q_1^2\right\}^{-\frac{1}{2}}.$$

The equations of motion may further be reduced to a system of the 6th order by the method of § 141, using the integral of energy

$$H = \text{Constant}$$

and eliminating the time. As the reduction is not required subsequently, it will not be given in detail here.

161. *The problem of three bodies in a plane.*

The motion of the three particles may be supposed to take place in a plane, instead of in three-dimensional space; this will obviously happen if the directions of the initial velocities of the bodies are in the plane of the bodies.

This case is known as the *problem of three bodies in a plane*: we shall now proceed to reduce the equations of motion to a Hamiltonian system of the lowest possible order.

Let (q_1, q_2) be the coordinates of m_1, (q_3, q_4) the coordinates of m_2, and (q_5, q_6) the coordinates of m_3, referred to any fixed axes Ox, Oy in the plane of the motion; and let $p_r = m_k \dot{q}_r$, where k denotes the greatest integer in $\frac{1}{2}(r+1)$. The equations of motion are (as in § 155)

$$\frac{dq_r}{dt} = \frac{\partial H}{\partial p_r}, \quad \frac{dp_r}{dt} = -\frac{\partial H}{\partial q_r} \qquad (r = 1, 2, \ldots, 6),$$

where

$$H = \frac{1}{2m_1}(p_1^2 + p_2^2) + \frac{1}{2m_2}(p_3^2 + p_4^2) + \frac{1}{2m_3}(p_5^2 + p_6^2) - m_2 m_3\{(q_3 - q_5)^2 + (q_4 - q_6)^2\}^{-\frac{1}{2}}$$

$$- m_3 m_1\{(q_5 - q_1)^2 + (q_6 - q_2)^2\}^{-\frac{1}{2}} - m_1 m_2\{(q_1 - q_3)^2 + (q_2 - q_4)^2\}^{-\frac{1}{2}}.$$

These equations will now be reduced from the 12th to the 8th order, by using the four integrals of motion of the centre of gravity. Perform on the variables the contact-transformation defined by the equations

$$q_r = \frac{\partial W}{\partial p_r}, \quad p_r' = \frac{\partial W}{\partial q_r'} \qquad (r = 1, 2, \ldots, 6),$$

where

$$W = p_1 q_1' + p_2 q_2' + p_3 q_3' + p_4 q_4' + (p_1 + p_3 + p_5) q_5' + (p_2 + p_4 + p_6) q_6'.$$

It is easily seen that (q_1', q_2') are the coordinates of m_1 relative to axes through m_3 parallel to the fixed axes, (q_3', q_4') are the coordinates of m_2 relative to the same axes, (q_5', q_6') are the coordinates of m_3 relative to the original axes, (p_1', p_2') are the components of momentum of m_1, (p_3', p_4') are the components of momentum of m_2, and (p_5', p_6') are the components of momentum of the system.

As in § 157, the equations for q_5', q_6', p_5', p_6' disappear from the system; and (suppressing the accents in the new variables) the equations of motion reduce to the system of the 8th order,

$$\frac{dq_r}{dt} = \frac{\partial H}{\partial p_r}, \quad \frac{dp_r}{dt} = -\frac{\partial H}{\partial q_r} \qquad (r = 1, 2, 3, 4),$$

where

$$H = \left(\frac{1}{2m_1} + \frac{1}{2m_3}\right)(p_1^2 + p_2^2) + \left(\frac{1}{2m_2} + \frac{1}{2m_3}\right)(p_3^2 + p_4^2) + \frac{1}{m_3}(p_1 p_3 + p_2 p_4)$$
$$- m_2 m_3 (q_3^2 + q_4^2)^{-\frac{1}{2}} - m_3 m_1 (q_1^2 + q_2^2)^{-\frac{1}{2}} + m_1 m_2 \{(q_1 - q_3)^2 + (q_2 - q_4)^2\}^{-\frac{1}{2}}.$$

Next, we shall shew that this system possesses an ignorable coordinate, which will make possible a further reduction through two units.

Perform on the system the contact-transformation defined by the equations

$$q_r = \frac{\partial W}{\partial p_r}, \quad p_r' = \frac{\partial W}{\partial q_r'} \qquad (r = 1, 2, 3, 4),$$

where

$$W = p_1 q_1' \cos q_4' + p_2 q_1' \sin q_4' + p_3 (q_2' \cos q_4' - q_3' \sin q_4') + p_4 (q_2' \sin q_4' + q_3' \cos q_4').$$

The physical interpretation of this transformation is as follows: q_1' is the distance $m_1 m_3$; q_2' and q_3' are the projections of $m_2 m_3$ on, and perpendicular to, $m_1 m_3$; q_4' is the angle between $m_3 m_1$ and the axis of x; p_1' is the component of momentum of m_1 along $m_3 m_1$; p_2' and p_3' are the components of momentum of m_2 parallel and perpendicular to $m_3 m_1$; and p_4' is the angular momentum of the system.

The differential equations, when expressed in terms of the new variables, become

$$\frac{dq_r'}{dt} = \frac{\partial H}{\partial p_r'}, \quad \frac{dp_r'}{dt} = -\frac{\partial H}{\partial q_r'} \qquad (r = 1, 2, 3, 4),$$

where

$$H = \left(\frac{1}{2m_1} + \frac{1}{2m_3}\right)\left\{p_1'^2 + \frac{1}{q_1'^2}(p_3' q_2' - p_2' q_3' - p_4')^2\right\} + \left(\frac{1}{2m_2} + \frac{1}{2m_3}\right)(p_2'^2 + p_3'^2)$$
$$+ \frac{1}{m_3}\left\{p_1' p_2' - \frac{p_3'}{q_1'}(p_3' q_2' - p_2' q_3' - p_4')\right\} - m_2 m_3 (q_2'^2 + q_3'^2)^{-\frac{1}{2}}$$
$$- m_3 m_1 q_1'^{-1} - m_1 m_2 \{(q_1' - q_2')^2 + q_3'^2\}^{-\frac{1}{2}}.$$

Since q_4' is not contained in H, it is an ignorable coordinate; the corresponding integral is $p_4' = k$, where k is a constant; this can be interpreted as the integral of angular momentum of the system. The equation $\dot{q}_4' = \partial H/\partial p_4'$ can be integrated by a quadrature when the rest of the equations have been integrated; and thus the equations for p_4' and q_4' disappear from the system.

Suppressing the accents on the new variables, the equations can therefore be written

$$\frac{dq_r}{dt} = \frac{\partial H}{\partial p_r}, \quad \frac{dp_r}{dt} = -\frac{\partial H}{\partial q_r} \quad (r = 1, 2, 3),$$

where

$$H = \left(\frac{1}{2m_1} + \frac{1}{2m_3}\right)\left\{p_1^2 + \frac{1}{q_1^2}(p_3 q_2 - p_2 q_3 - k)^2\right\} + \left(\frac{1}{2m_2} + \frac{1}{2m_3}\right)(p_2^2 + p_3^2)$$
$$+ \frac{1}{m_3}\left\{p_1 p_2 - \frac{p_3}{q_1}(p_3 q_2 - p_2 q_3 - k)\right\} - m_2 m_3 (q_2^2 + q_3^2)^{-\frac{1}{2}}$$
$$- m_3 m_1 q_1^{-1} - m_1 m_2 \{(q_1 - q_2)^2 + q_3^2\}^{-\frac{1}{2}}.$$

This is a system of the 6th order; it can be reduced to the 4th order by the process of § 141, making use of the integral of energy and eliminating the time.

162. *The restricted problem of three bodies.*

Another special case of the problem of three bodies, which has occupied a prominent place in recent researches, is the *restricted problem of three bodies*; this may be enunciated as follows:

Two bodies S and J revolve round their centre of gravity, O, in circular orbits, under the influence of their mutual attraction. A third body P, without mass (i.e. such that it is attracted by S and J, but does not influence their motion), moves in the same plane as S and J; the restricted problem of three bodies is to determine the motion of the body P, which is generally called the *planetoid*.

Let m_1 and m_2 be the masses of S and J, and write

$$F = \frac{m_1}{SP} + \frac{m_2}{JP}.$$

Take any fixed rectangular axes OX, OY, through O, in the plane of the motion; let (X, Y) be the coordinates, and (U, V) the components of velocity, of P. The equations of motion are

$$\frac{d^2X}{dt^2} = \frac{\partial F}{\partial X}, \quad \frac{d^2Y}{dt^2} = \frac{\partial F}{\partial Y},$$

or in the Hamiltonian form,

$$\frac{dX}{dt} = \frac{\partial H}{\partial U}, \quad \frac{dY}{dt} = \frac{\partial H}{\partial V}, \quad \frac{dU}{dt} = -\frac{\partial H}{\partial X}, \quad \frac{dV}{dt} = -\frac{\partial H}{\partial Y},$$

where
$$H = \tfrac{1}{2}(U^2 + V^2) - F.$$

Since F is a function not only of X and Y but also of t, the equation $H =$ Constant is *not* an integral of the system.

Perform on the variables the contact-transformation which is defined by the equations

$$X = \frac{\partial W}{\partial U}, \qquad Y = \frac{\partial W}{\partial V}, \qquad u = \frac{\partial W}{\partial x}, \qquad v = \frac{\partial W}{\partial y},$$

where $$W = U(x \cos nt - y \sin nt) + V(x \sin nt + y \cos nt),$$

and n is the angular velocity of SJ. The equations become

$$\frac{dx}{dt} = \frac{\partial K}{\partial u}, \qquad \frac{dy}{dt} = \frac{\partial K}{\partial v}, \qquad \frac{du}{dt} = -\frac{\partial K}{\partial x}, \qquad \frac{dv}{dt} = -\frac{\partial K}{\partial y},$$

where (§ 138) $$K = H - \frac{\partial W}{\partial t}$$

$$= \tfrac{1}{2}(u^2 + v^2) + n(uy - vx) - F;$$

it is at once seen that x and y are the coordinates of the planetoid referred to the moving line OJ as axis of x, and a line perpendicular to this through O as axis of y. F is now a function of x and y only, so K does not involve t explicitly, and

$$K = \text{Constant}$$

is an integral of the system; it is called the *Jacobian integral** of the restricted problem of three bodies.

Another form of the equations of motion is obtained by applying to the last system the contact-transformation

$$x = \frac{\partial W}{\partial u}, \qquad y = \frac{\partial W}{\partial v}, \qquad p_1 = \frac{\partial W}{\partial q_1}, \qquad p_2 = \frac{\partial W}{\partial q_2},$$

where $$W = q_1(u \cos q_2 + v \sin q_2).$$

The new variables may be defined directly by the equations

$$q_1 = OP, \qquad q_2 = POJ, \qquad p_1 = \frac{d}{dt}(OP), \qquad p_2 = OP^2 \frac{d}{dt}(POX),$$

and the equations of motion become

$$\frac{dq_r}{dt} = \frac{\partial H}{\partial p_r}, \qquad \frac{dp_r}{dt} = -\frac{\partial H}{\partial q_r} \qquad (r = 1, 2),$$

where $$H = \tfrac{1}{2}\left(p_1^2 + \frac{p_2^2}{q_1^2}\right) - np_2 - F.$$

Another form† is obtained by applying to these equations the contact-transformation

$$p_r = \frac{\partial W}{\partial q_r}, \qquad q_r' = \frac{\partial W}{\partial p_r'} \qquad (r = 1, 2),$$

where $$W = p_2' q_2 + \int_{p_1'\{p_1' - (p_1'^2 - p_2'^2)^{\frac{1}{2}}\}}^{q_1} \left\{ -\frac{p_2'^2}{u^2} + \frac{2}{u} - \frac{1}{p_1'^2} \right\}^{\frac{1}{2}} du,$$

* Jacobi, *Comptes Rendus*, III. (1836), p. 59.

† Adopted by Poincaré in his *Nouvelles méthodes de la Méc. Céleste.*

where u denotes a current variable of integration. These equations may be written

$$p_1 = \left(-\frac{p_2'^2}{q_1^2} + \frac{2}{q_1} - \frac{1}{p_1'^2}\right)^{\frac{1}{2}}, \qquad p_2 = p_2',$$

$$q_1' = \arccos\left\{\frac{1 - \dfrac{q_1}{p_1'^2}}{\left(1 - \dfrac{p_2'^2}{p_1'^2}\right)^{\frac{1}{2}}}\right\} - \left(-\frac{p_2'^2}{p_1'^2} + \frac{2q_1}{p_1'^2} - \frac{q_1^2}{p_1'^4}\right)^{\frac{1}{2}}, \quad q_2' = q_2 - \arccos\left\{\frac{\dfrac{p_2'^2}{q_1} - 1}{\left(1 - \dfrac{p_2'^2}{p_1'^2}\right)^{\frac{1}{2}}}\right\},$$

and it is easily seen that q_1' is the mean anomaly of the planetoid in the ellipse which it would describe about a fixed body of unit mass at O, if projected from its instantaneous position with its instantaneous velocity; q_2' is the longitude of the apse of this ellipse, measured from OJ; p_1' is $a^{\frac{1}{2}}$, and p_2' is $\{a(1-e^2)\}^{\frac{1}{2}}$, where a is the semi-major axis and e is the eccentricity of this ellipse. H does not involve t explicitly, so $H = \text{Constant}$ is an integral of the equations of motion, which are now

$$\frac{dq_r'}{dt} = \frac{\partial H}{\partial p_r'}, \qquad \frac{dp_r'}{dt} = -\frac{\partial H}{\partial q_r'} \qquad (r = 1, 2).$$

If we take the sum of the masses of S and J to be the unit of mass, and denote these masses by $1 - \mu$ and μ respectively, we have

$$H = \tfrac{1}{2}\left(\dot{p}_1^2 + \frac{p_2^2}{q_1^2}\right) - np_2 - \frac{1-\mu}{SP} - \frac{\mu}{JP}.$$

This is an analytic function of $p_1', p_2', q_1', q_2', \mu$, which is periodic in q_1' and q_2', with the period 2π. Moreover, to find the term independent of μ in H, we suppose μ to be zero; since SP now becomes q_1, we have

$$H = \tfrac{1}{2}\left(\frac{2}{q_1} - \frac{1}{p_1'^2}\right) - np_2' - \frac{1}{q_1} = -\frac{1}{2p_1'^2} - np_2'.$$

Thus finally, discarding the accents, *the equations of motion of the restricted problem of three bodies may be taken in the form*

$$\frac{dq_r}{dt} = \frac{\partial H}{\partial p_r}, \qquad \frac{dp_r}{dt} = -\frac{\partial H}{\partial q_r} \qquad (r = 1, 2),$$

where H can be expanded as a power-series in μ in the form

$$H = H_0 + \mu H_1 + \mu^2 H_2 + \ldots,$$

and

$$H_0 = -\frac{1}{2p_1^2} - np_2,$$

while $H_1, H_2, \ldots$ are periodic in q_1 and q_2, with the period 2π.

The equations of this 4th order system may be reduced to a Hamiltonian system of the second order by use of the integral $H = \text{Constant}$ and elimination of the time, as in § 141.

163. *Extension to the problem of n bodies.*

Many of the transformations which have been used in the present chapter in the reduction of the problem of three bodies can be extended so as to apply to the general problem of n bodies which attract each other according to the Newtonian law. In their original form, the equations of motion of the n bodies constitute a system of the $6n$th order; this can be reduced to the $(6n-12)$th order, by using the six integrals of motion of the centre of gravity, the three integrals of angular momentum, the integral of energy, the elimination of the time, and the elimination of the nodes.

The reduction has been performed by T. L. Bennett, *Mess. of Math.* (2) XXXIV. (1904), p. 113.

Miscellaneous Examples.

1. If in the problem of three bodies the units are so chosen that the energy integral is

$$\tfrac{1}{2}(v_1^2+v_2^2+v_3^2)=\frac{1}{r_{23}}+\frac{1}{r_{31}}+\frac{1}{r_{12}}-\frac{1}{r},$$

where r_{12} is the distance between the bodies whose velocities are v_1 and v_2, and if r is a positive constant, shew that the greatest possible value of the angular momentum of the system about its centre of gravity is $\frac{3}{2}\sqrt{2r}$.

(Camb. Math. Tripos, Part I, 1893.)

2. In the problem of three bodies, let Φ be Jacobi's function, let Ω be the angle between any fixed line in the invariable plane and the node of the plane of the three bodies on the invariable plane, let i be the inclination of the plane of the three bodies to the invariable plane, and let η be the area of the triangle formed by the three bodies. Shew that

$$\frac{d\Omega}{dt}=\frac{k}{\Phi},$$

$$\frac{1}{\sin i}\frac{di}{dt}=k\left\{\frac{M}{m_1m_2m_3\eta^2}-\frac{1}{\Phi^2}\right\}^{\frac{1}{2}},$$

where k is the angular momentum of the system round the normal to the invariable plane.

(De Gasparis.)

3. Let the problem of three bodies be replaced by the problem of two bodies μ and μ' as in § 160: let q_1 and q_2 be the distances of μ and μ' from the origin: let q_3 and q_4 be the angles made by q_1 and q_2 respectively with the intersection of the plane through the bodies and the invariable plane: let p_1 and p_2 denote $\mu\dot{q}_1$ and $\mu'\dot{q}_2$ respectively; and let p_3 and p_4 be the components of angular momentum of μ and μ' respectively, in the plane through the bodies and the origin. Shew that the equations of motion may be written

$$\frac{dq_r}{dt}=\frac{\partial H}{\partial p_r},\quad \frac{dp_r}{dt}=-\frac{\partial H}{\partial q_r}\qquad (r=1,2,3,4),$$

where $H=$ Constant is the integral of energy.

(Bour.)

4. Apply the contact-transformation defined by the equations

$$q_1'=\{(q_4-q_7)^2+(q_5-q_8)^2+(q_6-q_9)^2\}^{\frac{1}{2}},$$
$$q_2'=\{(q_7-q_1)^2+(q_8-q_2)^2+(q_9-q_3)^2\}^{\frac{1}{2}},$$
$$q_3'=\{(q_1-q_4)^2+(q_2-q_5)^2+(q_3-q_6)^2\}^{\frac{1}{2}},$$
$$q_4'=b_1(q_1+iq_2)+b_2(q_4+iq_5)+b_3(q_7+iq_8),$$
$$q_5'=c_1q_3+c_2q_6+c_3q_9,$$
$$q_6'=m_1q_1+m_2q_4+m_3q_7,$$
$$q_7'=m_1q_2+m_2q_5+m_3q_8,$$
$$q_8'=m_1q_3+m_2q_6+m_3q_9,$$
$$q_0'=\frac{a_1(q_1+iq_2)+a_2(q_4+iq_5)+a_3(q_7+iq_8)}{b_1(q_1+iq_2)+b_2(q_4+iq_5)+b_3(q_7+iq_8)},$$
$$p_r=\sum_{k=0}^{8}p_k'\frac{\partial q_k'}{\partial q_r} \qquad (r=0,\ 1,\ 2,\ \ldots,\ 8),$$

(where i stands for $\sqrt{-1}$ and a_1, a_2, a_3, b_1, b_2, b_3, c_1, c_2, c_3 are any nine constants which satisfy the equations

$$a_1+a_2+a_3=0, \quad b_1+b_2+b_3=0, \quad c_1+c_2+c_3=0, \quad a_2b_3-a_3b_2=1),$$

to the Hamiltonian system of the 18th order which (§ 155) determines the motion of the three bodies.

Shew that the integrals of motion of the centre of gravity are

$$q_6'=q_7'=q_8'=p_6'=p_7'=p_8'=0.$$

Shew further that when the invariable plane is taken as plane of xy, the variable p_5' is zero, and that the integral of angular momentum round the normal to the invariable plane is

$$p_4'q_4'=k, \qquad \text{where } k \text{ is a constant.}$$

Hence shew that the equations reduce to the 8th order system

$$\frac{dq_r'}{dt}=\frac{\partial H}{\partial p_r'}, \quad \frac{dp_r'}{dt}=-\frac{\partial H}{\partial q_r'} \qquad (r=0,\ 1,\ 2,\ 3),$$

where

$$H=\Sigma p_1'^2q_1'^2\cdot\frac{m_2+m_3}{2m_2m_3}+\Sigma\frac{p_2'p_3'}{q_2'q_3'}\cdot\frac{q_2'^2+q_3'^2-q_1'^2}{2m_1}$$
$$+\Sigma\frac{1}{m_1}\{p_0'(a_1-b_1q_0')+kb_1\}\left\{\frac{p_3'}{q_3'}(a_3-b_3q_0')-\frac{p_2'}{q_2'}(a_2-b_2q_0')\right\}-\Sigma\frac{m_2m_3}{q_1'}.$$

Reduce this to a system of the 6th order, by the theorem of § 141. (Bruns.)

CHAPTER XIV

THE THEOREMS OF BRUNS AND POINCARÉ

164. *Bruns' theorem.*

(i) *Statement of the theorem.*

We have seen (§ 155) that the problem of three bodies possesses 10 known integrals: namely the six integrals of motion of the centre of gravity, the three integrals of angular momentum, and the integral of energy; these are generally called the *classical* integrals of the problem. Each of them is an *algebraic* integral, i.e. is of the form

$$f(q_1, q_2, \ldots, q_9, p_1, p_2, \ldots, p_9, t) = \text{Constant},$$

where f is an algebraic function of the coordinates $(q_1, q_2, \ldots, q_9, p_1, \ldots, p_9)$ and of t.

Efforts have frequently been made to obtain other algebraic integrals of the problem of three bodies independent of these (i.e. not formed by combinations of them), but without success; and in 1887 Bruns* shewed that no such new algebraic integrals exist; in other words, *the classical integrals are the only independent algebraic integrals of the problem of three bodies.*

It may be remarked† that the non-existence of algebraic integrals does not necessarily imply great complexity in a system. One of the simplest of differential equations, namely the linear differential equation with constant coefficients

$$\ddot{x} - (\mu_1 + \mu_2)\dot{x} + \mu_1\mu_2 x = 0,$$

has no algebraic integral except when μ_1/μ_2 is a rational number, in which case the first integral

$$\frac{(\dot{x} - \mu_2 x)^{\mu_2}}{(\dot{x} - \mu_1 x)^{\mu_1}} = \text{Constant}$$

can be transformed into an algebraic integral.

(ii) *Expression of an integral in terms of the essential coordinates of the problem.*

We shall now proceed to a proof of Bruns' result, considering first those integrals which do not involve t explicitly.

* *Berichte der Kgl. Sächs. Ges. der Wiss.* 1887, pp. 1, 55; *Acta Math.* XI. p. 25. Cf. also Forsyth, *Theory of Differential Equations*, Vol. III. (1900), Ch. XVII.

† Cf. K. Bohlin, *Astron. Iakttagelser och Unders. å Stockholms Observ.* IX. (1908), Nr 1.

The equations of motion of the problem may (§ 160) be written in the form

$$\frac{dq_r}{dt} = \frac{\partial H}{\partial p_r}, \qquad \frac{dp_r}{dt} = -\frac{\partial H}{\partial q_r} \qquad (r = 1, 2, \ldots, 6) \ldots (1),$$

where

$$H = T - U,$$

$$T = \frac{1}{2\mu}(p_1^2 + p_2^2 + p_3^2) + \frac{1}{2\mu'}(p_4^2 + p_5^2 + p_6^2),$$

$$U = m_1 m_2 (q_1^2 + q_2^2 + q_3^2)^{-\frac{1}{2}}$$

$$+ m_1 m_3 \left\{q_4^2 + q_5^2 + q_6^2 + \frac{2m_2}{m_1 + m_2}(q_1 q_4 + q_2 q_5 + q_3 q_6) + \left(\frac{m_2}{m_1 + m_2}\right)^2 (q_1^2 + q_2^2 + q_3^2)\right\}^{-\frac{1}{2}}$$

$$+ m_2 m_3 \left\{q_4^2 + q_5^2 + q_6^2 - \frac{2m_1}{m_1 + m_2}(q_1 q_4 + q_2 q_5 + q_3 q_6) + \left(\frac{m_1}{m_1 + m_2}\right)^2 (q_1^2 + q_2^2 + q_3^2)\right\}^{-\frac{1}{2}},$$

$$\mu = \frac{m_1 m_2}{m_1 + m_2}, \qquad \mu' = \frac{m_3 (m_1 + m_2)}{m_1 + m_2 + m_3}.$$

We shall write $\mu_1 = \mu_2 = \mu_3 = \mu, \qquad \mu_4 = \mu_5 = \mu_6 = \mu'$,

so that $$T = \sum_{r=1}^{6} \frac{p_r^2}{2\mu_r}.$$

Let the coordinates of the three bodies be (q_1', q_2', q_3'), (q_4', q_5', q_6'), (q_7', q_8', q_9'), and let $m_k \dot{q}_r' = p_r'$, where k denotes the greatest integer contained in $\frac{1}{3}(r+2)$: the integrals whose existence we propose to discuss are of the form

$$\phi(q_1', q_2', \ldots, q_9', p_1', \ldots, p_9') = a,$$

where a is an arbitrary constant and ϕ is an algebraic function of its arguments. The formulae of § 160 enable us to express the variables $(q_1', q_2', \ldots, q_9', p_1', \ldots, p_9')$ as linear functions of $(q_1, q_2, \ldots, q_6, p_1, \ldots, p_6)$: we shall therefore, on making these substitutions in the integral, obtain an equation

$$f(q_1, q_2, \ldots, q_6, p_1, \ldots, p_6) = a \quad \ldots\ldots\ldots\ldots\ldots\ldots(2).$$

If the integral ϕ is compounded of the integrals of motion of the centre of gravity, f will evidently reduce to a constant; if not, f will be an algebraic function of the variables $(q_1, \ldots, q_6, p_1, \ldots, p_6)$. We have to enquire into the existence of integrals, such as (2), of the equations (1).

(iii) *An integral must involve the momenta.*

We shall first shew that an integral such as (2) must involve some of the quantities p, i.e. it cannot be a function of $(q_1, q_2, \ldots, q_6)$ only.

For suppose, if possible, that the integral, say

$$f(q_1, q_2, \ldots, q_6) = a,$$

does not involve $(p_1, p_2, \ldots, p_6)$. Differentiating with respect to t, we have

$$0 = \sum_{r=1}^{6} \frac{\partial f}{\partial q_r} \dot{q}_r = \sum_{r=1}^{6} \frac{\partial f}{\partial q_r} \frac{p_r}{\mu_r},$$

and therefore the equations

$$\frac{\partial f}{\partial q_r} = 0 \qquad (r = 1, 2, \dots, 6)$$

must be satisfied identically; that is, f does not involve $(q_1, q_2, \dots, q_6)$, and so is a mere constant.

(iv) *Only one irrationality can occur in the integral.*

As the mutual distances of the bodies are irrational functions of $(q_1, q_2, \dots, q_6)$, the function U will be an irrational function of these variables. Denoting by s the sum of the three mutual distances, it is easily seen that the mutual distances can be expressed as rational functions of the seven quantities $(q_1, q_2, \dots, q_6, s)$; in other words, the irrationalities involved in the mutual distances are all capable of being expressed by means of the irrationality of s; we may therefore suppose that U is expressed as a rational function of $(q_1, q_2, \dots, q_6, s)$.

Now the function f is algebraic, but not necessarily rational, in the variables $(q_1, \dots, q_6, p_1, \dots, p_6)$; let the equation (2) be rationalised, and let the resulting equation be arranged in powers of a, so that it becomes

$$a^m + a^{m-1}\phi_1(q_1, q_2, \dots, q_6, p_1, \dots, p_6) + a^{m-2}\phi_2(q_1, \dots, q_6, p_1, \dots, p_6) + \dots$$
$$+ \phi_m(q_1, \dots, q_6, p_1, \dots, p_6) = 0 \quad \dots(3),$$

where $\phi_1, \phi_2, \dots, \phi_m$ are rational functions of $(q_1, \dots, q_6, p_1, \dots, p_6)$. If this equation is reducible in the variables $(q_1, \dots, q_6, p_1, \dots, p_6, s)$, i.e. if it can be decomposed into other equations, each of the form

$$a^l + a^{l-1}\psi_1(q_1, \dots, q_6, p_1, \dots, p_6, s) + \dots + \psi_l(q_1, \dots, q_6, p_1, \dots, p_6, s) = 0 \quad \dots(4),$$

where $\psi_1, \psi_2, \dots, \psi$ are rational functions of $(q_1, \dots, q_6, p_1, \dots, p_6, s)$, then one of these last equations will give the value of a which corresponds to equation (2), and we shall consider this equation instead of (3). As the type of equation represented by (4) includes the type represented by (3) as a particular case, we shall suppose a to be given by an equation of the form (4), irreducible in $(q_1, \dots, q_6, p_1, \dots, p_6, s)$.

Differentiating with respect to t, and using equations (1), we have

$$a^{l-1}(\psi_1, H) + a^{l-2}(\psi_2, H) + \dots + (\psi_l, H) = 0 \quad \dots\dots\dots(5),$$

where (ψ_r, H) denotes as usual the Poisson-bracket of ψ_r and H.

We shall first suppose that the expressions (ψ_r, H), which are rational functions of $(q_1, \dots, q_6, p_1, \dots, p_6, s)$, are not all zero. Then equations (4) and (5) have one or more common roots a, and consequently equation (4) is reducible in $(q_1, q_2, \dots, q_6, p_1, \dots, p_6, s)$; but this equation is irreducible, and therefore this hypothesis is inadmissible, and the quantities (ψ_r, H) are all zero. This implies that all the coefficients $(\psi_1, \psi_2, \dots, \psi_l)$ in equation (4) are integrals of the equations (1): and hence *the integral f can be compounded algebraically from other integrals, which are rational functions of* $(q_1, \dots, q_6, p_1, \dots, p_6, s)$.

(v) *Expression of the integral as a quotient of two real polynomials.*

We need therefore henceforth only consider integrals of the type

$$f(q_1, q_2, \dots, q_6, p_1, \dots, p_6, s) = a \quad \dots\dots\dots\dots\dots\dots(6),$$

where f is a rational function of the arguments indicated. The form of f can be further restricted by the following observation. If in the equations of motion we replace q_r, p_r, t by $q_r k^2, p_r k^{-1}$, and tk^3, respectively, where k is any constant, the equations are unaltered. If therefore these substitutions are made in equation (6), this equation must still be an integral of the system, whatever k may be.

Now f is a rational function of its arguments: it can therefore be expressed as the quotient of two functions, each of which is a polynomial in $(q_1, q_2, \dots, q_6, p_1, \dots, p_6, s)$. When in these polynomials we replace q_r, p_r, s by $q_r k^2, p_r k^{-1}, sk^2$, respectively, the function f will (on multiplying its numerator and denominator by an appropriate power of k) take the form

$$f = \frac{A_0 k^p + A_1 k^{p-1} + \dots + A_p}{B_0 k^q + B_1 k^{q-1} + \dots + B_q},$$

where $(A_0, A_1, \dots, B_q)$ are polynomials in $(q_1, \dots, q_6, p_1, \dots, p_6, s)$. Since df/dt is zero, we have

$$(B_0 k^q + B_1 k^{q-1} + \dots + B_q)\left(\frac{dA_0}{dt} k^p + \dots + \frac{dA_p}{dt}\right)$$

$$- (A_0 k^p + A_1 k^{p-1} + \dots + A_p)\left(\frac{dB_0}{dt} k^q + \dots + \frac{dB_q}{dt}\right) = 0.$$

Now k is arbitrary, so the coefficients of successive powers of k in this equation must be zero; and therefore

$$\begin{cases} 0 = B_0 \dfrac{dA_0}{dt} - A_0 \dfrac{dB_0}{dt}, \\[2ex] 0 = B_1 \dfrac{dA_0}{dt} + B_0 \dfrac{dA_1}{dt} - A_1 \dfrac{dB_0}{dt} - A_0 \dfrac{dB_1}{dt}, \\ \dots\dots\dots\dots\dots\dots \\ \dots\dots\dots\dots\dots\dots \\ 0 = B_q \dfrac{dA_p}{dt} - A_p \dfrac{dB_q}{dt}. \end{cases}$$

These $(q + p + 1)$ equations are equivalent to the system

$$\frac{1}{A_0}\frac{dA_0}{dt} = \frac{1}{A_1}\frac{dA_1}{dt} = \dots = \frac{1}{A_p}\frac{dA_p}{dt} = \frac{1}{B_0}\frac{dB_0}{dt} = \dots = \frac{1}{B_q}\frac{dB_q}{dt},$$

from which it is evident that each of the quantities

$$\frac{A_1}{A_0}, \frac{A_2}{A_0}, \dots, \frac{A_p}{A_0}, \frac{B_0}{A_0}, \dots\dots, \frac{B_q}{A_0}$$

is an integral: and thus we have the result that *any integral such as f can be compounded from other integrals, which are of the form*

$$\frac{G_2(q_1, \ldots, q_6, p_1, \ldots, p_6, s)}{G_1(q_1, \ldots, q_6, p_1, \ldots, p_6, s)} = \text{Constant},$$

where each of the functions G_1, G_2 is a polynomial in its arguments, and is merely multiplied by a power of k when the variables q_r, p_r, s are replaced respectively by $q_r k^2$, $p_r k^{-1}$, sk^2. We need therefore only consider integrals of this form.

It may further be observed that the functions G_1 and G_2 may, without loss of generality, be taken to be free from imaginaries. For if P and iQ denote the real and imaginary parts of an integral

$$P + iQ = \text{Constant},$$

we have $$\frac{dP}{dt} + i\frac{dQ}{dt} = 0, \text{ identically.}$$

Since the differential equations are free from imaginaries, it follows that dP/dt and dQ/dt are free from imaginaries: and so dP/dt and dQ/dt must be zero separately. Hence P and Q are themselves integrals, and every complex integral can be compounded from real integrals. We shall therefore henceforth assume that G_1/G_2 is free from imaginaries.

(vi) *Derivation of integrals from the numerator and denominator of the quotient.*

It may be the case that the function G_1 is resoluble into a product of irresoluble polynomials in $(p_1, p_2, \ldots, p_6)$, the coefficients in these polynomials being rational functions of $(q_1, q_2, \ldots, q_6, s)$. Let ψ be such a polynomial, and suppose that it is repeated λ times in G_1: and let χ denote the remaining factors of G_1, so that

$$G_1 = \psi^\lambda \chi.$$

When G_1 is irreducible, we shall of course have $G_1 = \psi$, and $\chi = 1$.

The equation $$\frac{d}{dt}\left(\frac{G_1}{G_2}\right) = 0$$

gives $$\frac{\lambda}{\psi}\frac{d\psi}{dt} + \frac{1}{\chi}\frac{d\chi}{dt} - \frac{1}{G_2}\frac{dG_2}{dt} = 0,$$

or $$\frac{d\psi}{dt} = \psi\left(\frac{1}{\lambda G_2}\frac{dG_2}{dt} - \frac{1}{\lambda\chi}\frac{d\chi}{dt}\right).$$

Now $d\psi/dt$ is polynomial in $(p_1, \ldots, p_6)$, and ψ is also polynomial in $(p_1, \ldots, p_6)$, of order less by unity than the order of $d\psi/dt$. Also, ψ has no factor in common with G_2 or χ. Hence we see that

$$\frac{1}{\lambda G_2}\frac{dG_2}{dt} - \frac{1}{\lambda\chi}\frac{d\chi}{dt}$$

must be a polynomial in $(p_1, \ldots, p_6)$, of order unity: denote this polynomial by ω: then we have

$$\frac{d\psi}{dt} = \omega\psi.$$

It may be shewn in the same way that each of the other irreducible factors of G_1 satisfies an equation of this kind. Denote the various factors of G_1 by $\psi', \psi'', \ldots$, so that

$$G_1 = \psi'^{\mu} \psi''^{\nu} \ldots\ldots,$$

and let the equations they satisfy be

$$\frac{1}{\psi'}\frac{d\psi'}{dt} = \omega', \quad \frac{1}{\psi''}\frac{d\psi''}{dt} = \omega'', \ldots,$$

then we have

$$\frac{1}{G_1}\frac{dG_1}{dt} = \frac{\mu}{\psi'}\frac{d\psi'}{dt} + \frac{\nu}{\psi''}\frac{d\psi''}{dt} + \ldots = \mu\omega' + \nu\omega'' + \ldots = \omega \text{ say,}$$

where ω is a polynomial in $(p_1, \ldots, p_6)$, of order unity, and rational in $(q_1, \ldots, q_6, s)$. Thus G_1 satisfies the equation

$$\frac{dG_1}{dt} = \omega G_1,$$

and therefore (since G_1/G_2 is an integral), G_2 also satisfies the equation

$$\frac{dG_2}{dt} = \omega G_2.$$

As G_1 and G_2 satisfy the same differential equation, we shall in future use ϕ to denote either of them: so ϕ is a real polynomial in $(p_1, \ldots, p_6, q_1, \ldots, q_6, s)$, which satisfies the equation $\dot{\phi} = \omega\phi$.

Now ϕ is merely multiplied by a power of k when q_r, p_r, s are replaced respectively by $q_r k^2, p_r k^{-1}, sk^2$: since

$$\omega = \frac{1}{\phi}\frac{d\phi}{dt} = \sum_{r=1}^{6} \frac{1}{\phi}\left(\frac{\partial\phi}{\partial q_r}\frac{p_r}{\mu} + \frac{\partial\phi}{\partial p_r}\frac{\partial U}{\partial q_r}\right),$$

we see that ω is multiplied by k^{-3} when this substitution is made. It follows that ω cannot contain a term independent of $(p_1, \ldots, p_6)$, since such a term would be multiplied by an even power of k; ω is therefore of the form

$$\omega = \omega_1 p_1 + \omega_2 p_2 + \ldots + \omega_6 p_6,$$

where each of the quantities ω_r is homogeneous of degree -1 in the quantities $(q_1, \ldots, q_6, s)$.

Further, let one of the terms in ϕ be of order m in $(p_1, \ldots, p_6)$ and of order n in $(q_1, \ldots, q_6, s)$, while another term is of order m' in $(p_1, \ldots, p_6)$ and of order n' in $(q_1, \ldots, q_6, s)$: since these terms are multiplied by the same power of k when the above substitution is made, we have

$$-m + 2n = -m' + 2n',$$

so $m - m'$ is an even number. Hence ϕ can be arranged in the form

$$\phi = \phi_0 + \phi_2 + \phi_4 + \ldots,$$

where ϕ_0 denotes the terms of highest order in $(p_1, \ldots, p_6)$, ϕ_2 denotes terms of order less by two units in $(p_1, \ldots, p_6)$ than these, and so on: and each of the quantities ϕ_r is a polynomial in $(p_1, \ldots, p_6, q_1, \ldots, q_6, s)$, homogeneous in $(p_1, \ldots, p_6)$ and also in $(q_1, \ldots, q_6, s)$.

We shall now shew that *when ϕ_0 does not involve s, ϕ can be made into an integral by multiplying it by an appropriate rational function of $(q_1, \ldots, q_6)$.*

For suppose that ϕ_0 does not involve s: the equation

$$\frac{d\phi}{dt} = \omega\phi$$

or

$$\frac{d\phi_0}{dt} + \frac{d\phi_2}{dt} + \ldots\ldots = (\omega_1 p_1 + \omega_2 p_2 + \ldots + \omega_6 p_6)(\phi_0 + \phi_2 + \ldots)$$

gives, on equating the terms of highest degree in $(p_1, \ldots, p_6)$,

$$\sum_{r=1}^{6} \frac{p_r}{\mu_r}\frac{\partial \phi_0}{\partial q_r} = (\omega_1 p_1 + \ldots + \omega_6 p_6)\,\phi_0.$$

Now ϕ_0 may contain p_6 as a factor: in order to take account of this case, write $\phi_0 = p_6^k \phi_0'$, where ϕ_0' does not contain p_6 as a factor, and where as a special case we may have $k = 0$, $\phi_0' = \phi_0$. Substituting $p_6^k \phi_0'$ for ϕ_0 in the differential equation, it becomes

$$\sum_{r=1}^{6} \frac{p_r}{\mu_r}\frac{\partial \phi_0'}{\partial q_r} = (\omega_1 p_1 + \ldots + \omega_6 p_6)\,\phi_0'.$$

Let ϕ_0'' denote those terms in ϕ_0' which do not involve p_6; equating the terms which do not involve p_6 on the two sides of this equation. we have

$$\sum_{r=1}^{5} \frac{p_r}{\mu_r}\frac{\partial \phi_0''}{\partial q_r} = (\omega_1 p_1 + \ldots + \omega_5 p_5)\,\phi_0''.$$

It may be that ϕ_0'' is a mere function of $q_1, q_2, \ldots, q_6$, say equal to R; in this case we have

$$\frac{1}{\mu_r}\frac{\partial R}{\partial q_r} = \omega_r R \qquad (r = 1, 2, \ldots, 5)$$

or

$$\mu_r \omega_r = \frac{1}{R}\frac{\partial R}{\partial q_r} \qquad (r = 1, 2, \ldots, 5),$$

and therefore

$$\frac{\partial}{\partial q_s}(\mu_r \omega_r) = \frac{\partial}{\partial q_r}(\mu_s \omega_s) \qquad (r, s = 1, 2, \ldots, 5).$$

Supposing next that ϕ_0'' does involve some of the quantities $(p_1, \ldots, p_5)$, it may involve p_5 as a factor: to take account of this case, we write $\phi_0'' = p_5^{\lambda} \phi_0'''$, where ϕ_0''' does not involve p_5 as a factor. The equation now becomes

$$\sum_{r=1}^{5} \frac{p_r}{\mu_r}\frac{\partial \phi_0'''}{\partial q_r} = (\omega_1 p_1 + \ldots + \omega_5 p_5)\,\phi_0'''.$$

Let ϕ_0^{iv} denote the terms in ϕ_0''' which do not involve p_5: equating the terms which do not involve p_5 on the two sides of this equation, we have

$$\sum_{r=1}^{4} \frac{p_r}{\mu_r} \frac{\partial \phi_0^{\text{iv}}}{\partial q_r} = (\omega_1 p_1 + \ldots + \omega_4 p_4)\, \phi_0^{\text{iv}}.$$

Proceeding in this way, we ultimately arrive at the alternatives, that either

$$\frac{\partial}{\partial q_1}(\mu_2 \omega_2) = \frac{\partial}{\partial q_2}(\mu_1 \omega_1),$$

or else a function ψ exists, which is polynomial in $q_1, \ldots, q_6, p_1, p_2$, homogeneous in $q_1, \ldots, q_6$ and also in p_1, p_2, and is free from any factors which are mere powers of p_1 and p_2, and which satisfies the differential equation

$$\frac{p_1}{\mu_1}\frac{\partial \psi}{\partial q_1} + \frac{p_2}{\mu_2}\frac{\partial \psi}{\partial q_2} = (\omega_1 p_1 + \omega_2 p_2)\, \psi.$$

Now let $$\psi = a p_1^{l} + b p_2^{l} + c p_1^{l-1} p_2 + \ldots;$$

equating coefficients of p_1^{l+1} and p_2^{l+1} on the two sides of the last equation, we have

$$\omega_1 = \frac{1}{\mu_1 a}\frac{\partial a}{\partial q_1}, \qquad \omega_2 = \frac{1}{\mu_2 b}\frac{\partial b}{\partial q_2}.$$

The quantities $a, b, c, \ldots$ are polynomials in $(q_1, q_2, \ldots, q_6)$: they may have a common polynomial factor Q, so that

$$a = a'Q, \qquad b = b'Q, \text{ etc.}$$

Let $$\psi' = a' p_1^{l} + b' p_2^{l} + c' p_1^{l-1} p_2 + \ldots,$$

so that $$\psi = Q\psi'.$$

Then

$$\frac{1}{\psi'}\left(\frac{p_1}{\mu_1}\frac{\partial \psi'}{\partial q_1} + \frac{p_2}{\mu_2}\frac{\partial \psi'}{\partial q_2}\right) = \frac{1}{\psi}\left(\frac{p_1}{\mu_1}\frac{\partial \psi}{\partial q_1} + \frac{p_2}{\mu_2}\frac{\partial \psi}{\partial q_2}\right) - \frac{1}{Q}\left(\frac{p_1}{\mu_1}\frac{\partial Q}{\partial q_1} + \frac{p_2}{\mu_2}\frac{\partial Q}{\partial q_2}\right)$$

$$= \left(\omega_1 - \frac{1}{Q\mu_1}\frac{\partial Q}{\partial q_1}\right) p_1 + \left(\omega_2 - \frac{1}{Q\mu_2}\frac{\partial Q}{\partial q_2}\right) p_2$$

$$= \omega_1' p_1 + \omega_2' p_2, \text{ say,}$$

where $$\omega_1' = \frac{1}{\mu_1 a'}\frac{\partial a'}{\partial q_1}, \qquad \omega_2' = \frac{1}{\mu_2 b'}\frac{\partial b'}{\partial q_2},$$

so $$\frac{p_1}{\mu_1}\frac{\partial \psi'}{\partial q_1} + \frac{p_2}{\mu_2}\frac{\partial \psi'}{\partial q_2} = (\omega_1' p_1 + \omega_2' p_2)\, \psi'.$$

The left-hand side of this equation is a polynomial in $(q_1, q_2, \ldots, q_6, p_1, p_2)$; but if a' contains q_1, then ω_1' contains a', or some factor of it, as a denominator. Hence ψ' must contain a', or some factor of it, as a factor. But this is inconsistent with the supposition that $a', b', \ldots$ have no common factor. Hence a' cannot involve q_1; and therefore ω_1' is zero. Similarly ω_2' is zero.

Thus
$$\omega_1 = \frac{1}{Q\mu_1}\frac{\partial Q}{\partial q_1}, \qquad \omega_2 = \frac{1}{Q\mu_2}\frac{\partial Q}{\partial q_2},$$
and therefore
$$\frac{\partial}{\partial q_2}(\mu_1\omega_1) = \frac{\partial}{\partial q_1}(\mu_2\omega_2),$$
which is the same as the alternative previously noted: so this equation is true in any case

Similarly we can shew in general that
$$\frac{\partial}{\partial q_r}(\mu_s\omega_s) = \frac{\partial}{\partial q_s}(\mu_r\omega_r),$$
and hence we may write
$$\mu_r\omega_r = \frac{1}{R}\frac{\partial R}{\partial q_r},$$
where R is some rational function of $(q_1, q_2, \ldots, q_6)$.

Thus we have
$$\omega_1 p_1 + \omega_2 p_2 + \ldots + \omega_6 p_6 = \sum_{r=1}^{6} \frac{p_r}{\mu_r}\frac{1}{R}\frac{\partial R}{\partial q_r}$$
$$= \sum_{r=1}^{6} \frac{1}{R}\frac{\partial R}{\partial q_r}\frac{dq_r}{dt},$$
or
$$\frac{1}{\phi}\frac{d\phi}{dt} = \frac{1}{R}\frac{dR}{dt},$$
and therefore
$$\frac{\phi}{R} = \text{Constant}.$$

Thus ϕ *can be transformed into a constant, by multiplying it by an appropriate rational function of* $q_1, q_2, \ldots, q_6$, namely $1/R$; which is the required result.

If therefore the terms ϕ_0 in G_1 and G_2 do not involve s, we can transform G_1 and G_2 into integrals, by multiplying them by appropriate rational functions of $(q_1, q_2, \ldots, q_6)$; and hence, if it can be shewn that the terms ϕ_0 in G_1 and G_2 do not involve s, we shall have the result that any algebraic integral of the problem of three bodies can be compounded from integrals which are polynomial in $(p_1, p_2, \ldots, p_6)$ and rational in $q_1, q_2, \ldots, q_6, s$.

(vii) *Proof that* ϕ_0 *does not involve the irrationality* s.

The case in which ϕ_0 involves s is not included in the above investigation. We shall however now proceed to shew that no real function ϕ_0, which satisfies an equation
$$\sum_{r=1}^{6} \frac{p_r}{\mu_r}\frac{\partial \phi_0}{\partial q_r} = (\omega_1 p_1 + \ldots + \omega_6 p_6)\,\phi_0,$$
can involve s; and hence that the functions ϕ_0 occurring in our problem do not involve s, so that the above result is quite general.

For suppose that a function ϕ_0 exists, which involves s and satisfies the above differential equation. When the 8 values of s are substituted successively in ϕ_0, ϕ_0 will take a number of distinct values; let these values be denoted by ϕ_0', ϕ_0'', ...; they satisfy equations of the form

$$\sum_{\eta=1}^{6} \frac{p_r}{\mu_r} \frac{\partial \phi_0'}{\partial q_r} = \omega' \phi_0', \qquad \sum_{r=1}^{6} \frac{p_r}{\mu_r} \frac{\partial \phi_0''}{\partial q_r} = \omega'' \phi_0'', \ldots,$$

where ω', ω'', ... are the values of ω when the values of s corresponding to ϕ_0', ϕ_0'', ... respectively are substituted in it.

Let
$$\Phi = \phi_0' \phi_0'' \phi_0''' \ldots .$$

Then we have

$$\frac{1}{\Phi} \sum_{r=1}^{6} \frac{p_r}{\mu_r} \frac{\partial \Phi}{\partial q_r} = \sum_{r=1}^{6} \frac{p_r}{\mu_r} \left(\frac{1}{\phi_0'} \frac{\partial \phi_0'}{\partial q_r} + \frac{1}{\phi_0''} \frac{\partial \phi_0''}{\partial q_r} + \ldots \right)$$
$$= \omega' + \omega'' + \ldots$$
$$= \Omega,$$

where Ω is a linear function of $(p_1, p_2, \ldots, p_6)$, the coefficients being rational functions of $(q_1, q_2, \ldots, q_6)$.

Now Φ, from the manner of its formation, is a rational function of $(q_1, q_2, \ldots, q_6)$, not involving s: and it is clearly a polynomial in $(p_1, p_2, \ldots, p_6)$. So we can apply to Φ the results already obtained, which shew that (on multiplying Φ by some rational function of $q_1, q_2, \ldots, q_6$) Ω is zero, and therefore that Φ satisfies the equation

$$\sum_{r=1}^{6} \frac{p_r}{\mu_r} \frac{\partial \Phi}{\partial q_r} = 0.$$

This is a partial differential equation for Φ: there are 6 independent variables, and 5 independent solutions can at once be found, namely the quantities $\left(\frac{q_2 p_1}{\mu_1} - \frac{q_1 p_2}{\mu_2}\right), \ldots, \left(\frac{q_6 p_1}{\mu_1} - \frac{q_1 p_6}{\mu_6}\right)$. It follows that Φ is a function only of the quantities

$$\left(\frac{q_2 p_1}{\mu_1} - \frac{q_1 p_2}{\mu_2}\right), \ldots, \left(\frac{q_6 p_1}{\mu_1} - \frac{q_1 p_6}{\mu_6}\right), p_1, p_2, \ldots, p_6.$$

Now the factors of Φ differ from each other only in that different roots s are used in their formation: so when such a relation exists between $(q_1, q_2, \ldots, q_6)$ that two of these roots s become equal to each other, then two factors of Φ will become equal to each other; hence if $\Phi = 0$ be regarded as an equation in p_1, at least two roots will become equal to each other. When this relation

$$f(q_1, q_2, \ldots, q_6) = 0$$

exists between $(q_1, q_2, \ldots, q_6)$, we shall therefore have $\partial \Phi / \partial p_1 = 0$; and similarly $\partial \Phi / \partial p_2, \ldots, \partial \Phi / \partial p_6$ will each be zero.

Since Φ is homogeneous in $(p_1, p_2, \ldots, p_6)$, the equation

$$p_1 \frac{\partial \Phi}{\partial p_1} + p_2 \frac{\partial \Phi}{\partial p_2} + \ldots + p_6 \frac{\partial \Phi}{\partial p_6} = 0$$

is equivalent to $\Phi = 0$: so $\Phi = 0$ does not constitute an equation independent of the equations $\partial\Phi/\partial p_1 = 0, \ldots, \partial\Phi/\partial p_6 = 0$.

If small variations are given to the variables which satisfy the equation $\Phi = 0$, their increments are connected by the equation

$$\sum_{r=1}^{6} \left(\frac{\partial \Phi}{\partial q_r} \delta q_r + \frac{\partial \Phi}{\partial p_r} \delta p_r \right) = 0;$$

but if $(q_1, q_2, \ldots, q_6, p_1, \ldots, p_6)$ satisfy the equations $\partial\Phi/\partial p_r = 0$, this equation becomes

$$\sum_{r=1}^{6} \frac{\partial \Phi}{\partial q_r} \delta q_r = 0,$$

and this relation between the increments δq_r must therefore be equivalent to the relation

$$\sum_{r=1}^{6} \frac{\partial f}{\partial q_r} \delta q_r = 0.$$

Hence the equations

$$\frac{\partial f/\partial q_1}{\partial \Phi/\partial q_1} = \frac{\partial f/\partial q_2}{\partial \Phi/\partial q_2} = \ldots = \frac{\partial f/\partial q_6}{\partial \Phi/\partial q_6}$$

are consequences of the equations $\partial\Phi/\partial p_r = 0$; and so, since $\sum_{r=1}^{6} \frac{p_r}{\mu_r} \frac{\partial \Phi}{\partial q_r}$ is zero, we have (for sets of values of $q_1, q_2, \ldots, q_6, p_1, \ldots, p_6$ which satisfy these equations)

$$\sum_{r=1}^{6} \frac{p_r}{\mu_r} \frac{\partial f}{\partial q_r} = 0.$$

The equations $f = 0$ and $\sum_{r=1}^{6} \frac{p_r}{\mu_r} \frac{\partial f}{\partial q_r} = 0$ are therefore algebraically derivable from the equations $\partial\Phi/\partial p_r = 0$. Now the actual values of $(q_1, \ldots, q_6)$ are of no importance in this algebraical elimination; so we can replace q_r by $(q_r + p_r t/\mu_r)$ in all the equations: and thus we see that the equations

$$\left\{ \begin{aligned} & f\left(q_1 + \frac{p_1}{\mu_1} t, \ldots, q_6 + \frac{p_6}{\mu_6} t\right) = 0, \\ & \sum_{r=1} \frac{p_r}{\mu_r} \frac{\partial}{\partial q_r} f\left(q_1 + \frac{p_1}{\mu_1} t, \ldots, q_6 + \frac{p_6}{\mu_6} t\right) = 0, \end{aligned} \right.$$

are algebraical consequences of the equations

$$\frac{\partial}{\partial p_r} \Phi(q_1, \ldots, q_6, p_1, \ldots, p_6) - \frac{t}{\mu_r} \frac{\partial}{\partial q_r} \Phi(q_1, \ldots, q_6, p_1, \ldots, p_6) = 0 \qquad (r = 1, 2, \ldots, 6).$$

Hence the result of eliminating t between the equations

$$\begin{cases} f\left(q_1 + \frac{p_1}{\mu_1} t, \ldots, q_6 + \frac{p_6}{\mu_6} t\right) = 0, \\ \frac{\partial}{\partial t} f\left(q_1 + \frac{p_1}{\mu_1} t, \ldots, q_6 + \frac{p_6}{\mu_6} t\right) = 0, \end{cases}$$

must be an algebraic combination of the equations

$$\frac{\partial}{\partial p_r} \Phi(q_1, \ldots, q_6, p_1, \ldots, p_6) - \frac{t}{\mu_r} \frac{\partial}{\partial q_r} \Phi(q_1, \ldots, q_6, p_1, \ldots, p_6) = 0 \qquad (r = 1, 2, \ldots, 6).$$

Now one such algebraic combination of these equations is

$$\Phi(q_1, \ldots, q_6, p_1, \ldots, p_6) = 0;$$

for it can be derived by multiplying the equations by $(p_1, \ldots, p_6)$ in turn, and adding them. We shall shew that it is the eliminant which has just been mentioned.

For let the eliminant in question be denoted by Ψ; then the equation

$$\sum_{r=1}^{6} \left(\frac{\partial \Psi}{\partial q_r} \delta q_r + \frac{\partial \Psi}{\partial p_r} \delta p_r\right) = 0$$

must be a combination of the equations

$$\sum_{r=1}^{6} \frac{\partial f}{\partial q_r} \left(\delta q_r + \frac{t}{\mu_r} \delta p_r\right) = 0$$

and
$$\sum_{r=1}^{6} \frac{1}{\mu_r} \frac{\partial f}{\partial q_r} \delta p_r + \sum_{r=1}^{6} \sum_{s=1}^{6} \frac{p_r}{\mu_r} \frac{\partial^2 f}{\partial q_r \partial q_s} \left(\delta q_s + \frac{t}{\mu_s} \delta p_s + \frac{p_s}{\mu_s} \delta t\right) = 0.$$

Since the latter equation involves δt, we see that it cannot enter into the combination: and so we have

$$\frac{\partial \Psi / \partial q_1}{\partial f / \partial q_1} = \frac{\partial \Psi / \partial q_2}{\partial f / \partial q_2} = \ldots = \frac{\partial \Psi / \partial q_6}{\partial f / \partial q_6}, \quad \frac{\partial \Psi}{\partial p_r} = \frac{t}{\mu_r} \frac{\partial \Psi}{\partial q_r} \qquad (r = 1, 2, \ldots, 6).$$

The identity of these equations with those which have already been found for Φ shews that the equations $\Phi = 0$ and $\Psi = 0$ are equivalent. Hence $\Phi = 0$ is the eliminant of the equations

$$f\left(q_1 + \frac{p_1}{\mu_1} t, \ldots, q_6 + \frac{p_6}{\mu_6} t\right) = 0$$

and
$$\frac{\partial}{\partial t} f\left(q_1 + \frac{p_1}{\mu_1} t, \ldots, q_6 + \frac{p_6}{\mu_6} t\right) = 0.$$

Now the equations $f(q_1, q_2, \ldots, q_6) = 0$, which are the conditions that the equation for s may have equal roots, can easily be written down: and this result enables us then to find all possible polynomials Φ, and hence, by factorisation of Φ, to find all possible polynomials ϕ_0.

The eight roots s are the eight values of the expression $\pm r_1 \pm r_2 \pm r_3$, where r_1, r_2, r_3 denote the mutual distances: so we may have two roots s equal as a result of any one of the equations

$$r_1 = 0, \quad r_2 = 0, \quad r_3 = 0, \quad r_2 = \pm r_3, \quad r_3 = \pm r_1, \quad r_1 = \pm r_2, \quad r_1 \pm r_2 \pm r_3 = 0.$$

The equation $r_1 = 0$ gives

$$q_1^2 + q_2^2 + q_3^2 = 0;$$

and the eliminant of

$$\left(q_1 + \frac{p_1 t}{\mu_1}\right)^2 + \left(q_2 + \frac{p_2 t}{\mu_2}\right)^2 + \left(q_3 + \frac{p_3 t}{\mu_3}\right)^2 = 0,$$

and
$$\frac{d}{dt}\left\{\left(q_1 + \frac{p_1 t}{\mu_1}\right)^2 + \left(q_2 + \frac{p_2 t}{\mu_2}\right)^2 + \left(q_3 + \frac{p_3 t}{\mu_3}\right)^2\right\} = 0,$$

is
$$(q_1^2 + q_2^2 + q_3^2)\left(\frac{p_1^2}{\mu_1^2} + \frac{p_2^2}{\mu_2^2} + \frac{p_3^2}{\mu_3^2}\right) = \left(\frac{q_1 p_1}{\mu_1} + \frac{q_2 p_2}{\mu_2} + \frac{q_3 p_3}{\mu_3}\right)^2;$$

so the value of Φ arising in this connexion is

$$\Phi = \left(\frac{q_1 p_2}{\mu_2} - \frac{q_2 p_1}{\mu_1}\right)^2 + \left(\frac{q_2 p_3}{\mu_3} - \frac{p_2 q_3}{\mu_2}\right)^2 + \left(\frac{q_3 p_1}{\mu_1} - \frac{q_1 p_3}{\mu_3}\right)^2;$$

this expression is not resoluble into real factors, and therefore no real polynomials ϕ_0 can arise from this source.

A similar result can be deduced in connexion with the equations $r_2 = 0$ and $r_3 = 0$.

Consider next the equation

$$r_2 = \pm r_3;$$

it can be written in the form

$$q_4^2 + q_5^2 + q_6^2 + \frac{2m_2}{m_1 + m_2}(q_1 q_4 + q_2 q_5 + q_3 q_6) + \left(\frac{m_2}{m_1 + m_2}\right)^2 (q_1^2 + q_2^2 + q_3^2)$$
$$= q_4^2 + q_5^2 + q_6^2 - \frac{2m_1}{m_1 + m_2}(q_1 q_4 + q_2 q_5 + q_3 q_6) + \left(\frac{m_1}{m_1 + m_2}\right)^2 (q_1^2 + q_2^2 + q_3^2)$$

or
$$2(q_1 q_4 + q_2 q_5 + q_3 q_6) - \frac{m_1 - m_2}{m_1 + m_2}(q_1^2 + q_2^2 + q_3^2) = 0.$$

Replacing q_r by $(q_r + p_r t/\mu_r)$, and forming the discriminant with respect to t of the equation thus obtained, we find

$$\Phi = \left\{2(q_1 q_4 + q_2 q_5 + q_3 q_6) + \frac{m_1 - m_2}{m_1 + m_2}(q_1^2 + q_2^2 + q_3^2)\right\}$$
$$\times \left\{2\left(\frac{p_1 p_4}{\mu_1 \mu_4} + \frac{p_2 p_5}{\mu_2 \mu_5} + \frac{p_3 p_6}{\mu_3 \mu_6}\right) + \frac{m_1 - m_2}{m_1 + m_2}\left(\frac{p_1^2}{\mu_1^2} + \frac{p_2^2}{\mu_2^2} + \frac{p_3^2}{\mu_3^2}\right)\right\}$$
$$- \left\{\frac{q_1 p_4}{\mu_4} + \frac{q_4 p_1}{\mu_1} + \frac{q_2 p_5}{\mu_5} + \frac{q_5 p_2}{\mu_2} + \frac{q_3 p_6}{\mu_6} + \frac{q_6 p_3}{\mu_3} + \frac{m_1 - m_2}{m_1 + m_2}\left(\frac{q_1 p_1}{\mu_1} + \frac{q_2 p_2}{\mu_2} + \frac{q_3 p_3}{\mu_3}\right)\right\}^2.$$

This expansion cannot be factorised into polynomials ϕ_0, linear in $(p_1, p_2, \ldots, p_6)$; so no functions ϕ_0 can arise from this source.

Similarly it may be shewn that no polynomials ϕ_0 can arise in connexion with the equations $r_3 = \pm r_1$, $r_1 = \pm r_2$.

Lastly, the rationalised form of the equations

$$r_1 \pm r_2 \pm r_3 = 0$$

is

$$(r_3^2 - r_2^2 + r_1^2)^2 - 4r_3^2 r_1^2 = 0.$$

When r_1 is zero, this case reduces to that which was last discussed: and since the polynomial Φ is not resoluble in this special case, it cannot be resoluble in the general case.

Thus finally, *no real polynomials ϕ_0, involving s, can exist.*

Summarising the results obtained hitherto, we have shewn that any algebraic integral of the differential equations, which does not involve t, is an algebraic function of integrals ϕ, each of which can be written in the form

$$\phi_0 + \phi_2 + \phi_4 + \dots,$$

where ϕ_0 is a homogeneous polynomial in the variables p, say of degree k, and a homogeneous algebraic function of the variables q, say of degree l: ϕ_2 is a homogeneous polynomial in the variables p, of degree $(k-2)$, and a homogeneous algebraic function of the variables q, of degree $(l-1)$; ϕ_4 is a homogeneous polynomial in the variables p, of degree $(k-4)$, and a homogeneous algebraic function of the variables q, of degree $(l-2)$; and so on.

(viii) *Proof that ϕ_0 is a function only of the momenta and the integrals of angular momentum.*

We shall now proceed to shew that an integral ϕ, characterised by these properties, is an algebraic function of the classical integrals

The equation

$$\frac{d\phi}{dt} = 0, \quad \text{or} \quad \sum_{r=1}^{6} \left(\frac{p_r}{\mu_r} \frac{\partial \phi}{\partial q_r} + \frac{\partial U}{\partial q_r} \frac{\partial \phi}{\partial p_r} \right) = 0,$$

gives on replacing ϕ by $\phi_0 + \phi_2 + \phi_4 + \dots$, and equating terms of equal degree,

$$\begin{cases} 0 = \sum_{r=1}^{6} \dfrac{\partial \phi_0}{\partial q_r} \dfrac{p_r}{\mu_r}, \\ 0 = \sum_{r=1}^{6} \dfrac{\partial \phi_2}{\partial q_r} \dfrac{p_r}{\mu_r} + \dfrac{\partial \phi_0}{\partial p_r} \dfrac{\partial U}{\partial q_r}, \\ \dots\dots\dots\dots\dots\dots \\ \dots\dots\dots\dots\dots\dots \\ 0 = \sum_{r=1}^{6} \dfrac{\partial \phi_k}{\partial q_r} \dfrac{p_r}{\mu_r} + \dfrac{\partial \phi_{k-2}}{\partial p_r} \dfrac{\partial U}{\partial q_r}, \\ 0 = \sum_{r=1}^{6} \dfrac{\partial \phi_k}{\partial p_r} \dfrac{\partial U}{\partial q_r}. \end{cases}$$

The first of these equations is a linear partial differential equation for ϕ_0 which can at once be solved, and gives

$$\phi_0 = f_0(P_2, P_3, \ldots, P_6, p_1, p_2, \ldots, p_6),$$

where
$$P_r = \frac{q_r p_1}{\mu_1} - \frac{p_r q_1}{\mu_r} \qquad (r = 2, 3, \ldots, 6).$$

Let the expression of ϕ_2 in terms of the variables $q_1, P_2, \ldots, P_6, p_1, \ldots, p_6$ be
$$\phi_2 = f_2(q_1, P_2, P_3, \ldots, P_6, p_1, \ldots, p_6),$$

we have
$$\frac{\partial f_2}{\partial q_1} = \frac{\partial \phi_2}{\partial q_1} + \sum_{r=2}^{6} \frac{\partial \phi_2}{\partial q_r}\frac{\partial q_r}{\partial q_1}, \qquad \text{where } q_r = \frac{\mu_1 P_r}{p_1} + \frac{\mu_1 p_r q_1}{\mu_r p_1}$$
$$= \frac{\partial \phi_2}{\partial q_1} + \sum_{r=2}^{6} \frac{\partial \phi_2}{\partial q_r}\frac{\mu_1 p_r}{\mu_r p_1},$$

or
$$\frac{p_1}{\mu_1}\frac{\partial f_2}{\partial q_1} = \sum_{r=1}^{6} \frac{p_r}{\mu_r}\frac{\partial \phi_2}{\partial q_r} = -\sum_{r=1}^{6} \frac{\partial \phi_0}{\partial p_r}\frac{\partial U}{\partial q_r}.$$

Integrating we have
$$f_2 = \chi(P_2, P_3, \ldots, P_6, p_1, \ldots, p_6) - \int \frac{\mu_1}{p_1} \sum_{r=1}^{6} \frac{\partial \phi_0}{\partial p_r}\frac{\partial U}{\partial q_r}\, dq_1,$$
so there can be no logarithmic terms in $\int X\, dq_1$, where
$$X = \sum_{r=1}^{6} \frac{\partial \phi_0}{\partial p_r}\frac{\partial U}{\partial q_r}, \text{ expressed in terms of } q_1, P_2, \ldots, P_6, p_1, \ldots, p_6,$$
$$= \left(\frac{\partial f_0}{\partial p_1} + \sum_{s=2}^{6} \frac{\partial f_0}{\partial P_s}\frac{q_s}{\mu_1}\right)\frac{\partial U}{\partial q_1} + \sum_{r=2}^{6}\left(\frac{\partial f_0}{\partial p_r} - \frac{\partial f_0}{\partial P_r}\frac{q_1}{\mu_r}\right)\frac{\partial U}{\partial q_r}$$
$$= \sum_{r=1}^{6} \frac{\partial f_0}{\partial p_r}\frac{\partial U}{\partial q_r} + \sum_{r=2}^{6} \frac{\partial f_0}{\partial P_r}\left(\frac{q_r}{\mu_1}\frac{\partial U}{\partial q_1} - \frac{q_1}{\mu_r}\frac{\partial U}{\partial q_r}\right).$$

If V denotes the expression of U in terms of the variables
$$q_1, P_2, \ldots, P_6, p_1, \ldots, p_6,$$
we have
$$\frac{\partial U}{\partial q_r} = \frac{p_1}{\mu_1}\frac{\partial V}{\partial P_r} \quad (r > 1) \quad \text{and} \quad \frac{\partial U}{\partial q_1} = \frac{\partial V}{\partial q_1} - \sum_{r=2}^{6} \frac{\partial V}{\partial P_r}\frac{p_r}{\mu_r}.$$

The terms in X which may give rise to logarithmic terms in $\int X\, dq_1$ are now seen to be
$$\sum_{r=2}^{6} \frac{\partial f_0}{\partial P_r}\left\{\frac{p_r V}{\mu_r p_1} + \frac{p_r q_1}{\mu_r p_1}\sum_{s=2}^{6} \frac{\partial V}{\partial P_s}\frac{p_s}{\mu_s} + \frac{q_1 p_1}{\mu_r \mu_1}\frac{\partial V}{\partial P_r}\right\},$$

so the terms which may be logarithmic in $\int X\, dq_1$ are
$$\sum_{r=2}^{6} \frac{p_r}{\mu_r p_1}\frac{\partial f_0}{\partial P_r}\int V\, dq_1 + \sum_{r=2}^{6}\sum_{s=2}^{6} \frac{\partial f_0}{\partial P_r}\frac{p_r p_s}{\mu_r \mu_s p_1}\frac{\partial}{\partial P_s}\int q_1 V\, dq_1$$
$$+ \sum_{r=2}^{6} \frac{\partial f_0}{\partial P_r}\frac{p_1}{\mu_r \mu_1}\frac{\partial}{\partial P_r}\int q_1 V\, dq_1.$$

Now V is a sum of three terms, each of the form $(A+Bq_1+Cq_1^2)^{-\frac{1}{2}}$. Taking each of these terms separately, we have for the transcendental part of the last expression

$$\sum_{r=2}^{6}\frac{\partial f_0}{\partial P_r}\frac{p_r}{\mu_r p_1}\frac{1}{\sqrt{-C}}\arcsin\frac{2Cq_1+B}{(B^2-4AC)^{\frac{1}{2}}}$$
$$-\sum_{r=2}^{6}\sum_{s=2}^{6}\frac{\partial f_0}{\partial P_r}\frac{p_r p_s}{\mu_r\mu_s p_1}\frac{1}{2C\sqrt{-C}}\frac{\partial B}{\partial P_s}\arcsin\frac{2Cq_1+B}{(B^2-4AC)^{\frac{1}{2}}}$$
$$-\sum_{r=2}^{6}\frac{\partial f_0}{\partial P_r}\frac{p_1}{\mu_r\mu_1}\frac{1}{2C\sqrt{-C}}\frac{\partial B}{\partial P_r}\arcsin\frac{2Cq_1+B}{(B^2-4AC)^{\frac{1}{2}}}.$$

Thus for each of the fractions $(A+Bq_1+Cq_1^2)^{-\frac{1}{2}}$, we must have

$$C\sum_{r=2}^{6}\frac{\partial f_0}{\partial P_r}\frac{p_r}{\mu_r p_1}-\sum_{r=2}^{6}\sum_{s=2}^{6}\frac{\partial f_0}{\partial P_r}\frac{p_r p_s}{\mu_r\mu_s p_1}\frac{1}{2}\frac{\partial B}{\partial P_s}-\sum_{r=2}^{6}\frac{\partial f_0}{\partial P_r}\frac{p_1}{\mu_r\mu_1}\frac{1}{2}\frac{\partial B}{\partial P_r}=0.$$

Now for the first of these fractions, namely $(q_1^2+q_2^2+q_3^2)^{-\frac{1}{2}}$, we have

$$A=\frac{\mu_1^2}{p_1^2}(P_2^2+P_3^2),\quad \tfrac{1}{2}B=\frac{\mu_1^2P_2p_2}{\mu_2p_1^2}+\frac{\mu_1^2P_3p_3}{\mu_3p_1^2},\quad C=1+\frac{\mu_1^2p_2^2}{\mu_2^2p_1^2}+\frac{\mu_1^2p_3^2}{\mu_3^2p_1^2},$$

so the first of the three equations will be

$$\left(1+\frac{\mu_1^2p_2^2}{\mu_2^2p_1^2}+\frac{\mu_1^2p_3^2}{\mu_3^2p_1^2}\right)\sum_{r=2}^{6}\frac{\partial f_0}{\partial P_r}\frac{p_r}{\mu_r p_1}-\sum_{r=2}^{6}\frac{\partial f_0}{\partial P_r}\frac{p_r}{\mu_r p_1}\left(\frac{\mu_1^2p_2^2}{\mu_2^2p_1^2}+\frac{\mu_1^2p_3^2}{\mu_3^2p_1^2}\right)$$
$$-\left(\frac{\partial f_0}{\partial P_2}\frac{\mu_1p_2}{\mu_2^2p_1}+\frac{\partial f_0}{\partial P_3}\frac{\mu_1p_3}{\mu_3^2p_1}\right)=0,$$

or

$$\sum_{r=2}^{6}\frac{\partial f_0}{\partial P_r}\frac{p_r}{\mu_r}-\left(\frac{\partial f_0}{\partial P_2}\frac{\mu_1p_2}{\mu_2^2}+\frac{\partial f_0}{\partial P_3}\frac{\mu_1p_3}{\mu_3^2}\right)=0,$$

or (since $\mu_1=\mu_2=\mu_3$)

$$\sum_{r=4}^{6}p_r\frac{\partial f_0}{\partial P_r}=0 \quad\text{.............................(A)},$$

and on solving this equation we see that f_0 is a function of

$$p_1,\ p_2,\ \ldots,\ p_6,\ P_2,\ P_3,\ (p_4q_5-p_5q_4),\ \text{and}\ (p_4q_6-p_6q_4).$$

Since the three expressions $(A+Bq_1+Cq_1^2)$ are linear functions of the three quantities $(q_1^2+q_2^2+q_3^2)$, $(q_1q_4+q_2q_5+q_3q_6)$, $(q_4^2+q_5^2+q_6^2)$, we can for our present purpose replace them by these three quantities: so the second expression $(A+Bq_1+Cq_1^2)$ may be taken to be $(q_1q_4+q_2q_5+q_3q_6)$, or

$$q_1\left(\frac{\mu P_4}{p_1}+\frac{\mu p_4}{\mu' p_1}q_1\right)+\left(\frac{\mu P_2}{p_1}+\frac{p_2q_1}{p_1}\right)\left(\frac{\mu P_5}{p_1}+\frac{\mu p_5q_1}{\mu' p_1}\right)+\left(\frac{\mu P_3}{p_1}+\frac{p_3q_1}{p_1}\right)\left(\frac{\mu P_6}{p_1}+\frac{\mu p_6q_1}{\mu' p_1}\right),$$

so for this expression we have

$$B=\frac{\mu P_4}{p_1}+\frac{\mu P_5p_2}{p_1^2}+\frac{\mu^2P_2p_5}{\mu' p_1^2}+\frac{\mu P_6p_3}{p_1^2}+\frac{\mu^2P_3p_6}{\mu' p_1^2},$$
$$C=\frac{\mu p_4}{\mu' p_1}+\frac{\mu p_2p_5}{\mu' p_1^2}+\frac{\mu p_3p_6}{\mu' p_1^2},$$

and the corresponding equation is

$$\frac{\partial f_0}{\partial P_2}(p_2 p_4 - p_1 p_5) + \frac{\partial f_0}{\partial P_3}(p_3 p_4 - p_1 p_6) + \frac{\partial f_0}{\partial P_4}\left(\frac{\mu p_4^2}{\mu'} - p_1^2\right)$$
$$+ \frac{\partial f_0}{\partial P_5}\left(\frac{\mu p_4 p_5}{\mu'} - p_2 p_1\right) + \frac{\partial f_0}{\partial P_6}\left(\frac{\mu p_4 p_6}{\mu'} - p_1 p_3\right) = 0 \quad \text{...(B)}.$$

The third expression $(A + Bq_1 + Cq_1^2)$ may be taken to be $q_4^2 + q_5^2 + q_6^2$; the corresponding equation proves to be the same as equation (A), and may consequently be neglected. We have therefore only to consider equations (A) and (B): simplifying (B) by means of (A), they may be written

$$p_4 \frac{\partial f_0}{\partial P_4} + p_5 \frac{\partial f_0}{\partial P_5} + p_6 \frac{\partial f_0}{\partial P_6} = 0,$$

$$(p_2 p_4 - p_1 p_5)\frac{\partial f_0}{\partial P_2} + (p_4 p_3 - p_1 p_6)\frac{\partial f_0}{\partial P_3} - p_1\left(p_1 \frac{\partial f_0}{\partial P_4} + p_2 \frac{\partial f_0}{\partial P_5} + p_3 \frac{\partial f_0}{\partial P_6}\right) = 0;$$

these equations are obviously algebraically independent; and the Jacobian conditions of existence are satisfied identically for them, since the coefficients of the derivates $\frac{\partial f_0}{\partial P}$ do not involve the quantities P. These two equations therefore form a complete system, with 5 independent variables P_2, P_3, P_4, P_5, P_6: so there must be $5 - 2 = 3$ independent solutions, and any other solution will be a function of these three solutions and of $p_1, p_2, \ldots, p_6$.

It is easily verified that three independent solutions are

$$\begin{cases} P_2 p_3 - P_3 p_2 + P_5 p_6 - P_6 p_5, \\ P_3 p_1 \qquad\qquad + P_6 p_4 - P_4 p_6, \\ \qquad - P_2 p_1 + P_4 p_5 - P_5 p_4, \end{cases}$$

or

$$p_1 L/\mu, \quad p_1 M/\mu, \quad p_1 N/\mu,$$

where

$$\begin{cases} L = q_2 p_3 - q_3 p_2 + q_5 p_6 - q_6 p_5, \\ M = q_3 p_1 - q_1 p_3 + q_6 p_4 - q_4 p_6, \\ N = q_1 p_2 - q_2 p_1 + q_4 p_5 - q_5 p_4, \end{cases}$$

and the three equations

$$L = \text{Constant}, \quad M = \text{Constant}, \quad N = \text{Constant}$$

are the three integrals of angular momentum of the system. We have therefore the result that *ϕ_0 is a function of $L, M, N, p_1, p_2, \ldots, p_6$ only.*

(ix) *Proof that ϕ_0 is a function of T, L, M, N.*

Since ϕ_0, when expressed in terms of $q_1, q_2, \ldots, q_6, p_1, \ldots, p_6$, is a polynomial in $p_1, p_2, \ldots, p_6$, it is clear that ϕ_0 is a polynomial in its arguments $L, M, N, p_1, \ldots, p_6$. We shall write

$$\phi_0 = G(L, M, N, p_1, \ldots, p_6),$$

so we have $$\frac{d\phi_0}{dt} = \sum_{r=1}^{6} \frac{\partial G}{\partial p_r}\frac{dp_r}{dt} = \sum_{r=1}^{6} \frac{\partial G}{\partial p_r}\frac{\partial U}{\partial q_r};$$

and the equation for f_2 is

$$f_2 = \chi(P_2, \ldots, P_6, p_1, \ldots, p_6) - \frac{\mu_1}{p_1}\sum_{r=1}^{6} \frac{\partial G}{\partial p_r}\int Y_r dq_1,$$

where Y_r stands for $\partial U/\partial q_r$, supposed expressed in terms of $q_1, P_2, \ldots, P_6, p_1, \ldots, p_6$. We have therefore

$$\begin{aligned}\frac{\mu_1}{p_1}\sum_{r=1}^{6}\frac{\partial G}{\partial p_r}\int Y_r dq_1 &= \int \frac{\mu_1}{p_1}\left\{\frac{\partial G}{\partial p_1}\left(\frac{\partial V}{\partial q_1} - \Sigma \frac{\partial V}{\partial P_r}\frac{p_r}{\mu_r}\right) + \frac{p_1}{\mu_1}\frac{\partial V}{\partial P_2}\frac{\partial G}{\partial p_2} + \ldots + \frac{p_1}{\mu_1}\frac{\partial V}{\partial P_6}\frac{\partial G}{\partial p_6}\right\} dq_1\\ &= \frac{\mu_1}{p_1}\frac{\partial G}{\partial p_1}\int \frac{\partial V}{\partial q_1} dq_1 + \sum_{r=2}^{6}\int \frac{\partial V}{\partial P_r}\left(\frac{\partial G}{\partial p_r} - \frac{\mu_1 p_r}{\mu_r p_1}\frac{\partial G}{\partial p_1}\right) dq_1\\ &= \frac{\mu_1}{p_1}\frac{\partial G}{\partial p_1} V + \sum_{r=2}^{6}\left(\frac{\partial G}{\partial p_r} - \frac{\mu_1 p_r}{\mu_r p_1}\frac{\partial G}{\partial p_1}\right)\frac{\partial}{\partial P_r}\int V dq_1\\ &= \frac{\mu_1}{p_1}\frac{\partial G}{\partial p_1}\sum_A \frac{m_1 m_2}{(A + Bq_1 + Cq_1^2)^{\frac{1}{2}}}\end{aligned}$$

$$+ \sum_{r=2}^{6}\left(\frac{\partial G}{\partial p_r} - \frac{\mu_1 p_r}{\mu_r p_1}\frac{\partial G}{\partial p_1}\right)\sum_A \frac{m_1 m_2\left(-2A\dfrac{\partial B}{\partial P_r} - q_1 B\dfrac{\partial B}{\partial P_r} + B\dfrac{\partial A}{\partial P_r} + 2Cq_1\dfrac{\partial A}{\partial P_r}\right)}{(B^2 - 4AC)(A + Bq_1 + Cq_1^2)^{\frac{1}{2}}},$$

where the symbol $\sum\limits_A$ indicates summation over the three values of the expression $(A + Bq_1 + Cq_1^2)$.

Now the term $\chi(P_2, \ldots, P_6, p_1, \ldots, p_6)$ cannot give rise to terms involving $(A + Bq_1 + Cq_1^2)$ in the denominator: so the quantities multiplying each of the expressions $(A + Bq_1 + Cq_1^2)^{\frac{1}{2}}$ must themselves have the same character as ϕ_2, i.e. they must be polynomial in $(p_1, \ldots, p_6)$ when expressed in terms of $(q_1, q_2, \ldots, q_6, p_1, \ldots, p_6)$. We see therefore that the expression

$$\frac{\mu_1}{p_1}\frac{\partial G}{\partial p_1} + 2\sum_{r=2}^{6}\left(\frac{\partial G}{\partial p_r} - \frac{\mu_1 p_r}{\mu_r p_1}\frac{\partial G}{\partial p_1}\right)\frac{-A\dfrac{\partial B}{\partial P_r} - \frac{1}{2}q_1 B\dfrac{\partial B}{\partial P_r} + \frac{1}{2}B\dfrac{\partial A}{\partial P_r} + Cq_1\dfrac{\partial A}{\partial P_r}}{B^2 - 4AC}$$

must be a polynomial in $(p_1, \ldots, p_6)$, when expressed in terms of $(p_1, \ldots, p_6, q_1, \ldots, q_6)$. Taking first $A + Bq_1 + Cq_1^2 = q_1^2 + q_2^2 + q_3^2$, this expression becomes

$$\frac{\mu_1}{p_1}\frac{\partial G}{\partial p_1} - \sum_{r=2}^{3}\left(\frac{\partial G}{\partial p_r} - \frac{\mu_1 p_r}{\mu_r p_1}\frac{\partial G}{\partial p_1}\right)$$

$$\times \frac{-p_r\{\mu(P_2^2 + P_3^2) + q_1(P_2 p_2 + P_3 p_3)\} + P_r\{\mu(P_2 p_2 + P_3 p_3) + q_1(p_1^2 + p_2^2 + p_3^2)\}}{2\{p_1^2 P_2^2 + p_1^2 P_3^2 + (p_2 P_3 - p_3 P_2)^2\}},$$

or (omitting a factor μ)

$$\frac{1}{p_1}\frac{\partial G}{\partial p_1} - \sum_{r=2}^{3}\left(\frac{\partial G}{\partial p_r} - \frac{\mu_1 p_r}{\mu_r p_1}\frac{\partial G}{\partial p_1}\right)$$

$$\times \frac{-p_r\{p_1(q_2^2 + q_3^2) - p_2 q_1 q_2 - p_3 q_1 q_3\} + (q_r p_1 - p_r q_1)(p_1 q_1 + p_2 q_2 + p_3 q_3)}{2p_1\{(q_2 p_1 - p_2 q_1)^2 + (q_3 p_1 - p_3 q_1)^2 + (p_2 q_3 - p_3 q_2)^2\}},$$

or $\dfrac{1}{p_1}\dfrac{\partial G}{\partial p_1}$

$$\frac{\left(\frac{\partial G}{\partial p_2}-\frac{\mu_1 p_2}{\mu_2 p_1}\frac{\partial G}{\partial p_1}\right)(-p_2q_3^2-p_2q_1^2+p_1q_1q_2+p_3q_2q_3)+\left(\frac{\partial G}{\partial p_3}-\frac{\mu_1 p_3}{\mu_3 p_1}\frac{\partial G}{\partial p_1}\right)(-p_3q_2^2-p_3q_1^2+q_3q_1p_1+q_3q_2p_2)}{2\{(q_2p_1-q_1p_2)^2+(q_3p_1-q_1p_3)^2+(q_3p_2-q_2p_3)^2\}}$$

The last fraction must therefore represent a polynomial in $p_1, p_2, \ldots, p$ so the denominator must be a factor of the numerator.

Now G is a polynomial in L, M, N, so $\left(\dfrac{\partial G}{\partial p_2}-\dfrac{\mu_1 p_2}{\mu_2 p_1}\dfrac{\partial G}{\partial p_1}\right)$ and $\left(\dfrac{\partial G}{\partial p_3}-\dfrac{\mu_1 p_3}{\mu_3 p_1}\dfrac{\partial G}{\partial p_1}\right)$ are polynomials in L, M, N and involve q_1, q_2, q_3 only by means of L, M, N; so either they contain no terms in q_1, q_2, q_3—in which case the denominator cannot be a factor of the numerator—or else they contain some terms free from q_1, q_2, q_3—in which case also the denominator cannot be a factor of the numerator. The condition can therefore only be satisfied by supposing that

$$\frac{\partial G}{\partial p_2}-\frac{\mu_1 p_2}{\mu_2 p_1}\frac{\partial G}{\partial p_1}=0, \quad \frac{\partial G}{\partial p_3}-\frac{\mu_1 p_3}{\mu_3 p_1}\frac{\partial G}{\partial p_1}=0.$$

As might be expected from considerations of symmetry, the conditions arising from the other sets of values of A, B, C give

$$\frac{\partial G}{\partial p_r}-\frac{\mu_1 p_r}{\mu_r p_1}\frac{\partial G}{\partial p_1}=0 \qquad (r=4, 5, 6).$$

The function G therefore satisfies these five equations, which are evidently a complete system of five independent equations with six independent variables, and consequently possess only one independent solution; this solution is easily found to be

$$\sum_{s=1}^{6}\frac{p_s^2}{2\mu_s}, \text{ or } T.$$

The function G therefore involves $(p_1, \ldots, p_6)$ only by means of the expression T: and since G is polynomial in $(p_1, \ldots, p_6)$, it must also be polynomial in T.

Since ϕ_0 is homogeneous in $(q_1, q_2, \ldots, q_6)$, and also in $(p_1, p_2, \ldots, p_6)$, and the expressions (L, M, N) are each linear in $(q_1, \ldots, q_6)$, while T does not involve $(q_1, \ldots, q_6)$ and is of degree 2 in $(p_1, \ldots, p_6)$, it is clear that if T is involved in ϕ_0 at all, it must be as a factor: so we can write

$$\phi_0 = h(L, M, N)\, T^m,$$

where h is a homogeneous polynomial in its arguments.

(x) *Deduction of Bruns' theorem, for integrals which do not involve t.*

The equation which determines the function f_2 is

$$f_2 = \chi(P_2, \ldots, P_6, p_1, \ldots, p_6) - \frac{\mu_1}{p_1}\frac{\partial G}{\partial p_1}U.$$

But we have

$$\frac{\mu_1}{p_1}\frac{\partial G}{\partial p_1} = \frac{\mu_1}{p_1}\frac{\partial G}{\partial T}\frac{\partial T}{\partial p_1} = m\,h\,(L, M, N)\,T^{m-1},$$

and therefore

$$f_2 = \chi\,(P_2, \ldots, P_6, p_1, \ldots, p_6) - m\,h\,(L, M, N)\,T^{m-1}\,U.$$

Thus

$$\begin{aligned}\phi &= \phi_0 + \phi_2 + \phi_4 + \ldots \\ &= h(L, M, N)\,(T^m - mT^{m-1}U) + \chi\,(P_2, \ldots, P_6, p_1, \ldots, p_6) + \phi_4 + \phi_6 + \ldots.\end{aligned}$$

The integral ϕ can therefore be compounded from two other integrals, namely:

1° the integral $h\,(L, M, N)\,(T-U)^m$, which is itself compounded from the classical integrals,

and 2° the integral ϕ', where

$$\phi' = \phi_0' + \phi_2' + \phi_4' + \ldots$$

and

$$\phi_0' = \chi\,(P_2, \ldots, P_6, p_1, \ldots, p_6),$$

$$\phi_2' = \phi_4 - \frac{m\,(m-1)}{2\,!}\,h\,(L, M, N)\,T^{m-2}\,U^2,$$

$$\phi_4' = \phi_6 + \frac{m\,(m-1)\,(m-2)}{3\,!}\,h\,(L, M, N)\,T^{m-3}\,U^3,$$

....

But ϕ' is an integral of the same character as ϕ, except that its highest term, ϕ_0', is of order two degrees less in $(p_1, \ldots, p_6)$ than the highest term, ϕ_0, of ϕ. Now we have shewn that ϕ can be compounded from the classical integrals together with the integral ϕ'. Similarly ϕ' can be compounded from the classical integrals together with an integral ϕ'' which has the same character as ϕ, but is of order less by four units than ϕ in the variables p. Proceeding in this way, we see that ϕ can be compounded of the classical integrals together with an integral $\phi^{(n)}$, whose order in $(p_1, \ldots, p_6)$ is either unity or zero. If $\phi^{(n)}$ is of order unity in $(p_1, \ldots, p_6)$, then in the equation

$$\phi^{(n)} = \phi_0^{(n)} = h\,(L, M, N)\,T^k$$

we must evidently have $k=0$; in this case, therefore, $\phi^{(n)}$ is compounded of the classical integrals. If $\phi^{(n)}$ is of order zero in $(p_1, \ldots, p_6)$, it is a function of $(q_1, \ldots, q_6)$ only: but we have already shewn that such integrals do not exist: and so in any case ϕ can be compounded algebraically from the classical integrals. Hence we have Bruns' theorem, that *every algebraic integral of the differential equations of the problem of three bodies, which does not involve the time, can be compounded by purely algebraic processes from the classical integrals.*

(xi) *Extension of Bruns' result to integrals which involve the time.*

We now proceed to consider those algebraic integrals of the problem of three bodies which involve the time explicitly.

For this purpose we shall take the equations of motion as a system (§ 155) of the 18th order: we have therefore to investigate integrals of the form

$$f(q_1, q_2, \ldots, q_9, p_1, \ldots, p_9, t) = a,$$

where f is an algebraic function of its arguments, and a is a constant.

The function f is not necessarily rational in its arguments. Let the last equation be rationalised, so far as the variable t is concerned, so that it may be arranged in the form

$$a^m + a^{m-1}\phi_1(q_1, \ldots, q_9, p_1, \ldots, p_9, t) + a^{m-2}\phi_2(q_1, \ldots, q_9, p_1, \ldots, p_9, t) + \ldots \\ + \phi_m(q_1, \ldots, q_9, p_1, \ldots, p_9, t) = 0,$$

where the functions ϕ are rational functions of t and algebraic functions of their other arguments $(q_1, \ldots, q_9, p_1, \ldots, p_9)$. This equation may be supposed irreducible in t, i.e. such that it cannot be factorised into other equations which are of lower degree in a and are rational in t: for if it is reducible, we can suppose it replaced by that one of its irreducible factors which corresponds to the original equation $f = a$.

Differentiating with respect to t, we have

$$a^{m-1}\frac{d\phi_1}{dt} + a^{m-2}\frac{d\phi_2}{dt} + \ldots + \frac{d\phi_m}{dt} = 0.$$

Now the quantities $d\phi_r/dt$, when expressed in terms of $(q_1, q_2, \ldots, q_9, p_1, \ldots, p_9, t)$, are rational functions of t: so that the previous equation would be reducible in t if this equation did not vanish identically. It follows that this equation does vanish identically: that is,

$$\frac{d\phi_r}{dt} = 0 \qquad (r = 1, 2, \ldots, m).$$

The expressions ϕ_r are therefore themselves integrals: and hence *the integral f can be compounded from other integrals ϕ, which are rational functions of t and algebraic functions of $(q_1, \ldots, q_9, p_1, \ldots, p_9)$.*

Let such an integral ϕ be resolved into factors linear in t: so that it may be written

$$\frac{P(t-\phi_1)^{m_1}(t-\phi_2)^{m_2}\ldots(t-\phi_k)^{m_k}}{(t-\psi_1)^{n_1}(t-\psi_2)^{n_2}\ldots(t-\psi_l)^{n_l}},$$

where $(P, \phi_1, \phi_2, \ldots, \phi_k, \psi_1, \ldots, \psi_l)$ are algebraic functions of $(q_1, \ldots, q_9, p_1, \ldots, p_9)$. Since this expression is an integral, we have

$$\frac{1}{P}\frac{dP}{dt} + \frac{m_1}{t-\phi_1}\left(1 - \frac{d\phi_1}{dt}\right) + \ldots + \frac{m_k}{t-\phi_k}\left(1 - \frac{d\phi_k}{dt}\right) - \frac{n_1}{t-\psi_1}\left(1 - \frac{d\psi_1}{dt}\right) - \ldots \\ - \frac{n_l}{t-\psi_l}\left(1 - \frac{d\psi_l}{dt}\right) = 0.$$

When $\frac{dP}{dt}, \frac{d\phi_1}{dt}, \ldots, \frac{d\phi_k}{dt}, \frac{d\psi_1}{dt}, \ldots, \frac{d\psi_l}{dt}$ are replaced by their values (P, H), $(\phi_1, H), \ldots, (\psi_l, H)$, this equation must become an identity: but this can happen only if

$$\frac{dP}{dt} = 0, \quad 1 - \frac{d\phi_1}{dt} = 0, \ \ldots, \ 1 - \frac{d\phi_k}{dt} = 0, \quad 1 - \frac{d\psi_1}{dt} = 0, \ \ldots, \ 1 - \frac{d\psi_l}{dt} = 0,$$

i.e. if each of the expressions

$$P, \ t - \phi_1, \ t - \phi_2, \ \ldots, \ t - \phi_k, \ t - \psi_1, \ \ldots, \ t - \psi_l$$

is an integral. Hence *any algebraic integral of the problem of three bodies which involves t can be compounded* (1) *of algebraic integrals which do not involve t and* (2) *of integrals of the form*

$$t - \phi = \text{Constant},$$

where ϕ *is an algebraic function of* $(q_1, q_2, \ldots, q_9, p_1, \ldots, p_9)$.

Now it is known that

$$t - \frac{m_1 q_1 + m_2 q_4 + m_3 q_7}{p_1 + p_4 + p_7} = \text{Constant}$$

is an integral: hence any algebraic integral of the problem, which involves t, can be compounded of

(1) algebraic integrals which do not involve t;

(2) integrals of the form

$$\phi - \frac{m_1 q_1 + m_2 q_4 + m_3 q_7}{p_1 + p_4 + p_7} = \text{Constant},$$

where ϕ is an algebraic function of $(q_1, \ldots, q_9, p_1, \ldots, p_9)$; and

(3) the classical integral

$$t - \frac{m_1 q_1 + m_2 q_4 + m_3 q_7}{p_1 + p_4 + p_7}.$$

But the integrals in classes (1) and (2) are algebraic integrals which do not involve the time; and hence, by the result already obtained, they are combinations of the classical integrals.

Thus finally *every algebraic integral of the differential equations of the problem of three bodies, whether it involves the time or not, can be compounded from the classical integrals*

Bruns' theorem has been extended by Painlevé*, who has shewn that every integral of the problem of n bodies which involves the velocities algebraically (whether the coordinates are involved algebraically or not) is a combination of the classical integrals.

* *Bull. Astr.* xv. (1898), p. 81.

165. *Poincaré's theorem.*

We shall next establish another theorem on the non-existence of a certain type of integrals in the problem of three bodies, which is in many respects analogous to that of Bruns, and was discovered in 1889 by Poincaré*.

(i) *The equations of motion of the restricted problem of three bodies.*

In the restricted problem of three bodies, the equations of motion of the planetoid can (§ 162) be written in the form

$$\frac{dq_r}{dt} = \frac{\partial H}{\partial p_r}, \quad \frac{dp_r}{dt} = -\frac{\partial H}{\partial q_r} \qquad (r = 1, 2),$$

where
$$H = H_0 + \mu H_1 + \mu^2 H_2 + \ldots,$$

$$H_0 = -\frac{1}{2p_1^2} - np_2,$$

and $H_1, H_2, \ldots$ are periodic in q_1, q_2, with period 2π.

The Hessian

$$\begin{vmatrix} \dfrac{\partial^2 H_0}{\partial p_1^2} & \dfrac{\partial^2 H_0}{\partial p_1 \partial p_2} \\ \dfrac{\partial^2 H_0}{\partial p_1 \partial p_2} & \dfrac{\partial^2 H_0}{\partial p_2^2} \end{vmatrix}$$

is evidently zero: as this circumstance would prove inconvenient in the proof of Poincaré's theorem, we shall modify the form of the equations so as to obtain a system for which the corresponding Hessian is not zero.

Write $H^2 = K$, and let $H = h$ be the integral of energy; then we have

$$\frac{dq_r}{dt} = \frac{1}{2h}\frac{\partial K}{\partial p_r}, \quad \frac{dp_r}{dt} = -\frac{1}{2h}\frac{\partial K}{\partial q_r} \qquad (r = 1, 2),$$

and therefore, taking a new function H equal to $K/2h$, we can write the differential equations of the restricted problem of three bodies in the form

$$\frac{dq_r}{dt} = \frac{\partial H}{\partial p_r}, \quad \frac{dp_r}{dt} = -\frac{\partial H}{\partial q_r} \qquad (r = 1, 2),$$

where for sufficiently small values of μ, H can be expanded as a power-series in the parameter μ,

$$H = H_0 + \mu H_1 + \mu^2 H_2 + \ldots,$$

and
$$2hH_0 = \frac{1}{4p_1^4} + \frac{np_2}{p_1^2} + n^2 p_2^2;$$

the Hessian of H_0 is not now zero, and $(H_1, H_2, \ldots)$ are periodic in q_1, q_2, with period 2π.

* *Acta Math.* XIII. (1890), p. 259; *Nouv. Méth. de la Méc. Cél.* I. (1892), p. 233.

(ii) *Statement of Poincaré's theorem.*

Let Φ denote a function of $(q_1, q_2, p_1, p_2, \mu)$ which is one-valued and regular for all real values of q_1 and q_2, for values of μ which do not exceed a certain limit, and for values of p_1 and p_2 which form a domain D, which may be as small as we please; and suppose that Φ is periodic with respect to q_1 and q_2, having the period 2π. Under these conditions the function Φ can be expanded as a power-series in μ, say

$$\Phi = \Phi_0 + \mu\Phi_1 + \mu^2\Phi_2 + \dots,$$

where $\Phi_0, \Phi_1, \Phi_2, \dots$ are one-valued analytic functions of (q_1, q_2, p_1, p_2), periodic in q_1 and q_2. Poincaré's theorem is that *no integral of the restricted problem of three bodies exists (except the Jacobian integral of energy and integrals equivalent to it), which is of the form*

$$\Phi = \text{Constant},$$

where Φ is a function of this character. The proof which follows is applicable to any dynamical system whose equations of motion are of the same type as those of the restricted problem of three bodies.

The necessary and sufficient condition that $\Phi = \text{Constant}$ may be an integral is expressed by the vanishing of the Poisson-bracket (H, Φ); so that

$$(H_0, \Phi_0) + \mu\{(H_1, \Phi_0) + (H_0, \Phi_1)\} + \mu^2\{(H_2, \Phi_0) + (H_1, \Phi_1) + (H_0, \Phi_2)\} + \dots = 0,$$

and therefore
$$(H_0, \Phi_0) = 0,$$
$$(H_1, \Phi_0) + (H_0, \Phi_1) = 0.$$

(iii) *Proof that Φ_0 is not a function of H_0.*

We shall first shew that Φ_0 cannot be a function of H_0. For suppose a relation of the form $\Phi_0 = \psi(H_0)$ to exist. From the equation $H_0 = H_0(p_1, p_2)$ we have on solving for p_1 an equation of the form $p_1 = \theta(H_0, p_2)$, and θ will be a one-valued function of its arguments unless $\partial H_0/\partial p_1$ is zero in the domain D. Replacing p_1 by its value θ in the function $\Phi_0(q_1, q_2, p_1, p_2)$, we have an equation of the form

$$\Phi_0(q_1, q_2, p_1, p_2) = \psi(q_1, q_2, H_0, p_2);$$

and as Φ_0 is a one-valued function of its arguments ψ will be a one-valued function of (q_1, q_2, H_0, p_2); but by hypothesis, the function ψ depends only on H_0. It follows that ψ is a one-valued function of H_0, so long as the variables p_1, p_2 remain in the domain D, and provided $\partial H_0/\partial p_1$ is not zero in D; or more generally provided one of the derivates $\partial H_0/\partial p_1$ and $\partial H_0/\partial p_2$ is not zero in D, a condition which is evidently satisfied in general. Since ψ is a one-valued function, the equation $\psi(H) = \text{Constant}$ will be a one-valued integral of the differential equations, and therefore $\Phi - \psi(H) = \text{Constant}$ will also be a one-valued integral, and will be expansible as a power-series in μ: it will moreover be divisible by μ, since $\Phi_0 - \psi(H_0)$ is zero. If then we write

$$\Phi - \psi(H) = \mu\Phi',$$

the equation $\Phi' = $ Constant will be a one-valued analytic integral: writing

$$\Phi' = \Phi_0' + \mu\Phi_1' + \mu^2\Phi_2' + \ldots,$$

the function Φ_0' will not in general be a function of H_0: if however it is a function of H_0, we perform the same operation again, thus arriving at a third one-valued analytic integral, whose part independent of μ will not in general be a function of H_0; and so on. It is evident that in this way we shall ultimately obtain an integral which does not reduce to a function of H_0 when μ is zero, unless Φ is a function of H, in which case the two integrals H and Φ are not distinct.

If, therefore, there exists an integral Φ which is one-valued and analytic and distinct from H, but which is such that Φ_0 is a function of H_0, we can always derive from it another integral, of the same character but such that it does not reduce to a function of H_0 when μ vanishes. We can therefore always suppose that Φ_0 is not a function of H_0.

(iv) *Proof that Φ_0 cannot involve the variables q_1, q_2.*

If the function Φ_0 involves the variables q_1, q_2, then since it is periodic in these variables we can write

$$\Phi_0 = \sum_{m_1, m_2} A_{m_1, m_2} e^{i(m_1 q_1 + m_2 q_2)} = \sum_{m_1, m_2} A_{m_1, m_2} \zeta,$$

where m_1 and m_2 are positive or negative integers, i denotes $\sqrt{-1}$, the quantities A_{m_1, m_2} are functions of p_1, p_2, and ζ represents the exponential co-factor of A_{m_1, m_2}. Since H_0 does not involve q_1, q_2, we have

$$-(H_0, \Phi_0) = \frac{\partial H_0}{\partial p_1}\frac{\partial \Phi_0}{\partial q_1} + \frac{\partial H_0}{\partial p_2}\frac{\partial \Phi_0}{\partial q_2}.$$

But we have $\partial\Phi_0/\partial q_r = \sum_{m_1, m_2} i m_r A_{m_1, m_2} \zeta$, so the equation $(H_0, \Phi_0) = 0$ becomes

$$\sum_{m_1, m_2} A_{m_1, m_2}\left(m_1 \frac{\partial H_0}{\partial p_1} + m_2 \frac{\partial H_0}{\partial p_2}\right)\zeta = 0,$$

and therefore (as this equation is an identity)

$$A_{m_1, m_2}\left(m_1 \frac{\partial H_0}{\partial p_1} + m_2 \frac{\partial H_0}{\partial p_2}\right) = 0.$$

Hence we must have either

$$A_{m_1, m_2} = 0 \quad \text{or} \quad m_1 \partial H_0/\partial p_1 + m_2 \partial H_0/\partial p_2 = 0;$$

but the latter alternative is possible only when m_1 and m_2 are both zero, or when the Hessian of H_0 is zero, which is not the case. It follows that all the coefficients A_{m_1, m_2} are zero, except $A_{0,0}$; and consequently Φ_0 does not involve the variables q_1 and q_2.

(v) *Proof that the existence of a one-valued integral is inconsistent with the result of* (iii) *in the general case.*

Consider now the equation

$$(H_1, \Phi_0) + (H_0, \Phi_1) = 0,$$

or

$$\sum_{r=}^{2} \frac{\partial \Phi_0}{\partial p_r} \frac{\partial H_1}{\partial q_r} - \sum_{r=1}^{2} \frac{\partial H_0}{\partial p_r} \frac{\partial \Phi_1}{\partial q_r} = 0.$$

As the functions H_1 and Φ_1 are periodic with respect to q_1, q_2, they can be expanded in series of the form

$$H_1 = \sum_{m_1, m_2} B_{m_1, m_2} e^{i(m_1 q_1 + m_2 q_2)} = \sum_{m_1, m_2} B_{m_1, m_2} \zeta, \text{ say,}$$

$$\Phi_1 = \sum_{m_1, m_2} C_{m_1, m_2} e^{i(m_1 q_1 + m_2 q_2)} = \sum_{m_1, m_2} C_{m_1, m_2} \zeta,$$

where m_1 and m_2 are positive or negative integers, and the coefficients B_{m_1, m_2} and C_{m_1, m_2} depend only on p_1, p_2. We have therefore

$$\frac{\partial H_1}{\partial q_r} = i \sum_{m_1, m_2} B_{m_1, m_2} m_r \zeta, \qquad \frac{\partial \Phi_1}{\partial q_r} = i \sum_{m_1, m_2} C_{m_1, m_2} m_r \zeta,$$

so the equation

$$\sum_{r=1}^{2} \frac{\partial \Phi_0}{\partial p_r} \frac{\partial H_1}{\partial q_r} - \sum_{r=1}^{2} \frac{\partial H_0}{\partial p_r} \frac{\partial \Phi_1}{\partial q_r} = 0$$

becomes

$$\sum_{m_1, m_2} B_{m_1, m_2} \zeta \left(\sum_{r=1}^{2} m_r \frac{\partial \Phi_0}{\partial p_r} \right) - \sum_{m_1, m_2} C_{m_1, m_2} \zeta \left(\sum_{r=1}^{2} m_r \frac{\partial H_0}{\partial p_r} \right) = 0,$$

or (since this equation is an identity)

$$B_{m_1, m_2} \left(m_1 \frac{\partial \Phi_0}{\partial p_1} + m_2 \frac{\partial \Phi_0}{\partial p_2} \right) = C_{m_1, m_2} \left(m_1 \frac{\partial H_0}{\partial p_1} + m_2 \frac{\partial H_0}{\partial p_2} \right).$$

This equation is valid for all values of p_1, p_2: and therefore for values of p_1 and p_2 which satisfy the equation

$$m_1 \frac{\partial H_0}{\partial p_1} + m_2 \frac{\partial H_0}{\partial p_2} = 0,$$

we must have either

$$B_{m_1, m_2} = 0, \quad \text{or} \quad m_1 \partial \Phi_0 / \partial p_1 + m_2 \partial \Phi_0 / \partial p_2 = 0.$$

We shall say that a coefficient B_{m_1, m_2} becomes *secular* when p_1, p_2 have values such that $m_1 \partial H_0 / \partial p_1 + m_2 \partial H_0 / \partial p_2 = 0$.

As H is a given function, the coefficients B_{m_1, m_2} are given. In the general case of dynamical systems expressed by differential equations of the kind we are considering, no one of these coefficients will vanish when it becomes secular, and we shall take this case first: so that the equation

$$m_1 \partial \Phi_0 / \partial p_1 + m_2 \partial \Phi_0 / \partial p_2 = 0$$

is a consequence of the equation $m_1 \partial H_0 / \partial p_1 + m_2 \partial H_0 / \partial p_2 = 0$.

Now let k_1, k_2 be two integers: suppose that we give to p_1 and p_2 values such that the equation

$$\frac{\partial H_0}{k_1 \partial p_1} = \frac{\partial H_0}{k_2 \partial p_2}$$

is satisfied. We can find an infinite number of pairs of integers m_1, m_2 such that $m_1k_1 + m_2k_2$ is zero: and for each of these systems of integers the expression $m_1 \partial H_0/\partial p_1 + m_2 \partial H_0/\partial p_2$ is zero, and consequently

$$m_1 \partial\Phi_0/\partial p_1 + m_2 \partial\Phi_0/\partial p_2$$

is zero. Comparing these two equations, we have

$$\frac{\partial H_0/\partial p_1}{\partial\Phi_0/\partial p_1} = \frac{\partial H_0/\partial p_2}{\partial\Phi_0/\partial p_2},$$

so the Jacobian $\partial(H_0, \Phi_0)/\partial(p_1, p_2)$ is zero for all values of p_1, p_2 for which $\partial H_0/\partial p_1$ and $\partial H_0/\partial p_2$ are commensurable with each other. Thus in any domain, however small, there are an infinite number of systems of values of p_1, p_2 for which this Jacobian is zero: as the Jacobian is a continuous function, it must therefore vanish identically, and consequently Φ_0 must be a function of H_0. But this is contrary to what was proved in (iii), and therefore the fundamental assumption as to the existence of the integral Φ must be erroneous; that is, the Hamiltonian equations possess no one-valued analytic integral other than $H = h$, provided no one of the coefficients B_{m_1, m_2} vanishes when it becomes secular.

(vi) *Removal of the restrictions on the coefficients* B_{m_1, m_2}.

We have now to consider the case in which at least one of the coefficients B_{m_1, m_2} vanishes when it becomes secular. We shall say that two pairs of indices (m_1, m_2) and (m_1', m_2') belong to the same *class* when they satisfy the relation $m_1/m_1' = m_2/m_2'$, and that in this case the coefficients B_{m_1, m_2} and $B_{m_1', m_2'}$ belong to the same class.

We shall first shew that the result obtained in (v) as to the non-existence of one-valued integrals is true provided that in each of the classes there is *at least one* coefficient B_{m_1, m_2} which does not vanish on becoming secular. For suppose that the coefficient B_{m_1, m_2} is zero, but the coefficient $B_{m_1', m_2'}$ is not zero. If p_1, p_2 have values such that $m_1 \partial H_0/\partial p_1 + m_2 \partial H_0/\partial p_2$ is zero, we have $m_1' \partial H_0/\partial p_1 + m_2' \partial H_0/\partial p_2 = 0$, and consequently

$$B_{m_1, m_2}\left(m_1 \frac{\partial\Phi_0}{\partial p_1} + m_2 \frac{\partial\Phi_0}{\partial p_2}\right) = 0, \qquad B_{m_1', m_2'}\left(m_1 \frac{\partial\Phi_0}{\partial p_1} + m_2 \frac{\partial\Phi_0}{\partial p_2}\right) = 0,$$

and although the relation $m_1 \partial\Phi_0/\partial p_1 + m_2 \partial\Phi_0/\partial p_2 = 0$ cannot be inferred from the former of these equations, it can be inferred from the latter: the proof is in other respects the same as in (v).

Now a class is completely defined by the ratio of the indices m_1, m_2; let λ be any commensurable number, and let C be the class of indices for which $m_1/m_2 = \lambda$. We shall say for brevity that this class C belongs to a given domain, or is *in* this domain, if a set of values of p_1, p_2 can be found in this domain such that

$$\lambda \frac{\partial H_0}{\partial p_1} + \frac{\partial H_0}{\partial p_2} = 0.$$

We shall shew that the theorem is still true if in every domain δ, however small, which is contained in D, there are an infinite number of classes for which not all the coefficients of the class vanish when they become secular.

For take any set of values of p_1, p_2, such that for these values we have

$$\lambda \frac{\partial H_0}{\partial p_1} + \frac{\partial H_0}{\partial p_2} = 0.$$

Suppose that λ is commensurable, and that for the class which corresponds to this value of λ, all the coefficients of the class do not vanish when they become secular: the preceding reasoning then applies to this set of values, and so for these values of p_1 and p_2 the Jacobian $\partial(H_0, \Phi_0)/\partial(p_1, p_2)$ is zero. But, by hypothesis, there exists in every domain δ, however small, which is contained in D, an infinite number of such sets of values of p_1, p_2. The Jacobian consequently vanishes at all points of D, and therefore Φ_0 is a function of H_0; so, as before, there exists no one-valued integral distinct from H

(vii) *Deduction of Poincaré's theorem.*

In the four preceding sections, we have considered equations of the type

$$\frac{dq_r}{dt} = \frac{\partial H}{\partial p_r}, \quad \frac{dp_r}{dt} = -\frac{\partial H}{\partial q_r} \qquad (r = 1, 2),$$

in which H can be expanded in the form

$$H = H_0 + \mu H_1 + \mu^2 H_2 + \ldots,$$

where the Hessian of H_0 with respect to p_1 and p_2 is not zero, H_0 does not involve q_1 and q_2, and $H_1, H_2, \ldots$ are periodic functions of q_1, q_2: and we have shewn that no integral of these equations exists which is distinct from the equation of energy and is one-valued and regular for all real values of q_1 and q_2, for values of μ which do not exceed a certain limit, and for values of p_1 and p_2 which form a domain D; provided that in every domain, however small, contained in D, there are an infinite number of ratios m_1/m_2 for which not all the corresponding coefficients B_{m_1, m_2} vanish when they become secular.

This result can be applied at once to the restricted problem of three bodies: for we have seen in (i) that the equations of motion in this problem are of the character specified, and on determining the function H_1 by actual expansion we find that the last condition is satisfied. *Poincaré's theorem is thus established.*

Poincaré's theorem establishes the non-existence of integrals *uniform with respect to the Keplerian variables*, which implies uniformity in the neighbourhood of all the trajectories which have the same osculating ellipse. This however does not exclude the existence of integrals which are uniform in domains of a different character. Cf. Levi-Civita, *Acta Math.* XXX. (1905), p. 305.

The theorem has been extended by Poincaré to the general problem of three bodies: cf. *Nouv. Méth. de la Méc. Cél.* I. p. 253; it has also been extended by Painlevé, *C.R.* CXXX. (1900), p. 1699. See also Cherry, *Proc. Camb. Phil. Soc.* XXII. (1924). p. 287.

CHAPTER XV

THE GENERAL THEORY OF ORBITS

166. *Introduction.*

We shall now pass to the study of the general form and disposition of the orbits of dynamical systems. For simplicity we shall in the present chapter chiefly consider the motion of a particle which is free to move in a plane under the action of conservative forces, but many of the results obtained can be readily extended to more general dynamical systems.

It has already been observed (§ 104) that the determination of the motion of a particle with two degrees of freedom under the action of conservative forces is reducible to the problem of finding the geodesics on a surface with a given line-element; an account of the properties of geodesics might therefore be regarded as falling within the scope of the discussion. Many of these properties are however of no importance for our present purpose: and as the theory of geodesics is fully treated in many works on Differential Geometry, we shall only consider those theorems which are of general dynamical interest.

167. *Periodic solutions.*

Great interest has attached in recent years to the investigation of those particular modes of motion of dynamical systems in which the same configuration of the system is repeated at regular intervals of time, so that the motion is purely periodic. Such modes of motion are called *periodic solutions.* The term *periodic solution* is also used in cases where a relative rather than an absolute configuration is periodically repeated: thus in the problem of three bodies, a solution is said to be periodic if the mutual distances of the bodies are periodic functions of the time, although the bodies may not necessarily have the same orientation at the end of a period as at its beginning.

The periodic orbits described by a spherical pendulum have been studied by A. Emch, *Proc. Nat. Ac. Sci.* IV. (1918), p. 218 and *Tôhoku Math. J.* XV. (1919), p. 146.

168. *A criterion for the discovery of periodic orbits.*

We shall now shew that the existence and position of periodic orbits can be determined by a theorem* analogous to those theorems which furnish the

* Whittaker, *Monthly Notices R.A.S.* LXII. (1902), p. 186. Cf. A. Signorini, *Rend. d. Lincei,* XXI. (1912), p. 36; *Rend. d. Palermo,* XXXIII. (1912), p. 187; L. Tonelli, *Rend. d. Lincei,* XXI. (1912), pp. 251, 332; *Mem. Ist. Bologna,* I. (1924), p. 21; Birkhoff, *Trans. Amer. Math. Soc.* XVIII. (1917), p. 224; Onicescu, *Math. Ann.* XCVIII. (1927), p. 576; Damköhler, *Ann. Sc. Norm. Sup. Pisa,* V. (1936), p. 127.

position of the roots of an algebraic equation by considerations depending on the sign of expressions connected with the equation. We shall for simplicity suppose the dynamical problem considered to be that of the motion of a particle of unit mass in a plane under the action of conservative forces: the result can be extended to more general systems without difficulty*.

Let (x, y) be the coordinates of the particle at time t, referred to any fixed rectangular axes in the plane, and let $V(x, y)$ be its potential energy function, so that the equation of energy is

$$\tfrac{1}{2}(\dot{x}^2+\dot{y}^2)+V(x,y)=h,$$

where h is the constant energy.

The differential equations of motion of the particle form a system of the fourth order, and their general solution consequently involves four arbitrary constants. One of these constants is, however, merely the constant additive to t, which determines the epoch in the orbit, so there are only ∞^3 really distinct orbits. This triple infinity of orbits can be arranged in sets, each containing a double infinity of orbits, by associating together those orbits for which the constant of energy has the same value h: such a set of ∞^2 orbits may be defined analytically by the principle of least action (§ 100), namely that the orbit between two given points (x_0, y_0) and (x_1, y_1) is such as to make the value of the expression

$$\int \{h-V(x,y)\}^{\frac{1}{2}} \{(dx)^2+(dy)^2\}^{\frac{1}{2}}$$

stationary as compared with other curves joining the given terminal points (x_0, y_0) and (x_1, y_1)†.

Consider any simple closed curve C in the plane of xy; and let another simple closed curve C' be drawn, enclosing C and differing only slightly from it. We may regard C' as defined by an equation of the form

$$\delta p=\phi(\gamma),$$

where δp is the normal distance between the curves C and C' (measured outwards from C, and consequently always positive) and γ is the inclination of this normal distance to the axis of x. Then if I be the value of the integral

$$\int \{h-V(x,y)\}^{\frac{1}{2}} \{(dx)^2+(dy)^2\}^{\frac{1}{2}},$$

when the integration is taken round the curve C, and if $I+\delta I$ denote the value of the same integral when the integration is taken round the curve C'

* For the extension to the restricted problem of three bodies, cf. *Monthly Notices R.A.S.* LXII. (1902), p. 346.

† Following Painlevé, *Liouville's Journal*, x. (1894), it is customary to call a family of orbits which have the same constant of energy a *natural* family.

(so that the symbol δ denotes an increment obtained in passing from C to C'), we have

$$\delta I = \int \{(dx)^2 + (dy)^2\}^{\frac{1}{2}} \delta \{h - V(x, y)\}^{\frac{1}{2}} + \int \{h - V(x, y)\}^{\frac{1}{2}} \delta \{(dx)^2 + (dy)^2\}^{\frac{1}{2}}.$$

But we have

$$\delta \{h - V(x, y)\}^{\frac{1}{2}} = -\tfrac{1}{2} \{h - V(x, y)\}^{-\frac{1}{2}} \left(\frac{\partial V}{\partial x} \delta x + \frac{\partial V}{\partial y} \delta y\right)$$

$$= -\tfrac{1}{2} \{h - V(x, y)\}^{-\frac{1}{2}} \left(\frac{\partial V}{\partial x} \cos \gamma + \frac{\partial V}{\partial y} \sin \gamma\right) \delta p,$$

and
$$\delta \{(dx)^2 + (dy)^2\}^{\frac{1}{2}} = \delta p : d\gamma = \frac{\delta p}{\rho} \{(dx)^2 + (dy)^2\}^{\frac{1}{2}},$$

where ρ is the radius of curvature of the curve C at the point (x, y).

Thus we have

$$\delta I = \int \{(dx)^2 + (dy)^2\}^{\frac{1}{2}} \{h - V(x, y)\}^{-\frac{1}{2}} \left\{\frac{h - V(x, y)}{\rho} - \tfrac{1}{2} \cos \gamma \frac{\partial V}{\partial x} - \tfrac{1}{2} \sin \gamma \frac{\partial V}{\partial y}\right\} \delta p.$$

This equation shews that if the quantity

$$\frac{h - V(x, y)}{\rho} - \tfrac{1}{2} \cos \gamma \frac{\partial V}{\partial x} - \tfrac{1}{2} \sin \gamma \frac{\partial V}{\partial y}$$

is negative at all points of C, then δI is negative, and so the integral I has its value diminished when any curve surrounding C and adjacent to C is taken instead of C as the path of integration.

Now suppose that another simple closed curve D can be drawn enclosing C, and such that at all points of D the quantity

$$\frac{h - V}{\rho} - \tfrac{1}{2} \left(\cos \gamma \frac{\partial V}{\partial x} + \sin \gamma \frac{\partial V}{\partial y}\right)$$

is positive. Then, in the same way, it can be shewn that the integral I is diminished when any simple closed curve D', enclosed by D and adjacent to D, is taken instead of D as the path of integration.

When, therefore, we consider the aggregate of all simple closed curves situated in the ring-shaped space bounded by C and D—which is assumed to contain no singularity of the function $V(x, y)$—it is clear that the curve which furnishes the least value of I cannot be C or D, and cannot coincide with C or D for any part of its length. There exist, therefore, among the simple closed curves of this aggregate, one or more curves K for which the value of I is less than for all other curves of the aggregate. Since K does not coincide with C or D along any part of its length, it follows that the curves adjacent to K are all members of the aggregate in question, and hence that the curve K furnishes a stationary value of I as compared with all the curves adjacent to it. The curve K is therefore an orbit in the dynamical system. We have thus arrived at the theorem: *If one closed curve be enclosed by another closed curve, and if the quantity*

$$\frac{h - V(x, y)}{\rho} - \tfrac{1}{2}\cos\gamma\,\frac{\partial V}{\partial x} - \tfrac{1}{2}\sin\gamma\,\frac{\partial V}{\partial y}$$

be negative at all points of the inner curve and positive at all points of the outer curve, then in the ring-shaped space between the two curves there exists a periodic orbit of the dynamical system, for which the constant of energy is h. As the quantity

$$\frac{h - V(x, y)}{\rho} - \tfrac{1}{2}\cos\gamma\,\frac{\partial V}{\partial x} - \tfrac{1}{2}\sin\gamma\,\frac{\partial V}{\partial y}$$

can be calculated immediately for every point on the curves C and D, depending as it does only on the potential-energy function and the curves themselves, this result furnishes a means of detecting the presence of periodic orbits.

169. *Asymptotic solutions.*

It sometimes happens in dynamical systems that motion in one orbit is *asymptotic* to motion in another, i.e., continually approaches more and more closely to coincidence with it as the time increases, just as a particle describing the curve whose equation in polar coordinates is

$$r = a\,\frac{\theta}{\theta - 1}$$

continually approximates to motion in the circle $r = a$, as θ increases indefinitely. In particular, we may have orbits which are asymptotic to periodic orbits, so that the motion, which originally bears no resemblance to periodic motion, approximates more and more nearly to periodic motion as the time tends to infinity. Such a motion is called an *asymptotic solution**.

We can of course have also a second kind of asymptotic solution which differs widely from a periodic solution when $t \to +\infty$, but approximates to it for $t \to -\infty$: in fact, if a periodic orbit is unstable, the path of a particle which is slightly disturbed from the periodic orbit will evidently be an asymptotic solution of this kind.

As we shall see in the next article, solutions also exist which belong at the same time to both of the classes of asymptotic solutions, i.e., they are very nearly periodic when $t \to +\infty$ and $t \to -\infty$, but differ widely from periodic orbits in the meantime. These are called *doubly-asymptotic solutions*†.

170. *The orbits of planets in the relativity-theory.*

Excellent illustrations of periodic and asymptotic orbits are provided by the paths which small particles ("planets") describe in the gravitational field

* Cf. Poincaré, *Méth. Nouv.* I. Ch. VII: Picard, *Traité d'Analyse*, III. Ch. VII.

† Poincaré, *Acta Math.* XIII. p. 225; *Méth. Nouv.* III. Ch. XXXIII.

due to a single attracting mass (the "sun"), when the Newtonian law of gravitation is replaced by the more accurate and philosophical laws belonging to the general relativity-theory.

A very full discussion of these orbits has been published by Y. Hagihara, *Jap. Journ. of Ast. and Geoph.* VIII. (1931), pp. 67–176.

Take the origin of coordinates at the sun and let the position of the planet in the plane of motion be specified by two coordinates (r, θ) which may be plotted as ordinary polar coordinates (though strictly speaking the distance of a point from the origin is represented not by r but by a certain function of r). Denote the time by t, and let ds^2 be defined as equal to the quadratic differential form

$$c^2\left(1-\frac{\alpha}{r}\right)dt^2-\frac{dr^2}{1-\frac{\alpha}{r}}-r^2d\theta^2,$$

where c denotes the velocity of light in free space and α is a constant depending on the mass of the sun. Let

$$T=\tfrac{1}{2}c^2\left(1-\frac{\alpha}{r}\right)t'^2-\tfrac{1}{2}\frac{r'^2}{1-\frac{\alpha}{r}}-\tfrac{1}{2}r^2\theta'^2,$$

where accents denote derivatives with respect to s: then it is shewn in treatises on Relativity that the equations which determine the orbit of the planet are the Lagrangian equations

$$\begin{cases}\dfrac{d}{ds}\left(\dfrac{\partial T}{\partial r'}\right)-\dfrac{\partial T}{\partial r}=0,\\[2ex] \dfrac{d}{ds}\left(\dfrac{\partial T}{\partial \theta'}\right)-\dfrac{\partial T}{\partial \theta}=0,\\[2ex] \dfrac{d}{ds}\left(\dfrac{\partial T}{\partial t'}\right)-\dfrac{\partial T}{\partial t}=0.\end{cases}$$

The last two of these equations give at once $\dfrac{\partial T}{\partial \theta'}=$ Constant and $\dfrac{\partial T}{\partial t'}=$ Constant,

or

$$r^2\frac{d\theta}{ds}=\frac{1}{\sqrt{\beta}},$$

$$\left(1-\frac{\alpha}{r}\right)\frac{dt}{ds}=-\frac{k}{c^2\sqrt{\beta}}$$

where k and β are constants.

Substituting from these equations in the equation

$$ds^2=c^2\left(1-\frac{\alpha}{r}\right)dt^2-\frac{dr^2}{1-\frac{\alpha}{r}}-r^2\,d\theta^2,$$

we have

$$\frac{dr^2}{1-\frac{\alpha}{r}}=\left\{-\beta r^4+\frac{k^2r^4}{c^2\left(1-\frac{\alpha}{r}\right)}-r^2\right\}d\theta^2,$$

so writing u for $1/r$, the differential equation of the orbit of the planet is

$$\left(\frac{du}{d\theta}\right)^2 = \frac{k^2}{c^2} - (1 - \alpha u)(\beta + u^2).$$

Since the expression on the right-hand side is a cubic polynomial in u, this differential equation may be integrated in terms of elliptic functions: putting

$$u = \frac{4}{\alpha} U + \frac{1}{3\alpha},$$

the equation becomes

$$\left(\frac{dU}{d\theta}\right)^2 = 4U^3 - g_2 U - g_3,$$

where

$$g_2 = \frac{1}{12} - \frac{\alpha^2 \beta}{4}, \qquad g_3 = \frac{1}{216} + \frac{\alpha^2\beta}{24} - \frac{\alpha^2 k^2}{16c^2}.$$

The integral is therefore

$$U = \wp(\theta + C)$$

where C denotes a constant of integration; so *the equation of the orbit in terms of the coordinates r and θ is*

$$\frac{\alpha}{4r} = \frac{1}{12} + \wp(\theta + C).$$

Among these orbits we shall consider specially the following:

(i) *Quasi-elliptic orbits.*

When g_2 and g_3 are real and the discriminant $\Delta = g_2^3 - 27g_3^2$ is positive, the three roots e_1, e_2, e_3, are all real: let them be arranged in the order $e_1 > e_3 > e_2$. Then

$$\omega_1 = \int_{e_1}^{\infty} \frac{dz}{(4z^3 - g_2 z - g_3)^{\frac{1}{2}}} \text{ is purely real,}$$

and

$$\omega_2 = i\int_{-e_2}^{\infty} \frac{dz}{(4z^3 - g_2 z - g_3)^{\frac{1}{2}}} \text{ is purely imaginary.}$$

Under these circumstances, for the orbit whose equation is

$$\frac{\alpha}{4r} = \frac{1}{12} + \wp(\theta - \omega_3),$$

we have

$$\text{when } \theta = 0, \quad \frac{\alpha}{4r} = \frac{1}{12} + e_3, \quad \frac{d}{d\theta}\left(\frac{1}{r}\right) = 0, \quad \frac{d^2}{d\theta^2}\left(\frac{1}{r}\right) < 0,$$

$$\text{and wnen } \theta = \omega_1, \quad \frac{\alpha}{4r} = \frac{1}{12} + e_2, \quad \frac{d}{d\theta}\left(\frac{1}{r}\right) = 0, \quad \frac{d^2}{d\theta^2}\left(\frac{1}{r}\right) > 0,$$

so there is a pericentre at $\theta = 0$ and an apocentre at $\theta = \omega_1$, provided $\frac{1}{12} + e_2$ is positive, a condition which is readily found to be equivalent to $c^2\beta > k^2$. For such orbits, the radius vector r oscillates between the two fixed finite values $\dfrac{\alpha}{\frac{1}{3} + 4e_3}$ and $\dfrac{\alpha}{\frac{1}{3} + 4e_2}$, so that the planet's orbit is comprised between two concentric circles whose centres are at the sun: the motion of the planet in fact has a general resemblance to elliptic motion under the Newtonian law,

except that there is a progressive motion of the apsides: between the consecutive pericentres or apocentres, the increment in the angle θ is not 2π as it is under the Newtonian law, but $2\omega_1$, which differs from 2π. In our actual planetary system, however, the difference $2\omega_1 - 2\pi$ is too small to be observed except in the case of the planet Mercury.

Evidently these "*quasi-elliptic orbits*" as we may call them, *are periodic whenever* ω_1 *is commensurable with* π. Thus we obtain a family of ∞^2 periodic orbits which have a pericentre on the line $\theta = 0$, or taking account of the fact that rotations round the origin transfer orbits into orbits, *we have* ∞^3 *periodic orbits of this class.*

(ii) *Orbits doubly-asymptotic to circles.*

Suppose now that the constants k and β (which depend on the initial

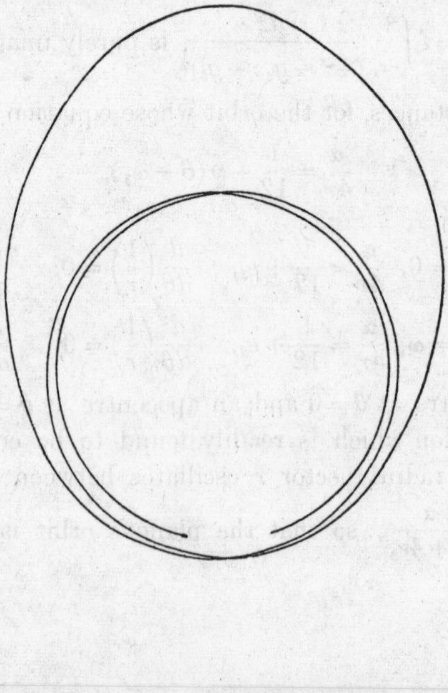

conditions) are such that the discriminant Δ is zero, so that two of the roots e_1, e_2, e_3 are equal. Let e be the repeated root, so the other root is $-2e$; when e is positive, putting $3e = n^2$, we find readily that the differential equation is satisfied by

$$\frac{\alpha}{4r} = \frac{1}{12} + \frac{n^2}{3} - \frac{n^2}{\cosh^2 n\theta}.$$

There is an apocentre at $\theta = 0$ if this gives a positive value for r, that is, if $8n^2 < 1$, a condition which is readily found to be equivalent to $\alpha^2\beta > \frac{1}{4}$.

When θ increases or decreases indefinitely, the orbit approaches spirally to the asymptotic circle

$$\frac{\alpha}{4r} = \frac{1}{12} + \frac{n^2}{3}$$

which is interior to the orbit. As n^2 lies between 0 and $\frac{1}{8}$, the radius of the asymptotic circle lies between 3α and 2α.

(iii) *Circular periodic orbits.*

If in the quasi-elliptic orbits we suppose the aphelion distance to be equal to the perihelion distance, we obtain a circular orbit. In this case $e_2 = e_3$, where e_2 and e_3 are the smaller roots of the cubic: the discriminant Δ is zero, and the repeated root of the cubic is negative, so $\frac{\alpha}{4r} < \frac{1}{12}$; and therefore, *the radius of the circular orbit, if it is a limiting case of a quasi-elliptic orbit, must be* $> 3\alpha$.

We can however have a different class of circular orbits, for which the radius is $< 3\alpha$, namely the circles to which the asymptotic orbits discussed under (ii) are asymptotic: for these the repeated root of the cubic is positive. *These circular orbits for which $r < 3\alpha$ are unstable,* since orbits exist which are spirally asymptotic to them.

171. *The motion of a particle on an ellipsoid under no external forces.*

As a second example illustrating the general theory of orbits, we shall consider the motion of a particle on the surface of an ellipsoid under no external forces. As we have seen in §54, the particle describes a geodesic on the surface, so the theory of the orbits is simply the theory of the geodesics on an ellipsoid, and the periodic solutions are simply those geodesics which are closed curves. Now for a geodesic on an ellipsoid we have Joachimstal's equation

$$pd = \text{Constant},$$

where p denotes the perpendicular from the centre of the ellipsoid on the tangent-plane at the point, and d is the diameter parallel to the tangent to the geodesic at the point. The same equation holds for the lines of curvature on the ellipsoid; so that every geodesic may be associated with a line of curvature, namely, that line of curvature for which pd has the same value as it has for the geodesic. We shall speak of the geodesic as "belonging to" the line of curvature. There is only one line of curvature having a prescribed value for pd, but there is an infinite number of geodesics having this value for pd, so that an infinite number of geodesics "belong to" each line of curvature. Now the line of curvature consists of two closed curves on the ellipsoid (being in fact the intersection of the ellipsoid with a confocal quadric): the region between these two portions of the line of curvature is a belt extending round the ellipsoid: and all the geodesics which belong to this line of curvature are comprised within this belt*, and touch the two portions of the line of curvature alternately. The matter is represented schematically in the diagram, where ABCDEF and PQRSTU are the two portions of the line of curvature,

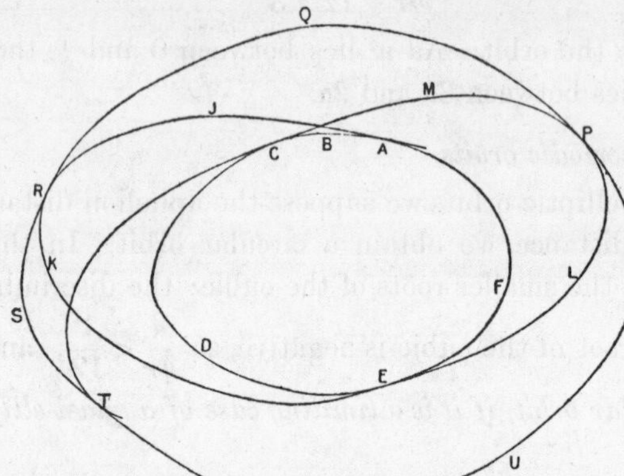

and AJRKELPMCT is an arc of one of the geodesics belonging to it, touching one of the portions of the line of curvature at A, C, E, and touching the other portion at R, P, T.

In order that the geodesic may be closed, it is necessary (as in all poristic problems) that a certain parameter (depending in this case on the value of the constant pd of the line of curvature) should be a rational number: the geodesic is unclosed if this parameter is an irrational number. If it is closed, then there are ∞^1 other geodesics which belong to the same line of curvature and which are also closed; but if it is not closed, then no other geodesic belonging to this particular line of curvature can be a closed geodesic†.

Now consider the connection between the ∞^1 members of the family of geodesics which belong to the same line of curvature. It is known (§ 144) that if

$$\phi(q_1, q_2, p_1, p_2) = \text{Constant}$$

* Ignoring the exceptional case of those geodesics which pass through an umbilicus.

† This is obvious in the case when the ellipsoid is of revolution: for then the two portions of the line of curvature are parallel circles on the surface, and the ∞^1 geodesics which belong to this line of curvature are obtained from each other by mere rotation about the axis of symmetry.

is an integral of a dynamical system, then the infinitesimal contact-transformation which is defined by the equations

$$\delta q_1 = \epsilon \frac{\partial \phi}{\partial p_1}, \quad \delta q_2 = \epsilon \frac{\partial \phi}{\partial p_2}, \quad \delta p_1 = -\epsilon \frac{\partial \phi}{\partial q_1}, \quad \delta p_2 = -\epsilon \frac{\partial \phi}{\partial q_2}$$

(where ϵ is a small constant) transforms any trajectory into an adjacent curve which is also a trajectory. If we apply this theorem to the motion on the ellipsoid, we find without much difficulty that the infinitesimal transformation which corresponds to the integral

$$pd = \text{Constant}$$

transforms any geodesic into another geodesic which belongs to the same line of curvature.

Summing up, we see that *the* ∞^2 *orbits of a particle moving on an ellipsoid under no external forces may be classified into* ∞^1 *families, each family consisting of* ∞^1 *orbits: the members of any one family are either all closed or all unclosed: and a certain continuous group of transformations, which is closely associated with the integral* pd = *Constant, transforms any orbit into all the orbits which belong to the same family.*

172. *Ordinary and Singular periodic solutions.*

Still considering the motion of a particle on an ellipsoid under no external forces, we now observe that besides the geodesics which can be arranged in families, there are on the ellipsoid three other closed geodesics, namely, the three principal sections of the ellipsoid. These have quite a different character they are solitary, instead of belonging to families: and the infinitesimal transformation which has just been mentioned transforms them not into other geodesics but into themselves—that is, they are invariant under the transformation. This last property suggests a resemblance with the theory of "singular solutions" of ordinary differential equations of the first order: for if a differential equation of the first order admits a particular infinitesimal transformation, then this infinitesimal transformation changes the ordinary integral-curves into each other, but it leaves invariant the singular integral-curve. On account of this resemblance a periodic solution (of a dynamical system with two degrees of freedom) is called* *ordinary* if it belongs to a continuous family of ∞^1 periodic solutions for which the constant of energy has the same value, and which are transformed into each other by the infinitesimal transformation belonging to a certain integral (this is specified more closely later on); but a periodic solution is to be called *singular* if there is no periodic solution adjacent to it which corresponds to the same value of the constant of energy: the above-mentioned infinitesimal transformation leaves the singular periodic solutions invariant.

It should be noticed that we have inserted the condition "for which the

* Whittaker, *Proc. R. S. Edinburgh*, xxxvii. (1916), p. 95.

constant of energy has the same value." If we suppose the constant of energy to vary, an "ordinary" periodic solution of a dynamical system with two degrees of freedom is in general a member of a continuous family of ∞^2 periodic solutions, over which the period varies continuously, being constant over singly-infinite sub-families. A "singular" periodic solution is a member of a family of ∞^1 periodic solutions*, over which the period varies continuously.

In the problem of the motion of a planet about a single attracting mass in general relativity (§ 170), the periodic "quasi-elliptic" orbits are "ordinary" periodic orbits, while the circular periodic orbits are "singular."

There are marked differences between the properties of "ordinary" and those of "singular" periodic solutions. For instance, the "asymptotic solutions" of § 169 can exist only in connection with *singular* periodic solutions, and not in connection with ordinary periodic solutions: an illustration of this is again afforded by the theory of geodesics on quadrics; for the only asymptotic solutions among the geodesics of quadrics are those geodesics which wind round and round the hyperboloid of one sheet, becoming ultimately asymptotic to the principal elliptic section of the hyperboloid: and this elliptic section is a singular periodic solution.

The words "ordinary" and "singular" suggest that the "ordinary" periodic orbits occur most frequently, while the "singular" orbits are exceptional: and indeed we find this to be true, so long as we confine ourselves to studying the *soluble* problems of dynamics. It is therefore with some surprise that we learn—what was first shewn by T. M. Cherry in 1927—that for Hamiltonian systems with two degrees of freedom *in general* there are no "ordinary" periodic solutions: all the periodic solutions are "singular." The explanation of the apparent paradox is that a Hamiltonian system is in general *insoluble*, and for insoluble systems the periodic orbits are of the "singular" type.

The paradox may be compared to a familiar one in the theory of partial differential equations. If a partial differential equation of the first order

$$f(x, y, z, p, q) = 0$$

is given, the student is told that by eliminating p and q between the three equations

$$f = 0, \quad \frac{\partial f}{\partial p} = 0, \quad \frac{\partial f}{\partial q} = 0$$

he will obtain in general an equation $F(x, y, z) = 0$ which represents the singular solution of the partial differential equation: and he finds that, for the particular partial differential equations which are given in text-books, this statement is true. Nevertheless, as was shewn in 1883 by Darboux, a partial differential equation of the first order *in general* does not possess a singular solution: the equation $F(x, y, z) = 0$ in general represents the locus of the cusps of the characteristics. The apparent contradiction between the student's experience and Darboux's theorem is explained by the circumstance that the equations given as examples in the text-books have mostly been manufactured by starting with a complete

* The case of motion on a surface under no external forces is exceptional, as in it the value of the constant of energy is immaterial.

integral which represents a family of surfaces, and eliminating the two arbitrary constants: and a partial differential equation formed in this way *does* in general possess a singular solution, namely the envelope of the surfaces represented by the complete integral.

Cherry's theorem on periodic orbits and Darboux's theorem on partial differential equations alike warn us not to take the properties of the more familiar "soluble" systems as representative of the properties of systems in general.

173. *Characteristic exponents.*

The stability of types of motion of dynamical systems may be discussed by the aid of certain constants to which Poincaré has given the name *characteristic exponents**.

Consider any set of differential equations

$$\frac{dx_i}{dt} = X_i \qquad (i = 1, 2, \ldots, n),$$

where $(X_1, X_2, \ldots, X_n)$ are functions of $(x_1, x_2, \ldots, x_n)$ and possibly also of t, having a period T in t; and suppose that a periodic solution of these equations is known, defined by the equations

$$x_i = \phi_i(t) \qquad (i = 1, 2, \ldots, n),$$

where

$$\phi_i(t+T) = \phi_i(t) \qquad (i = 1, 2, \ldots, n).$$

In order to investigate solutions adjacent to this, we write

$$x_i = \phi_i(t) + \xi_i \qquad (i = 1, 2, \ldots, n),$$

where $(\xi_1, \xi_2, \ldots, \xi_n)$ are supposed to be small, and are given by the variational equations (§ 112)

$$\frac{d\xi_i}{dt} = \sum_{k=1}^{n} \xi_k \frac{\partial X_i}{\partial x_k} \qquad (i = 1, 2, \ldots, n).$$

As these are linear differential equations, with coefficients periodic in the independent variable t, it is known from the general theory of linear differential equations that each of the variables ξ_i will be of the form

$$\sum_{k=1}^{n} e^{\alpha_k t} S_{ik},$$

where the quantities S_{ik} denote periodic functions of t with the period T, and the n quantities α_k are constants, which are called the *characteristic exponents* of the periodic solution.

If all the characteristic exponents are purely imaginary, the functions $(\xi_1, \xi_2, \ldots, \xi_n)$ can evidently be expressed as sums and products of purely periodic terms; while this is evidently not the case if the characteristic exponents are not all purely imaginary. Hence *a necessary condition for stability of the periodic orbit is that all the characteristic exponents must be purely imaginary.*

* *Acta Math.* XIII. (1890), p. 1; *Nouv. Méth. de la Méc. Cél.* I. (1892). On the general problem of stability the reader should consult the extensive memoir of A. Liapounoff, originally published in 1892 by the Math. Soc. of Kharkow, and translated into French by E. Davaux, *Annales de Toulouse* (2), IX. (1907), p. 203.

We shall now find the equation which determines the characteristic exponents of a given solution.

In one of the orbits adjacent to the given periodic orbit, let $(\beta_1, \beta_2, \ldots, \beta_n)$ denote the initial values of $(\xi_1, \xi_2, \ldots, \xi_n)$ and let $\beta_i + \psi_i$ be the value of ξ_i after the lapse of a period. As the quantities $(\psi_1, \psi_2, \ldots, \psi_n)$ are one-valued functions of $(\beta_1, \beta_2, \ldots, \beta_n)$, which are zero when $(\beta_1, \beta_2, \ldots, \beta_n)$ are all zero, we have by Taylor's theorem (neglecting squares and products of $\beta_1, \beta_2, \ldots, \beta_r$

$$\psi_i = \frac{\partial \psi_i}{\partial \beta_1}\beta_1 + \frac{\partial \psi_i}{\partial \beta_2}\beta_2 + \ldots + \frac{\partial \psi_i}{\partial \beta_n}\beta_n \qquad (i = 1, 2, \ldots, n).$$

If α_k is one of the characteristic exponents, one of the adjacent orbits will be defined by equations of the form

$$\xi_1 = e^{\alpha_k t} S_{1k}, \qquad \xi_2 = e^{\alpha_k t} S_{2k}, \quad \ldots, \qquad \xi_n = e^{\alpha_k t} S_{nk},$$

so that

$$\beta_i + \psi_i = e^{\alpha_k T} S_{ik}(0) = e^{\alpha_k T} \beta_i \qquad (i = 1, 2, \ldots, n),$$

and consequently a set of values of $\beta_1, \beta_2, \ldots, \beta_n$ exists for which the equations

$$\frac{\partial \psi_i}{\partial \beta_1}\beta_1 + \frac{\partial \psi_i}{\partial \beta_2}\beta_2 + \ldots + \left(\frac{\partial \psi_i}{\partial \beta_i} + 1 - e^{\alpha_k T}\right)\beta_i + \ldots + \frac{\partial \psi_i}{\partial \beta_n}\beta_n = 0$$
$$(i = 1, 2, \ldots, n)$$

are satisfied: the quantity α_k must therefore be a root of the equation in α

$$\begin{vmatrix} \frac{\partial \psi_1}{\partial \beta_1} + 1 - e^{\alpha T} & \frac{\partial \psi_1}{\partial \beta_2} & \cdots\cdots & \frac{\partial \psi_1}{\partial \beta_n} \\ \frac{\partial \psi_2}{\partial \beta_1} & \frac{\partial \psi_2}{\partial \beta_2} + 1 - e^{\alpha T} & \cdots\cdots & \frac{\partial \psi_2}{\partial \beta_n} \\ \cdots\cdots & \cdots\cdots & \cdots\cdots & \cdots\cdots \\ \cdots\cdots & \cdots\cdots & \cdots\cdots & \cdots\cdots \\ \frac{\partial \psi_n}{\partial \beta_1} & \frac{\partial \psi_n}{\partial \beta_2} & \cdots\cdots & \cdots\cdots \end{vmatrix} = 0.$$

*The characteristic exponents are therefore the roots of this determinantal equation**.

174. *Characteristic exponents when t does not occur explicitly.*

When t is not contained explicitly in the functions $(X_1, X_2, \ldots, X_n)$, it is evident that if

$$x_i = \phi_i(t) \qquad (i = 1, 2, \ldots, n)$$

is a solution of the equations, then

$$x_i = \phi_i(t + \epsilon) \qquad (i = 1, 2, \ldots, n)$$

is also a solution, where ϵ is an arbitrary constant. The equations

$$\xi_i = \frac{\partial}{\partial \epsilon}\phi_i(t + \epsilon) \qquad (i = 1, 2, \ldots, n)$$

* Cf. H. F. Baker, *Proc. Camb. Phil. Soc.* xx. (1920), p. 181.

therefore define a particular solution of the variational equations; but as $\partial\phi_i(t+\epsilon)/\partial\epsilon$ is evidently a periodic function of t, it follows that the coefficient $e^{\alpha_k t}$ reduces in this case to unity: and hence *when t is not contained explicitly in the original differential equations, one of the characteristic exponents of every periodic solution is zero.*

175. *The characteristic exponents of a system which possesses a one-valued integral.*

Suppose next that the system possesses an integral of the form

$$F(x_1, x_2, \ldots, x_n) = \text{Constant}$$

where F is a one-valued function of $(x_1, x_2, \ldots, x_n)$ and does not involve t. In the notation of the last article, we have

$$F\{\phi_i(0)+\beta_i+\psi_i\} = F\{\phi_i(0)+\beta_i\}$$

where for brevity $F(x_i)$ is written in place of $F(x_1, x_2, \ldots, x_n)$. Differentiating this equation with respect to β_i, we have

$$\frac{\partial F}{\partial x_1}\frac{\partial \psi_1}{\partial \beta_i} + \frac{\partial F}{\partial x_2}\frac{\partial \psi_2}{\partial \beta_i} + \ldots + \frac{\partial F}{\partial x_n}\frac{\partial \psi_n}{\partial \beta_i} = 0 \qquad (i=1, 2, \ldots, n),$$

where in $\partial F/\partial x_1$, $\partial F/\partial x_2$, etc., the quantities $(x_1, x_2, \ldots, x_n)$ are to be replaced by $\phi_1(0), \phi_2(0), \ldots, \phi_n(0)$. From these equations it follows that either the Jacobian $\dfrac{\partial(\psi_1, \psi_2, \ldots, \psi_n)}{\partial(\beta_1, \beta_2, \ldots, \beta_n)}$ is zero, or else the quantities $\dfrac{\partial F}{\partial x_1}, \dfrac{\partial F}{\partial x_2}, \ldots, \dfrac{\partial F}{\partial x_n}$ are all zero when $t=0$.

Now if the latter alternative is correct, we see that (since the origin of time is arbitrary) the equations

$$\frac{\partial F}{\partial x_1}=0, \quad \frac{\partial F}{\partial x_2}=0, \quad \ldots, \quad \frac{\partial F}{\partial x_n}=0 \quad \ldots\ldots\ldots\ldots\ldots\ldots (1)$$

must be satisfied at all points of the periodic solution. This actually happens, e.g., in the case of the circular periodic orbits of a planet round a single attracting mass in general relativity (§ 170 (iii)): for the general differential equations connecting the radius r with the time t in motion round the attracting mass are readily seen from § 170 to be $\left(\text{writing } \rho \text{ for } \dfrac{1}{1-(\alpha/r)}\dfrac{dr}{dt}\right)$

$$\begin{cases} \dfrac{d\rho}{dt} = \dfrac{\alpha\rho^2}{2r^2} - \tfrac{1}{2}\dfrac{\alpha c^2}{r^2} + \dfrac{\left(1-\dfrac{\alpha}{r}\right)^2}{r^3}\dfrac{c^4}{k^2} \\ \dfrac{dr}{dt} = \rho\left(1-\dfrac{\alpha}{r}\right) \end{cases}$$

and these equations possess the one-valued integral

$$\frac{\rho^2}{c^4\left(1-\dfrac{\alpha}{r}\right)} - \frac{1}{c^2\left(1-\dfrac{\alpha}{r}\right)} + \frac{1}{k^2 r^2} = -\frac{\beta}{k^2}.$$

If this integral is written $F(r, \rho) = $ Constant, it is easily verified that for a circular orbit we have

$$\frac{\partial F}{\partial r} = 0, \quad \text{and} \quad \frac{\partial F}{\partial \rho} = 0$$

which shews that for these circular periodic orbits the conditions (1) are satisfied.

However, it is evidently only in quite exceptional cases such as this that the conditions (1) are satisfied at all points of the periodic solution, and the other alternative, namely that the Jacobian $\partial(\psi_1, \psi_2, \ldots, \psi_n) / \partial(\beta_1, \beta_2, \ldots, \beta_n)$ vanishes, must be in general the true one: but when the Jacobian is zero, the determinantal equation for the characteristic exponents is evidently satisfied by the value $e^{\alpha T} = 1$, i.e. by $\alpha = 0$: so that one of the characteristic exponents is zero. Thus *in general, if the differential equations possess a one-valued integral, one of the characteristic exponents is zero.*

Another way of arriving at this theorem, in the case when the equations are those of a dynamical system, is as follows: by § 144, if the system admits the integral

$$\phi(q_1, q_2, \ldots, q_n, p_1, \ldots, p_n, t) = \text{Constant},$$

then the variational equations are satisfied by

$$\delta q_r = \epsilon \frac{\partial \phi}{\partial p_r}, \quad \delta p_r = -\epsilon \frac{\partial \phi}{\partial q_r} \qquad (r = 1, 2, \ldots, n),$$

where ϵ is a small constant: but these values of δq_r and δp_r are periodic when the orbit whose variations are studied is a periodic orbit; and therefore the corresponding characteristic exponent is zero, which proves the theorem. The only exception is when the periodic orbit is transformed into *itself* by the infinitesimal transformation corresponding to the integral,—which again gives us the exceptional case of "singular" periodic orbits. These are, in fact, the Levi-Civita particular solutions (§ 149) belonging to the integral.

Example. If the differential equations do not involve the time explicitly, and possess p one-valued integrals $F_1, \ldots, F_p$ which do not involve t, shew that either $(p+1)$ characteristic exponents are zero, or that all the determinants contained in the array

$$\left\| \frac{\partial F_\iota}{\partial x_\kappa} \right\| \qquad (\iota = 1, 2, \ldots, p;\ \kappa = 1, 2, \ldots, n)$$

are zero at all points of the periodic solution considered. (Poincaré.)

176. *The theory of matrices.*

In the next article it will be necessary to make use of matrices: the elementary notions of the theory of matrices will therefore be given here for the benefit of readers who have no previous acquaintance with it.

Consider any square array

$$\begin{pmatrix} a_{11} & a_{12} & a_{13} \cdots\cdots & a_{1n} \\ a_{21} & a_{22} & a_{23} \cdots\cdots & a_{2n} \\ \cdots & \cdots & \cdots & \cdots \\ \cdots & \cdots & \cdots & \cdots \\ a_{n1} & a_{n2} & a_{n3} \cdots\cdots & a_{nn} \end{pmatrix}$$

formed of ordinary (real or complex) numbers a_{pq}, which will be called the *elements.* This array will be called a *matrix*, and may be denoted by a single letter A or by the notation (a_{pq}). It may be thought of as representing an operation, namely the operation of performing the linear substitution

$$\begin{cases} y_1 = a_{11}x_1 + a_{12}x_2 + \ldots\ldots + a_{1n}x_n, \\ y_2 = a_{21}x_1 + a_{22}x_2 + \ldots\ldots + a_{2n}x_n, \\ \ldots\ldots\ldots\ldots\ldots\ldots\ldots\ldots\ldots\ldots \\ \ldots\ldots\ldots\ldots\ldots\ldots\ldots\ldots\ldots\ldots \\ y_n = a_{n1}x_1 + a_{n2}x_2 + \ldots\ldots + a_{nn}x_n: \end{cases}$$

but the theory of matrices is based on the idea that a matrix is a *number*, in the general sense of the word number, so that matrices may be added, multiplied, etc. Two matrices $A \equiv (a_{pq})$ and $B \equiv (b_{pq})$ are said to be *equal* when their elements are equal, each to each: that is, $a_{pq} = b_{pq}$ for $p, q = 1, 2, \ldots, n$. The product BA of any two matrices $B \equiv (b_{pq})$ and $A \equiv (a_{pq})$ is defined to be the matrix which has

$$b_{p1}a_{1q} + b_{p2}a_{2q} + b_{p3}a_{3q} + \ldots + b_{pn}a_{nq}$$

as the element of the pth row and q^{th} column. The matrix BA corresponds to the operation of performing the substitutions A and B in succession.

Matrix multiplication does not in general satisfy the commutative law: that is, AB and BA are in general two different matrices. But matrix multiplication satisfies the associative law

$$A(BC) = (AB)C.$$

The matrix

$$\begin{pmatrix} 1 & 0 & 0 & 0\ldots \\ 0 & 1 & 0 & 0\ldots \\ 0 & 0 & 1 & 0\ldots \\ \ldots & \ldots & \ldots & \ldots \\ \ldots & \ldots & \ldots & \ldots \end{pmatrix}$$

is called the *unit matrix* and is denoted by E. The matrix B such that $BA = E$ is called the *reciprocal* matrix of A, and is denoted by A^{-1}. The matrix which is obtained from A by interchanging its rows and columns is called the *conjugate* matrix of A, and is denoted by A'.

If $A \equiv (a_{pq})$ is a matrix, the roots of the determinantal equation in r

$$\begin{vmatrix} a_{11} - r & a_{12} & a_{13} \ldots\ldots\ldots\ldots \\ a_{21} & a_{22} - r & a_{23} \ldots\ldots\ldots\ldots \\ \ldots\ldots & \ldots\ldots & \ldots\ldots\ldots\ldots \\ \ldots\ldots & \ldots\ldots & \ldots\ldots\ldots\ldots \\ a_{n1} \ldots\ldots & \ldots\ldots & \ldots\ldots a_{nn} - r \end{vmatrix} = 0$$

are called the *latent roots* of the matrix A. One of the chief theorems of the theory of matrices is *Sylvester's law of latency*, viz., that *if $r_1, r_2, \ldots, r_n$ are the latent roots of a matrix A, then the latent roots of any function of A, say $f(A)$, are $f(r_1), f(r_2), \ldots, f(r_n)$.* In particular, *the latent roots of the matrix A^{-1} are the reciprocals of the latent roots of A.*

If A be a matrix, and S another matrix, the matrix SAS^{-1} has the same latent roots as A.

(It will be understood that the above theorems have been stated in their general form without referring to exceptional cases.)

177. *The characteristic exponents of a Hamiltonian system.*

Consider now in particular a conservative dynamical system which for simplicity we shall suppose to have two degrees of freedom. The equations of motion may be written

$$\frac{dq_1}{dt}=\frac{\partial H}{\partial p_1},\quad \frac{dq_2}{dt}=\frac{\partial H}{\partial p_2},\quad \frac{dp_1}{dt}=-\frac{\partial H}{\partial q_1},\quad \frac{dp_2}{dt}=-\frac{\partial H}{\partial q_2},$$

where H is a given function of (q_1, q_2, p_1, p_2). Suppose that this system admits a periodic solution

$$q_1=Q_1(t),\quad q_2=Q_2(t),\quad p_1=P_1(t),\quad p_2=P_2(t),$$

and let an adjacent solution be represented by

$$q_1=Q_1+\xi_1,\quad q_2=Q_2+\xi_2,\quad p_1=P_1+\varpi_1,\quad p_2=P_2+\varpi_2.$$

Then the equations to determine $(\xi_1, \xi_2, \varpi_1, \varpi_2)$ are evidently

$$\frac{d\xi_1}{dt}=\xi_1\frac{\partial^2 H}{\partial p_1\partial q_1}+\xi_2\frac{\partial^2 H}{\partial p_1\partial q_2}+\varpi_1\frac{\partial^2 H}{\partial p_1{}^2}+\varpi_2\frac{\partial^2 H}{\partial p_1\partial p_2}\quad\ldots\ldots\ldots\ldots(1),$$

and three similar equations.

Let $(\xi_1', \xi_2', \varpi_1', \varpi_2')$ be another solution of these equations, so

$$\frac{d\xi_1'}{dt}=\xi_1'\frac{\partial^2 H}{\partial p_1\partial q_1}+\xi_2'\frac{\partial^2 H}{\partial p_1\partial q_2}+\varpi_1'\frac{\partial^2 H}{\partial p_1{}^2}+\varpi_2'\frac{\partial^2 H}{\partial p_1\partial p_2}\quad\ldots\ldots\ldots(2),$$

and three similar equations.

Multiply equations (1) by $\varpi_1', \varpi_2', -\xi_1', -\xi_2'$, respectively, and the equations (2) by $-\varpi_1, -\varpi_2, \xi_1, \xi_2$, respectively, and add. We obtain

$$\frac{d}{dt}(\xi_1\varpi_1'+\xi_2\varpi_2'-\xi_1'\varpi_1-\xi_2'\varpi_2)=0,$$

and hence the integrated equation

$$\xi_1\varpi_1'+\xi_2\varpi_2'-\xi_1'\varpi_1-\xi_2'\varpi_2=\text{Constant}\quad\ldots\ldots\ldots\ldots\ldots(3).$$

Now the values, which $(\xi_1, \xi_2, \varpi_1, \varpi_2)$ acquire when t is increased by a complete period T, are linear functions of their values at the beginning of the period: let the transformation which gives the values at the end of a period, $(\bar{\xi}_1, \bar{\xi}_2, \bar{\varpi}_1, \bar{\varpi}_2)$, in terms of the values at the beginning of the period, $(\xi_1, \xi_2, \varpi_1, \varpi_2)$, be

$$\begin{aligned}
\bar{\xi}_1&=r_{11}\xi_1+r_{12}\xi_2+r_{13}\varpi_1+r_{14}\varpi_2,\\
\bar{\xi}_2&=r_{21}\xi_1+r_{22}\xi_2+r_{23}\varpi_1+r_{24}\varpi_2,\\
\bar{\varpi}_1&=r_{31}\xi_1+r_{32}\xi_2+r_{33}\varpi_1+r_{34}\varpi_2,\\
\bar{\varpi}_2&=r_{41}\xi_1+r_{42}\xi_2+r_{43}\varpi_1+r_{44}\varpi_2.
\end{aligned}$$

Denote the transformation or matrix (r_{pq}) by R. Then by (3), R transforms $\xi_1\varpi_1'+\xi_2\varpi_2'-\xi_1'\varpi_1-\xi_2'\varpi_2$ into itself, and therefore

$$\begin{aligned}
&\xi_1\varpi_1'+\xi_2\varpi_2'-\xi_1'\varpi_1-\xi_2'\varpi_2\\
&\quad=(r_{11}\xi_1+r_{12}\xi_2+r_{13}\varpi_1+r_{14}\varpi_2)(r_{31}\xi_1'+r_{32}\xi_2'+r_{33}\varpi_1'+r_{34}\varpi_2')\\
&\qquad+\text{three similar products.}
\end{aligned}$$

Equating coefficients of $\xi_1 \xi_1'$ etc. in this equation, we have a set of equations which are comprehended in the matrix-equation

$$\begin{pmatrix} 0 & 0 & 1 & 0 \\ 0 & 0 & 0 & 1 \\ -1 & 0 & 0 & 0 \\ 0 & -1 & 0 & 0 \end{pmatrix}$$

$$= \begin{pmatrix} 0 & -r_{31}r_{12}-r_{41}r_{22}+r_{11}r_{32}+r_{21}r_{42} & -r_{31}r_{13}-r_{41}r_{23}+r_{11}r_{33}+r_{21}r_{43} & \cdots \\ -r_{32}r_{11}-r_{42}r_{21}+r_{12}r_{31}+r_{22}r_{41} & 0 & -r_{32}r_{13}-r_{42}r_{23}+r_{12}r_{33}+r_{22}r_{43} & \cdots \\ \cdots & \cdots & \cdots & \cdots \\ \cdots & \cdots & \cdots & \cdots \end{pmatrix}$$

The matrix on the right-hand side is equal to

$$\begin{pmatrix} -r_{31} & -r_{41} & r_{11} & r_{21} \\ -r_{32} & -r_{42} & r_{12} & r_{22} \\ -r_{33} & -r_{43} & r_{13} & r_{23} \\ -r_{34} & -r_{44} & r_{14} & r_{24} \end{pmatrix} \begin{pmatrix} r_{11} & r_{12} & r_{13} & r_{14} \\ r_{21} & r_{22} & r_{23} & r_{24} \\ r_{31} & r_{32} & r_{33} & r_{34} \\ r_{41} & r_{42} & r_{43} & r_{44} \end{pmatrix}$$

or to

$$\begin{pmatrix} r_{11} & r_{21} & r_{31} & r_{41} \\ r_{12} & r_{22} & r_{32} & r_{42} \\ r_{13} & r_{23} & r_{33} & r_{43} \\ r_{14} & r_{24} & r_{34} & r_{44} \end{pmatrix} \begin{pmatrix} 0 & 0 & 1 & 0 \\ 0 & 0 & 0 & 1 \\ -1 & 0 & 0 & 0 \\ 0 & -1 & 0 & 0 \end{pmatrix} \begin{pmatrix} r_{11} & r_{12} & r_{13} & r_{14} \\ r_{21} & r_{22} & r_{23} & r_{24} \\ r_{31} & r_{32} & r_{33} & r_{34} \\ r_{41} & r_{42} & r_{43} & r_{44} \end{pmatrix}$$

Therefore denoting the matrix

$$\begin{pmatrix} 0 & 0 & 1 & 0 \\ 0 & 0 & 0 & 1 \\ -1 & 0 & 0 & 0 \\ 0 & -1 & 0 & 0 \end{pmatrix}$$

by S, we have $S = R'SR$, where R' denotes the matrix conjugate to R; and therefore $(R')^{-1} = SRS^{-1}$.

This equation shews that $(R')^{-1}$ has the same latent roots as R, and therefore that R^{-1} has the same latent roots as R, and therefore that *the set of the latent roots of R is the same as the set of their reciprocals*: the reciprocal of every latent root of R is itself a latent root of R. Now if there exists a linear combination $\eta = \alpha\xi_1 + \beta\xi_2 + \gamma\varpi_1 + \delta\varpi_2$ of $(\xi_1, \xi_2, \varpi_1, \varpi_2)$, such that its value $\bar{\eta}$ after completing a period satisfies the equation

$$\bar{\eta} = \lambda\eta \quad \text{.................................(4)},$$

then on writing down the equations which express this condition, it follows at once that λ must be a latent root of the matrix R. But as we have seen in the last article, any solution of the equations (1) is of the form

$$\sum_k e^{a_k t} S_k(t) \quad \text{...............................(5)},$$

where the α_k's are the characteristic exponents and the S_k's are periodic functions of t of period T; and equation (4) can be satisfied by an expression of the form (5) only when η involves *a single one* of the characteristic exponents, say

$$\eta = e^{\alpha_k t} S_k(t) \qquad \text{.......(6).}$$

Substituting from (6) in (4), we have

$$\lambda = e^{\alpha_k T},$$

so *the latent roots of the matrix R are the quantities $e^{\alpha_k T}$, where the α_k's are the characteristic exponents.*

Thus the result that the set of latent roots of R is the same as the set of their reciprocals now yields the theorem that *when the differential equations form a Hamiltonian system, the characteristic exponents of any periodic solution may be arranged in pairs, the exponents of each pair being equal in magnitude but opposite in sign**.

From the results of this article and § 174 it is evident that *in the motion of a particle in a plane under the action of conservative forces, the characteristic exponents of any periodic orbit are* $(0, 0, \alpha, -\alpha)$, *where α is some number.*

Example 1. *Shew that, in the motion of a particle in a plane under the action of conservative forces, if u_n, u_{n+1}, u_{n+2} denote the normal displacements in an orbit adjacent to a known periodic orbit in three consecutive revolutions, the ratio $k = (u_{n+2} + u_n)/u_{n+1}$ has a constant value, which is the same for all adjacent orbits.*

(Korteweg, *Wiener Sitzungsber.* XCIII. (1886).)

This number k is called the *index of stability* of the periodic orbit. Let the characteristic exponents of the periodic orbit be $(0, 0, a, -a)$: then the characteristic exponent a is connected with the index of stability k and the period T by the equation

$$k = 2 \cosh aT.$$

Example 2. *Shew that* (so far as the question of stability is determined by considering small displacements) *a periodic orbit is stable or not according as the associated index of stability is less or greater* (*in absolute value*) *than two.*

This of course corresponds to the fact that the orbit is stable or unstable according as a is purely imaginary or not.

Example 3. Discuss the transitional case in which the index of stability has one of the values ± 2: shewing that the equation of the adjacent orbits is of one of the forms

$$u = K_1 \{\phi(s) + s\psi(s)\} + K_2 \psi(s),$$

$$u = K_1 \phi(s) + K_2 \psi(s),$$

where ϕ and ψ either have the period S (denoting by s the arc in the periodic orbit, when S denotes its complete length), or satisfy the equations

$$\phi(s+S) = -\phi(s), \quad \psi(s+S) = -\psi(s),$$

and that the known orbit may be either stable or unstable. (Korteweg.)

* This theorem is due to Poincaré, *Méc. Cél.* I. (1892), p. 193.

178. *The asymptotic solutions of* § 170 *deduced from the theory of characteristic exponents.*

We shall now shew how those orbits of planets in the general relativity-theory, which (§ 170) approach spirally to an asymptotic circular orbit, may be obtained by considering the characteristic exponents of this circular orbit. Writing down the equations of motion, namely (§ 170)

$$\frac{d}{ds}\left(\frac{r'}{1-\frac{\alpha}{r}}\right)+\frac{1}{2}\frac{\alpha c^2}{r^2}t'^2+\frac{\alpha r'^2}{2r^2\left(1-\frac{\alpha}{r}\right)^2}-r\theta'^2=0 \quad \text{............(1)},$$

$$\left(1-\frac{\alpha}{r}\right)t'=-\frac{k}{c^2\sqrt{\beta}} \quad \text{......(2)},$$

$$r^2\theta'=\frac{1}{\sqrt{\beta}} \quad \text{.........(3)},$$

and eliminating s and θ between them, we obtain

$$\frac{1}{1-\frac{\alpha}{r}}\frac{d^2r}{dt^2}-\frac{3\alpha}{2r^2\left(1-\frac{\alpha}{r}\right)^2}\left(\frac{dr}{dt}\right)^2+\frac{1}{2}\frac{\alpha c^2}{r^2}-r\left(\frac{d\theta}{dt}\right)^2=0 \quad \text{.........(4)},$$

while from (2) and (3) we have

$$\frac{r^2}{1-\frac{\alpha}{r}}\frac{d\theta}{dt}=-\frac{c^2}{k},$$

and differentiating this,

$$r^2\frac{d^2\theta}{dt^2}+2r\frac{dr}{dt}\frac{d\theta}{dt}-\frac{\alpha}{1-\frac{\alpha}{r}}\frac{dr}{dt}\frac{d\theta}{dt}=0 \quad \text{..................(5)}.$$

Equations (4) and (5) are the equations of motion when r and θ are taken as dependent variables and t as independent variable. If $r=r_0(t)$, $\theta=\theta_0(t)$ are the equations of an orbit, then in order to determine the small oscillations about this orbit we put

$$r=r_0(t)+\xi, \quad \theta=\theta_0(t)+\eta,$$

and neglect the squares and products of ξ and η. The characteristic exponents are found by solving the differential equations in ξ and η.

In particular, if the orbit is circular and of radius r_0, we must have

$$\frac{dr}{dt}=0 \text{ and } \frac{d^2r}{dt^2}=0, \text{ when } r=r_0,$$

and therefore by (4)

$$r_0^3\,\dot{\theta}_0^2=\tfrac{1}{2}\alpha c^2.$$

The equation in ξ now becomes, after a simple reduction,

$$\frac{d^2\xi}{dt^2}+\frac{\alpha c^2}{2r_0^3}\left(1-\frac{3\alpha}{r_0}\right)\xi=0 \quad \text{..........................(6)}.$$

It is evident from this equation that *a circular orbit can be stable only when its radius is greater than* 3α, as we found in § 170. When the radius is less than 3α, the solution of equation (6) is of the form

$$\xi = Ae^{t\left\{\frac{\alpha c^2}{2r_0^3}\left(\frac{3\alpha}{r_0}-1\right)\right\}^{\frac{1}{2}}} + Be^{-t\left\{\frac{\alpha c^2}{2r_0^3}\left(\frac{3\alpha}{r_0}-1\right)\right\}^{\frac{1}{2}}} \quad \text{..............(7).}$$

The characteristic exponents being $\pm\left\{\frac{\alpha c^2}{2r_0^3}\left(\frac{3\alpha}{r_0}-1\right)\right\}^{\frac{1}{2}}$, *the two terms on the right-hand side of* (7) *correspond to the two types of asymptotic solution to the circular orbit, namely the type which approaches the circle by a right-handed spiral and the type which approaches by a left-handed spiral.* It appeared from § 170 that a planet moving away from the circular orbit in a right-handed spiral, when it attains a certain distance from the attracting centre, passes through an aphelion after which it moves in again asymptotically towards the circular orbit in a left-handed spiral: but of course this fact cannot be inferred by the mere consideration of characteristic exponents.

179. *The characteristic exponents of "ordinary" and "singular" periodic solutions.*

We have seen (§ 177) that in a dynamical system with two degrees of freedom, the characteristic exponents of any periodic orbit are $(0, 0, \alpha, -\alpha)$, where α is some number; and in the last article we have seen that for the circular periodic orbits of a planet in general relativity-theory, which are "singular" orbits, the quantity α is not zero. On the other hand for the quasi-elliptic orbits (§ 170), which are "ordinary" orbits, and for which consequently the period is constant over a singly-infinite sub-family, α is zero. This result is true for Hamiltonian systems with two degrees of freedom in general: *for "ordinary" periodic solutions, all the characteristic exponents are zero**: *but for "singular" periodic solutions, two exponents are zero and the other two are* $\pm\alpha$, *where* α *is not zero and varies continuously over the family of singular periodic solutions.*

180. *Lagrange's three particles.*

We shall now consider specially certain periodic solutions of the problem of three bodies.

Let the equations of motion of the problem be taken in the reduced form obtained in § 160, and let us first enquire whether these equations admit of a particular solution in which the mutual distances of the bodies are invariable throughout the motion.

* Cf. Poincaré, *Méth. Nouv.* II. § 148.

The mutual distances are

$$q_1, \quad \left\{q_2^2 - \frac{2m_2 q_1 q_2}{m_1+m_2}\left(\cos q_3 \cos q_4 - \frac{k^2 - p_3^2 - p_4^2}{2p_3 p_4}\sin q_3 \sin q_4\right) + \frac{m_2^2}{(m_1+m_2)^2} q_1^2\right\}^{\frac{1}{2}},$$

and $$\left\{q_2^2 + \frac{2m_1 q_1 q_2}{m_1+m_2}\left(\cos q_3 \cos q_4 - \frac{k^2 - p_3^2 - p_4^2}{2p_3 p_4}\sin q_3 \sin q_4\right) + \frac{m_1^2}{(m_1+m_2)^2} q_1^2\right\}^{\frac{1}{2}};$$

it follows that, in the particular solution considered, the quantities

$$q_1, \quad q_2, \quad \text{and} \quad \cos q_3 \cos q_4 - \frac{k^2 - p_3^2 - p_4^2}{2p_3 p_4}\sin q_3 \sin q_4$$

must be constant, and hence the functions $U, \partial U/\partial q_1, \partial U/\partial q_2$ must be constant, where $U = \Sigma m_1 m_2 r_{12}^{-1}$.

The equations

$$0 = \dot{q}_1 = \frac{\partial H}{\partial p_1} = \frac{p_1}{\mu}, \qquad 0 = \dot{q}_2 = \frac{\partial H}{\partial p_2} = \frac{p_2}{\mu}$$

shew that p_1 and p_2 must be permanently zero: while the equations

$$0 = \dot{p}_1 = -\frac{\partial H}{\partial q_1} = \frac{p_3^2}{\mu q_1^3} + \frac{\partial U}{\partial q_1}, \qquad 0 = \dot{p}_2 = -\frac{\partial H}{\partial q_2} = \frac{p_4^2}{\mu' q_2^3} + \frac{\partial U}{\partial q_2}$$

shew that p_3 and p_4 must be constant.

Moreover, the equations

$$0 = \dot{p}_3 = -\frac{\partial H}{\partial q_3}, \qquad 0 = \dot{p}_4 = -\frac{\partial H}{\partial q_4}$$

shew that the expressions

$$\frac{\partial}{\partial q_3}\left(\cos q_3 \cos q_4 - \frac{k^2 - p_3^2 - p_4^2}{2p_3 p_4}\sin q_3 \sin q_4\right)$$

and $$\frac{\partial}{\partial q_4}\left(\cos q_3 \cos q_4 - \frac{k^2 - p_3^2 - p_4^2}{2p_3 p_4}\sin q_3 \sin q_4\right)$$

are zero, so we have

$$\tan q_3 \cot q_4 = \cot q_3 \tan q_4 = \frac{p_3^2 + p_4^2 - k^2}{2p_3 p_4},$$

and therefore $$p_3^2 + p_4^2 - k^2 = \pm 2p_3 p_4,$$

or $$k^2 = (p_3 \pm p_4)^2,$$

an equation which shews that the instantaneous planes of motion of the bodies μ and μ' coincide with the plane through these bodies and the origin: in other words, the motion of μ and μ' takes place in a plane: and therefore *the motion of m_1, m_2, m_3 takes place in a plane.*

It follows that, the centre of gravity O of the system being supposed at rest, the particles m_1, m_2, m_3 (which we shall denote by P, Q, R) must move in circular orbits round O. We have now to see if such a motion is possible.

One condition which must obviously be satisfied is that the resultant attraction of any two of the particles on the third must act in the line joining the third particle to the centre of gravity. This condition is satisfied if the three particles are in the same straight line. If they are not in a straight line, it gives

$$\frac{m_1}{PR^2}\sin PRO = \frac{m_2}{QR^2}\sin QRO, \text{ and two similar equations.}$$

But since O is the centre of gravity of the particles, we have

$$\frac{m_1 \sin PRO}{m_2 \sin QRO} = \frac{\sin QPR}{\sin PQR} = \frac{QR}{PR},$$

and this combined with the preceding equation gives $PR = QR$: similarly we find $PR = PQ$.

Hence *either the bodies must be collinear, or else the triangle formed by them must be equilateral.*

Considering first the collinear case, let the distances of the bodies from their centre of gravity (measured positively in the same direction) be a_1, a_2, a_3 respectively: we shall suppose that $a_1 < a_2 < a_3$, which does not lessen the generality of the discussion. Since the force acting on P must be that corresponding to circular motion round O, we have

$$n^2 a_1 = -m_2(a_2 - a_1)^{-2} - m_3(a_3 - a_1)^{-2},$$

where n is the angular velocity of the line PQR; and similarly

$$n^2 a_2 = -m_3(a_3 - a_2)^{-2} + m_1(a_2 - a_1)^{-2}, \quad n^2 a_3 = m_1(a_3 - a_1)^{-2} + m_2(a_3 - a_2)^{-2}.$$

From these equations we readily find

$$m_1 k^2\{(1+k)^3 - 1\} + m_2(1+k)^2(k^3 - 1) + m_3\{k^3 - (1+k)^3\} = 0,$$

where k denotes the ratio $(a_3 - a_2)/(a_2 - a_1)$.

This is a quintic equation in k, with real coefficients. Since the left-hand side of the equation is negative when k is zero, and positive when $k \to +\infty$, there is at least one positive real root; such a root determines uniquely real values for the ratios $a_1 : a_2 : a_3$; and if n is given, the distances a_1, a_2, a_3 can be completely determined. It follows that *there are an infinite number of solutions of the problem of three bodies, in which the bodies remain always in a straight line at constant distances from each other; the straight line rotates uniformly, and when its angular velocity has been (arbitrarily) assigned, the mutual distances of the bodies are determinate.*

Considering next the equilateral case, let a be the length of one side of the triangle formed by the bodies, and let n be its angular velocity. Since the force acting on m_3 is that which corresponds to a circular orbit round O, we have

$$\frac{m_1}{a^2}\cos PRO + \frac{m_2}{a^2}\cos QRO = n^2 . OR,$$

a condition which reduces to

$$m_1 + m_2 + m_3 = n^2 a^3.$$

The conditions relating to the motion of Q and of R reduce to the same equation: and hence a motion of the kind indicated is possible, provided n and a are connected by this relation. Hence *there are an infinite number of solutions of the problem of three bodies, in which the triangle formed by the bodies remains equilateral and of constant size, and rotates uniformly in the plane of the motion: the angular velocity of its rotation can be arbitrarily assigned, and the size of the triangle is then determinate.*

The two particular types of motion which have now been found will be called *Lagrange's collinear particles* and *Lagrange's equidistant particles* respectively*.

For more than a century after Lagrange's discovery, its interest was supposed to be purely theoretical. But in 1906 a new minor planet, 588 Achilles, was found to have a mean distance equal to that of Jupiter: and it was soon realised that the Sun, Jupiter, and Achilles constitute, approximately at any rate, an example of the Lagrangian equilateral-triangular configuration. Shortly afterwards came the discovery of three other Asteroids, 617 Patroclus, 624 Hector, and 659 Nestor, which are in the same case†, and in the next 25 years, six more. Of this "Trojan group," five are in longitude Jupiter $-60°$, and the other five in longitude Jupiter $+60°$.

Example. Shew that particular solutions of the problem of three bodies exist, in which the bodies are always collinear or always equidistant, although the mutual distances are not constant but are periodic functions of the time.

These are evidently *periodic solutions* of the problem, and include Lagrange's particles as a limiting case.

181. *Stability of Lagrange's particles: periodic orbits in the vicinity.*

It has been observed that in the neighbourhood of any configuration of stable equilibrium or steady motion there exists in general a family of periodic solutions, namely the normal vibrations about the position of equilibrium or steady motion. We shall now apply this idea to the case of the Lagrange's-particle solution of the restricted problem of three bodies, and thereby obtain certain families of periodic orbits of the planetoid.

Let S and J be the bodies of finite mass, m_1 and m_2 their masses, O their centre of gravity, n the angular velocity of SJ, x and y the coordinates of the

* They were discovered by Lagrange in 1772: *Oeuvres de Lagrange*, VI. p. 229. For references to extensions of these results to the problem of n bodies, cf. my article in the *Encyklopädie d. math. Wiss.* VI. 2, 12, p. 529; to the papers there mentioned may be added E. O. Lovett, *Annali di Mat.* (3), XI. (1904), p. 1; W. R. Longley, *Bull. Amer. Math. Soc.* XIII. (1907), p. 324, and F. R. Moulton, *Annals of Math.* XII. (1910), p. 1; C. Carathéodory, *Sitz. Bayer. Ak. Wiss.* H 2 (1933), p. 257.

† Cf. F. J. Linders, *Arkiv för Mat.* IV. (1908), No. 20.

planetoid P when O is taken as origin and OJ as axis of x. The equations of motion of the planetoid are (§ 162)

$$\frac{dx}{dt}=\frac{\partial K}{\partial u},\quad \frac{dy}{dt}=\frac{\partial K}{\partial v},\quad \frac{du}{dt}=-\frac{\partial K}{\partial x},\quad \frac{dv}{dt}=-\frac{\partial K}{\partial y},$$

where $$K=\tfrac{1}{2}(u^2+v^2)+n(uy-vx)-m_1/SP-m_2/JP.$$

Let (a, b) denote the values of (x, y) in the position of relative equilibrium considered, so that for the collinear case we have $b=0$, and for the equidistant case we have $a=\frac{1}{2}(m_1-m_2)\,l/(m_1+m_2)$, $b=\frac{1}{2}\sqrt{3}l$, where l denotes the distance SJ, so that (§ 46)

$$m_1+m_2=n^2l^3.$$

The values of u, v in the position of relative equilibrium are easily seen to be $-nb$ and na respectively.

Write $\quad x=a+\xi,\quad y=b+\eta,\quad u=-nb+\theta,\quad v=na+\phi,$

where ξ, η, θ, ϕ are supposed to be small quantities: neglecting a constant term, we have

$$K=\tfrac{1}{2}(\theta^2+\phi^2)+n(\eta\theta-\xi\phi)-n^2(a\xi+b\eta)$$
$$-m_1\left\{\left(a+\frac{m_2 l}{m_1+m_2}+\xi\right)^2+(b+\eta)^2\right\}^{-\frac{1}{2}}-m_2\left\{\left(a-\frac{m_1 l}{m_1+m_2}+\xi\right)^2+(b+\eta)^2\right\}^{-\frac{1}{2}}.$$

On expanding and retaining only terms of the second order in the small quantities, we obtain an expression for K with which the equations for the vibrations about relative equilibrium can be formed: we shall for definiteness consider vibrations about the equidistant configuration: in this case the expression for K becomes

$$K=\tfrac{1}{2}(\theta^2+\phi^2)+n(\eta\theta-\xi\phi)$$
$$-\frac{n^2}{8(m_1+m_2)}\{4(m_1+m_2)(\xi^2+\eta^2)-3m_1(\xi+\sqrt{3}\eta)^2-3m_2(\xi-\sqrt{3}\eta)^2\}.$$

The equations of motion are

$$\dot{\xi}=\frac{\partial K}{\partial\theta},\quad \dot{\eta}=\frac{\partial K}{\partial\phi},\quad \dot{\theta}=-\frac{\partial K}{\partial\xi},\quad \dot{\phi}=-\frac{\partial K}{\partial\eta}.$$

Solving these equations in the manner described in Chapter VII, we find that the period of a normal vibration is $2\pi/\lambda$, where λ is a root of the equation

$$\lambda^4-n^2\lambda^2+(\tfrac{27}{16}-k^2)n^4=0,\text{ where } k=\frac{3\sqrt{3}}{4}\cdot\frac{m_1-m_2}{m_1+m_2}.$$

The two values of λ^2 given by this equation will be positive provided they are real, since $(\frac{27}{16}-k^2)$ is positive: and they will be real provided $4(\frac{27}{16}-k^2)<1$, or $(m_1+m_2)^2>27m_1m_2$; a relation which is satisfied provided one of the masses S, J is at least about 25 times as large as the other. *When this condition is satisfied, there exist two families of periodic orbits of the planetoid in the vicinity of its equidistant configuration of relative equilibrium*; the periods are, to a first approximation, $2\pi/\lambda_1$ and $2\pi/\lambda_2$, where λ_1^2 and λ_2^2 are the roots of the equation in λ^2,

$$\lambda^4 - n^2\lambda^2 + (\tfrac{27}{16} - k^2)\, n^4 = 0.$$

These orbits have been computed by various investigators by numerical integration.

Example. Shew that, for one of the modes of normal vibration of the planetoid in the vicinity of the equidistant configuration, the constant of relative energy is greater than in the configuration of relative equilibrium, while for the other mode the constant is less than in the configuration of relative equilibrium. (Charlier.)

A similar discussion leads to the result that *the collinear Lagrange's-particle configurations are unstable; but the equation for the periods of normal modes of vibration has always one real root, and consequently in the neighbourhood of a position of relative equilibrium of the planetoid on the line SJ there exists a family of unstable periodic orbits.*

These orbits have been computed by numerical integration.

Let S and J be the bodies of finite mass, and P the "point of libration"

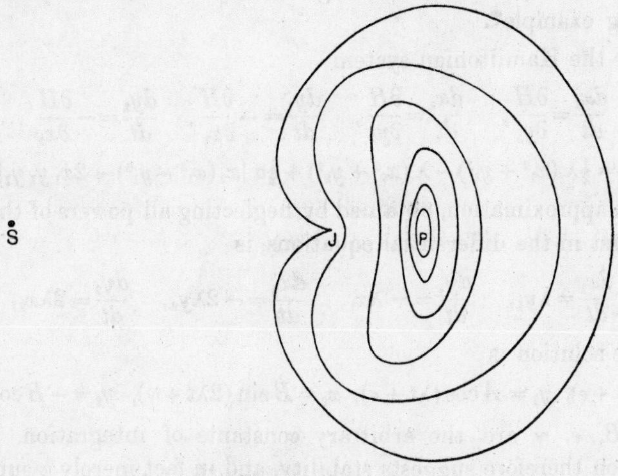

beyond J, at which the planetoid could remain at relative rest. Then there is a sequence of periodic solutions as in the above diagram, leading (as the orbit enlarges) to an "orbit of ejection" discovered by Burrau, in which the planetoid collides with J and rebounds from it. Beyond the orbit of ejection we find orbits which have loops round J; and, as was shewn in 1924 by Strömgren of the Copenhagen Observatory and his co-workers, after many changes of type we come back to the original simple orbit round P, so that the sequence returns into itself and we obtain a continuous complete closed set of periodic orbits.

For further work on the subject of orbits in the neighbourhood of the Lagrange's-particle solutions and the restricted problem of three bodies generally, cf. the memoirs referred to on page 530 of my article in the *Encyklopädie*, and also E. O. Lovett, *Astr.*

Nach. CLIX. (1902), p. 281; F. R. Moulton, *Proc. L.M.S.* (2), XI. (1912), p. 367, *Math. Ann.* LXXIII. (1912), p. 441, *Proc. Inter. Cong. of Math.* Cambridge, 1912, II. p. 182, and *Periodic Orbits* (Washington, 1920), which gives an account of researches by Moulton himself and by W. D. McMillan, T. Buck, D. Buchanan, W. R. Longley, and F. L. Griffin; various memoirs by E. Haerdtl, G. Pavanini, L. A. H. Warren, J. Chazy, L. Amoroso, J. Fischer-Petersen, P. Pedersen, K. Bohlin, W. W. Heinrich, A. Wintner, K. Popoff, T. Matukuma, M. Martin; N. Moïsseiev, *Publ. of the Sternberg State Astron. Inst.* VII. (1936): and a long series of papers by E. Strömgren and his school in the *Publikationer fra Köbenhavns Observatorium*, the results of which are collected in No. 100 (1935).

182. *The stability of orbits as affected by terms of higher order in the displacement.*

In the present chapter, in our treatment of orbits adjacent to a given orbit or to a position of equilibrium, we have considered only an *approximate* solution of the equations of the varied orbit, since we have neglected the influence of all powers of the displacement above the first. The effect of these neglected terms may however be of great importance, as may be seen from the following example*.

Consider the Hamiltonian system

$$\frac{dx_1}{dt}=\frac{\partial H}{\partial y_1},\quad \frac{dx_2}{dt}=\frac{\partial H}{\partial y_2},\quad \frac{dy_1}{dt}=-\frac{\partial H}{\partial x_1},\quad \frac{dy_2}{dt}=-\frac{\partial H}{\partial x_2} \quad \text{.........(1)},$$

where $H=\frac{1}{2}\lambda\,(x_1^2+y_1^2)-\lambda\,(x_2^2+y_2^2)+\frac{1}{2}\alpha\,\{x_2(x_1^2-y_1^2)-2x_1\,y_1\,y_2\}$.

The first approximation, obtained by neglecting all powers of the variables above the first in the differential equations, is

$$\frac{dx_1}{dt}=\lambda y_1,\quad \frac{dy_1}{dt}=-\lambda x_1,\quad \frac{dx_2}{dt}=-2\lambda y_2,\quad \frac{dy_2}{dt}=2\lambda x_2,$$

of which the solution is

$$x_1=A\sin(\lambda t+\epsilon),\;\; y_1=A\cos(\lambda t+\epsilon),\;\; x_2=B\sin(2\lambda t+\gamma),\;\; y_2=-B\cos(2\lambda t+\gamma),$$

where A, B, ϵ, γ are the arbitrary constants of integration. This first approximation therefore suggests stability, and in fact merely a superposition of two simple-harmonic oscillations.

It can however be verified without difficulty that the differential equations (1) admit the solution

$$x_1=\frac{\sqrt{2}}{\alpha(t+\epsilon)}\sin(\lambda t+\gamma),\qquad y_1=\frac{\sqrt{2}}{\alpha(t+\epsilon)}\cos(\lambda t+\gamma),$$

$$x_2=\frac{1}{\alpha(t+\epsilon)}\sin(2\lambda t+\gamma),\qquad y_2=\frac{-1}{\alpha(t+\epsilon)}\cos(2\lambda t+\gamma),$$

where ϵ and γ are arbitrary constants; and these equations represent orbits which are indefinitely near the origin when $t\to\infty$ and $t\to-\infty$, but which have infinite branches, all the coordinates becoming infinite as t approaches

* Due to T. M. Cherry, *Trans. Camb. Phil. Soc.* XXIII. (1925), p. 199.

the arbitrary value $-\epsilon$. The position of equilibrium at the origin is therefore unstable, in spite of being stable to the first order.

The general effect of the neglected terms of higher order on stability in dynamical systems has been considered by T. Levi-Civita* and A. R. Cigala†: and in a series of memoirs by D. J. Korteweg‡ and his pupil H. J. E. Beth§. Considering the oscillations about equilibrium of a system with any number of degrees of freedom, Korteweg shewed that if $s_1, s_2, \ldots$ denote the frequencies corresponding to infinitesimal oscillations in the different normal coordinates, then when

$$p_1 s_1 + p_2 s_2 + \ldots$$

is zero or very small (where $p_1, p_2, \ldots$ are small integers, positive or negative) certain vibrations of higher order, which are usually of small intensity compared with the principal vibrations, may acquire an abnormally great intensity. The most important cases occur when

$$|p_1| + |p_2| + \ldots \leqslant 4;$$

these cases have been fully discussed by Beth.

183. *Attractive and repellent regions of a field of force.*

The general character of the motion of a conservative holonomic system is illustrated by a theorem which was given by Hadamard‖ in 1897. For simplicity, we shall suppose that the system consists of a particle of unit mass, which is free to move on a given smooth surface under forces derivable from a potential energy function V; a similar result will readily be seen to hold for more complex systems.

Let (u, v) be two parameters which specify the position of the particle on the surface, and let the line-element on the surface be given by the equation

$$ds^2 = E\,du^2 + 2F\,du\,dv + G\,dv^2,$$

where (E, F, G) are given functions of u and v The kinetic energy of the particle is

$$T = \tfrac{1}{2}(E\dot{u}^2 + 2F\dot{u}\dot{v} + G\dot{v}^2),$$

and the Lagrangian equations of motion are

$$\frac{d}{dt}\left(\frac{\partial T}{\partial \dot{u}}\right) - \frac{\partial T}{\partial u} = -\frac{\partial V}{\partial u}, \quad \frac{d}{dt}\left(\frac{\partial T}{\partial \dot{v}}\right) - \frac{\partial T}{\partial v} = -\frac{\partial V}{\partial v},$$

which can be written

$$(EG - F^2)\ddot{u} = -G\frac{\partial V}{\partial u} + F\frac{\partial V}{\partial v} + \dot{u}^2\left(F\frac{\partial F}{\partial u} - \tfrac{1}{2}G\frac{\partial E}{\partial u} - \tfrac{1}{2}F\frac{\partial E}{\partial v}\right)$$
$$+ \dot{u}\dot{v}\left(F\frac{\partial G}{\partial u} - G\frac{\partial E}{\partial v}\right) + \dot{v}^2\left(\tfrac{1}{2}G\frac{\partial G}{\partial u} + \tfrac{1}{2}F\frac{\partial G}{\partial v} - G\frac{\partial F}{\partial v}\right),$$

* *Annali di Mat.* v. (1901), p. 221. † *Annali di Mat.* xi. (1904), p. 67.

‡ *Verhand. d. K. Akad. v. Wetensch.* v. No. 8 (1897): *Archives Néerland.* (2), i. (1897), p. 229.

§ *Amsterdam Proc.* xii. (1910), p. 618 and p. 735: xiii. (1911), p. 742: *Archives Néerland.* (2), xv. (1910), p. 246: (3a), i. (1912), p. 185: *Phil. Mag.* (6), xxvi. (1913), p. 268.

‖ *Journ. de Math.* (5), iii. p. 331. Cf. Synge, *Trans. Amer. Math. Soc.* xxxiv. (1932), p. 481.

$$(EG-F^2)\ddot{v}=F\frac{\partial V}{\partial u}-E\frac{\partial V}{\partial v}+\dot{u}^2\left(\tfrac{1}{2}E\frac{\partial E}{\partial v}+\tfrac{1}{2}F\frac{\partial E}{\partial u}-E\frac{\partial F}{\partial u}\right)$$
$$+\dot{u}\dot{v}\left(F\frac{\partial E}{\partial v}-E\frac{\partial G}{\partial u}\right)+\dot{v}^2\left(F\frac{\partial F}{\partial v}-\tfrac{1}{2}E\frac{\partial G}{\partial v}-\tfrac{1}{2}F\frac{\partial G}{\partial u}\right).$$

We have, by differentiation,

$$\dot{V}=\frac{\partial V}{\partial u}\dot{u}+\frac{\partial V}{\partial v}\dot{v},$$
$$\ddot{V}=\frac{\partial V}{\partial u}\ddot{u}+\frac{\partial V}{\partial v}\ddot{v}+\frac{\partial^2 V}{\partial u^2}\dot{u}^2+2\frac{\partial^2 V}{\partial u\partial v}\dot{u}\dot{v}+\frac{\partial^2 V}{\partial v^2}\dot{v}^2.$$

Substituting for $\ddot{u}$ and $\ddot{v}$ their values from the preceding equations, we have

$$\ddot{V}=-(EG-F^2)^{-1}\left\{E\left(\frac{\partial V}{\partial v}\right)^2-2F\frac{\partial V}{\partial u}\frac{\partial V}{\partial v}+G\left(\frac{\partial V}{\partial u}\right)^2\right\}+\Phi(\dot{u},\dot{v}),$$

where

$$\Phi(\dot{u},\dot{v})=\left[\frac{\partial^2 V}{\partial u^2}+(EG-F^2)^{-1}\left\{\frac{\partial V}{\partial u}\left(F\frac{\partial F}{\partial u}-\tfrac{1}{2}G\frac{\partial E}{\partial u}-\tfrac{1}{2}F\frac{\partial E}{\partial v}\right)\right.\right.$$
$$\left.\left.+\frac{\partial V}{\partial v}\left(\tfrac{1}{2}E\frac{\partial E}{\partial v}+\tfrac{1}{2}F\frac{\partial E}{\partial u}-E\frac{\partial F}{\partial u}\right)\right\}\right]\dot{u}^2$$
$$+\left[2\frac{\partial^2 V}{\partial u\partial v}+(EG-F^2)^{-1}\left\{\frac{\partial V}{\partial u}\left(F\frac{\partial G}{\partial u}-G\frac{\partial E}{\partial v}\right)\right.\right.$$
$$\left.\left.+\frac{\partial V}{\partial v}\left(F\frac{\partial E}{\partial v}-E\frac{\partial G}{\partial u}\right)\right\}\right]\dot{u}\dot{v}$$
$$+\left[\frac{\partial^2 V}{\partial v^2}+(EG-F^2)^{-1}\left\{\frac{\partial V}{\partial u}\left(\tfrac{1}{2}G\frac{\partial G}{\partial u}+\tfrac{1}{2}F\frac{\partial G}{\partial v}-G\frac{\partial F}{\partial v}\right)\right.\right.$$
$$\left.\left.+\frac{\partial V}{\partial v}\left(F\frac{\partial F}{\partial v}-\tfrac{1}{2}E\frac{\partial G}{\partial v}-\tfrac{1}{2}F\frac{\partial G}{\partial u}\right)\right\}\right]\dot{v}^2.$$

The quantities occurring in this equation can be expressed in terms of *deformation-covariants**. The principal deformation-covariants connected with the surface whose line-element is given by the equation

$$ds^2=Edu^2+2F\,du\,dv+G\,dv^2$$

are the differential parameters

$$\Delta(\phi,\psi)=(EG-F^2)^{-1}\left\{E\frac{\partial\phi}{\partial v}\frac{\partial\psi}{\partial v}-F\left(\frac{\partial\phi}{\partial u}\frac{\partial\psi}{\partial v}+\frac{\partial\phi}{\partial v}\frac{\partial\psi}{\partial u}\right)+G\frac{\partial\phi}{\partial u}\frac{\partial\psi}{\partial u}\right\}$$
$$\Delta_1(\phi)=(EG-F^2)^{-1}\left\{E\left(\frac{\partial\phi}{\partial v}\right)^2-2F\frac{\partial\phi}{\partial u}\frac{\partial\phi}{\partial v}+G\left(\frac{\partial\phi}{\partial u}\right)^2\right\}$$
$$\Delta_2(\phi)=(EG-F^2)^{-\frac{1}{2}}\left[\frac{\partial}{\partial u}\left\{(EG-F^2)^{\frac{1}{2}}\left(G\frac{\partial\phi}{\partial u}-F\frac{\partial\phi}{\partial v}\right)\right\}\right.$$
$$\left.+\frac{\partial}{\partial v}\left\{(EG-F^2)^{\frac{1}{2}}\left(-F\frac{\partial\phi}{\partial u}+E\frac{\partial\phi}{\partial v}\right)\right\}\right],$$

where ϕ and ψ are arbitrary functions of the variables u and v.

* The definition of a deformation-covariant is given in the footnote on page 111.

With this notation, the preceding equation becomes

$$\ddot{V} = -\Delta_1(V) + \Phi(\dot{u}, \dot{v}).$$

Utilising the equation of energy

$$E\dot{u}^2 + 2F\dot{u}\dot{v} + G\dot{v}^2 = 2(h - V),$$

and observing that the expression

$$\frac{\Phi(\dot{u}, \dot{v})}{E\dot{u}^2 + 2F\dot{u}\dot{v} + G\dot{v}^2} - \frac{\Phi(\partial V/\partial v, -\partial V/\partial u)}{E(\partial V/\partial v)^2 - 2F(\partial V/\partial v)(\partial V/\partial u) + G(\partial V/\partial u)^2}$$

contains the quantity $(\dot{u}\partial V/\partial u + \dot{v}\partial V/\partial v)$ as a factor, we can write

$$\ddot{V} = -\Delta_1(V) + \frac{2(h - V)I_V}{\Delta_1(V)} + (\lambda\dot{u} + \mu\dot{v})\dot{V},$$

where λ and μ contain in their denominators only the quantity

$$E\left(\frac{\partial V}{\partial v}\right)^2 - 2F\frac{\partial V}{\partial v}\frac{\partial V}{\partial u} + G\left(\frac{\partial V}{\partial u}\right)^2,$$

and where I_V denotes the expression

$$\Phi(\partial V/\partial v, -\partial V/\partial u)/(EG - F^2);$$

we readily find that I_V can be expressed in the form

$$I_V = \Delta_1(V)\Delta_2(V) - \tfrac{1}{2}\Delta\{V, \Delta_1(V)\}.$$

Consider, on the orbit of the particle, a point at which V has a minimum value; at such a point $\dot{V}$ is zero and $\ddot{V}$ is positive: as $\Delta_1(V)$ is essentially positive (since the line-element of the surface is a positive definite form), it follows that $I_V \geqslant 0$, the inequality becoming an equality only when $\Delta_1(V)$ is zero, i.e. at an equilibrium-position of the particle.

As the particle describes any trajectory, the function V will either have an infinite number of successive maxima and minima (this is the general case) or (in exceptional cases) the function will, after passing some point of the orbit, vary continually in the same sense. Suppose first that the former of these alternatives is the true one: then if we divide the given surface into two regions, in which I_V is positive and negative respectively, it follows from what has been proved above that the former of these regions contains all the points of the orbit at which V has a minimum value, i.e. it contains in general an infinite number of distinct parts of the orbit, each of finite length; whereas in the other region, for which I_V is negative, the particle cannot remain permanently. These two parts of the surface are on this account called the *attractive* and *repellent* regions. Each of these regions exists in general, for it is easily found that any isolated point of the surface at which V is a minimum (i.e. any point where stable equilibrium is possible) is in an attractive region, and any point at which V is a maximum is in a repellent region.

It is interesting to compare this result with that which corresponds to it in the motion of a particle with one degree of freedom, e.g. a particle which is free to move on a curve under the action of a force which depends only on the position of the particle. In this case the particle either ultimately travels an indefinite distance in one direction or oscillates about a position of stable equilibrium. The attractive region, in motion with two degrees of freedom, corresponds to the position of stable equilibrium in motion with one degree of freedom.

Consider next the alternative supposition, namely that after some definite instant the variation of V is always in the same sense. We shall suppose that the surface has no infinite sheets and is regular at all points, and that V is an everywhere regular function of position on the surface; so that, since the variation of V is always in the same sense, V must tend toward some definite finite limit, $\dot{V}$ and $\ddot{V}$ tending to the limit zero. Considering the equation

$$\ddot{V} = -\Delta_1(V) + 2(h - V) I_V/\Delta_1(V) + (\lambda\dot{u} + \mu\dot{v})\dot{V},$$

we see that if $\Delta_1(V)$ is not very small, λ and μ are finite and the last term on the right-hand side of the equation is infinitesimal; and consequently either there exist values of t as large as we please for which I_V is positive (in which case the part of the orbit described in the attractive region is of length greater than any assignable quantity) or else $\Delta_1(V)$ tends to zero. But $\Delta_1(V)$ can be zero only when $\partial V/\partial u$ and $\partial V/\partial v$ are zero; if therefore (as is in general the case) the surface possesses only a finite number of equilibrium positions, the particle will tend to one of these positions, with a velocity which tends to zero. A position of equilibrium thus approached asymptotically must be a position of unstable equilibrium: for the asymptotic motion reversed is a motion in which the particle, being initially near the equilibrium position with a small velocity, does not remain in the neighbourhood of the equilibrium position; and this is inconsistent with the definition of stability.

Thus finally we obtain Hadamard's theorem, which may be stated as follows: *If a particle is free to move on a surface which is everywhere regular and has no infinite sheets, the potential energy function being regular at all points of the surface and having only a finite number of maxima and minima on it, either the part of the orbit described in the attractive region is of length greater than any assignable quantity, or else the orbit tends asymptotically to one of the positions of unstable equilibrium.*

Example. If all values of t from $-\infty$ to $+\infty$ are considered, shew that the particle must for part of its course be in the attractive region.

184. *Application of the energy integral to the problem of stability.*

A simple criterion for determining the character of a given form of motion of a dynamical system is often furnished by the equation of energy of the system. Considering the case of a single particle of unit mass which moves in a plane under the influence of forces derived from a potential energy function $V(x, y)$, the equation of energy can be written

$$\tfrac{1}{2}(\dot{x}^2 + \dot{y}^2) = h - V(x, y).$$

Now the branches of the curve $V(x, y) = h$ separate the plane into regions for which $\{V(x, y) - h\}$ is respectively positive and negative; but as $(\dot{x}^2 + \dot{y}^2)$ is essentially positive, an orbit for which the total energy is h can only exist in the regions for which $V(x, y) < h$. If then the particle is at any time in the interior of a closed branch of the curve $V(x, y) = h$, it must always remain within this region. The word *stability* is often applied to characterise types of motion in which the moving particle is confined to certain limited regions, and in this sense we may say that the motion of the particle in question is stable.

The above method has been used by Hill*, Bohlin†, and Darwin‡, chiefly in connexion with the restricted problem of three bodies.

185. *Application of integral-invariants to investigations of stability.*

The term *stability* was applied in a different sense by Poisson to a system which, in the lapse of time, returns infinitely often to positions indefinitely near to its original position, the intervening oscillations being of any magnitude. It has been shewn by Poincaré that the theory of integral-invariants may be applied to the discussion of Poisson stability.

Considering a system of differential equations

$$\frac{dx_r}{dt} = X_r(x_1, x_2, \ldots, x_n) \qquad (r = 1, 2, \ldots, n),$$

for which

$$\iiint \ldots \int \delta x_1 \delta x_2 \ldots \delta x_n$$

is an integral-invariant, we regard these equations as defining the trajectory in n dimensions of a point P whose coordinates are $(x_1, x_2, \ldots, x_n)$. If the trajectories have no branches receding to an infinite distance from the origin, it may be shewn§ that if any small region R is taken in the space, there exist trajectories which traverse R infinitely often: and, in fact, the probability that a trajectory issuing from a point of R does not traverse this region infinitely often is zero, however small R may be. Poincaré has given several extensions of this method, and has shewn that under certain conditions it is applicable in the restricted problem of three bodies.

The development of this line of thought led to the *ergodic* theorem, for which see Birkhoff, *Bull. Amer. Math. Soc.* XXXVIII. (1932), p. 361, where references are given to the work of Birkhoff himself and of Koopman, Hopf, Wintner, and J. v. Neumann. Cf. also T. Carleman, *Ark. Mat. Astr. Fys.* 22 B, Nr 7 (1932).

186. *Synge's "Geometry of Dynamics."*

A brief account will now be given of recent work by Synge‖; in which dynamical problems are treated by aid of the tensor calculus.

The motion of a dynamical system whose configurations are specified by N coordinates $(q^1, q^2, \ldots, q^N)$ may be thought of as the motion of a point in a manifold of N dimensions—the "manifold of configurations." If the kinetic energy of the system is given by $T = \frac{1}{2} a_{mn} \dot{q}^m \dot{q}^n$ (where the repetition of an index in a product implies summation of that index from 1 to N), the infinitesimal

$$ds^2 = 2T dt^2 = a_{mn} dq^m dq^n$$

* *Amer. J. Math.* I. (1878), p. 75.

† *Acta Math.* X. (1887), p. 109.

‡ *Acta Math.* XXI. (1897), p. 99.

§ Poincaré, *Acta Math.* XIII. (1890), p. 67; *Nouv. Méch.* III. Ch. XXVII: on Poincaré's work on stability *à la Poisson* cf. E. Picard, *Bull. des sc. math.* (2), XXXVIII. (1914), p. 320.

‖ J. L. Synge, "On the Geometry of Dynamics," *Phil. Trans.* A, 226 (1926), pp. 31–106. For the essentials of the tensor calculus the reader is referred to Levi-Civita's *Absolute Differential Calculus* (English translation), or to Eisenhart's *Riemannian Geometry*.

is an invariant, defined by two adjacent configurations. The square root of this quantity may be called the *distance* between the two configurations and the above form the *kinematical line-element.* In the case of a particle of unit mass moving in space or on a surface the kinematical line-element so defined is precisely the ordinary geometrical line-element.

The vector whose components are $\dot{q}^r$ is called the *velocity vector* and the vector whose components are f^r, where

$$f^r = \ddot{q}^r + \{^{m\ n}_{\ r}\}\dot{q}^m\dot{q}^n,$$

is called the *acceleration vector.* It is easily seen (p. 39) that the equations of motion may be written $f^r = Q^r$, i.e. acceleration = force. Just as in particle dynamics the acceleration vector can be resolved into components along the tangent and principal normal to the trajectory, the components in these directions being respectively $v dv/ds$ and $v^2\kappa$, where v is the magnitude of the velocity ($v^2 = a_{mn}\dot{q}^m\dot{q}^n = 2T$), and κ is the first curvature. This leads at once to a generalisation of Bonnet's Theorem (p. 94).

The purely geometrical concept of the relative curvature of two curves in Riemannian space, due to Lipka (*Bull. Amer. Math. Soc.* XXIX. (1923), p. 345), throws further light on the question of Least Curvature (Chap. IX) and leads to the theorem: *When a holonomic conservative system is subjected to constraints, holonomic or non-holonomic, the natural constrained trajectory has, relative to the unconstrained natural trajectory with the same velocity vector, a smaller curvature than any other curve having the same tangent and satisfying the conditions of constraint.*

When the methods of the tensor calculus are applied to non-holonomic systems, a determinate form of Lagrange's equations is obtained. A system of M equations of constraint implies that the velocity vector must be perpendicular to a certain element of M dimensions. It is always possible to choose M orthogonal unit vectors in this element; if these are denoted by $B^r_{(1)}, B^r_{(2)}, \ldots, B^r_{(M)}$, the equations of motion may be written

$$\begin{aligned}\frac{d}{dt}\left(\frac{\partial T}{\partial \dot{q}^r}\right) - \frac{\partial T}{\partial q^r} = Q_r &- (B_{(1)m}Q^m + B_{(1)mn}Q^mQ^n)\,B^r_{(1)}\\ &- (B_{(2)m}Q^m + B_{(2)mn}Q^mQ^n)\,B^r_{(2)}\\ &- \ldots\\ &- (B_{(M)m}Q^m + B_{(M)mn}Q^mQ^n)\,B^r_{(M)},\end{aligned}$$

where $B_{(1)mn}$, for example, is the covariant derivative of $B_{(1)m}$,

$$B_{(1)mn} = \frac{\partial B_{(1)m}}{\partial q^n} - \{^{m\ n}_{\ s}\}B_{(1)s}.$$

The application of the method is perhaps most interesting in treating the problem of the stability of a state of motion. Two motions, the undisturbed and the disturbed, take place along neighbouring curves in the manifold. If the distance between simultaneous configurations of the undisturbed and the disturbed motions remains permanently small, we say that the motion is stable *in the kinematical sense.* The infinitesimal disturbance vector joining the undisturbed configuration to the corresponding disturbed configuration being denoted by η^r, it is found that η^r satisfies the equation

$$\hat{\ddot{\eta}}^r + G^r_{\ msn}\eta^s\dot{q}^m\dot{q}^n - Q_s^{\ r}\eta^s = 0,$$

where

$$G^r_{\ msn} = \frac{\partial}{\partial q^s}\{^{m\ n}_{\ r}\} - \frac{\partial}{\partial q^n}\{^{m\ s}_{\ r}\} + \{^{m\ n}_{\ t}\}\{^{s\ t}_{\ r}\} - \{^{m\ s}_{\ t}\}\{^{n\ t}_{\ r}\},$$

the mixed curvature tensor of the manifold, and

$$Q_s^{\ r} = \frac{\partial Q^r}{\partial q^s} + \{^{s\ n}_{\ r}\}Q^n;$$

the accent $\wedge$ denotes the *contravariant time-flux*, viz.

$$\hat{\eta}^r = \frac{d\eta^r}{dt} + \{^{m\,n}_{\ r}\}\,\eta^m \frac{dq^n}{dt}, \qquad \hat{\hat{\eta}}^r = \frac{d\hat{\eta}^r}{dt} + \{^{m\,n}_{\ r}\}\,\hat{\eta}^m \frac{dq^n}{dt}.$$

The magnitude of the disturbance vector, defined as $\eta = (a_{mn}\eta^m\eta^n)^{\frac{1}{2}}$, satisfies the equation

$$\ddot{\eta} + \eta\,(G_{mnst}\,\mu^m \dot{q}^n \mu^s \dot{q}^t - \hat{\mu}^2 - Q_{mn}\,\mu^m\,\mu^n) = 0,$$

where μ^r is the unit vector codirectional with η^r, $\hat{\mu}$ the magnitude of $\hat{\mu}^r$ and Q_{mn} the covariant derivative of Q_m.

When the correspondence between configurations of the undisturbed and disturbed trajectories is not that of simultaneity but is defined by the condition that the disturbance vector η^r shall be normal to the undisturbed trajectory, permanent smallness of η for this correspondence is called stability in the *kinematico-statical sense*. In this case also differential equations for the components and magnitude of the disturbance vector have been given in the case of a conservative system. In the case of two degrees of freedom we obtain

$$\ddot{\beta} + \beta(v^2K + V_{mn}\,\nu^m\,\nu^n + 3v^2\kappa^2) + 2\kappa\,\delta h = 0,$$

where β is the magnitude of the disturbance vector (but here counted positive on one side of the undisturbed trajectory and negative on the other side), K the Gaussian curvature of the manifold, ν^r the unit vector along the normal, κ the first curvature of the trajectory and δh the excess of the total energy in the disturbed motion over that in the undisturbed motion.

As long as we confine our attention to conservative systems and, in considering stability, think only of disturbances which do not change the total energy, the geometrical statement of dynamical problems assumes a simpler form when, instead of the kinematical line-element, we employ the *action line-element*,

$$ds^2 = 2\,(h - V)\,Tdt^2 = (h - V)\,a_{mn}\,dq^m\,dq^n = g_{mn}\,dq^m\,dq^n,$$

where h denotes the total energy of the system.

For this line-element the curves of natural motion are geodesics (curves of stationary length) by the Principle of Least Action (cf. p. 251), and their equations may be written

$$\frac{d^2q^r}{ds^2} + \{^{m\,n}_{\ r}\}\,\frac{dq^m}{ds}\,\frac{dq^n}{ds} = 0.$$

In discussing questions of stability we have now to deal with the separation of two neighbouring geodesics. Corresponding points being those at equal (action) distances from assigned points on the two geodesics, the disturbance vector satisfies the equation

$$\bar{\bar{\eta}}^r + G^r{}_{msn}\,\eta^s\,\frac{dq^m}{ds}\,\frac{dq^n}{ds} = 0,$$

where
$$\bar{\eta}^r = \frac{d\eta^r}{ds} + \{^{m\,n}_{\ r}\}\,\eta^m\,\frac{dq^n}{ds}, \qquad \bar{\bar{\eta}}^r = \frac{d\bar{\eta}^r}{ds} + \{^{m\,n}_{\ r}\}\,\bar{\eta}^m\,\frac{dq^n}{ds}.$$

The Christoffel symbols and curvature tensor are, of course, to be calculated for the action line-element. For the magnitude of the disturbance vector we have

$$\frac{d^2\eta}{ds^2} + \eta\left(G_{mnst}\,\mu^m\,\frac{dq^n}{ds}\,\mu^s\,\frac{dq^t}{ds} - \bar{\mu}^2\right) = 0.$$

In the case of a system with two degrees of freedom this gives

$$\frac{d^2\beta}{ds^2} + K\beta = 0,$$

where β is the magnitude of the disturbance vector (taken without loss of generality along the normal and affected with sign) and K the Gaussian curvature of the manifold calculated with respect to the action line-element.

187. *Connexion with the theory of surface transformations.*

The questions of integrability, stability, and the classification of various types of motion of a dynamical system with two degrees of freedom, have been studied by Poincaré and Birkhoff by a method which depends on the theory of the transformation of surfaces into themselves: a full account of this method is given in two papers by Birkhoff, "Dynamical systems with two degrees of freedom," *Trans. Amer. Math. Soc.* XVIII. (1917), p. 199 and "Surface transformations and their dynamical applications," *Acta Math.* XLIII. (1920).

The topics studied in the present chapter are treated more fully in Birkhoff's valuable work *Dynamical Systems* (*Amer. Math. Soc. Coll. Pub.* IX. (1927)) and in Husson's *Les trajectoires de la dynamique* (Paris, 1932).

MISCELLANEOUS EXAMPLES.

1. Shew that the motion of a particle in an ellipse under the influence of two fixed Newtonian centres of force is stable. (Novikoff.)

2. A particle of unit mass is free to move in a plane under the action of several centres of force which attract it according to the Newtonian law of the inverse square of the distance: denoting the resulting potential energy of the particle by $V(x, y)$, shew that the integral

$$\frac{1}{2\pi}\iint\left[\left(\frac{\partial^2}{\partial x^2}+\frac{\partial^2}{\partial y^2}\right)\log\{h-V(x,y)\}\right]dx\,dy,$$

where the integration is taken over the interior of any periodic orbit for which the constant of energy has the value h (the centres of force being excluded from the field of integration by small circles of arbitrary infinitesimal radius), is equal to the number of centres of force enclosed by the orbit, diminished by two. (*Monthly Notices R.A.S.* LXII. p. 186.)

3. Let a family of orbits in a plane be defined by a differential equation

$$\frac{d^2y}{dx^2}=\phi(x,y),$$

where (x, y) are the current rectangular coordinates of a point on an orbit of the family; and let δn denote the normal distance from the point (x, y) to some definite adjacent orbit of the family. Shew that δn satisfies the equation

$$\frac{d^2\delta n}{dt^2}+I\delta n=0,$$

where

$$I=\left\{1+\left(\frac{dy}{dx}\right)^2\right\}\left(-\frac{\partial\phi}{\partial y}+\frac{dy}{dx}\frac{\partial\phi}{\partial x}\right)+\{\phi(x,y)\}^2,$$

and t is a variable defined by the equation

$$\frac{dx}{dt}=1+\left(\frac{dy}{dx}\right)^2.$$

(Sheepshanks Astron. Exam.)

4. A particle moves under the influence of a repulsive force from a fixed centre: shew that the path is always of a hyperbolic character, and never surrounds the centre of force; that the asymptotes do not pass through the centre in the cases when the work, which has to be done against the force in order to bring the particle to its position from an infinite distance, has a finite value; but that when this work is infinitely great, the asymptotes pass through the centre, and the duration of the whole motion may be finite. (Schouten.)

5. Shew that in the motion of a particle on a fixed smooth surface under the influence of gravity, the curve of separation between the attractive and repellent regions of the surface is formed by the apparent horizontal contour of the surface, together with the locus of points at which an asymptotic direction is horizontal.

6. A particle moves freely in space under the influence of two Newtonian centres of attraction; shew that when its constant of energy is negative, it describes a spiral curve round the line joining the centres, remaining within a tubular region bounded by two ellipsoids of rotation and two hyperboloids of rotation, whose foci are the centres of force: and that when the constant of energy is zero or positive, the particle describes a spiral path within a region which is bounded by an ellipsoid and two infinite sheets of hyperboloids of the same confocal system. (Bonacini.)

7. The necessary and sufficient condition in order that a two-parameter family of curves defined by a differential equation

$$y'' = \phi(x, y, y')$$

may be orbits of a system defined by equations of motion

$$\begin{cases} \ddot{x} + \lambda(x, y)\,\dot{y} = \dfrac{\partial \mu(x, y)}{\partial x} \\ \ddot{y} - \lambda(x, y)\,\dot{x} = \dfrac{\partial \mu(x, y)}{\partial y} \end{cases}$$

is that

$$\frac{\partial \phi}{\partial y'} - \frac{3\phi y'}{1 + y'^2}$$

shall be the total derivative of a function of x and y with respect to x.

(P. Frank and K. Ogura.)

8. In the motion of a particle in a plane under forces which depend only on its position, a one-parameter family of trajectories is obtained by starting particles at a given point in a given direction with all possible velocities. Shew that the locus of the foci of the osculating parabolas is a circle passing through the point. If the initial direction is now varied, shew that the locus of the centres of the ∞^1 circles obtained is a conic having the given point as focus: and if the forces are conservative, this degenerates into a straight line counted twice.

9. In order that a system of ∞^5 space-curves, of which ∞^1 pass through every point in every direction, may be identifiable with the system of trajectories of a particle in an arbitrary positional field of force, it is necessary (but not sufficient) that the system should have the following properties:

(α) The osculating planes of the ∞^2 curves passing through a given point form a pencil: that is, all the planes pass through a fixed direction.

(β) The osculating spheres of the ∞^1 curves passing through a given point in a given direction form a pencil: their centres thus lie on a straight line.

10. Shew that the ∞^2 curves of a natural family which meet any surface orthogonally are orthogonal to ∞^1 surfaces, that is, form a normal congruence. (The surfaces in question are the surfaces of equal Action.) (Hamilton.)

11. Shew that the property referred to in Ex. 10 belongs exclusively to natural families.

12. In order that a family of ∞^4 curves in space may constitute a natural family of orbits, two properties are necessary and sufficient, viz.:

(α) If the osculating circles of those curves of the family which pass through a given point p are constructed at that point, they have a second point P in common, and thus form a *bundle*. Consequently, three of the circles in such a bundle have four-point contact with the corresponding curves.

(β) These three hyperosculating circles will be mutually orthogonal.

13. The only point-transformations which convert every natural family into a natural family are those belonging to the conformal group.

[Examples 8, 9, 11, 12, 13 above are taken from memoirs by E. Kasner in the *Transactions of the Amer. Math. Soc.*, 1906–1909. For further work in this direction the reader is referred to Professor Kasner's Princeton Colloquium lectures on *Differential Geometric Aspects of Dynamics.*]

14. Two sets of ∞^1 curves in a plane, which form an orthogonal system, are orbits in a certain conservative field of force. If U denote the Action at any point (x, y) of a particle considered as moving on one of the first set of orbits, and V denote the Action at (x, y) when the particle is considered as moving on one of the second set of orbits, shew that U and V are conjugate functions of x and y: and that the families of curves U=constant, V=constant, are identical with the orbits.

(P. G. Tait and K. Ogura.)

CHAPTER XVI

INTEGRATION BY SERIES

188. *The need for series which converge for all values of the time; Poincaré's series.*

We have already observed (§ 32) that the differential equations of motion of a dynamical system can be solved in terms of series of ascending powers of the time measured from some fixed epoch; these series converge in general for values of t within some definite circle of convergence in the t-plane, and consequently will not furnish the values of the coordinates except for a limited interval of time. By means of the process of analytic continuation* it would be possible to derive from these series successive sets of other power-series, which would converge for values of the time outside this interval; but the process of continuation is too cumbrous to be of much use in practice, and the series thus derived give no insight into the general character of the motion, or indication of the remote future of the system. The efforts of investigators have therefore been directed to the problem of expressing the coordinates of a dynamical system by means of expansions which converge for all values of the time. One method of achieving this result† is to apply a transformation to the t-plane. Assuming that the motion of the system is always regular (i.e. that there are no collisions or other discontinuities, and that the coordinates are always finite), there will be no singularities of the system at points on the real axis in the t-plane, and the divergence of the power-series in $t - t_0$ after a certain interval of time must therefore be due to the existence of singularities of the solution in the finite part of the t-plane but not on the real axis. Suppose that the singularity which is nearest to the real axis is at a distance h from the real axis; and let τ be a new variable defined by the equation

$$t - t_0 = \frac{2h}{\pi} \log \frac{1+\tau}{1-\tau}.$$

A band which extends to a distance h on either side of the real axis in the t-plane evidently corresponds to the interior of the circle $|\tau| = 1$ in the τ-plane; the coordinates of the dynamical system are therefore regular functions of τ at all points in the interior of this circle, and consequently they can be

* Cf. Whittaker and Watson, *Modern Analysis*, § 5·5.

† Due to Poincaré, *Acta Math.* IV. (1884), p 211

expressed as power-series in the variable τ, convergent within this circle. These series will therefore converge for all real values of τ between -1 and 1, i.e. for all real values of t between $-\infty$ and $+\infty$. Thus *these series are valid for all values of the time.*

189. *The regularisation of the Problem of Three Bodies.*

In the last article we made the reservation that there are to be no collisions or other discontinuities for real values of t. The importance of collisions in the mathematical theory of the Problem of Three Bodies was first indicated by Painlevé*, who shewed that the motion of the bodies is regular (i.e. their coordinates are holomorphic functions of t) for all time, provided the initial conditions are not such that after a finite interval of time two of the bodies collide. The relations which must subsist between the initial values of the variables in order that a collision may ultimately happen between two of the three bodies have been discussed by Levi-Civita† for the restricted problem of three bodies (when there is one such relation) and by Bisconcini‡ for the general problem, when there are two relations: these relations are analytic, but they are expressed by somewhat complicated infinite series, and are not directly applicable except when the interval of time between the initial instant and the collision is sufficiently short.

A considerable advance was made when K. F. Sundman§ shewed that the singularity of the differential equations which corresponds to a collision of two of the bodies is not of an essential character, and that it may in fact be removed altogether by making a suitable change of the independent variable: that is to say, it is possible to choose the variables which specify the motion, and the independent variable, in such a way that the differential equations of motion are regular even when two of the three bodies occupy coincident positions‖. It is thus possible to obtain a real prolongation of the motion after the collision¶: the coordinates can be specified for all values of the time t from $-\infty$ to $+\infty$, whether collisions take place or not: and a positive lower bound ι can be assigned to the two greater of the mutual distances. There

* *Leçons sur la théorie anal. des éq. diff.*, Paris, 1897, p. 583.

† *Annali di Mat.* (3) IX. (1903), p. 1; *Comptes Rendus*, CXXXVI. (1903), pp. 82, 221.

‡ *Acta Math.* XXX. (1905), p. 49. Cf. also H. Block, *Medd. från Lunds Obs.*, Series II., No. 6 (1909); *Arkiv f. Mat. Astr. och Fys.* V. (1909), No. 9.

§ *Acta Math.* XXXVI. (1912), p. 105. The essential features of the work were originally published in *Acta Societatis Scient. Fennicae* in 1906 and 1909. It seems to have been inspired largely by Poincaré's theory of the uniformisation of analytic functions: cf. *Acta Math.* XXXI. (1907), p. 1.

‖ Levi-Civita regularised the differential equations of the restricted problem of three bodies by an elementary transformation in *Acta Math.* XXX. (1906), p. 306; and in a later paper, *Rend. d. Lincei*, XXIV. (1915), p. 61, he extended this to the problem of three bodies in a plane. Cf. also his paper in *Acta Math.* XLII. (1917), p. 99.

¶ The variables can be expanded in ascending powers of $(t_1 - t)^{\frac{1}{3}}$, where t_1 represents the instant of collision: the orbits have cusps at the point of impact.

is only one case of exception, namely when all three bodies collide simultaneously: but this can happen only in a very special type of motion, in which all the constants of angular momentum are zero together*.

Disregarding this case of triple collision, Sundman introduced a new independent variable w defined by the equation

$$dt = (1 - e^{-\frac{r_0}{l}})(1 - e^{-\frac{r_1}{l}})(1 - e^{-\frac{r_2}{l}})\, dw\dagger,$$

where r_0, r_1, r_2 denote the three mutual distances, and l is the lower bound already mentioned. The coordinates of the bodies, and the time, are then holomorphic functions of w within a band of finite breadth 2Ω in the w-plane, bounded by two lines parallel to the real axis and on either side of it. There exists a continuous one-to-one correspondence between the real values of t and the real values of w, so that when t varies from $-\infty$ to $+\infty$, w likewise varies from $-\infty$ to $+\infty$.

Lastly, Sundman applied Poincaré's transformation

$$w = \frac{2\Omega}{\pi} \log \frac{1+\tau}{1-\tau},$$

in order to transform the band in the w-plane into a circle of radius unity in the plane of a new variable τ. The coordinates of the three bodies, and the time, are now holomorphic functions of τ everywhere within the unit circle in the τ-plane: and therefore *they can be expanded as convergent series of powers of τ for all real values of the time*, whether there are collisions or not: the case of triple collision alone being excepted‡.

190. *Trigonometric series.*

The series discussed in the preceding articles are all open to the objection that they give no evident indication of the nature of the motion of the system after the lapse of a great interval of time: they also throw no light on the number and character of the distinct types of motion which are possible in the problem: and the actual execution of the processes described is attended with great difficulties. Under these circumstances we are led to investigate expansions of an altogether different type.

If in the solution of the problem of the simple pendulum (§ 44) we consider the oscillatory type of motion, and replace the elliptic function by its expansion as a trigonometric series§, we have

$$\sin \tfrac{1}{2}\theta = \frac{2\pi}{K} \sum_{s=1}^{\infty} \frac{q^{\frac{1}{2}(2s-1)}}{1 - q^{2s-1}} \sin \frac{(2s-1)\,\pi\mu\,(t - t_0)}{2K},$$

* This last fact had been known to Weierstrass: cf. *Acta Math.* xxxv. p. 55. The motion is then in one plane.

† A simpler equation available in the restricted problem of three bodies was given by G. Armellini, *Comptes Rendus*, CLVIII. (1914), p. 253.

‡ Cf. K. Popoff, *Comptes Rendus*, CLXXXIII. (1926), p. 472, and M. Kiveliovitch, *Thèse*, Paris, 1932.

§ Cf. Whittaker and Watson, *Modern Analysis*, § 22·6.

where θ denotes the inclination of the pendulum to the vertical at time t; K and t_0 may be regarded as the two arbitrary constants of the solution, and μ is a definite constant, while q denotes $e^{-\pi K'/K}$, where K' is the complete elliptic integral complementary to K. This expansion, each term of which is a trigonometric function of t, is valid for all time. Moreover, when the constant q is not large, the first few terms of the series give a close approximation to the motion for all values of t. The circulatory type of motion of the pendulum may be similarly expressed by a trigonometric series of the same general character.

Turning now to Celestial Mechanics, we find that series of trigonometric terms have long been recognised as the most convenient method of expressing the coordinates of the members of the solar system; these series are of the type

$$\Sigma a_{n_1, n_2, \ldots, n_k} \cos (n_1 \theta_1 + n_2 \theta_2 + \ldots + n_k \theta_k),$$

where the summation is taken over positive and negative integer values of $n_1, n_2, \ldots, n_k$, and θ_r is of the form $\lambda_r t + \epsilon_r$; the quantities a, λ, and ϵ being constants. Delaunay* shewed in 1860 that the coordinates of the moon can be expressed in this way; Newcomb† in 1874 obtained a similar result for the coordinates of the planets, and several later writers‡ have designed processes for the solution of the general Problem of the Three Bodies in this form; these processes are also applicable to other dynamical systems whose equations of motion are of a certain type resembling those of the Problem of Three Bodies. In the following articles we shall give a method§ which is applicable to all dynamical systems and leads to solutions in the form of trigonometric series.

191. *Removal of terms of the first degree from the energy function.*

Consider then a dynamical system, whose equations of motion are

$$\frac{dq_r}{dt} = \frac{\partial H}{\partial p_r}, \quad \frac{dp_r}{dt} = -\frac{\partial H}{\partial q_r} \quad (r = 1, 2, \ldots, n),$$

where the energy function H does not involve the time t explicitly.

The algebraic solution of the $2n$ simultaneous equations

$$\frac{\partial H}{\partial p_r} = 0, \quad \frac{\partial H}{\partial q_r} = 0 \quad (r = 1, 2, \ldots, n)$$

will furnish in general one or more sets of values $(a_1, a_2, \ldots, a_n, b_1, b_2, \ldots, b_n)$ for the variables $(q_1, q_2, \ldots, q_n, p_1, \ldots, p_n)$; and each of these sets of values will correspond to a form of equilibrium or (if the above equations are those of a reduced system) steady motion of the system.

* *Théorie du mouvement de la lune.* Paris, 1860. † *Smithsonian Contributions*, 1874.

‡ e.g. Lindstedt, Tisserand, and Poincaré.

§ Whittaker, *Proc. Lond. Math. Soc.* XXXIV. (1902), p. 206. *Proc. R.S.E.* XXXVII. (1916), p. 95.

Let any one of these set of values $(a_1, a_2, \ldots, a_n, b_1, b_2, \ldots, b_n)$ be selected; we shall shew how to find expansions which represent the solution of the problem when the motion is of a type terminated by this form of equilibrium or steady motion. Thus if the system considered were the simple pendulum, and the form of equilibrium chosen were that in which the pendulum hangs vertically downwards at rest, our aim would be to find series which would represent the solution of the pendulum problem when the motion is of the oscillatory type.

Take then new variables $(q_1', q_2', \ldots, q_n', p_1', p_2', \ldots, p_n')$, defined by the equations

$$q_r = a_r + q_r', \qquad p_r = b_r + p_r' \qquad (r = 1, 2, \ldots, n);$$

the equations of motion become

$$\frac{dq_r'}{dt} = \frac{\partial H}{\partial p_r'}, \qquad \frac{dp_r'}{dt} = -\frac{\partial H}{\partial q_r'} \qquad (r = 1, 2, \ldots, n),$$

and for sufficiently small values of the new variables the function H can be expanded as a multiple power series in the form

$$H = H_0 + H_1 + H_2 + H_3 + \ldots,$$

where H_k denotes terms homogeneous of the kth degree in the variables

$$(q_1', q_2', \ldots, q_n', p_1', \ldots, p_n').$$

Since H_0 does not contain any of the variables, it may be omitted: and the fact that the differential equations are satisfied when $(q_1', q_2', \ldots, q_n', p_1', \ldots, p_n')$ are permanently zero requires that H_1 should vanish identically. The expansion of H therefore begins with the terms H_2, which (suppressing the accents of the new variables) may be written in the form

$$H_2 = \tfrac{1}{2}\Sigma\,(a_{rr}q_r^2 + 2a_{rs}q_rq_s) + \Sigma b_{rs}q_rp_s + \tfrac{1}{2}\Sigma\,(c_{rr}p_r^2 + 2c_{rs}p_rp_s),$$

where $\qquad a_{rs} = a_{sr}, \qquad c_{rs} = c_{sr},$

but b_{rs} is not necessarily equal to b_{sr}. If the terms $H_3, H_4, \ldots$ were neglected in comparison with H_2, the equations would become those of a vibrational problem (Chapter VII).

192. *Determination of the normal coordinates by a contact-transformation.*

We shall now apply a contact-transformation to the system in order to express H_2 in a simpler form*,—in fact, to obtain variables which correspond to normal coordinates for small vibrations of the system.

Consider the set of $2n$ equations

$$\left.\begin{aligned} sy_r + \frac{\partial}{\partial x_r} H_2(x_1, x_2, \ldots, x_n, y_1, \ldots, y_n) = 0 \\ -sx_r + \frac{\partial}{\partial y_r} H_2(x_1, x_2, \ldots, x_n, y_1, \ldots, y_n) = 0 \end{aligned}\right\} \qquad (r = 1, 2, \ldots, n)$$

* In obtaining the transformation of this article a method is used which was suggested to the author by Dr Bromwich, and which furnishes the transformation more directly than the method originally devised.

or
$$\left.\begin{aligned} -sy_r &= a_{r1}x_1 + a_{r2}x_2 + \ldots + a_{rn}x_n + b_{r1}y_1 + \ldots + b_{rn}y_n \\ sx_r &= b_{1r}x_1 + b_{2r}x_2 + \ldots + b_{nr}x_n + c_{r1}y_1 + \ldots + c_{rn}y_n \end{aligned}\right\} (r = 1, 2, \ldots, n).$$

On solving these equations, we obtain for s the determinantal equation which in § 84 was denoted by $f(s) = 0$: we shall suppose that H_2 is a positive definite form, and (as in § 84) we shall denote the roots of the equation by $\pm is_1, \pm is_2, \ldots, \pm is_n$; the quantities $s_1, s_2, \ldots, s_n$ are all real, and for simplicity we shall suppose no two of them to be equal.

To each root there will correspond a set of values for the ratios of the quantities $(x_1; x_2, \ldots, x_n, y_1, \ldots, y_n)$; let the set which correspond to the root is_r be denoted by $({}_rx_1, {}_rx_2, \ldots, {}_rx_n, {}_ry_1, \ldots, {}_ry_n)$, and let the set which correspond to the root $-is_r$ be denoted by $({}_{-r}x_1, {}_{-r}x_2, \ldots, {}_{-r}x_n, {}_{-r}y_1, \ldots, {}_{-r}y_n)$, so that we have

$$\begin{aligned} -is_r\, {}_ry_p &= a_{p1}\, {}_rx_1 + a_{p2}\, {}_rx_2 + \ldots + a_{pn}\, {}_rx_n + b_{p1}\, {}_ry_1 + \ldots + b_{pn}\, {}_ry_n, \\ is_r\, {}_rx_p &= b_{1p}\, {}_rx_1 + b_{2p}\, {}_rx_2 + \ldots + b_{np}\, {}_rx_n + c_{p1}\, {}_ry_1 + \ldots + c_{pn}\, {}_ry_n. \end{aligned}$$

Multiply these equations by ${}_kx_p$ and ${}_ky_p$ respectively, add them, and sum with respect to p; we thus obtain the equation

$$is_r \sum_{p=1}^{n} ({}_rx_p\, {}_ky_p - {}_kx_p\, {}_ry_p) = H(r, k),$$

where

$$H(r, k) = a_{11}\, {}_rx_1\, {}_kx_1 + a_{12}({}_rx_1\, {}_kx_2 + {}_kx_1\, {}_rx_2) + \ldots + b_{11}({}_rx_1\, {}_ky_1 + {}_kx_1\, {}_ry_1) + \ldots + c_{11}\, {}_ry_1\, {}_ky_1 + \ldots,$$

so that $H(r, k)$ is symmetrically related to r and k.

Interchanging r and k, we have

$$is_k \sum_{p=1}^{n} ({}_kx_p\, {}_ry_p - {}_rx_p\, {}_ky_p) = H(r, k),$$

and therefore
$$(s_r + s_k) \sum_{p=1}^{n} ({}_kx_p\, {}_ry_p - {}_rx_p\, {}_ky_p) = 0.$$

So, unless $s_r + s_k$ is zero, we have

$$\sum_{p=1}^{n} ({}_rx_p\, {}_ky_p - {}_kx_p\, {}_ry_p) = 0,$$

and consequently $H(r, k)$ is zero: if $s_r + s_k$ is zero, we have ${}_kx_p = {}_{-r}x_p$, ${}_ky_p = {}_{-r}y_p$, and therefore

$$is_r \sum_{p=1}^{n} ({}_rx_p\, {}_{-r}y_p - {}_{-r}x_p\, {}_ry_p) = H(r, -r).$$

If now we define new variables $(q_1', q_2', \ldots, q_n', p_1', \ldots, p_n')$ by the equations

$$\left.\begin{aligned} q_r &= {}_1x_r q_1' + {}_2x_r q_2' + \ldots + {}_nx_r q_n' + {}_{-1}x_r p_1' + \ldots + {}_{-n}x_r p_n' \\ p_r &= {}_1y_r q_1' + {}_2y_r q_2' + \ldots + {}_ny_r q_n' + {}_{-1}y_r p_1' + \ldots + {}_{-n}y_r p_n' \end{aligned}\right\} \quad (r = 1, 2, \ldots, n),$$

and if δ and Δ denote any two independent modes of variation, it is evident that the coefficient of $\delta q_r' \Delta p_k'$ in $\sum_{l=1}^{n} (\delta q_l \Delta p_l - \Delta q_l \delta p_l)$ is $\sum_{l=1}^{n} ({}_rx_l\, {}_{-k}y_l - {}_{-k}x_l\, {}_ry_l)$, which is zero when r is not equal to k. Thus $\sum_{l=1}^{n} (\delta q_l \Delta p_l - \Delta q_l \delta p_l)$ contains no terms except such as $(\delta q_r' \Delta p_r' - \Delta q_r' \delta p_r')$, and the coefficient of this term is $\sum_{l=1}^{n} ({}_r x_l\, {}_{-r}y_l - {}_{-r}x_l\, {}_r y_l)$. Now hitherto the actual values of ${}_r x_l$, ${}_r y_l$ have not been fixed, as only their ratios are determined from their equations of definition; we may therefore choose their values so that

$$\sum_{l=1}^{n} ({}_r x_l\, {}_{-r}y_l - {}_{-r}x_l\, {}_r y_l) = 1 \qquad (r = 1, 2, \ldots, n),$$

and then we shall have

$$\sum_{l=1}^{n} (\delta q_l \Delta p_l - \Delta q_l \delta p_l) = \sum_{r=1}^{n} (\delta q_r' \Delta p_r' - \Delta q_r' \delta p_r'),$$

so that (§ 128) the transformation from the variables $(q_1, q_2, \ldots, q_n, p_1, \ldots, p_n)$ to the variables $(q_1', q_2', \ldots, q_n', p_1', \ldots, p_n')$ is a contact-transformation. Moreover, if in H_2 we substitute for $(q_1, q_2, \ldots, q_n, p_1, \ldots, p_n)$ in terms of $(q_1', q_2', \ldots, q_n', p_1', \ldots, p_n')$, we obtain

$$H_2 = \sum_{r=1}^{n} H(r, -r)\, q_r' p_r'$$

or

$$H_2 = i \sum_{r=1}^{n} s_r q_r' p_r'.$$

Now apply to the variables $(q_1', q_2', \ldots, q_n', p_1', \ldots, p_n')$ the contact-transformation defined by the equations

$$q_r'' = \frac{\partial W}{\partial p_r''}, \qquad p_r' = \frac{\partial W}{\partial q_r'} \qquad (r = 1, 2, \ldots, n),$$

where

$$W = \sum_{r=1}^{n} \left(p_r'' q_r' + \tfrac{1}{2} \frac{i p_r''^2}{s_r} - \tfrac{1}{4} i s_r q_r'^2 \right),$$

which gives

$$H_2 = \tfrac{1}{2} \sum_{r=1}^{n} (p_r''^2 + s_r^2 q_r''^2).$$

As all the transformations concerned have been linear, we see that $H_3, H_4, \ldots$ will be homogeneous polynomials of degrees 3, 4, ... in the new variables: and thus, omitting the accents, we have the result that *the equations of motion of the dynamical system have been brought to the form*

$$\frac{dq_r}{dt} = \frac{\partial H}{\partial p_r}, \qquad \frac{dp_r}{dt} = -\frac{\partial H}{\partial q_r} \qquad (r = 1, 2, \ldots, n);$$

where

$$H = H_2 + H_3 + H_4 + \ldots,$$

in which H_r is a homogeneous polynomial of degree r in the variables, and in particular

$$H_2 = \tfrac{1}{2} \sum_{r=1}^{n} (p_r^2 + s_r^2 q_r^2).$$

It is clear that if we neglect $H_3, H_4, \ldots$ in comparison with H_2, and integrate the equations, the solution obtained will be identical with that found in § 84.

193. *Transformation to the trigonometric form of H.*

The system will now be further transformed by applying to it a contact-transformation from the variables $(q_1, q_2, \ldots, q_n, p_1, \ldots, p_n)$ to new variables $(q_1', q_2', \ldots, q_n', p_1', \ldots, p_n')$, defined by the equations

$$p_r = \frac{\partial W}{\partial q_r'}, \quad q_r = \frac{\partial W}{\partial p_r} \qquad (r = 1, 2, \ldots, n),$$

where
$$W = \sum_{r=1}^{n} \left[q_r' \arcsin \frac{p_r}{(2s_r q_r')^{\frac{1}{2}}} + \frac{p_r}{2s_r} \{2s_r q_r' - p_r^2\}^{\frac{1}{2}} \right]$$

so that

$$p_r = (2s_r q_r')^{\frac{1}{2}} \sin p_r', \qquad q_r = (2q_r')^{\frac{1}{2}} s_r^{-\frac{1}{2}} \cos p_r', \qquad (r = 1, 2, \ldots, n).$$

The differential equations become

$$\frac{dq_r'}{dt} = \frac{\partial H}{\partial p_r'}, \quad \frac{dp_r'}{dt} = -\frac{\partial H}{\partial q_r'} \qquad (r = 1, 2, \ldots, n),$$

where
$$H = s_1 q_1' + s_2 q_2' + \ldots + s_n q_n' + H_3 + H_4 + \ldots,$$

and now H_r denotes an aggregate of terms which are homogeneous of degree $\frac{1}{2}r$ in the quantities q_r', and homogeneous of degree r in the quantities $\cos p_r$, $\sin p_r'$.

Since a product of powers of $\cos p_r'$, $\sin p_r'$ can be expressed as a sum of sines and cosines of angles of the form $(n_1 p_1' + n_2 p_2' + \ldots + n_n p_n')$, where $n_1, n_2, \ldots, n_n$ have integer or zero values, it follows that H_r can be expressed as the sum of a finite number of terms, each of the form

$$q_1'^{m_1} q_2'^{m_2} \ldots q_n'^{m_n} \begin{matrix}\sin\\ \cos\end{matrix} (n_1 p_1' + n_2 p_2' + \ldots + n_n p_n'),$$

where
$$m_1 + m_2 + \ldots + m_n = \tfrac{1}{2}r, \qquad |n_r| \leqslant 2m_r,$$

and therefore
$$|n_1| + |n_2| + \ldots + |n_n| \leqslant r.$$

The function H is thus expressed in the form

$$H = \Sigma A_{n_1, n_2, \ldots, n_n}^{m_1, m_2, \ldots, m_n} q_1'^{m_1} q_2'^{m_2} \ldots q_n'^{m_n} \begin{matrix}\sin\\ \cos\end{matrix} (n_1 p_1' + n_2 p_2' + \ldots + n_n p_n'),$$

where for each term we have

$$|n_1| + |n_2| + \ldots + |n_n| \leqslant 2(m_1 + m_2 + \ldots + m_n),$$

and clearly the series is absolutely convergent for all values of $p_1', p_2', \ldots, p_n'$, provided $q_1', q_2', \ldots, q_n'$ do not exceed certain limits of magnitude.

To avoid unnecessary complexity, we shall ignore the terms in

$$\sin(n_1 p_1' + n_2 p_2' + \ldots + n_n p_n'),$$

as they are to be treated in the same way as the terms in

$$\cos(n_1 p_1' + n_2 p_2' + \ldots + n_n p_n'),$$

and their presence complicates, but does not in any important respect modify, the later developments.

194. *Other types of motion which lead to equations of the same form.*

The equations which have now been obtained have been shewn to be applicable when the motion is of a type not far removed from a steady motion or an equilibrium-configuration, e.g. the oscillatory motion of the simple pendulum, or those types of motion of the Problem of Three Bodies which have been studied in § 181. But these equations may be shewn to be applicable also to motion which is not of this character, and in particular to motion such as that of the planets round the sun, or the moon round the earth*.

For let the equations of motion of the Problem of Three Bodies be taken in the form obtained in § 160; and let the contact-transformation which is defined by the equations

$$p_r = \frac{\partial W}{\partial q_r}, \quad p_r' = -\frac{\partial W}{\partial q_r'} \qquad (r=1, 2, 3, 4)$$

be applied to this system, where

$$W = q_1' q_3 + q_2' q_4 + \int^{q_1} \left\{ -\frac{\mu^2 m_1^2 m_2^2}{q_3'^2} + \frac{2\mu m_1 m_2}{q_1} - \frac{q_1'^2}{q_1^2} \right\}^{\frac{1}{2}} dq_1 + \int^{q_2} \left\{ -\frac{\mu'^2 m_1^2 m_3^2}{q_4'^2} + \frac{2\mu' m_1 m_3}{q_2} - \frac{q_2'^2}{q_2^2} \right\}^{\frac{1}{2}} dq_2.$$

The new variables can be interpreted in the following way. Suppose that at the instant t all the forces acting on the particle μ cease, except a force of magnitude $m_1 m_2/q_1^2$ directed to the origin; and let a be the semi-major axis and e the eccentricity of the ellipse described after this instant: then

$$q_1' = \{m_1 m_2 \mu a (1 - e^2)\}^{\frac{1}{2}}, \qquad q_3' = \{m_1 m_2 \mu a\}^{\frac{1}{2}}.$$

Further, if the lower limits of the integrals are suitably chosen, $p_1' + q_3$ is the true anomaly of μ in its ellipse, and $-p_3'$ is the mean anomaly. The variables q_2', q_4', p_2', p_4' stand in a corresponding relation to the particle μ'.

The equations of motion now take the form

$$\frac{dq_r'}{dt} = \frac{\partial H}{\partial p_r'}, \quad \frac{dp_r'}{dt} = -\frac{\partial H}{\partial q_r'} \qquad (r=1, 2, 3, 4);$$

when the particles m_2 and m_3 are supposed to be of small mass compared with m_1, and are describing orbits of a planetary character about m_1, it is readily found that H can be expanded in terms of the new variables in the form

$$H = a_{0,0,0,0} + \Sigma a_{n_1, n_2, n_3, n_4} \cos (n_1 p_1' + n_2 p_2' + n_3 p_3' + n_4 p_4'),$$

where the coefficients a are functions of (q_1', q_2', q_3', q_4') only, the summation extends over positive and negative integer and zero values of n_1, n_2, n_3, n_4, and the coefficient $a_{0,0,0,0}$ is much the most important part of the series. As this expansion of H is of the same character as that obtained in § 193, it follows that *the method of solution given in the following articles is applicable either to motion of the planetary type or to motion of the type studied in* § 181.

* Delaunay, *Théorie de la Lune*; Tisserand, *Annales de l'Obs. de Paris, Mémoires*, XVIII. (1885).

195. *The problem of integration.*

For simplicity we shall suppose in what follows that the system has only two degrees of freedom: so that the equations to be integrated are

$$\frac{dq_1}{dt}=\frac{\partial H}{\partial p_1},\quad \frac{dq_2}{dt}=\frac{\partial H}{\partial p_2},\quad \frac{dp_1}{dt}=-\frac{\partial H}{\partial q_1},\quad \frac{dp_2}{dt}=-\frac{\partial H}{\partial q_2} \quad \text{..........(1)},$$

where the Hamiltonian function H can be expanded as an infinite series proceeding in powers of $\sqrt{q_1}$ and $\sqrt{q_2}$, and in trigonometric functions of multiples of p_1 and p_2: that is to say, in terms of the type

$$q_1^{\frac{1}{2}m}q_2^{\frac{1}{2}n}\cos(ip_1+jp_2),$$

where m and n are integers (positive or zero) and i and j are integers (positive or negative or zero): moreover, if we call $(m+n)$ the "order" of a term, the terms of lowest order are linear in q_1 and q_2 and free from p_1 and p_2, so that they may be written $(s_1q_1+s_2q_2)$, where s_1 and s_2 are constants. Further, $m-|i|$ is zero or an even integer, and $n-|j|$ is also zero or an even integer.

The Hamiltonian function H may therefore be expanded in the form

$$H=s_1q_1+s_2q_2+H_3+H_4+H_5+\ldots \quad \text{.................(2)},$$

where H_r denotes the terms of order r, so we may write

$$\begin{aligned}H_3=&q_1^{\frac{3}{2}}(U_1\cos p_1+U_2\cos 3p_1)+q_1q_2^{\frac{1}{2}}\{U_3\cos p_2+U_4\cos(2p_1+p_2)+U_5\cos(2p_1-p_2)\}\\&+q_1^{\frac{1}{2}}q_2\{U_6\cos p_1+U_7\cos(2p_2+p_1)\,U_8\cos(2p_2-p_1)\}+q_2^{\frac{3}{2}}\{U_9\cos p_2+U_{10}\cos 3p_2\},\end{aligned}$$

and

$$\begin{aligned}H_4=&q_1^2(X_1+X_2\cos 2p_1+X_3\cos 4p_1)\\&+q_1^{\frac{3}{2}}q_2^{\frac{1}{2}}\{X_4\cos(p_1+p_2)+X_5\cos(p_1-p_2)+X_6\cos(3p_1+p_2)+X_7\cos(3p_1-p_2)\}\\&+q_1q_2\{X_8+X_9\cos 2p_1+X_{10}\cos 2p_2+X_{11}\cos(2p_1+2p_2)+X_{12}\cos(2p_1-2p_2)\}\\&+q_1^{\frac{1}{2}}q_2^{\frac{3}{2}}\{X_{13}\cos(p_1+p_2)+X_{14}\cos(p_1-p_2)+X_{15}\cos(p_1+3p_2)+X_{16}\cos(p_1-3p_2)\}\\&+q_2^2\{X_{17}+X_{18}\cos 2p_2+X_{19}\cos 4p_2\},\end{aligned}$$

the coefficients $U_1, U_2, \ldots U_{10}, X_1, X_2, \ldots X_{19}$ being constants.

We know one integral of the differential equations (1), namely the integral of energy

$$H(q_1,\ q_2,\ p_1,\ p_2)=\text{Constant}:$$

and we know (§ 121) that if we can find one other integral, the system can be completely integrated.

It was shewn by the author in 1916* that this other integral can be found, but that *it cannot in general be represented by a single analytical expression which is valid for all values of the ratio s_1/s_2, or even for any finite range of values of s/s_2*: in fact, it is necessary to distinguish three cases:

* *Proc. R.S. Edin.* XXXVII. (1916), p. 95.

CASE I. The ratio s_1/s_2 is an irrational number.

CASE II. The ratio s_1/s_2 is a rational number, say equal to m/n (where m and n are integers and the fraction m/n is in its lowest terms) and terms in $\cos(np_1 - mp_2)$ are absent from H_3.

CASE III. The ratio s_1/s_2 is a rational number, say equal to m/n, and terms in $\cos(np_1 - mp_2)$ are present in H_3.

The integral which we are seeking (and which we shall call the *adelphic* integral, for reasons which will appear later) always exists, but its analytic expression is different in these three cases; so that as s_1/s_2 is continuously varied, the form of the adelphic integral is abruptly changed whenever s_1/s_2 passes from a rational to an irrational value, or conversely. This is the ultimate fact which lies at the basis of Poincaré's celebrated theorem that the series of Celestial Mechanics, if they converge at all, cannot converge uniformly for all values of the time on the one hand, and on the other hand for all values of the constants comprised between certain limits.

We shall now determine the adelphic integral in each of these cases in turn.

196. *Determination of the adelphic integral in Case I.*

Let us then first suppose that the Hamiltonian function is expanded as in § 195, and that the ratio s_1/s_2 is an irrational number. We shall now shew how to set up formally a series which, if it converges, is an integral of the system.

If $\phi(q_1, q_2, p_1, p_2) = \text{Constant}$ is an integral, we must have (from the equations of motion)

$$\frac{\partial\phi}{\partial q_1}\frac{\partial H}{\partial p_1} + \frac{\partial\phi}{\partial q_2}\frac{\partial H}{\partial p_2} - \frac{\partial\phi}{\partial p_1}\frac{\partial H}{\partial q_1} - \frac{\partial\phi}{\partial p_2}\frac{\partial H}{\partial q_2} = 0 \quad \text{..............(1)},$$

an equation which we may write $(\phi, H) = 0$.

Let us see if this equation can be satisfied formally by a series proceeding in ascending powers of $\sqrt{q_1}$ and $\sqrt{q_2}$ and trigonometric functions of p_1 and p_2 (like the series for H), whose terms of lowest order are $(s_1 q_1 - s_2 q_2)$: so that we may write

$$\phi \equiv s_1 q_1 - s_2 q_2 + \phi_3 + \phi_4 + \phi_5 + \ldots,$$

where ϕ_r denotes the terms which are of degree r in $\sqrt{q_1}$ and $\sqrt{q_2}$.

Substituting in equation (1), and equating to zero the terms of lowest order, we have

$$s_1\frac{\partial\phi_3}{\partial p_1} + s_2\frac{\partial\phi_3}{\partial p_2} = s_1\frac{\partial H_3}{\partial p_1} - s_2\frac{\partial H_3}{\partial p_2}.$$

This evidently implies that to any term $A\cos(mp_1 + np_2)$ in H_3, there corresponds a term $\dfrac{s_1 m - s_2 n}{s_1 m + s_2 n} A\cos(mp_1 + np_2)$ in ϕ_3: so the value of ϕ_3

may be written down at once. Having thus determined ϕ_3, we equate to zero the terms in equation (1) which are of order 4 in $\sqrt{q_1}$ and $\sqrt{q_2}$: this gives the equation

$$s_1\frac{\partial\phi_4}{\partial p_1}+s_2\frac{\partial\phi_4}{\partial p_2}=s_1\frac{\partial H_4}{\partial p_1}-s_2\frac{\partial H_4}{\partial p_2}+(\phi_3,\,H_3).$$

As the quantities on the right-hand side are all known, we can solve this equation for ϕ_4* in the same way as the preceding equation was solved for ϕ_3: and thus combining our results we obtain for our integral series ϕ

$$\begin{aligned}
\text{Constant}=\phi\equiv{}& s_1q_1-s_2q_2+q_1^{\frac{3}{2}}(U_1\cos p_1+U_2\cos 3p_1)\\
&+q_1q_2^{\frac{1}{2}}\left\{-U_3\cos p_2+\frac{2s_1-s_2}{2s_1+s_2}U_4\cos(2p_1+p_2)+\frac{2s_1+s_2}{2s_1-s_2}U_5\cos(2p_1-p_2)\right\}\\
&+q_1^{\frac{1}{2}}q_2\left\{U_6\cos p_1+\frac{s_1-2s_2}{s_1+2s_2}U_7\cos(2p_2+p_1)+\frac{s_1+2s_2}{s_1-2s_2}U_8\cos(2p_2-p_1)\right\}\\
&+q_2^{\frac{3}{2}}\left\{-U_9\cos p_2-U_{10}\cos 3p_2\right\}\\
&+q_1^2\left[\left\{\frac{1}{2s_1+s_2}U_3U_4+\frac{1}{2s_1-s_2}U_3U_5+X_2\right\}\cos 2p_1+\left\{\frac{s_2}{(2s_1+s_2)(2s_1-s_2)}U_4U_5\right.\right.\\
&\qquad\left.\left.+X_3\right\}\cos 4p_1\right]\\
&+q_1^{\frac{3}{2}}q_2^{\frac{1}{2}}\left[\frac{\cos(p_1+p_2)}{s_1+s_2}\left\{\frac{-2s_1}{s_1+2s_2}U_3U_7-\frac{6s_1s_2}{(2s_1-s_2)(s_1-2s_2)}U_5U_8+U_3U_6\right.\right.\\
&\qquad\left.+\frac{s_2}{2s_1+s_2}U_4U_6-U_1U_3+\frac{4s_2}{2s_1+s_2}U_1U_4+\frac{6s_2}{2s_1-s_2}U_2U_5+(s_1-s_2)X_4\right\}\\
&\quad+\frac{\cos(p_1-p_2)}{s_1-s_2}\left\{\frac{-6s_1s_2U_4U_7}{(2s_1+s_2)(s_1+2s_2)}+\frac{2s_1}{s_1-2s_2}U_3U_8-U_3U_6\right.\\
&\qquad\left.+\frac{s_2}{2s_1-s_2}U_5U_6-U_1U_3-\frac{4s_2}{2s_1-s_2}U_1U_5-\frac{6s_2}{2s_1+s_2}U_2U_4+(s_1+s_2)X_5\right\}\\
&\quad+\frac{\cos(3p_1+p_2)}{3s_1+s_2}\left\{\frac{+10s_1s_2}{(2s_1-s_2)(s_1+2s_2)}U_5U_7+\frac{s_2}{2s_1+s_2}U_4U_6-3U_2U_3\right.\\
&\qquad\left.+\frac{2s_2}{2s_1+s_2}U_1U_4+(3s_1-s_2)X_6\right\}\\
&\quad+\frac{\cos(3p_1-p_2)}{3s_1-s_2}\left\{\frac{+10s_1s_2}{(2s_1+s_2)(s_1-2s_2)}U_4U_8+\frac{s_2}{2s_1-s_2}U_5U_6-3U_2U_3\right.\\
&\qquad\left.\left.-\frac{2s_2}{2s_1-s_2}U_1U_5+(3s_1+s_2)X_7\right\}\right.
\end{aligned}$$

* The equation for ϕ_4 will not have a solution of the desired form unless on the right the coefficients of q_1^2, q_1q_2, q_2^2 are all zero, since terms of this form in ϕ_4 are annulled by the operator $s_1\frac{\partial}{\partial p_1}+s_2\frac{\partial}{\partial p_2}$. It is easy to verify that these terms are in fact absent from $s_1\frac{\partial H_4}{\partial p_1}-s_2\frac{\partial H_4}{\partial p_2}$ and from (ϕ_3, H_3). The same point arises in the equations for ϕ_6, ϕ_8, ... : for a general discussion of these "critical" terms, cf. T. M. Cherry, *Proc. Camb. Phil. Soc.* XXII. (1924), pp. 325, 510.

$$
\begin{aligned}
&+ q_1 q_2 \Bigg[\cos 2p_1 \left\{ \frac{-2}{2s_1+s_2} U_4 U_9 + \frac{2}{2s_1-s_2} U_5 U_9 + \frac{8s_2}{(s_1+2s_2)(s_1-2s_2)} U_7 U_8 \right. \\
&\qquad\qquad \left. - \frac{2}{2s_1+s_2} U_3 U_4 - \frac{2}{2s_1-s_2} U_3 U_5 + X_9 \right\} \\
&\quad + \cos 2p_2 \left\{ \frac{+2}{s_1+2s_2} U_6 U_7 - \frac{2}{s_1-2s_2} U_6 U_8 + \frac{2}{s_1+2s_2} U_1 U_7 + \frac{2}{s_1-2s_2} U_1 U_8 \right. \\
&\qquad\qquad \left. + \frac{8s_1}{(2s_1-s_2)(2s_1+s_2)} U_4 U_5 - X_{10} \right\} \\
&\quad + \frac{\cos(2p_1+2p_2)}{2s_1+2s_2} \left\{ \frac{-2s_1}{2s_1+s_2} U_4 U_9 + \frac{6s_1}{2s_1-s_2} U_5 U_{10} + \frac{4s_2}{s_1+2s_2} U_6 U_7 \right. \\
&\qquad\qquad \left. + \frac{2s_2}{s_1+2s_2} U_1 U_7 + \frac{6s_2}{s_1-2s_2} U_2 U_8 - \frac{4s_1}{2s_1+s_2} U_3 U_4 + (2s_1-2s_2) X_{11} \right\} \\
&\quad + \frac{\cos(2p_1-2p_2)}{2s_1-2s_2} \left\{ \frac{+2s_1}{2s_1-s_2} U_5 U_9 - \frac{6s_1}{2s_1+s_2} U_4 U_{10} + \frac{4s_2}{s_1-2s_2} U_6 U_8 \right. \\
&\qquad\qquad \left. - \frac{2s_2}{s_1-2s_2} U_1 U_8 - \frac{6s_2}{s_1+2s_2} U_2 U_7 - \frac{4s_1}{2s_1-s_2} U_3 U_5 + (2s_1+2s_2) X_{12} \right\} \Bigg] \\
&+ q_1^{\frac{1}{2}} q_2^{\frac{3}{2}} \Bigg[\frac{\cos(p_1+p_2)}{s_1+s_2} \left\{ + U_6 U_9 - \frac{4s_1}{s_1+2s_2} U_7 U_9 + \frac{6s_1}{s_1-2s_2} U_8 U_{10} - U_3 U_6 \right. \\
&\qquad\qquad \left. - \frac{s_1}{s_1+2s_2} U_3 U_7 + \frac{2s_2}{2s_1+s_2} U_4 U_6 + \frac{6s_1 s_2}{(2s_1-s_2)(s_1-2s_2)} U_5 U_8 + (s_1-s_2) X_{13} \right\} \\
&\quad + \frac{\cos(p_1-p_2)}{s_1-s_2} \left\{ - U_6 U_9 - \frac{6s_1}{s_1+2s_2} U_7 U_{10} + \frac{4s_1}{s_1-2s_2} U_8 U_9 - U_3 U_6 \right. \\
&\qquad\qquad \left. - \frac{s_1}{s_1-2s_2} U_3 U_8 - \frac{6s_1 s_2}{(s_1+2s_2)(2s_1+s_2)} U_4 U_7 - \frac{2s_2}{2s_1-s_2} U_5 U_6 + (s_1+s_2) X_{14} \right\} \\
&\quad + \frac{\cos(p_1+3p_2)}{s_1+3s_2} \left\{ + 3 U_6 U_{10} - \frac{2s_1}{s_1+2s_2} U_7 U_9 - \frac{s_1}{s_1+2s_2} U_3 U_7 \right. \\
&\qquad\qquad \left. + \frac{10 s_1 s_2}{(s_1-2s_2)(2s_1+s_2)} U_4 U_8 + (s_1-3s_2) X_{15} \right\} \\
&\quad + \frac{\cos(p_1-3p_2)}{s_1-3s_2} \left\{ - 3 U_6 U_{10} + \frac{2s_1}{s_1-2s_2} U_8 U_9 - \frac{s_1}{s_1-2s_2} U_3 U_8 \right. \\
&\qquad\qquad \left. - \frac{10 s_1 s_2}{(s_1+2s_2)(2s_1-s_2)} U_5 U_7 + (s_1+3s_2) X_{16} \right\} \Bigg] \\
&+ q_2^2 \Bigg[\left\{ \frac{1}{s_1+2s_2} U_6 U_7 + \frac{1}{s_1-2s_2} U_6 U_8 - X_{18} \right\} \cos 2p_2 \\
&\qquad\qquad + \left\{ \frac{s_1}{(s_1-2s_2)(s_1+2s_2)} U_7 U_8 - X_{19} \right\} \cos 4p_2 \Bigg] \\
&+ \text{terms of the 5th and higher orders in } \sqrt{q_1} \text{ and } \sqrt{q_2} \quad \text{.........(2).}
\end{aligned}
$$

The terms of higher order in the series may be determined in the same way as the terms in ϕ_3 and ϕ_4, and we thus obtain the complete expansion

We may note that instead of assuming $(s_1 q_1 - s_2 q_2)$ as the lowest term of our integral, we might have assumed q_1, or q_2, or any linear function of q_1 and q_2; the integral then obtained would be merely a linear combination of our integral (2) with the integral of energy, whose lowest terms are $(s_1 q_1 + s_2 q_2)$.

We may further note that in the above process, when finding ϕ_4, we may if we please add to ϕ_4 any terms of the form $\alpha q_1^2 + \beta q_1 q_2 + \gamma q_2^2$, where α, β, γ are constants; for these terms are annulled by the operator $\left(s_1 \frac{\partial}{\partial p_1} + s_2 \frac{\partial}{\partial p_2}\right)$, and therefore ϕ_4 satisfies its differential equation just as well when these terms are present as when they are absent. The introduction of these terms into ϕ_4 will cause changes in the terms of higher order—in ϕ_5, ϕ_6, etc.: and the sum total of all the changes will merely amount to adding to our function ϕ a quadratic function of the two integrals which we know, namely, the integral of energy and the integral (2) itself.

Similarly we may add any terms of the form $(\alpha q_1^3 + \beta q_1^2 q_2 + \gamma q_1 q_2^2 + \delta q_2^3)$ to ϕ_6: the ultimate effect is merely to add to our integral a cubic function of itself and the integral of energy. There is evidently nothing to be gained by doing this, and we may therefore omit these arbitrary terms in $\phi_4, \phi_6, \phi_8, \ldots$.

197. *An example of the adelphic integral in Case I.*

As an example, consider the motion of a particle of unit mass in a field when the potential energy is (with rectangular coordinates x and y)

$$-\frac{1+3x}{3(1+2x+x^2+y^2)^{\frac{3}{2}}}$$

or (expanding) $x^2 + \frac{1}{2}y^2 - \frac{8}{3}x^3 - xy^2 + 5x^4 - \frac{5}{8}y^4 +$ terms of the 5th order, so that the origin is a position of equilibrium. We shall study orbits near to this equilibrium position.

If we make the contact-transformation

$$\dot{x} = 2^{\frac{3}{4}} q_1^{\frac{1}{2}} \sin p_1, \qquad x = 2^{\frac{1}{4}} q_1^{\frac{1}{2}} \cos p_1,$$

$$\dot{y} = 2^{\frac{1}{2}} q_2^{\frac{1}{2}} \sin p_2, \qquad y = 2^{\frac{1}{2}} q_2^{\frac{1}{2}} \cos p_2,$$

the system is now specified by the Hamiltonian function

$$H = 2^{\frac{1}{2}} q_1 \sin^2 p_1 + q_2 \sin^2 p_2 - \frac{1 + 3 \cdot 2^{\frac{1}{4}} q_1^{\frac{1}{2}} \cos p_1}{3\,(1 + 2^{\frac{5}{4}} q_1^{\frac{1}{2}} \cos p_1 + 2^{\frac{1}{2}} q_1 \cos^2 p_1 + 2 q_2 \cos^2 p_2)^{\frac{3}{2}}},$$

or expanding,

$$H = 2^{\frac{1}{2}} q_1 + q_2 + 2^{\frac{7}{4}} q_1^{\frac{3}{2}} (-\cos p_1 - \tfrac{1}{3}\cos 3p_1)$$
$$+ 2^{-\frac{3}{4}} q_1^{\frac{1}{2}} q_2 \{-2\cos p_1 - \cos(p_1 + 2p_2) - \cos(p_1 - 2p_2)\} + \ldots \quad \ldots(1).$$

The corresponding adelphic integral, obtained by substituting in formula (2) of the last article, is

$$\text{Constant} = \phi \equiv 2^{\frac{1}{2}} q_1 - q_2 + 2^{\frac{7}{4}} q_1^{\frac{3}{2}} (-\cos p_1 - \tfrac{1}{3} \cos 3p_1)$$
$$+ 2^{-\frac{3}{4}} q_1^{\frac{1}{2}} q_2 \{-2 \cos p_1 + (1 - \sqrt{2})^2 \cos (p_1 + 2p_2) + (1 + \sqrt{2})^2 \cos (p_1 - 2p_2)\} + \dots \qquad \dots\dots\dots(2).$$

Now it may be verified readily by differentiation that this dynamical system possesses the integral

$$\text{Constant} = (q_2^{\frac{1}{2}} \sin p_2 + 2^{\frac{1}{4}} q_1^{\frac{1}{2}} q_2^{\frac{1}{2}} \sin p_2 \cos p_1 - 2^{\frac{3}{4}} q_1^{\frac{1}{2}} q_2^{\frac{1}{2}} \sin p_1 \cos p_2)^2$$
$$- \frac{1 + 2^{\frac{1}{4}} q_1^{\frac{1}{2}} \cos p_1}{(1 + 2^{\frac{5}{4}} q_1^{\frac{1}{2}} \cos p_1 + 2^{\frac{1}{2}} q_1 \cos^2 p_1 + 2 q_2 \cos^2 p_2)^{\frac{1}{2}}},$$

which when expanded takes the form

$$\text{Constant} = q_2 + 2^{-\frac{1}{4}} (1 - \sqrt{2}) q_1^{\frac{1}{2}} q_2 \cos (p_1 + 2p_2)$$
$$- 2^{\frac{1}{4}} (1 + \sqrt{2}) q_1^{\frac{1}{2}} q_2 \cos (p_1 - 2p_2) + \dots \qquad \dots\dots\dots(3).$$

It is evident, on comparing the series, that the series (2) is what would be obtained by subtracting twice the series (3) from the series (1), which represents the integral of energy. This shews that for the particular dynamical system we are considering, the ϕ-series (2) is identical with the expansion, formed by ordinary algebraic and trigonometric processes under conditions which ensure convergence, of a known integral: and the convergence of the series (2), for sufficiently small values of $\sqrt{q_1}$ and $\sqrt{q_2}$, is thereby established for this particular system.

198. *The question of convergence.*

For particular dynamical systems such as that considered in the last article, the convergence of the adelphic-integral series (2) of § 196 can be proved, for sufficiently small values of q_1 and q_2, so long as the ratio s_1/s_2 is an irrational number. No proof of convergence applicable to the most general case, has yet been devised, and the procedure of § 196 must therefore be regarded, in the present state of the subject, merely as a method of constructing a formal series, whose convergence must be investigated separately in the case of each particular dynamical system to which the method is applied. In all such particular systems examined hitherto, the series is convergent: and the following considerations may be adduced in support of the opinion that the convergence is general.

Since the ratio s_1/s_2 is an irrational number, none of the denominators $(s_1 + s_2)$, $(s_1 - s_2)$, $(2s_1 + s_2)$, $(2s_1 - s_2)$, $(s_1 + 2s_2)$, $(3s_1 + s_2)$, ... can vanish, and therefore no term of the series can be infinite. The series is a power-series in $\sqrt{q_1}$ and $\sqrt{q_2}$, and it has been derived from another absolutely convergent power-series in $\sqrt{q_1}$ and $\sqrt{q_2}$, namely, the series for H, by operations which are of an ordinary algebraical and trigonometrical combinatory character, except as regards the operation of introducing the divisors of the type (ms_1+ns_2) (where m and n are positive or negative integers) in the integrations. We may therefore expect that the series will converge for sufficiently small values of $\sqrt{q_1}$ and $\sqrt{q_2}$, unless the smallness of some of these divisors causes the series to diverge for all values of $\sqrt{q_1}$ and $\sqrt{q_2}$, however small. Now the values of the integers m and n may indeed be so chosen that the divisor (ms_1+ns_2) may be as small as we please: but $|m|$ and $|n|$ are then large, and since $|m|$ and $|n$ are not greater than the order of the term, this small divisor can occur only in a term of high order, where it will be more or less neutralised by the high powers of $\sqrt{q_1}$ and $\sqrt{q_2}$: and it was in fact shewn many years ago by Bruns* that this state of things is consistent with the absolute convergence of a series. The example given by Bruns was the series

$$\sum_{m=1}^{\infty}\sum_{n=1}^{\infty}\frac{q_1^{\,m}q_2^{\,n}}{m-nA},$$

where q_1 and q_2 are proper fractions and A is a positive irrational number which is an algebraic number, i.e. a root of an irreducible algebraic equation

$$A^s+G_1A^{s-1}+G_2A^{s-2}+\ldots+G_n=0,$$

with integer coefficients G. If we multiply the numerator and denominator of any term in Bruns' series by

$$(m-nA')(m-nA'')\ldots,$$

where A', A'', ..., are the other roots of the algebraic equation, then the denominator becomes a polynomial in m and n with integer coefficients: and as it is never zero, it must be at least equal to unity: while in the numerator we now have a polynomial in m and n of degree $(s-1)$: whence it follows at once that Bruns' series converges.

The adelphic-integral series (2) of § 196 is much more complicated than Bruns' series: and although the analogy so far as it goes is favourable to the convergence of (2), yet our opinion must rest mainly on the undoubted convergence of (2) in the case of particular systems where a test is possible.

199. *Use of the adelphic integral in order to complete the integration.*

Still restricting ourselves to Case I, in which the ratio s_1/s_2 is an irrational number, we now know two integrals of the dynamical system, namely, the integral of energy (which is obtained by equating the Hamiltonian function to a constant) and the adelphic integral expressed by equation (2) of § 196. Now it is known (§ 121) that if, in any conservative holonomic dynamical system with two degrees of freedom, we know one integral in addition to the integral of energy, the system can be completely integrated, i.e. we can find expressions for the coordinates and momenta (q_1, q_2, p_1, p_2) in terms of the

* *Astr. Nach.* CIX. (1884), p. 215. Cf. also W. J. MacMillan, *Proc. Nat. Ac. Sc.* I. (1915), p. 437 and *Bull. Am. M. S.* XXII. (1915), p. 26.

time and 4 arbitrary constants of integration. We shall now perform this process.

If we add the integral of energy to the adelphic integral (2) of § 196, and divide throughout by $2s_1$, we obtain

$$l_1 = q_1 + q_1^{\frac{3}{2}} \left\{ \frac{1}{s_1} U_1 \cos p_1 + \frac{1}{s_1} U_2 \cos 3p_1 \right\}$$

$$+ q_1 q_2^{\frac{1}{2}} \left\{ \frac{2}{2s_1 + s_2} U_4 \cos (2p_1 + p_2) + \frac{2}{2s_1 + s_2} U_5 \cos (2p_1 - p_2) \right\}$$

$$+ q_1^{\frac{1}{2}} q_2 \left\{ \frac{1}{s_1} U_6 \cos p_1 + \frac{1}{s_1 + 2s_2} U_7 \cos (2p_2 + p_1) + \frac{1}{s_1 - 2s_2} U_8 \cos (2p_2 - p_1) \right\}$$

+ terms of the 4th and higher orders,

where l_1 denotes an arbitrary constant.

Similarly by subtracting the adelphic integral from the integral of energy, and dividing by s_2, we obtain

$$l_2 = q_2 + q_1 q_2^{\frac{1}{2}} \left\{ \frac{1}{s_2} U_3 \cos p_2 + \frac{1}{2s_1 + s_2} U_4 \cos (2p_1 + p_2) - \frac{1}{2s_1 - s_2} U_5 \cos (2p_1 - p_2) \right\}$$

$$+ q_1^{\frac{1}{2}} q_2 \left\{ \frac{2}{s_1 + 2s_2} U_7 \cos (2p_2 + p_1) - \frac{2}{s_1 - 2s_2} U_8 \cos (2p_2 - p_1) \right\}$$

$$+ q_2^{\frac{3}{2}} \left\{ \frac{1}{s_2} U_9 \cos p_2 + \frac{1}{s_2} U_{10} \cos 3p_2 \right\} + \text{terms of the 4th and higher orders,}$$

where l_2 represents a second arbitrary constant.

It is an easy matter to obtain q_1 and q_2 from these equations in terms of (l_1, l_2, p_1, p_2) by successive approximation: in fact, the first approximation gives $q_1 = l_1, q_2 = l_2$, and the second approximation gives

$$q_1 = l_1 - l_1^{\frac{3}{2}} \left\{ \frac{1}{s_1} U_1 \cos p_1 + \frac{1}{s_1} U_2 \cos 3p_1 \right\}$$

$$- l_1 l_2^{\frac{1}{2}} \left\{ \frac{2}{2s_1 + s_2} U_4 \cos (2p_1 + p_2) + \frac{2}{2s_1 - s_2} U_5 \cos (2p_1 - p_2) \right\}$$

$$- l_1^{\frac{1}{2}} l_2 \left\{ \frac{1}{s_1} U_6 \cos p_1 + \frac{1}{s_1 + 2s_2} U_7 \cos (2p_2 + p_1) + \frac{1}{s_1 - 2s_2} U_8 \cos (2p_2 - p_1) \right\}$$

+ terms of the 4th and higher order in $\sqrt{l_1}$ and $\sqrt{l_2}$,

$$q_2 = l_2 - l_1 l_2^{\frac{1}{2}} \left\{ \frac{1}{s_2} U_3 \cos p_2 + \frac{1}{2s_1 + s_2} U_4 \cos (2p_1 + p_2) - \frac{1}{2s_1 - s_2} U_5 \cos (2p_1 - p_2) \right\}$$

$$- l_1^{\frac{1}{2}} l_2 \left\{ \frac{2}{s_1 + 2s_2} U_7 \cos (2p_2 + p_1) - \frac{2}{s_1 - 2s_2} U_8 \cos (2p_2 - p_1) \right\}$$

$$- l_2^{\frac{3}{2}} \left\{ \frac{1}{s_2} U_9 \cos p_2 + \frac{1}{s_2} U_{10} \cos 3p_2 \right\}$$

+ terms of the 4th and higher order in $\sqrt{l_1}$ and $\sqrt{l_2}$.

Now we know (§ 121) that the expressions thus found for q_1 and q_2 must be the partial differential coefficients with respect to p_1 and p_2 of some function of (l_1, l_2, p_1, p_2): and, in fact, we have obviously

$$q_1 = \frac{\partial W}{\partial p_1}, \quad q_2 = \frac{\partial W}{\partial p_2},$$

where

$$\begin{aligned} W = l_1 p_1 + l_2 p_2 - l_1^{\frac{3}{2}} \left(\frac{1}{s_1} U_1 \sin p_1 + \frac{1}{3s_1} U_2 \sin 3p_1 \right) \\ - l_1 l_2^{\frac{1}{2}} \left\{ \frac{1}{s_2} U_3 \sin p_2 + \frac{1}{2s_1 + s_2} U_4 \sin (2p_1 + p_2) + \frac{1}{2s_1 - s_2} U_5 \sin (2p_1 - p_2) \right\} \\ - l_1^{\frac{1}{2}} l_2 \left\{ \frac{1}{s_1} U_6 \sin p_1 + \frac{1}{2s_2 + s_1} U_7 \sin (2p_2 + p_1) + \frac{1}{2s_2 - s_1} U_8 \sin (2p_2 - p_1) \right\} \\ - l_2^{\frac{3}{2}} \left\{ \frac{1}{s_2} U_9 \sin p_2 + \frac{1}{3s_2} U_{10} \sin 3p_2 \right\} \\ + \text{terms of the 4th and higher orders in } \sqrt{l_1} \text{ and } \sqrt{l_2}. \end{aligned}$$

The terms in which p_1 and p_2 occur otherwise than in the arguments of trigonometric functions are

$$\begin{aligned} p_1 (l_1 + \text{terms of the 4th and higher order in } \sqrt{l_1} \text{ and } \sqrt{l_2}) \\ + p_2 (l_2 + \quad ,, \quad ,, \quad ,, \quad ,, \quad). \end{aligned}$$

Denote the coefficients of p_1 and p_2 in this expression by α_1 and α_2 respectively: express l_1 and l_2 in terms of α_1 and α_2 by reversion of series, and replace l_1 and l_2 throughout in the series for W by these values in terms of α_1 and α_2; so that we now have

$$\begin{aligned} W = \alpha_1 p_1 + \alpha_2 p_2 - \alpha_1^{\frac{3}{2}} \left(\frac{1}{s_1} U_1 \sin p_1 + \frac{1}{3s_1} U_2 \sin 3p_1 \right) \\ - \alpha_1 \alpha_2^{\frac{1}{2}} \left\{ \frac{1}{s_2} U_3 \sin p_2 + \frac{1}{2s_1 + s_2} U_4 \sin (2p_1 + p_2) + \frac{1}{2s_1 - s_2} U_5 \sin (2p_1 - p_2) \right\} \\ - \alpha_1^{\frac{1}{2}} \alpha_2 \left\{ \frac{1}{s_1} U_6 \sin p_1 + \frac{1}{2s_2 + s_1} U_7 \sin (2p_2 + p_1) + \frac{1}{2s_2 - s_1} U_8 \sin (2p_2 - p_1) \right\} \\ - \alpha_2^{\frac{3}{2}} \left\{ \frac{1}{s_2} U_9 \sin p_2 + \frac{1}{3s_2} U_{10} \sin 3p_2 \right\} \\ + \text{terms of the 4th and higher orders in } \sqrt{\alpha_1} \text{ and } \sqrt{\alpha_2} \ \ldots\ldots\ldots\ldots (1), \end{aligned}$$

and now p_1 and p_2 do not occur except in the arguments of trigonometric functions and in the terms $(\alpha_1 p_1 + \alpha_2 p_2)$.

Now the equations

$$q_1 = \frac{\partial W}{\partial p_1}, \quad q_2 = \frac{\partial W}{\partial p_2}, \quad \beta_1 = \frac{\partial W}{\partial \alpha_1}, \quad \beta_2 = \frac{\partial W}{\partial \alpha_2}$$

define a contact-transformation from the variables (q_1, q_2, p_1, p_2) to the variables $(\alpha_1, \alpha_2, \beta_1, \beta_2)$: so in terms of these new variables the differential equations take the form

$$\frac{d\alpha_1}{dt}=\frac{\partial H}{\partial \beta_1}, \quad \frac{d\alpha_2}{dt}=\frac{\partial H}{\partial \beta_2}, \quad \frac{d\beta_1}{dt}=-\frac{\partial H}{\partial \alpha_1}, \quad \frac{d\beta_2}{dt}=-\frac{\partial H}{\partial \alpha_2}.$$

But we know that $\alpha_1 = \text{Constant}$ and $\alpha_2 = \text{Constant}$ are two of the integrals of the system, since l_1 and l_2 are constant: and therefore

$$\frac{\partial H}{\partial \beta_1}=0, \quad \frac{\partial H}{\partial \beta_2}=0,$$

so when H is expressed in terms of $(\alpha_1, \alpha_2, \beta_1, \beta_2)$, it will be found to involve α_1 and α_2 only *; and then the equations

$$\frac{d\beta_1}{dt}=-\frac{\partial H}{\partial \alpha_1}, \quad \frac{d\beta_2}{dt}=-\frac{\partial H}{\partial \alpha_2}$$

give

$$\left.\begin{aligned} \beta_1 &= -\frac{\partial H(\alpha_1,\alpha_2)}{\partial \alpha_1}t+\epsilon_1 \\ \beta_2 &= -\frac{\partial H(\alpha_1,\alpha_2)}{\partial \alpha_2}t+\epsilon_2 \end{aligned}\right\} \quad \ldots\ldots\ldots\ldots\ldots\ldots\ldots\ldots(2),$$

where ϵ_1 and ϵ_2 are arbitrary constants.

Thus we have the complete solution of the dynamical system expressed by the equations

$$\frac{\partial W}{\partial p_1}=q_1, \quad \frac{\partial W}{\partial p_2}=q_2,$$

$$\frac{\partial W}{\partial \alpha_1}=-\frac{\partial H(\alpha_1,\alpha_2)}{\partial \alpha_1}t+\epsilon_1, \quad \frac{\partial W}{\partial \alpha_2}=-\frac{\partial H(\alpha_1,\alpha_2)}{\partial \alpha_2}t+\epsilon_2,$$

where W is given by equation (1), and the four arbitrary constants of integration are $(\alpha_1, \alpha_2, \epsilon_1, \epsilon_2)$. On referring to the form of W, we see that these equations enable us to express q_1 and q_2 as purely trigonometric series, the arguments of the trigonometric functions being of the form

$$m\beta_1+n\beta_2,$$

where m and n are integers (positive, negative, or zero) and where β_1 and β_2 are linear functions of the time, given by equations (2). *We have thus obtained expressions for the coordinates in terms of the time, by means of series in which the time occurs only in the arguments of trigonometric functions: each coordinate is, in fact, represented by a series whose terms are of the form*

$$a_{mn}\cos(m\beta_1+n\beta_2),$$

where m and n are integers (positive, negative, or zero); the coefficients a_{mn} are

* We may remark that $H=s_1\alpha_1+s_2\alpha_2+\frac{1}{2}A\alpha_1^2+H\alpha_1\alpha_2+\frac{1}{2}B\alpha_2^2+\ldots$, there being no terms of order $\alpha^{\frac{1}{2}}$.

functions of two of the constants of integration, α_1 and α_2 only; and the angles β_1 and β_2 are defined by equations

$$\beta_1 = \mu_1 t + \epsilon_1, \quad \beta_2 = \mu_2 t + \epsilon_2,$$

where μ_1 and μ_2 are functions of α_1 and α_2 only, and ϵ_1 and ϵ_2 are the two remaining constants of integration.

200. *The fundamental property of the adelphic integral.*

Periodic solutions of the dynamical system evidently arise when the constants α_1 and α_2 are such that μ_1 is commensurable with μ_2: the period of the solution is then $2\pi/\nu$, where ν is the largest quantity of which μ_1 and μ_2 are integer multiples.

Suppose then that α_1 and α_2 have such values. Then if the constant ϵ_1 be varied continuously, we obtain a family of periodic solutions, each having the same period (since this does not depend on ϵ_1). The constant of energy depends only on α_1 and α_2, and is therefore the same for each of these periodic solutions. The family is therefore a family of "ordinary" periodic solutions (§ 172).

It might hastily be supposed that by varying ϵ_2 as well as ϵ_1 we should get a family of ∞^2 periodic solutions. But it is easily seen that the transformation which is obtained by varying ϵ_2 may be obtained by combining the transformation which consists in varying ϵ_1 with that which consists in adding a small constant to t. Now this latter transformation merely transforms every orbit into itself (each point being displaced in the direction of the tangent to the orbit), and so may be disregarded. The ϵ_1 and ϵ_2 transformations are therefore to be regarded as not distinct from each other*.

Consider, then, those infinitesimal transformations which change each trajectory of the system into an adjacent trajectory, *in such a way that every ordinary periodic solution is changed into an adjacent periodic solution of the same family*, i.e. having the same period and the same constant of energy. In the notation we have just been using, this transformation corresponds to a small change in ϵ_1. This transformation will be called the *adelphic* transformation†. The adelphic transformation changes any solution of the dynamical system, whether periodic or not, into one of ∞^1 other solutions which stand in a particularly close relation to it, being in fact derived from it by a change of the constant ϵ_1 only.

Now, referring to the formulae of the last article, it is evident that a change in ϵ_1, in which the other constants of integration (ϵ_2, α_1, α_2) are left unaltered, does not affect either of the constants l_1 and l_2 (since these depend only on α_1

* The only case of exception is when *all* the orbits of the system are periodic.

† From ἀδελφικός, *brotherly*, because these orbits stand in very close relation to each other, and also because the integral corresponding to the transformation stands in a much closer relation to the integral of energy than do the other integrals of the system.

and α_2) and therefore does not affect the constant of the adelphic integral or the constant of energy: this shews that all the orbits, which differ from each other only in having different values of the constant ϵ_1, have the same values for the constant of the adelphic integral and the constant of energy: and hence that the infinitesimal transformation, which corresponds (§ 144) to the adelphic integral, transforms these orbits into each other: that is to say, *the adelphic integral is the integral which corresponds to the adelphic transformation.* This is the fundamental property of the adelphic integral.

As there is only one really distinct adelphic transformation of a given dynamical system with two degrees of freedom, so there is only one really distinct adelphic integral: all other adelphic integrals may be obtained from this by combining it in various ways with the integral of energy*.

In practically all the known soluble problems of dynamics with two degrees of freedom, the integral which enables us to effect the solution is an adelphic integral. Thus, when the trajectories are the geodesics on an ellipsoid, the adelphic integral is the equation $pd =$ Constant. When the problem is that of two centres of gravitation, the adelphic integral is Euler's integral of that problem. When the solubility of the problem is due to the presence of an ignorable coordinate, say q_2, the corresponding integral (namely $p_2 =$ Constant) is adelphic.

201. *Determination of the adelphic integral in Case II.*

We now proceed to the discussion of "Case II," in which the ratio s_1/s_2 is a rational number (say equal to m/n), but no term in $\cos(np_1 - mp_2)$ is present among the third-order terms in the Hamiltonian function H. Certain terms of the series (2) of § 196 now contain in their denominators the factor $(ns_1 - ms_2)$, which vanishes since $s_1/s_2 = m/n$: and therefore the series (2) as it stands cannot converge in Case II, unless the terms which have zero denominators have numerators which also vanish. We have here come upon the real root of the principal difficulty of Celestial Mechanics: by removing it here, so as to obtain a valid adelphic integral in Cases II and III, we shall be enabled to remove it from the whole subject.

To fix ideas, we shall suppose that $s_1 = 2$, $s_2 = 1$, so that s_1/s_2 has the rational value 2, and the denominator $(s_1 - 2s_2)$, which occurs frequently in the series (2) of § 196, is zero.

In this case the equation for ϕ_3 becomes

$$2\frac{\partial \phi_3}{\partial p_1} + \frac{\partial \phi_3}{\partial p_2} = 2\frac{\partial H_3}{\partial p_1} - \frac{\partial H_3}{\partial p_2},$$

* The integral of energy corresponds to that infinitesimal transformation which changes every orbit into itself, each point of an orbit being displaced in the direction of the tangent to the orbit.

and indeed the equation for any one of the functions $\phi_3, \phi_4, \phi_5, \ldots$ takes the form

$$2\frac{\partial\phi_r}{\partial p_1}+\frac{\partial\phi_r}{\partial p_2}=\text{a known sum of terms of the type } q_1^{\frac{1}{2}m}q_2^{\frac{1}{2}n}\sin(kp_1+lp_2).$$

Now in integrating the differential equations for $\phi_3, \phi_4, \ldots$ in § 196, we used only the "particular integral," which corresponds term-by-term to the known function on the right-hand side of the equation: so that, e.g., the integral of the equation

$$s_1\frac{\partial\phi_3}{\partial p_1}+s_2\frac{\partial\phi_3}{\partial p_2}=q_1^{\frac{3}{2}}\sin p_1$$

would be taken to be

$$\phi_3=-\frac{q_1^{\frac{3}{2}}}{s_1}\cos p_1.$$

The reason for this was that the "complementary function," or arbitrary part of the solution of the differential equation, is a function of $(s_2p_1-s_1p_2)$, and so does not contain terms of the type appropriate to ϕ_3. But when $s_1=2$, $s_2=1$, the arbitrary part of the solution of the differential equation *does* contain terms of the type proper to ϕ_3, and these must be taken account of; so that we must take the integral of the equation

$$2\frac{\partial\phi_3}{\partial p_1}+\frac{\partial\phi_3}{\partial p_2}=q_1^{\frac{3}{2}}\sin p_1$$

to be

$$\phi_3=-\tfrac{1}{2}q_1^{\frac{3}{2}}\cos p_1+\alpha q_1^{\frac{1}{2}}q_2\cos(p_1-2p_2),$$

where α is an arbitrary constant. *In this way we obtain terms with arbitrary coefficients in $\phi_3, \phi_4, \phi_5, \ldots$: and these arbitrary coefficients must be chosen in such a way as to remove terms with vanishing denominators from the subsequently determined part of ϕ.* This principle enables us to obtain, in Case II, an adelphic integral free from vanishing denominators.

202. *An example of the adelphic integral in Case II.*

We shall now illustrate the working of this principle by an example. Consider the dynamical system which is specified by the Hamiltonian function

$$\left.\begin{aligned}H=2q_1\sin^2 p_1+q_2\sin^2 p_2&+\frac{1}{2\left(1+2q_1^{\frac{1}{2}}\cos p_1+q_1\cos^2 p_1+2q_2\cos^2 p_2\right)^2}\\&-\frac{1+q_1^{\frac{1}{2}}\cos p_1}{\left(1+2q_1^{\frac{1}{2}}\cos p_1+q_1\cos^2 p_1+2q_2\cos^2 p_2\right)^{\frac{3}{2}}}\end{aligned}\right\}\ldots(1)$$

If this be expanded in ascending powers of $\sqrt{q_1}$ and $\sqrt{q_2}$, we obtain

$$\begin{aligned}H=2q_1+q_2&+q_1^{\frac{3}{2}}\left(-\tfrac{9}{2}\cos p_1-\tfrac{3}{2}\cos 3p_1\right)+q_1^2\left(\tfrac{75}{16}+\tfrac{25}{4}\cos 2p_1+\tfrac{25}{16}\cos 4p_1\right)\\&+q_1q_2\left\{-3-3\cos 2p_1-3\cos 2p_2-\tfrac{3}{2}\cos(2p_1+2p_2)-\tfrac{3}{2}\cos(2p_1-2p_2)\right\}\\&+q_2^2\left\{-\tfrac{9}{16}-\tfrac{3}{4}\cos 2p_2-\tfrac{3}{16}\cos 4p_2\right\}\\&+\text{terms of the 5th and higher order in } \sqrt{q_1} \text{ and } \sqrt{q_2},\end{aligned}$$

so that in this case $s_1=2$, $s_2=1$.

As explained at the end of § 196, we may assume that the lowest term of the adelphic integral is simply q_2. Then if we write

$$\phi = q_2 + \phi_3 + \phi_4 + \phi_5 + \ldots$$

the equation to determine ϕ_3 is

$$2\frac{\partial \phi_3}{\partial p_1} + \frac{\partial \phi_3}{\partial p_2} = 0,$$

so by § 201,

$$\phi_3 = \alpha q_1^{\frac{1}{2}} q_2 \cos(p_1 - 2p_2),$$

where α is an arbitrary constant.

The equation for ϕ_4 now becomes

$$2\frac{\partial \phi_4}{\partial p_1} + \frac{\partial \phi_4}{\partial p_2}$$
$$= q_1 q_2 \left\{\left(6 + \frac{9\alpha}{2}\right)\sin 2p_2 + \left(3 + \frac{9\alpha}{4}\right)\sin(2p_1 + 2p_2) - \left(3 + \frac{9\alpha}{4}\right)\sin(2p_1 - 2p_2)\right\}$$
$$+ q_2^2\left(\tfrac{3}{2}\sin 2p_2 + \tfrac{3}{2}\sin 4p_2\right),$$

of which the integral is

$$\phi_4 = q_1 q_2 \left\{-\left(3 + \frac{9\alpha}{4}\right)\cos 2p_2 - \left(\frac{1}{2} + \frac{3\alpha}{8}\right)\cos(2p_1 + 2p_2) + \left(\frac{3}{2} + \frac{9\alpha}{8}\right)\cos(2p_1 - 2p_2)\right\}$$
$$+ q_2^2\left(-\tfrac{3}{4}\cos 2p_2 - \tfrac{3}{16}\cos 4p_2\right)$$

The equation to determine ϕ_5 is now

$$2\frac{\partial \phi_5}{\partial p_1} + \frac{\partial \phi_5}{\partial p_2} = \frac{\partial H_5}{\partial p_2} + (\phi_3, H_4) + (\phi_4, H_3),$$

where, as usual, (ϕ_3, H_4) denotes the expression

$$\frac{\partial \phi_3}{\partial q_1}\frac{\partial H_4}{\partial p_1} + \frac{\partial \phi_3}{\partial q_2}\frac{\partial H_4}{\partial p_2} - \frac{\partial \phi_3}{\partial p_1}\frac{\partial H_4}{\partial q_1} - \frac{\partial \phi_3}{\partial p_2}\frac{\partial H_4}{\partial q_2};$$

and we have to choose α so as to annul the terms in $\sin(p_1 - 2p_2)$ on the right-hand side of this equation. On calculating these terms, we find

$\left(\text{from } \frac{\partial H_5}{\partial p_2}\right)$	$\frac{39}{2} q_1^{\frac{3}{2}} q_2 \sin(p_1 - 2p_2)$
(from (ϕ_4, H_3))	$-\frac{45}{2}\left(1 + \frac{3\alpha}{4}\right) q_1^{\frac{3}{2}} q_2 \sin(p_1 - 2p_2)$
(from (ϕ_3, H_4))	$+\frac{123}{8}\alpha q_1^{\frac{3}{2}} q_2 \sin(p_1 - 2p_2)$.

The quantity α must therefore satisfy the equation

$$\tfrac{39}{2} - \tfrac{45}{2}\left(1 + \frac{3\alpha}{4}\right) + \tfrac{123}{8}\alpha = 0,$$

which gives

$$\alpha = -2.$$

Substituting this value of α in ϕ_3 and ϕ_4, our integral becomes

$$\begin{aligned}\text{Constant} = q_2 - 2q_1^{\frac{1}{2}} q_2 \cos(p_1 - 2p_2) \\ + q_1 q_2 \{\tfrac{3}{2}\cos 2p_2 + \tfrac{1}{4}\cos(2p_1 + 2p_2) - \tfrac{3}{4}\cos(2p_1 - 2p_2)\} \\ + q_2^2(-\tfrac{3}{4}\cos 2p_2 - \tfrac{3}{16}\cos 4p_2) \\ + \text{terms of the 5th and higher orders in } \sqrt{q_1} \text{ and } \sqrt{q_2} \quad \ldots\ldots(2).\end{aligned}$$

Now it may be verified by differentiation that the dynamical system specified by equation (1) possesses the integral

$$\text{Constant} = \tfrac{1}{2}\{\sqrt{2q_2}\sin p_2 + q_1^{\frac{1}{2}}\sqrt{2q_2}\cos p_1 \sin p_2 - 2\sqrt{2q_1 q_2}\sin p_1 \cos p_2\}^2 - \frac{1 + q_1^{\frac{1}{2}}\cos p_1}{(1 + 2q_1^{\frac{1}{2}}\cos p_1 + q_1\cos^2 p_1 + 2q_2\cos^2 p_2)^{\frac{1}{2}}} \quad \ldots\ldots\ldots\ldots(3),$$

and this integral is adelphic, as may be shewn by completing the solution, or more simply by remarking that the integral (3) is a function of the variables $(\sqrt{q_1}, \sqrt{q_2}, p_1, p_2)$ which is one-valued and free from singularities for a certain range of values, and therefore the infinitesimal transformation corresponding to it will also be one-valued and free from singularities, and so must transform closed orbits into closed orbits.

But *on expanding this integral* (3) *in ascending powers of* $\sqrt{q_1}$ *and* $\sqrt{q_2}$ *by the multinomial theorem, we arrive at the series* (2). This shews that, *for the dynamical system we are considering, the series obtained by the process of the last article converges for all real values of* p_1 *and* p_2 *so long as* $|q_1|$ *and* $|q_2|$ *are inferior to certain fixed quantities, and that the series represents the adelphic integral of the dynamical system.*

203. *Determination of the adelphic integral in Case III.*

The principle for the removal of vanishing divisors from the adelphic integral, which has been explained and illustrated in the last two articles, is not sufficient for the purpose if the Hamiltonian function contains, among its third-order terms, a term in $\cos(s_2 p_1 - s_1 p_2)$: for this term gives rise to a vanishing divisor in ϕ_3, and the arbitrary terms which are used in order to annul terms with vanishing divisors do not come into operation early enough to remove vanishing divisors from ϕ_3.

In this "Case III" we must make use of another principle (concurrently with the principle of § 201) which may be explained thus: Suppose that an integral of a system of differential equations in variables (q_1, q_2, p_1, p_2) is of the form

$$f(q_1, q_2, p_1, p_2) + \frac{g(q_1, q_2, p_1, p_2)}{\mu} = \gamma,$$

where γ is the *arbitrary* constant and μ is a *definite* constant formed of quantities occurring in the differential equations. The integral in this form ceases to have a meaning when μ tends to zero. But we may derive from it an integral which has a meaning when $\mu \to 0$, by merely supposing first that μ is different from zero, and multiplying the equation throughout by μ, so that it becomes

$$\mu f(q_1, q_2, p_1, p_2) + g(q_1, q_2, p_1, p_2) = \mu\gamma$$

and then making $\mu \to 0$; the equation becomes

$$g(q_1, q_2, p_1, p_2) = c,$$

where c denotes $\underset{\mu \to 0}{\mathrm{Lt}}(\mu\gamma)$. This is the desired form of the integral when μ is zero.

Our case is not so simple as this, since the vanishing divisor occurs not only in the inverse first power, but in an infinite series containing all the inverse powers. The method we follow, which will be illustrated in the next article, is really equivalent to using the principle of § 201 in order to remove all inverse powers of the small divisor except the first, and then using the principle of this article in order to remove this inverse first power.

204. *An example of the adelphic integral in Case III.*

We shall now shew by considering a particular dynamical system how the principle just mentioned is applied in order to obtain an adelphic integral free from vanishing divisors in "Case III."

Consider the dynamical system whose Hamiltonian function is

$$H = q_1 - 2q_2 + q_1^{\frac{3}{2}} U_1 \cos p_1 + q_1 q_2^{\frac{1}{2}} U_4 \cos(2p_1 + p_2).$$

Now if the Hamiltonian function is

$$H = s_1 q_1 + s_2 q_2 + q_1^{\frac{3}{2}} U_1 \cos p_1 + q_1 q_2^{\frac{1}{2}} U_4 \cos(2p_1 + p_2),$$

where s_1 and s_2 are arbitrary, the adelphic-integral series to which we are led by the method of § 196 is

$$\begin{aligned}\text{Constant} = {} & s_1 q_1 - s_2 q_2 + q_1^{\frac{3}{2}} U_1 \cos p_1 + \frac{2s_1 - s_2}{2s_1 + s_2} q_1 q_2^{\frac{1}{2}} U_4 \cos(2p_1 + p_2) \\ & + \frac{s_2}{2s_1 + s_2} U_1 U_4 q_1^{\frac{3}{2}} q_2^{\frac{1}{2}} \left\{ \frac{4\cos(p_1 + p_2)}{s_1 + s_2} + \frac{2\cos(3p_1 + p_2)}{3s_1 + s_2} \right\} \\ & + U_1^2 U_4 q_1^2 q_2^{\frac{1}{2}} \frac{s_2}{2s_1 + s_2} \left\{ -\frac{3}{3s_1 + s_2} \frac{\cos(4p_1 + p_2)}{4s_1 + s_2} - \frac{6}{3s_1 + s_2} \frac{\cos(2p_1 + p_2)}{2s_1 + s_2} \right. \\ & \left. \qquad - \frac{6}{s_1 + s_2} \frac{\cos p_2}{s_2} \right\} \\ & + U_1 U_4^2 q_1^{\frac{3}{2}} q_2 \frac{s_2}{2s_1 + s_2} \left\{ \frac{2(9s_1 + s_2)}{(s_1 + s_2)(3s_1 + s_2)} \frac{\cos p_1}{s_1} + \frac{4}{s_1 + s_2} \frac{\cos(3p_1 + 2p_2)}{3s_1 + 2s_2} \right\} \\ & + U_1 U_4^2 q_1^{\frac{5}{2}} \frac{s_2}{2s_1 + s_2} \cdot \frac{5s_1 + s_2}{(3s_1 + s_2)(s_1 + s_2)} \cdot \frac{\cos p_1}{s_1} \\ & + \text{terms of the 6th and higher orders in } \sqrt{q_1} \text{ and } \sqrt{q_2} \quad \ldots(1).\end{aligned}$$

In our problem $s_1 = 1$, $s_2 = -2$, so $2s_1 + s_2$ is a vanishing denominator. This denominator makes its appearance in the fourth term of the above expression, and occurs in every subsequent term, being squared in the coefficient of the fifth-order term $q_1^2 q_2^{\frac{1}{2}} \cos(2p_1 + p_2)$. We must now modify this series (1) so as to obtain an integral which has no vanishing denominators.

In the first place, we apply the principle of § 203: the lowest term which is affected by the vanishing denominator is the term

$$\frac{2s_1 - s_2}{2s_1 + s_2} q_1 q_2^{\frac{1}{2}} U_4 \cos(2p_1 + p_2):$$

we therefore try to form an integral whose lowest term (discarding the non-essential factors $(2s_1 - s_2)$ and U_4) shall be

$$q_1 q_2^{\frac{1}{2}} \cos(2p_1 + p_2).$$

If then we suppose this integral to be

$$\text{Constant} = \phi \equiv q_1 q_2^{\frac{1}{2}} \cos(2p_1 + p_2) + \phi_4 + \phi_5 + \phi_6 + \ldots,$$

where ϕ_r denotes the terms of degree r in $\sqrt{q_1}$ and $\sqrt{q_2}$, and substitute in the equation $(\phi, H) = 0$, we find on equating to zero the terms of order 4 that ϕ_4 is to be determined from the equation

$$\frac{\partial \phi_4}{\partial p_1} - 2\frac{\partial \phi_4}{\partial p_2} = q_1^{\frac{3}{2}} q_2^{\frac{1}{2}} U_1 \{2 \sin(p_1 + p_2) + \sin(3p_1 + p_2)\}.$$

The integral of this is

$$\phi_4 = q_1^{\frac{3}{2}} q_2^{\frac{1}{2}} U_1 \{2 \cos(p_1 + p_2) - \cos(3p_1 + p_2)\},$$

to which, however, we may add terms of the type

$$\alpha q_1^2 + \beta q_1 q_2 + \gamma q_2^2 \quad \ldots\ldots\ldots\ldots\ldots\ldots\ldots\ldots (2),$$

where α, β, γ are arbitrary constants, since these terms satisfy the differential equation and are of the type proper to ϕ_4. It should be noticed that these terms are not now superfluous, as they were in the general case studied in § 196; for in the general case the addition of such terms to ϕ_4 would merely be equivalent to adding on an arbitrary quadratic function of the integral of energy and the adelphic integral: but in our present case the adelphic integral does not begin with terms linear in q_1 and q_2, and therefore a quadratic function of it does not account for terms like those in (2). The arbitrary constants in (2) are to be determined in such a way as to make terms with vanishing denominators disappear from the higher-order terms of ϕ. Thus, writing now

$$\phi_4 = q_1^{\frac{3}{2}} q_2^{\frac{1}{2}} U_1 \{2 \cos(p_1 + p_2) - \cos(3p_1 + p_2)\} + \alpha q_1^2,$$

and substituting in the differential equation satisfied by ϕ_5 which is

$$\frac{\partial \phi_5}{\partial p_1} - 2\frac{\partial \phi_5}{\partial p_2} = (\phi_4, H_3) \quad \ldots\ldots\ldots\ldots\ldots\ldots\ldots\ldots (3),$$

we find that on the right-hand side of (3) the terms involving $\sin(2p_1+p_2)$ (which would lead to vanishing denominators on integration) are

$$-3q_1^2q_2^{\frac{1}{2}}U_1^2\sin(2p_1+p_2)-4\alpha q_1^2q_2^{\frac{1}{2}}U_4\sin(2p_1+p_2),$$

and these will collectively vanish provided

$$\alpha=-\frac{3}{4}\frac{U_1^2}{U_4}.$$

In this way, by repeated application of the principle of § 201, we are able to remove all terms with vanishing denominators and obtain an adelphic integral free from them.

205. *Completion of the integration of the dynamical system in Cases II and III.*

Having now in §§ 201–204 overcome the difficulty caused by the presence of terms with vanishing divisors in the adelphic integral in Cases II and III, we can use this integral in order to integrate the dynamical system completely, just as was done for Case I in § 199. We thus obtain expansions for the coordinates in terms of the time in all cases: but these expansions are completely different in form, according as the dynamical system falls under Case I, II, or III. This result supplies the underlying explanation of Poincaré's theorem that the series of Celestial Mechanics cannot converge uniformly over any continuous range of values of the constants: for the series to which he was referring were of the kind which we have classified under Case I, and we have seen that when the constants s_1, s_2 are continuously varied, these series must be replaced by the series appropriate to Case II or Case III, whenever the ratio s_1/s_2 passes from an irrational to a rational value. The advantage of solving by means of the adelphic integral is that the forms of the adelphic integral corresponding to the three cases can be readily determined: and thus the difficulty is removed before the adelphic integral is used in order to obtain the complete expressions for the coordinates in terms of the time.

For further recent researches on the general solution of the equations of dynamical systems, reference may be made to the important series of memoirs by T. M. Cherry, published in the years 1924–27 in the *Proc. Camb. Phil. Soc.*, *Trans. Camb. Phil. Soc.*, and *Proc. Lond. Math. Soc.*: and also to papers by B. B. Baker and E. B. Ross in *Proc. Edin. Math. Soc.* Vols. XXXIX.–XLI. (1921–23).

Miscellaneous Examples.

1. Let ϕ denote any function of the variables $q_1, q_2, \ldots, q_n, p_1, \ldots, p_n$ of a dynamical system which possesses an integral of energy

$$H(q_1, q_2, \ldots, q_n, p_1, \ldots, p_n)=\text{Constant};$$

let $a_1, a_2, \ldots, a_n, b_1, \ldots, b_n$ be the values of $q_1, q_2, \ldots, q_n, p_1, \ldots, p_n$ respectively at the instant $t=t_0$; and let $\{f, g\}$ denote the value of the Poisson-bracket (f, g) when the quantities $q_1, q_2, \ldots, q_n, p_1, \ldots, p_n$ occurring in it are replaced respectively by $a_1, a_2, \ldots, a_n, b_1, \ldots, b_n$.

Shew that

$$\phi(q_1, q_2, \ldots, q_n, p_1, \ldots, p_n) = \phi(a_1, a_2, \ldots, a_n, b_1, \ldots, b_n) + (t-t_0)\{\phi, H\} + \frac{(t-t_0)^2}{2!}\{\{\phi, H\}, H\} + \ldots.$$

2. Shew that the dynamical system whose equations of motion are

$$\frac{dq}{dt} = \frac{\partial H}{\partial p}, \quad \frac{dp}{dt} = -\frac{\partial H}{\partial q},$$

where

$$H = \tfrac{1}{2}p^2 + \frac{l^4k^2}{2q^2} - \frac{l^3k^2}{q},$$

possesses a family of solutions represented by the expansion (retaining only terms of order less than $a^{\frac{3}{2}}$)

$$q = l + \frac{3a}{kl} + \left(\frac{2a}{k}\right)^{\frac{1}{2}} \cos\beta - \frac{3a}{2kl}\cos 2\beta,$$

where

$$\beta = -\left(k + \frac{\alpha a}{2b^2}\right)t + \epsilon,$$

and a and ϵ are arbitrary constants.

INDEX OF AUTHORS QUOTED

(The numbers refer to the pages.)

INDEX OF TERMS EMPLOYED

(The numbers refer to the pages, where the term occurs for the first time in the book or is defined.)